Medical Intelligence Unit 26

New Vaccine Technologies

Ronald W. Ellis, Ph.D.
BioChem Pharma
(to become Shire Biologics after soon-expected merger)
Northborough, Massachusetts, U.S.A.

CRC Press
Taylor & Francis Group
Boca Raton London New York

CRC Press is an imprint of the
Taylor & Francis Group, an **informa** business

New Vaccine Technologies

Medical Intelligence Unit

designed by Lana K. Moore

CRC Press
Taylor & Francis Group
6000 Broken Sound Parkway NW, Suite 300
Boca Raton, FL 33487-2742

First issued in hardback 2019

ISBN-13: 978-1-58706-050-2 (hbk)
ISBN-13: 978-1-58706-080-9 (pbk)

Visit the Taylor & Francis Web site at
http://www.taylorandfrancis.com

and the CRC Press Web site at
http://www.crcpress.com

Library of Congress Cataloging-in-Publication Data

New Vaccine Technologies / [edited by] Ronald W. Ellis
p.;cm.-- (Biotechnology intelligence unit)
Includes bibliographical references and index.
ISBN 1-58706-050-7 (alk. paper)
1. Vaccines--Biotechnology. I. Ellis, Ronald W. II. Series.
[DNLM: 1. Vaccines. 2. Biotechnology. QW 805 N5324 2001]
TP248.65.V32N49 2001
615'372--dc21 00-063630

Dedication

Dedicated to my wife Danielle and children Jacob and Miriam for their love, patience and support, and to the memory of my father.

CONTENTS

EDITOR

Ronald W. Ellis, Ph.D.
BioChem Pharma
(to become Shire Biologics after soon-expected merger)
Northborough, Massachussetts, U.S.A.
Chapter 1

CONTRIBUTORS

Charles J. Arntzen
Boyce Thompson Institute for Plant Research
Ithaca, New York, U.S.A.
Chapter 15

Remco M. P. Brandt
Cardinal Bernardin Cancer Center
Loyola University Chicago
Maywood, Illinois, U.S.A.
Chapter 12

Michel Bublot
Arbovirus Unit
Griffin Lab
Slingerlands, New York, U.S.A.
Chapter 7

Julianne C.M. Clifford
Office of Vaccines Research and Review
Division of Vaccines and Related Products Applications
Rockville, Maryland, U.S.A.
Chapter 3

Victor J. DiRita
Department Microbiology and Immunology
Unit for Laboratory Animal Medicine
University of Michigan Medical School
Ann Arbor, Michigan, U.S.A.
Chapter 9

Rafaat Fahim
Connaught Laboratories, Ltd.
Willowdale, Ontario, Canada
Chapter 11

Stephen Freese
Wyeth-Lederle Vaccines
West Henrietta, New York, U.S.A.
Chapter 13

Russell R. Gettig
Arbovirus Unit
Griffin Lab
Slingerlands, New York, U.S.A.
Chapter 7

Gregory M. Glenn
IOMAI Corporation
Washington, District of Columbia, U.S.A.
Chapter 17

Marion Gruber
Office of Vaccines Research and Review
Division of Vaccines and Related Products Applications
Rockville, Maryland, U.S.A.
Chapter 3

Luc Hessel
Medical Department
Pasteur Merieux MSD
Lyon Cedex, France
Chapter 2

Gregory E. Holt
Cardinal Bernardin Cancer Center
Loyola University Chicago
Maywood, Illinois, U.S.A.
Chapter 12

Ke-Fei Hu
Department of Virology
National Veterinary Institute
Uppsala, Sweden
Chapter 16

W. Martin Kast
Cardinal Bernardin Cancer Center
Loyola University Chicago
Maywood, Illinois, U.S.A.
Chapter 12

Elizabeth Kauffman
Arbovirus Unit
Griffin Lab
Slingerlands, New York, U.S.A.
Chapter 7

Kevin Killeen
AVANT Immunotherapeutics Inc.
Needham, Massachusetts, U.S.A.
Chapter 9, 10

Sims K. Kochi
AVANT Immunotheraputics, Inc.
Needham, Massachusetts, U.S.A.
Chapter 10

Eugene D. Kwon
Cardinal Bernardin Cancer Center
Loyola University Chicago
Maywood, Illinois, U.S.A.
Chapter 12

Shaw-guang Lee
Wyeth-Lederle Vaccines and Pediatrics
Malvern, Pennsylvania, U.S.A.
Chapter 14

Keith J. Limbach
Arbovirus Unit
Griffin Lab
Slingerlands, New York, U.S.A.
Chapter 7

Sheena M. Loosmore
Connaught Laboratories, Ltd.
Willowdale, Ontario, Canada
Chapter 11

Daniel McCallus
Wyeth-Lederle Vaccines and Pediatrics
Malvern, Pennsylvania, U.S.A.
Chapter 14

Bror Morein
Department of Virology
National Veterinary Institute
Uppsala, Sweden
Chapter 16

Andrew Murdin
Aventis Pasteur Ltd.
Willowdale, Ontario, Canada
Chapter 8

Catherine J. Pachuk
Wyeth-Lederle Vaccines and Pediatrics
Malvern, Pennsylvania, U.S.A.
Chapter 14

Steven E Pincus
Arbovirus Unit
Griffin Lab
Slingerlands, New York, U.S.A.
Chapter 7

Paul C. Richman
Office of Vaccines Research and Review
Division of Vaccines and Related
Products Applications
Rockville, Maryland, U.S.A.
Chapter 3

Benjamin Rovinski
Aventis Pasteur Ltd.
Willowdale, Ontario, Canada
Chapter 8

Michael P. Rudolf
Cardinal Bernardin Cancer Center
Loyola University Chicago
Maywood, Illinois, U.S.A.
Chapter 12

Suryaprakash Sambhara
Aventis Pasteur Ltd.
Willowdale, Ontario, Canada
Chapter 8

C. Satishchandran
Wyeth-Lederle Vaccines and Pediatrics
Malvern, Pennsylvania, U.S.A.
Chapter 14

Alan Shaw
Virus and Cell Biology
Merck Research Laboratory
West Point, Pennsylvania, U.S.A.
Chapter 5

Richard Spaete
Aviron
Mountain View, California, U.S.A.
Chapter 6

Dale Spriggs
Project Planning and Management
BioChem Pharma, Inc.
Northborough, Massachussetts, U.S.A.
Chapter 4

Jill Taylor
Arbovirus Unit
Griffin Lab
Slingerlands, New York, U.S.A.
Chapter 7

Markwin P. Velders
Cardinal Bernardin Cancer Center
Loyola University Chicago
Maywood, Illinois, U.S.A.
Chapter 12

Amanda Walmsely
Boyce Thompson Institute for Plant Research
Ithaca, New York, U.S.A.
Chapter 15

Damu Yang
Cardinal Bernardin Cancer Center
Loyola University Chicago
Maywood, Illinois, U.S.A.
Chapter 12

Gavin R. Zealey
Connaught Laboratories, Ltd.
Willowdale, Ontario, Canada
Chapter 11

PREFACE

Vaccines are one of the most cost-effective interventions in health-care. Vaccination is estimated to have been responsible for 10-15 years of the increase in the average human lifespan during the 20th century, an increase probably second in impact only to that of clean water. In addition to considerable morbidity, there are over 10 million deaths annually worldwide attributable to infectious diseases. A large number of these deaths can be prevented by wider use of existing vaccines, while most of these deaths would be preventable by the development of effective new vaccines.

There is an increasingly broad array of new technologies that are being employed for developing vaccines. Such technologies are based on breakthrough discoveries in the fields of immunology, biochemistry, molecular biology and related areas. The broad applications of such discoveries should result in the development of many new vaccines that have not been feasible previously. Alternatively it may be possible to improve existing vaccines in terms of their safety and efficacy. There are about 40 new vaccines (not including competing versions of the same product) that were developed and introduced during the 20th century. It is noteworthy that almost half of these new vaccines were introduced during the 1980s and 1990s, with many of these based on new technologies such as recombinant proteins and conjugates. Therefore, the development of new vaccine technologies offers yet further potential for considerably reducing worldwide mortality and morbidity from infectious diseases.

Beyond the applications of vaccines to infectious diseases, it should be noted that there are increasing efforts to develop vaccines for the treatment or prevention of chronic diseases such as cancer, autoimmunity and allergy. There currently is one vaccine licensed for the treatment of a cancer (BCG vaccine for bladder cancer), with numerous other cancer vaccines of multiple designs in various stages of preclinical and clinical development. Such therapeutic vaccines, in conjunction with other therapeutic modalities, offer the prospect for improving health and recovery from a range of chronic diseases.

There are two general categories for vaccines, active and passive. Active vaccines stimulate the production of both antibodies and/or of immune system cells with memory and effector functions (*e.g.*, cytotoxic T-cells). Passive vaccines are antibody preparations that are used in cases where developing an active vaccine is not feasible or where there is a need for immediate immunity due to acute exposure to a virus or bacteria. Passive vaccines, which do not stimulate immunological memory, historically have been polyclonal human antibodies from individuals with the requisite antibody specificities. However, most new passive vaccines are based on monoclonal antibodies that are human or humanized, given recent advances in molecular biology that have enabled the production of such antibodies.

This book focuses upon the applications of new technologies to active vaccines for the prevention of human infectious diseases, which represent all but one of the available licensed vaccines.

There are many challenges in fully applying and developing new vaccine technologies. Most of these technologies can be divided into five general categories: 1) discovery of new leads and candidate antigens; 2) production; 3) design of the overall vaccine; 4) formulation of the final product; and 5) administration modality for human use.

Vaccine discovery has relied historically on a range of technologies. Live attenuated vaccines have been based on isolating and growing the virus or bacteria in vitro. In the case of inactivated vaccines, the in vitro-cultivated microorganism is chemically treated to destroy its infectivity. Vaccine antigens have been identified through an approach akin to proteomics, viz., the study of the proteins (and polysaccharides and other antigens) associated with viruses and bacteria. These antigens may be identified by means of antibodies raised against the whole microorganism or in acute or convalescent sera following infection. Alternatively, the microorganism is grown and biochemically fractionated to identify antigens. Such approaches also have been taken to identifying candidate antigens for diseases such as cancer and allergy. More recently, genomics-based approaches have been applied to identifying vaccine antigens, whereby the complete sequence of the microorganism is derived and annotated. Candidate vaccine antigens then are identified by homology to known vaccine antigens or by structures (hydrophobic signal sequence) that would direct the antigen to the cell surface. This approach has been applied to diverse bacteria such as non-typeable *Haemophilus influenzae*, *Helicobacter pylori*, and *Neisseria meningitidis*, from which novel antigens have been identified and then validated in animal studies as candidate vaccine antigens. Based on human genomics and the study of disease-specific gene expression, novel candidate vaccine antigens are being discovered and developed for cancer and other diseases.

There have been tremendous advances in production technologies for vaccines. Highly productive recombinant expression systems have been developed and optimized for a broad range of prokaryotic and eukaryotic cells. Large-scale fermentation equipment and processes as well as growth media have been developed that enable the attaining of high cell densities for high levels of accumulation of viruses or recombinant antigens. New large-scale filtration and chromatography modalities have enabled the efficient processing of large biomasses in order to isolate highly purified antigens. Advances in biochemistry and analytical chemistry/biochemistry have enabled macromolecules to be very well characterized and stably formulated. Continued technical advances in all these areas offer the prospect for even more efficient and reproducible large-scale production of vaccines.

Vaccines can be divided into three general categories: live, subunit/inactivated, and DNA (Chapter 1). There are several subcategories of specific designs within each of these three general groups, as described in Chapters 5-15. One or more of these specific designs may be applicable for developing a vaccine for a particular virus, bacteria or disease. Each design has different potential advantages and disadvantages in terms of production, immunobiology, potential safety and efficacy, and ability to be analytically and biologically characterized. All of these factors need to be weighed when selecting a design early in a development program. Given the long timeframe and large expense for development, such decisions assume significant weight. Table 1.1 lists the status of development of vaccines made by each specific approach, whether licensed or in clinical or preclinical evaluations.

The clinical development plan for a vaccine based on novel technologies is similar to that of a traditional vaccine (Chapter 2). Nevertheless, there are several clinical issues that must be considered for evaluating vaccines made by new technologies, especially approaches such as live vectors, DNA, adjuvants and delivery systems. Certain approaches may present new safety-related issues that require significant monitoring, especially during initial clinical studies. It is also important that a new vaccine technology be validated for proof-of-principle early in the clinical development program before significant resources are applied to its development. The clinical endpoint and surrogate markers of protection should be understood in order to facilitate development, especially for a new vaccine target.

Regulatory issues may present special challenges for technologies with which there has been little or no experience (Chapter 3), since specific standards for criteria such as safety, purity and potency of new vaccines may not exist. In that sense, the review by regulatory agencies of vaccines based on new technologies often is done case-by-case in an indication-based and product-specific fashion. Another important consideration is the risk (perceived or actual) relative to potential benefit for the particular vaccine, one which differs for (e.g.) a prophylactic vaccine for infants vs. a therapeutic cancer vaccine. Guidance Documents, such as Points-to-Consider monographs, prepared by the US FDA and the International Conference on Harmonization may provide useful guidelines regarding regulatory needs or desires to groups preparing IND and license applications.

It has been increasingly uncommon that any single organization has all the technologies at its disposal to be able to develop a safe and effective vaccine of a particular type. Therefore, in-licensing and business development have been areas of increasing activity for vaccines (Chapter 4). The licensor is usually an academic or government laboratory or a small biotech company. The technology available for licensing may be an antigen, vector, method for discovery or screening of antigens, production method, adjuvant or delivery system, formulation, device, or combination of such inventions for which the necessary financial, technical and physical resources are not available in that group. In order to receive the best value in a licensing agreement, it is important that the licensor develop its technology and associated patent portfolio to the point where it has added as much value to it as possible within a useful timeframe.

There are three general categories of live viral vaccines (Chapters 5-7). Live attenuated vaccines are derived by means of the passage of a virus in cell culture until its pathogenicity has been sufficiently attenuated for humans, but with the retention of sufficient infectivity in vivo to stimulate protective immunity (Chapter 5). Such passaging is empirical in terms of the number of passages and cell types used for attenuation. Furthermore, the mutations found to be associated with such attenuated viruses are generally random. In most cases, the precise mutations responsible for the attenuation phenotype are unknown. Since many human viruses lack useful animal models for virus replication and virulence, it is necessary to test such vaccines extensively in humans for safety until the vaccine virus is judged to be sufficiently attenuated. Nevertheless, these vaccines have been very successful in terms of control of disease. Smallpox, the first human disease ever eliminated from the earth, was eradicated through the use of a live vaccine. This also will be the case for polio, which should be eradicated within the next few years, and possibly for measles in the following decade.

In order to make the technique of attenuation less empirical in nature, recombinant technology can be used to introduce mutations or deletions in key genes responsible for pathogenicity. Such live recombinant viral vaccines have well-defined molecular changes (Chapter 6). These mutations may exert their attenuating effects by limiting *in vivo* replication potential or considerably reducing or eliminating virulence. By making multiple changes, one can assure that there is no possibility or a very low probability for the vaccine virus to revert to virulence. These mutations also can be exploited as immunological or molecular markers for distinguishing the mutated virus from its wild-type counterpart. Even though attenuated human vaccines of this type have been only in early clinical evaluations, a live attenuated recombinant animal vaccine was licensed in the 1980s for the prevention of pseudorabies infections of pigs.

Live viral vaccines can be engineered to express that encode vaccines from other pathogens (usually viruses), thereby functioning as live vectored vaccines. Several classes of viruses have been developed as viral vaccine vectors (Chapter 7), including poxviruses, adenoviruses, herpesviruses, and alphaviruses. The advantage of such live vectors is that the vaccine antigen encoded by the transgene (inserted into the viral genome) is processed intracellularly as part of a live virus infection, by which it may stimulate both antibody and cellular immunity. The main challenges to successful development include achieving appropriate expression levels of the transgene, assuring adequate attenuation of the virus vector while retaining sufficient infectivity, and obviating potential host immunity to the vector. While it is possible that this development can yield a dual vaccine against both the vectored virus and the virus encoding the transgene, most of the common vectors are not vaccine targets in their own right. Virus vectors also may be used to prime the immune system, to be followed by a booster with a recombinant protein as in the case of HIV vaccines in clinical studies.

Many viruses can be grown to high titer in cell cultures. Such viruses become the starting material for purification and inactivation. Many such inactivated viral vaccines (Chapter 8) have multiple repeat surface epitopes, which are composed of repeat units of viral structural proteins. As a consequence, these vaccines are among the most potent immunogens ever characterized. For example, immunization with a single 50-ng dose of a hepatitis A vaccine was shown to protect against clinical disease. Many of these inactivated vaccines (e.g., influenza, polio) have been used for decades, with an excellent track record of safety and efficacy.

There are very few examples of empirically attenuated bacterial vaccines. The only two such licensed vaccines are the BCG vaccine for tuberculosis and bladder cancer (attenuated by >200 passages in vitro) and *Salmonella typhi* vaccine for typhoid fever (attenuated by random chemical mutagenesis). The mutations associated with these attenuations remain unknown. Thus, recombinant technology has been applied to making defined mutations in bacterial genes responsible for pathogenesis in order to derive live recombinant vaccines (Chapter 9). Two or more mutations are made in order to assure the lack of reversion to pathogenicity. While only one such vaccine has been licensed to date (cholera), this approach continues to be applied, especially for enteric pathogens.

Several bacterial species have been developed into live vectored bacterial vaccines (Chapter 10) expressing proteins from other pathogens (usually bacteria) according to similar principles as for live viral vectors. Appropriate attenuation of the live bacterial vector involves both the reduction of pathogenicity and the maintenance of sufficient in vivo infection/replication potential to assure effective immunization against the protein antigen encoded by the transgene. The transgene may be integrated into the bacterial chromosome or may be encoded by a plasmid. As a result, expression of the vaccine antigen is in the context of that for the whole bacteria, thus providing for a potentially broader immune response to the bacteria encoding the transgene. The promoter for expression of the transgene may be prokaryotic (in which case expression of the protein is as a typical bacterial protein) or eukaryotic (in which case the protein may be expressed as for a DNA vaccine [Chapter 14]). Such vaccines are in early clinical studies.

Subunit vaccines consist of proteins, peptides or polysaccharides that carry protective epitopes. While the first examples of licensed vaccines were with proteins isolated directly from bacteria (e.g., diphtheria, tetanus and pertussis) or viruses (e.g., hepatitis B and pertussis), most recent applications have involved the recombinant expression of such proteins (Chapter 11). Recombinant protein antigens that have been developed into licensed vaccines have been expressed in diverse cells such as *Saccharomyces cerevisiae*, *Vibrio cholerae*, *Bordetella pertussis*, and *Escherichia coli*, with new candidate vaccines also being expressed in mammalian and insect cells as well as in whole plants (Chapter 15) or animals. Hybrid or chimeric recombinant protein antigens also have been designed and developed as candidate vaccines. Recombinant vaccine antigens are isolated to a high level of purity and typically are very well characterized analytically. These antigens require multiple doses for eliciting both protective immunity and immunological memory. While some proteins are sufficiently immunogenic to be formulated on their own, most require adjuvants (Chapter 16) for being sufficiently immunogenic to elicit protective immunity.

There are cases where the full-length polypeptide with protective epitope(s) is not optimal as a vaccine antigen. For instance, the polypeptide may have immunodominant epitopes that do not elicit effective immunity, or there may be a need to focus the immune response toward a particular protective epitope. In such instances, a peptide-derived vaccine may enable the immune response to be focused on a single key epitope (Chapter 12). Some peptide vaccines are based on a B-cell epitope that stimulates a protective antibody response. In this case, the B-cell epitope peptide is linked to a T-cell epitope peptide or a carrier protein that provides for T-cell help for the immune response. Other peptide vaccines are based on a T-cell epitope that stimulates a cell-mediated immune response such as cytotoxic T lymphocytes (CTL), as is being applied to novel vaccines for the prevention or therapy of cancer or chronic infections.

There are many bacteria (both Gram$^-$ and Gram$^+$) that are encapsulated with polysaccharides. These capsular polysaccharides carry the major seroreactivity of the bacterial species or subspecies and as such are the object of protective antibodies, such that the polysaccharides are effective vaccine antigens (Chapter 13). In a few cases, there is a single polysaccharide serotype for the particular pathogen (*e.g.*, *Haemophilus influenzae* type *b* [Hib] for invasive *H. influenzae* type *b* meningitis), such that a single polysaccharide type can be developed into a monovalent vaccine. However, in most cases, there are multiple capsular polysaccharide serotypes (about 90 for *Streptococcus pneumoniae*), and such vaccines need to be multivalent in order to have a high enough rate of overall efficacy. The first generation of these vaccines consisted of purified polysaccharides. These vaccines usually are effective in eliciting protective immunity in adults and children over about 2 years of age. For preventing diseases in <2-year-old children or in adults with other underlying diseases, polysaccharides are conjugated to carrier proteins for increasing their immunogenicity. Both monovalent (Hib) and multivalent (*S. pneumoniae*) polysaccharide conjugate vaccines have been developed and licensed.

The most recent of the major technologies for vaccine design is DNA (Chapter 14). This field began with the serendipitous observation that purified plasmid DNA injected intramuscularly could stimulate antibody- and cell-based immune responses. This field has evolved further in terms of the development of formulations to improve DNA uptake and expression, the optimized design of plasmid molecules, the exploration of new routes of administration, and the use of nonreplicating viruses or bacteria to deliver DNA to cells. Clinical studies have been performed both for DNA vaccines *per se* as well as for such vaccines as priming doses followed by boosting with protein antigen-based vaccines.

A wide range of prokaryotic and eukaryotic cell types have been developed into host cells for the expression of recombinant proteins as vaccine antigens. One of the most recent such developments has been the engineering of whole plants as recombinant expression systems (Chapter 15). In some cases, the recombinant protein may be purified from the plant and formulated as a vaccine antigen. This approach offers the advantage of the relatively inexpensive production of a large biomass as feedstock for the purification of the vaccine antigen. In other cases, it may be possible to eat the recombinant plant itself, e.g., tomato or spinach, as a user-friendly route of immunization. Initial clinical studies have been conducted with such vaccines.

Adjuvants and delivery systems are the basis of most of the new technologies in vaccine formulation (Chapter 16). Aluminum salts have been used as vaccine adjuvants (which modulate immune responses) throughout the 20th century. However, these salts often are not potent enough for adjuvanting protein-based vaccines to elicit strong enough immune responses. Therefore, a range of novel chemical and biochemical molecules as well as proteins have been evaluated as adjuvants, many of which have advanced to clinical studies. While many of these adjuvants are more potent than aluminum salts in animal studies and some more potent in clinical studies, their tolerability has not always been good enough to permit full clinical development. Nevertheless, within the last year an oil-in-water emulsion became the first new approved adjuvant (MF59 for inactivated influenza vaccine). This development augurs well for the development and approval of other new adjuvants. Delivery systems, which generally do not modulate immune responses, provide for the physical targeting of the active vaccine component to particular cells of the immune system. These vehicles may function by mechanisms such as depot effects, slow or pulsatile release, and presentation to mucosal surfaces.

One of the key considerations in the administration of vaccines is the route of uptake, for which there have been investigations of new routes besides injection in order to increase the rate of compliance as well as to potentially induce mucosal immune responses more efficiently. The only route other than injection used in any currently licensed vaccines is oral, as employed for whole virus or bacteria vaccines (polio, cholera). There have been clinical investigations into formulations of inactivated or subunit antigens in delivery systems for oral or nasal administration. Live attenuated cold-adapted influenza vaccine has been developed for intranasal administration and has been shown to be well-tolerated and efficacious in large clinical trials.

Transcutaneous immunization (Chapter 17) has been demonstrated for several different types of vaccines, including proteins, viruses and DNA. The coadministration on skin of a vaccine active-component with a mucosally-active toxoid (cholera toxin or *E. coli* heat-labile toxin) results in transcutaneous uptake and recognition by antigen-presenting cells. This mode of administration can result in the stimulation of both serum and mucosal immune responses, as has been observed in initial clinical studies. If this technology proves to be successful clinically, it would provide for relative ease of administration and consequent improved compliance with vaccination programs.

Combination vaccines, which are covered in depth in other books and reviews, are an important technology for administration. Combination vaccines are defined as the physical mixture of one or more vaccines during the manufacturing process or at the time of administration. In cases where vaccines indicated for the same age-group can be combined, the use of combination vaccines would result in fewer needlesticks. This would make multiple immunizations easier for subjects (and the parents of immunized children!) as well as for health-care practitioners. In this way, combination vaccines represent another technology for improving compliance with vaccination programs.

While recent technologies have expanded the horizons for new and improved vaccines, considerable financial and staff resources must be available to support full vaccine development. From the time that an initial lead has been identified, it can take 10 years and well over $100 million to develop a new vaccine. Furthermore, the success rate from the time of entry to development to availability on the market is only ca. 10-15%. Therefore, given this long timeframe, large cost and high risk, it is very important to design and implement a Product Development Plan early during this time-period in order to map out all the technologies and resources (money, people, facilities) necessary for optimizing the likelihood of success of the program.

I hope that *New Vaccine Technologies* will serve as a comprehensive reference on the major aspects of new approaches to developing vaccines. Since vaccination remains the most cost-effective and one of the most practical ways for preventing infectious diseases (and potentially for treating some diseases), the development and widespread applications of new technologies should spawn new vaccines that have not been approachable technically, with consequent impact on reducing morbidity and mortality worldwide. This book should prove useful for scientists, developers of vaccines and biotechnology products, clinicians, regulators, and health-care practitioners.

I am very grateful for the many collaborations I have been fortunate to have had over the last 17 years with innumerable coworkers in vaccines in BioChem Pharma, Merck, and Astra as well as with many colleagues in diverse collaborating groups. The loving support and encouragement of my wife Danielle and children Jacob and Miriam have been very important to me throughout my career and preparation of this book. Most importantly, I thank all the authors for their outstanding contributions that should make this book a key reference in the field of vaccine technologies.

Ronald W. Ellis

CHAPTER 1

New Technologies for Making Vaccines

Ronald W. Ellis

Introduction

The past two decades have witnessed an explosion in the number of technological and immunological approaches for making new vaccines. These developments have flowed from advances in a broad range of scientific fields. Some of the earliest applications of the newer technologies were to improving previously existing vaccines. However, most recent applications have been directed toward the development of new vaccines for diseases not previously approachable. The protective immunity elicited by a vaccine ideally would be life-long and robust after one or a few doses with minimal side effects (reactogenicity). Available vaccines and those under development fall short of this ideal, thus stimulating new research in the field.

There are two broad categories of vaccines, active and passive. An active vaccine stimulates the host's immune system to produce specific antibodies or cellular immune responses or both, which would protect against or eliminate a disease. A passive vaccine is a preparation of antibodies that neutralizes a pathogen and is administered before or around the time of known or potential exposure. Most references to the term **vaccine** are to active vaccines, which are the object of the vast majority of research and development activities in the field as well as the subject of this chapter. Although it is desirable or essential to administer a passive vaccine in specific instances (particularly if no active vaccine is available or sometimes for immunocompromised individuals), establishing lasting immunity through the administration of an active vaccine is a very important means of preventive medicine.

This chapter summarizes the major technologies, key issues and immunological objectives for making different kinds of active vaccines. The status of development of vaccines made by each approach is identified, whether licensed or in clinical or preclinical evaluations (Table 1.1). Only a few salient examples of each approach along with most common licensed vaccines are given with one or two accompanying references. While most examples are prophylactic vaccines for viruses and bacteria, there also is research into prophylactic vaccines for parasites and fungi and therapeutic vaccines for infectious diseases, cancer and autoimmunity. The technologies and examples presented should provide a strong framework for the reader to appreciate the diverse approaches to the research and development of new vaccines.

There are three general categories of active vaccines. A **live** vaccine is a microorganism that can replicate in the host or can infect cells, thereby functioning as an immunogen without causing its natural disease. A **subunit** or **inactivated** vaccine is an immunogen that cannot replicate in the host. A (DNA) **nucleic acid** vaccine, which cannot replicate in humans, is taken up by cells, in which it directs the synthesis of vaccine antigen(s).

Table 1.1. Status of development of representative human vaccines made by different technologies

Type of Vaccine**	Status of development*: Preclinical Evaluation***	Clinical Evaluation§	Licensed Product§§	Example#	Reference
A. Live					
1.Classical strategies -Viral					
a. Attenuation in cell culture			x	Poliovirus	2
			x	Measles virus	3
			x	Mumps virus	4
			x	Rubella virus	5
			x	Varicella-zoster virus (VZV)	6
b. Variants from other species			x	Smallpox (vaccinia virus)	7
		x		Rotavirus	8
c. Reassorted genomes			x	Rotavirus [a]	9,10
d. Temperature-selected mutants		x		Influenza virus	13
2. Recombinant virus		x		Herpes simplex virus (HSV)	16
3. Recombinant viral vector		x		Vaccinia virus [b]	18,19
		x		Adenovirus [c]	21
4. Classical strategies-Bacterial			x	Tuberculosis (*bacille Calmette-Guérin (BCG)*)	1
			x	Typhoid fever 14,15 (*Salmonella typhi*)	
5. Recombinant bacteria			x	Cholera (*Vibrio cholerae*)	17
6. Recombinant bacterial vector		x		*Salmonella typhi* [d]	23
		x		*V. cholerae* [d]	24
	x			*Shigella flexneri* [d]	25
B. Subunit/inactivated vaccines					
1. Whole pathogen					
a. Inactivated bacteria			x	Pertussis (*Bordetella pertussis*)	26
			x	Cholera	27
		x		Enterotoxigenic *Escherichia coli*	28
b. Inactivated virus			x	Poliovirus	30
			x	Influenza virus	31
			x	Rabies virus	32
			x	Japanese encephalitis virus	33
			x	Hepatitis A virus	35
2. Protein-based					
a. Natural			x	Hepatitis B virus (HBV)	39
			x	Pertussis	40-42
b. Chemically inactivated			x	Tetanus (*Clostridium tetani*)	43
			x	Diphtheria (*Corynebacterium diphtheriae)*	44
			x	Pertussis	45

continued on next page

Table 1.1. Status of development of representative human vaccines made by different technologies (continued)

Type of Vaccine**	Status of development*: Preclinical Evaluation***	Clinical Evaluation§	Licensed Product§§	Example#	Reference
c. Genetically inactivated			x	Pertussis	46
		x		Diphtheria	47
d. Recombinant			x	HBV	48
polypeptide			x	Lyme disease (*Borrelia burgdorferi*)	50
		x		HSV	51
		x		Human papilloma virus	52
		x		Yeast Ty [e]	55
	x			Rotavirus	53
3. Peptide based					
a. Fusion protein		x		Malaria [f]	58
b. Conjugate		x		Malaria [g]	60
	x			*Pseudomonas aeruginosa* [g]	57
c. Complex peptide	x			HIV	61
d. T-cell epitope		x		HBV	62
		x		cancer	63
4. Polysaccharide-based					
a. Plain polysaccharide			x	*Haemophilus influenza type b* (Hib)	64
			x	Meningococcal (*Neisseria meningitidis*)	65
			x	Pneumococcal (*Streptococcus pneumoniae*)	66
b. Conjugate			x	Hib [h]	67
			x	Pn [h]	68
5. Anti-idiotype (antibodies)		x		Cancer [i]	71,72
	x			HBV	70
C. DNA-based					
1. DNA- naked		x		Influenza	74
2. Facilitated DNA		x		Hepatitis B	76
		x		HIV	78
3. Viral delivery		x		Canarypox virus [j]	81
4. Bacterial delivery	x			*S. flexneri*	83
	x			*S. typhimurium*	84

continued on next page

Table 1.1. Status of development of representative human vaccines made by different technologies (continued)

*These categories are presented in the same outline as in the text.
**This denotes the single most advanced status achieved by each example.
***Not yet evaluated in a human clinical trial.
§ In clinical trial but not yet licensed.
§§ Licensed in one or more major countries in the world.
# These are representative examples for each vaccine strategy, with one or two key illustrative references.
a Licensed then withdrawn from distribution by the manufacturer.
b Expressing more than 50 different foreign polypeptides.
c Expressing at least six different foreign polypeptides.
d Examples of foreign polypeptides include toxoids from *E. coli, V. cholerae,* and *C. tetani.*
e Examples of foreign polypeptides include those encoded by HIV-1 *gag* and *env* genes.
f Fusion partner is HBsAg.
g Conjugate carrier is TT.
h Conjugate carriers are TT, DT, CRM_{197} and OMPC.
i Specificities are for human tumor carbohydrate and a human colorectal carcinoma antigen.
j Expressing rabies virus glycoproteins.

The strategic decision for developing a live, subunit/inactivated or nucleic acid-based vaccine should be made after considering the epidemiology, pathogenesis and immunobiology of the infection or disease in question as well as the technical feasibility of the various approaches. Epidemiology dictates the target population for the vaccine. The age and state of health of this population usually favors certain strategies as more appropriate for eliciting protective immunity. For example, minimal reactogenicity is very important for a vaccine intended for healthy infants, and certain types of vaccines are useless for infants because they do not elicit protective immunity. However, the degree of reactogenicity is less important in cases such as a therapeutic cancer vaccine. Knowledge of immunobiology can aid in identifying the nature of protective immunity that should be elicited by the vaccine; certain immune responses may be protective and others useless to the prevention or treatment of a particular infection. For example, the clearance of the natural infection may correlate with the appearance of antibodies against a particular microbial antigen; this would define that antigen as a candidate vaccine immunogen. Alternatively, the study of immunobiology is greatly facilitated or enabled by developing an experimental animal model, the availability of which enables candidate vaccines to be tested and optimized for protective efficacy before bringing the best one(s) forward for clinical evaluation. Historically, only a limited range of technical approaches has been feasible for a particular vaccine. Nevertheless, considering the expanding number of technical approaches, it may be possible in the future to custom-design many vaccines for optimal efficacy and tolerability.

Live Vaccines

Some live vaccines come very close to meeting the criteria for an ideal vaccine by being able to elicit lifelong protection with minimal reactogenicity using one or two doses. Such vaccines may be feasible in cases where the natural infection confers lifelong protection on the host. These vaccines consist of microorganisms (usually viruses) that replicate in the host in a fashion like that of the natural microorganism so that the vaccine may elicit an immune response similar to that elicited by the natural infection. The live vaccine is attenuated, meaning that its disease-causing capacity is eliminated by biological or technical manipulations. Care should be taken to ensure that the live vaccine is neither underattenuated (retaining pathogenicity even to a limited extent) nor overattenuated (no longer infectious enough to function as a vaccine). Live vaccines usually elicit both humoral immunity (antibodies) as well as cellular immunity (e.g., cytotoxic T lymphocytes (CTL)).

Although these properties per se might make live vaccines highly desirable, this is not technically feasible for most vaccines currently under development. A live vaccine may be incompletely attenuated and consequently cause its natural disease at a low frequency or be completely attenuated and incompletely immunogenic. Because a live vaccine can replicate, it may be possible for it to revert to its more naturally pathogenic form. Live vaccine strains can be transmitted from the vaccine to an unvaccinated individual, which can be quite serious if the recipient is immunodeficient or is undergoing cancer chemotherapy. In some cases, the natural viral infection per se fails to produce a protective immune response, such that an attenuated virus (without further engineering) would not be expected to produce a protective response.

Classical Strategies

The term **classical** refers to technical strategies that do not utilize rDNA technology. The production of live viral vaccines relies on propagating the virus efficiently in cell culture.

Attenuation in vitro

It has not been readily possible to develop live attenuated bacterial vaccines by classical strategies because there has been relatively little success with in vitro culture of bacteria for attenuation while maintaining immunogenicity. There also may be little competitive or selective pressure for bacteria to become less virulent during in vitro passage; bacteria could stop expressing virulence factors in vitro, then turn on their expression in vivo. The one widely available live bacterial vaccine based on serial in vitro passage is for tuberculosis. This vaccine consists of a live attenuated strain of *Mycobacterium bovis*, known as bacille Calmette-Guérin (BCG),[1] which was attenuated by 231 successive in vitro subculturings over 13 years. The available BCG vaccines vary in tolerability, immunogenicity and rate of protective efficacy in clinical trials. BCG vaccines have been inoculated into more than 1 billion people worldwide and have generally acceptable tolerability profiles. One would anticipate that the techniques of rDNA technology would be applied to attenuating a new bacterial strain. Therefore, by current technical and regulatory standards, it seems highly unlikely that a new live bacterial vaccine attenuated by a classical strategy alone will be developed.

The first classical strategy for viruses became possible during the 1950s with the ability to propagate viruses in cell culture. The approach is empirical, in that the wild-type virus isolated from a human infection is passaged in vitro through one or more cell types with the goal of attenuating its pathogenicity. In such cases, there may be selective pressure to produce less damage to cells. The mechanism by which mutation(s) are introduced during the course of attenuation is not well understood. In some cases (e.g., poliovirus[2]), it has been possible to demonstrate attenuation in a primate species, whereas attenuation has been proven in most cases only through the course of extensive clinical trials. The success of this empirical approach, which has been applied to both an oral vaccine (oral poliovirus vaccine[2] (OPV)) and to injected (parenteral) vaccines (measles,[3] mumps,[4] rubella,[5] varicella[6]), has been born out by the number of available licensed vaccines. The reactogenicity of such vaccines has been low enough that some of them (polio, measles) are widely accepted worldwide for routine pediatric use. By means of intensive immunization programs with OPV, polio is well on its way to worldwide eradication.

Variants from Other Species

An animal virus that causes a veterinary disease similar to a human disease can be isolated and cultivated, as was done for smallpox vaccine vaccine (derived from cowpox virus). The anticipated outcome is that the animal virus will be attenuated for humans yet will be sufficiently related immunologically to the natural human virus to elicit protective immunity. The immunization program was applied worldwide using vaccinia virus[7] and resulted in the complete eradication of smallpox worldwide by the mid-1970s, the only infectious disease ever eradicated. This program is a tribute to an effective control strategy and to the tireless efforts of countless individuals. Based on this model, first-generation vaccines for rotavirus consisted of animal-derived viruses.[8] However, these rotavirus vaccines were not reproducibly efficacious as human vaccines.

Reassorted Genomes

A reassortant virus derived following coinfection of a culture with two different viruses with segmented genomes contains genes from both parental viruses. To improve the efficacy of animal rotaviruses, reassortant rotaviruses were isolated containing mostly animal rotavirus genes, which confer the attenuation phenotype for humans, as well as the gene(s) for a human rotavirus surface protein, which elicits serotype-specific neutralizing antibodies for human rotavirus.[9,10] These reassortant rotaviruses have elicited higher efficacy rates as vaccine candidates than their parental animal viruses. A quadrivalent reassortant rhesus rotavirus vaccine was licensed in 1998. However, due to an increased rate of intussuseption (1:10,000) observed immediately following immunization, the vaccine was withdrawn from use. This withdrawal highlights safety as a key challenge for the development of new live vaccines. The same approach has been applied to influenza vaccines, in which a newly chosen influenza virus provides the genes that encode the immunogenic surface glycoproteins (hemagglutinin and neuraminidase), and an attenuated virus provides all other genes and, with them, the attenuation phenotype.[11] Such reassortant influenza viruses can be adapted to grow in mammalian cell lines such as MDCK[12] as a cell substrate to replace the use of chicken eggs.

Temperature-Sensitive Mutants

Viral mutants can be selected according to their growth properties at different temperatures. These viruses have been referred to as temperature-sensitive (ts), being unable to grow at elevated temperatures, or cold-adapted (ca), having been selected for growth in vitro at lower than physiological (37°C) temperatures, i.e., down to 25°C. The idea behind this approach is that ca viruses will be less vigorous in their in vivo growth than their wild-type parental virus, hence less virulent and phenotypically attenuated. A ca influenza vaccine has been developed.[13]

Chemical Mutagenesis

Another technique for creating an attenuated strain has been chemical mutagenesis followed by selection. The Ty21a strain of *Salmonella typhi* was derived in this fashion[14] and licensed for preventing typhoid fever based on its record of safety and efficacy over several years.[15]

Recombinant Microorganisms

Viral

The increased stability of the attenuation phenotype results from making the modifications or deletions in viral genes extensive enough that reversion through back-mutation is impossible or highly unlikely. In contrast, attenuated viruses derived by classical strategies may have only point mutations and therefore the capability to revert.

A deletion was made in a herpes simplex virus (HSV) gene encoding a glycoprotein required for infectivity. This glycoprotein in supplied to the virus in trans by the cell line during in vitro cultivation so that the resultant virus can initiate infection in vivo but not spread, which provides for its molecular attenuation.[16]

Recombinant Bacteria

The engineering of bacteria for attenuation is more complex than for viruses, given the much larger size of bacterial genomes. The strategy is to identify the gene(s) responsible for the bacterial virulence or colonization and survival and to either eliminate the gene (preferred) or to abolish or modulate its in vivo expression. As for viruses, there can be a balance between virulence and activity as a vaccine, which means that it is possible to overattenuate a bacterial strain to the point that it no longer replicates sufficiently to elicit an effective immune response.

Attenuation of *V. cholerae* strains has been accomplished by the rDNA-directed deletion of genes that encode virulence factors (such as cholera toxin (CT)).[17] Live attenuated cholera vaccine candidates prepared in this fashion have been evaluated clinically and one has been licensed. In order to assure attenuation by reducing the probability of reversion, it is desirable to delete two or more independent genes or genetic loci that contribute to virulence.

Recombinant Vectors

The second application of rDNA technology to the development of new live vaccines has been the engineering of viruses as vectors for "foreign" polypeptides from other pathogens. The goal of creating such vectors is to present the foreign antigen to the immune system in the context of a live infection so that the immune system responds to the antigen as a live immunogen and thereby develops broader immunity (humoral and cellular) to the corresponding human pathogen. The recombinant polypeptide is expressed within the infected cell and either is transported to the cell surface to stimulate antibody production or is broken down into peptide fragments that are transported to the cell surface where they elicit CTL responses. This strategy also has the potential advantage of amplification of the immunogenic signal when the live vector replicates.

Viral

The prototype viral vector is vaccinia virus. Dozens of different recombinant polypeptides have been expressed in vaccinia virus.[18] At least 25 models for different infections have shown that vaccination of animals can protect against the pathogen encoding the recombinant polypeptide. Recombinant vaccinia viruses expressing tumor antigens also have been shown to be protective in rodent tumor model challenge studies. Given the known sequelae to immunization for smallpox, which are more serious in immunocompromised individuals, vaccinia virus itself has been engineered to reduce its virulence without compromising its efficacy as a live viral vector.[19] Cytokines can influence the nature of magnitude of the immune response. In order to selectively manipulate the type of immune response to a vaccine antigen in the context of a live vector vaccination, a recombinant vector has been constructed which expresses both a cytokine as well as a recombinant vaccine antigen.[20] Fowlpox and canarypox viruses are being developed as live vectors that can infect human cells but not produce infectious viral progeny. This inability to spread makes these viral vectors also classifiable as DNA-based vaccines (see *Viral Delivery* in this Chapter).

Other mammalian viruses have been engineered into live vectors. Adenovirus strains, which have been used extensively as vaccines in military recruits to prevent acute respiratory disease, have been engineered to express foreign polypeptides and have elicited protective immunity in several viral challenge models in animals.[21] Optimizing recombinant polypeptide expression remains an important technical objective for all these live viral vectors.

RNA viruses can be engineered in similar fashion. Sindbis and other alphaviruses have received extensive attention due to their broad host range, ability to infect nondividing cells, and potential high-level expression per cell.[22] On this basis, Sindbis has been developed into a nucleic acid-based vaccine (see *Viral Delivery* in this Chapter).

Bacterial

Pathogenic bacteria can be engineered into live recombinant vectors for the expression of foreign polypeptide antigens. The most common applications have been to engineer enteric pathogens so that they can induce mucosal immunity against the foreign polypeptide upon oral delivery. In the field of developing live bacterial vectors, *S. typhi* has been the focus of the most effort in terms of strain development, immunology, molecular development; and clinical testing.[23] *V. cholerae*,[24] and *S. flexneri*[25] also have been engineered into oral recombinant vectors for clinical evaluations. The challenges for these live attenuated vectors are both to retain sufficient virulence for replication in the gut and expression of appropriate levels of foreign polypeptides as well as to achieve sufficient attenuation to assure good tolerability. The ability of some of these bacterial species to replicate intracellularly may augment the ability of expressed foreign polypeptides to elicit cellular immune responses against their respective pathogens.

Subunit/Inactivated Vaccines

Such vaccines have advantages that relate to their inability to multiply within the host. Generally they are well tolerated, especially for the majority of such vaccines that undergo purification to remove other macromolecules. Given the broad range of available approaches, it also is generally more feasible technically to produce a subunit or inactivated vaccine. Immuno-genicity may be enhanced by its administration with an adjuvant or delivery system (see *Formulation of Antigens* in this Chapter). Nevertheless, a development program should be undertaken with the realization that multiple doses, often followed by booster doses, most often are necessary for attaining long-term protective immunity. These vaccines usually function by stimulating humoral immune responses as well as by priming for immunological memory. In certain cases, especially when administered with certain adjuvants and delivery systems, nonlive vaccines may stimulate CTL immunity.

Whole Pathogen

The earliest approach to making inactivated vaccines relied on the use of whole bacteria or viruses with the objective of eliciting the formation of antibodies to many antigens, some of which would neutralize the pathogen.

Bacteria

These vaccines are prepared by cultivating the bacteria, collecting the cells, and inactivating them with heat or with chemical agents such as thimerosal or phenol.[26] The final vaccine does not undergo further purification. Owing to their biochemically highly crude nature, which includes virtually all bacterial cellular components, the reactogenicity of such vaccines when given parenterally (e.g., *Bordetella pertussis*) is usually greater than that of other types of vaccines. On the other hand, inactivated whole-cell *V. cholerae*[27] and enterotoxigenic *Escherichia coli* (ETEC)[28] vaccines have been well-tolerated by the oral route. Oral inactivated whole-cell cholera (WCC) vaccine, which lacks CT (and its toxic effects), has been shown to be very well tolerated and to have a rate of efficacy of ca. 60% for three years in a high-risk population.[27] In order to elicit antibodies that would neutralize CT and increase efficacy, the recombinant B subunit of CT (CTB) which lacks toxin activity is independently expressed, purified, and added back to the WCC vaccine. This combined WCC + rCTB vaccine was shown to have a somewhat higher rate of efficacy than WCC vaccine alone.[29]

Virus

Some inactivated viral vaccines have been available for decades and are generally very well tolerated. Because viruses generally are shed into the cell culture media when grown in vitro, cell-free media from infected cultures are collected. The large size of the virus particles relative to other macromolecules in the media enables the particles to be enriched readily by simple purification techniques that exploit the size of the particles. Examples include poliovirus,[30] influenza virus,[31] rabies virus[32] and Japanese encephalitis virus.[33] Alternatively in the case of killed hepatitis A virus (HAV) vaccine, infected cells are lysed and virus particles are purified.[34] The virus particles are inactivated chemically, typically by treatment with formalin, and then may be adjuvanted by an aluminum salt. The key epitope(s) on the surface of many nonenveloped small viruses that elicits a protective immune response (protective epitope) is often conformational, being formed by the highly ordered assembly of structural proteins into precise structures. For most of the listed viruses for which inactivated vaccines have been developed and licensed, it has not been possible to readily mimic the conformation of such epitopes by other technologies, e.g., recombinant polypeptides. Inactivated viral vaccines tend to be highly potent immunologically, e.g., one dose of hepatitis A vaccine is protective at a dosage of 50 ng.[35] Thus, this classical strategy, which has had an excellent track record of producing well-tolerated and efficacious vaccines, remains the technology of choice for many viral vaccines.

Protein-Based

Developing a protein-based vaccine is a preferred strategy for many pathogens in which a polypeptide contains protective epitopes, given the abovementioned issues regarding inactivated vaccines. Protein-based approaches have relied on genetic, biochemical, and immunological techniques to identify protective epitopes and their corresponding polypeptides as candidate vaccine antigens.

More recently, genomics technology has enabled the identification of new vaccine antigens in lieu of prior available biochemical or antigen data. Once the complete sequence (or portions thereof) of the genomic DNA or RNA are available, open reading frames (ORFs) are identified. The derived amino acid sequence can be inspected for structural features, such as homologies with proteins from other related pathogens that are vaccine candidates or a hydrophobic N-terminal sequence that suggests surface localization. The genes are expressed in a recombinant host cell (typically *E. coli*) and the recombinant polypeptide is purified and used to immunize animals to derive polyclonal antibodies for identifying whether the hypothetical protein is produced by the pathogen. Antisera also can be used in biological assays (neutralization of viruses, opsonization of bacteria) to see whether the protein may be an attractive vaccine candidate. The new protein also can be used for immunization and challenge in an animal model. Some of the earliest applications of genomics technology to viruses were for the discovery of hepatitis C virus (HCV)[36] and hepatitis E virus (HEV),[37] which resulted in the direct definition of candidate vaccine antigens. The genomic approach was applied to *Neisseria meningitidis* in which a number of candidate vaccine antigens were defined.[38]

Natural

The first protein-based vaccines relied on natural sources of antigens. In this regard, the first-generation hepatitis B vaccine was unique among active vaccines in that it utilized a human tissue source (plasma) for the vaccine antigen. Liver cells of individuals chronically infected with hepatitis B virus (HBV) shed excess viral surface protein, i.e., hepatitis B surface antigen (HBsAg), into blood as 22 nm virus-like particles (VLP) with protective epitopes. To develop a vaccine, plasma was harvested from long-term chronic carriers of hepatitis B, HBsAg purified and the final preparation subjected to 1-3 inactivation techniques (depending on the manufacturer) to kill HBV and any other potential human pathogens.[39]

Proteins purified from cultures of *B. pertussis* are combined to formulate acellular pertussis ($_a$P) vaccines, which eventually should replace whole-cell pertussis vaccine for routine pediatric vaccinations in many developed countries. Depending on the number of different protein antigens, these $_a$P vaccines are referred to as one-, two-, three-, four-, or five-component vaccines and have been licensed based on recent efficacy studies.[40-42] These vaccines all contain pertussis toxoid (PT) as a component, whose preparation is described below.

Chemical Inactivation

Many bacteria produce protein toxins that are responsible for the pathogenesis of infection. It had been recognized for many decades that, when a toxin was pathogenic after infection, antitoxins (antisera enriched in toxin-specific antibodies) that were effective in neutralizing toxin activity in vivo could prevent or ameliorate symptoms of certain bacterial infections. This precedent established the basis for bacterial toxins to be formulated as active vaccines. The toxin molecules are purified from bacterial cultures (e.g., *Corynebacterium diphtheriae* (D), *Clostridium tetani* (T), *B. pertussis* (P)) and then detoxified by incubation with a chemical such as formalin or glutaraldehyde. Detoxified toxins, referred to as toxoids, thus represent two of the vaccines (D,T) in the diphtheria, tetanus and pertussis (DTP) combination vaccine.[43,44] PT[45] combined with other pertussis antigens comprise the $_a$P vaccines.

Genetic Inactivation

The chemical toxoiding procedure has possible disadvantages, including the alteration of protective epitopes with ensuing reduced immunogenicity and potential reversion to a biologically active toxin. To produce a stable PT, codons for amino acids required for toxin bioactivity (adenosine diphosphate (ADP) ribosyl transferase) were mutated. The altered gene was substituted for the native gene in the parental organism, which then produces immunogenic but stably inactivated PT. As a refinement of this strategy, two mutations were introduced into PT to assure the lack to reversion;[46] this double mutant PT (which also is treated with formalin under milder conditions to improve its immunogenicity or stability) is a component of a $_a$P vaccine.[40] In a related application, mutated cultures of *C. diphtheriae* were screened for the secretion of enzymatically inactive yet antigenic toxin molecules. Subsequent cloning and sequencing of one such mutated toxin gene identified a single amino acid mutation at the enzymatic active site (also an ADP-ribosyl transferase). This genetic toxoid (CRM_{197})[47] is the protein carrier for a licensed *H. influenzae* type *b* (Hib) conjugate vaccine (Section II.D). This technology also has been applied to *V. cholerae* toxin (CT) and ETEC heat-labile toxin (LT) to produce candidate mucosal adjuvants (see Formulation of Antigens-Adjuvants p. 15).

Recombinant Polypeptides

The first application of rDNA technology to the production of a vaccine was for hepatitis B. Given the precedent of plasma-derived HBsAg as a well-tolerated and efficacious vaccine, the S gene encoding HBsAg was expressed in bakers' yeast *S. cerevisiae*,[48] which express 22-nm HBsAg particles within cells. HBsAg is a VLP in that its surface is similar to that of HBV virions. The yeast-derived vaccine, which is available worldwide in large supply, has largely supplanted the equally efficacious and well-tolerated plasma-derived vaccine. HBsAg also has been expressed in transgenic tobacco leaves and potato tubers; the purified HBsAg was immunogenic.[49]

There are innumerable ongoing research and development applications of rDNA technology to produce proteins as vaccine candidates. The major *Borrelia burgdorferi* surface protein (OspA), expressed in *E. coli* as a recombinant lipoprotein,[50] has been licensed as a vaccine for Lyme disease. Recombinant-derived HSV glycoproteins expressed in Chinese hamster ovary (CHO) cells and formulated as vaccines were tested in clinical trials.[51]

Large particles most often are more immunogenic than individual polypeptides. Furthermore, as in the case of HBsAg VLPs, particles usually elicit antibodies to conformational epitopes on the particle, while isolated surface polypeptides of the particle might not elicit the production of such antibodies. The human papilloma virus (HPV) virion is a highly ordered structure whose major protein is L1. Expression of L1 in eukaryotic cells (e.g., *S. cerevisiae*) results in the formation of L1 VLPs, which after immunization elicit antibodies that bind to virions.[52] Recombinant rotavirus[53] and parvovirus[54] VLPs also have been expressed as potential parenteral vaccines.

Many host cells have been used for the expression of heterologous recombinant genes. In addition to the previously mentioned (*E. coli*, *S. cerevisiae* and CHO), expression systems have been developed for cells from other bacterial and yeast species and other mammalian continuous cell lines (CCLs), e.g., African green monkey kidney (Vero). Whole animals and plants also can be employed as hosts for recombinant expression. In general, smaller proteins that do not require posttranslational modifications can be expressed efficiently in authentic form in microbial expression systems. In contrast, polypeptides that require posttranslational modifications for immunogenicity such as glycosylation for proper immunogenicity are expressed in mammalian CCLs capable of correctly performing such modifications.

Carrier

A novel approach to recombinant vaccines is the use of yeast Ty particles as killed carriers for foreign proteins. Yeast Ty is a particle assembled in *S. cerevisiae* that cannot replicate in mammals. It is possible to express a gene encoding a foreign protein in conjunction with Ty genes such that the foreign proteins assemble with Ty proteins into mixed particles.[55] Because the foreign proteins are expressed on the surface of these large particles, their immunogenicity as vaccine antigens might be enhanced.

Peptide-Based

In many cases, it has been possible to identify B-cell epitopes within a polypeptide against which neutralizing antibodies are directed. Many B-cell epitopes are conformational, being formed by the juxtaposition in three-dimensional space of amino acid residues from different portions of the polypeptide, which means that such epitopes require the full polypeptide for their proper immunogenic presentation. In contrast, other peptide epitopes are linear in nature, being fully antigenic as short linear sequences in the range of 6-20 consecutive amino acid residues in the polypeptide. Some linear epitopes are only weakly immunogenic when presented in the context of the full polypeptide. In other cases, natural peptides would be effective vaccine antigens if they were rendered sufficiently immunogenic. Linear B-cell epitopes of this type have been defined for the malarial circumsporzoite (CS) protein (repetitive 4-amino acid sequence)[56] and for the *Pseudomonas aeruginosa* pilus protein.[57] Both of these polypeptides contain linear epitopes that are recognized by antibodies that neutralize the respective pathogens, yet the whole polypeptides elicit such antibodies only weakly. It is interesting to speculate that this may represent a mechanism by which these and other pathogens have evolved to escape immunological surveillance by rendering their neutralization epitopes less immunogenic.

The application of the following strategies (fusion protein, conjugate, complex peptide) to weakly immunogenic linear epitopes has resulted in immunogenic presentations that elicit substantially increased titers of neutralizing antibody compared with those elicited by the epitope presented in the context of its natural full-length polypeptide. Nevertheless, the most effective strategy in terms of ultimate clinical utility remains to be established on a case-by-case basis.

Fusion Protein

The immunogenicity of linear epitopes can be increased by making a genetic fusion of defined epitopes to a carrier protein that forms a large particle to improve the immune presentation of the peptide. Two commonly used protein fusion partners of this type are HBsAg[58] and hepatitis B core antigen,[59] a 28-nm particle encoded by hepatitis B virus. Fusions have been made at the N-terminus, the C-terminus, or the internal portion of the polypeptide sequence of the protein partner, depending on which location affords the best immunogenic presentation while maintaining efficient particle formation.

Conjugate

The peptide can be chemically conjugated to a carrier protein. The peptide sequence is synthesized chemically with a reactive amino acid residue through which conjugation occurs to the carrier protein. The most commonly used carrier proteins in conjugates are bacterial proteins that humans commonly encounter such as tetanus toxoid (TT), for which a conjugate with the malarial CS epitope has been tested clinically.[60]

Complex Peptide

Multimers of the peptide sequence can be synthesized for linkage together in repeated arrays, as applied to the malarial CS and HIV-1 gp120 peptide epitopes.[61]

T-Cell Epitopes

Peptide epitopes recognized by CTL may be useful immunogens for the prophylaxis of infections by agents such as HIV or immunotherapy for chronic diseases such as hepatitis B. CTL peptide epitopes generally are poor immunogens. Thus, for an immunotherapeutic hepatitis B vaccine, a CTL epitope from the HBV core protein was modified by covalent linkage to a T-helper epitope (from tetanus toxoid) as well as two palmitic acid molecules.[62] This vaccine was shown in clinical studies to be immunogenic for eliciting HBV-specific CTL and memory CTL. Melanoma-specific T-cell epitopes as peptides have been used to pulse dendritic cells in vitro for delivery to the patient, with some observations of tumor regression.[63]

Polysaccharide-Based

There are many bacteria with an outer polysaccharide (Ps) capsule. In many if not most of the encapsulated bacteria studied, antibodies directed against capsular Ps are protective against infection. These observations have established capsular Ps as vaccine antigens.

Plain Ps

Native capsular Ps contain up to hundreds of repeat units distinctive for each bacterial species and antigenic subtype in which each monomer consists of a combination of monosaccharides, phosphate groups and small organic moieties. The Ps is shed by the organism during its growth and is harvested from the culture medium. These Ps preparations are usually immunogenic in adults and children over 2 years of age and elicit antibodies that may mediate the opsonization of the organism thereby protecting against infection. Ps vaccines have been licensed for Hib[64] (monovalent for serotype b), *Neisseria meningitidis*[65] (quadrivalent) and *Streptococcus pneumoniae*[66] (23-valent). The shortcoming of these vaccines is that Ps, being T-cell–independent (TI) immunogens, are poorly immunogenic or nonimmunogenic in children younger than 2 years, and they do not elicit immunological memory in older children and adults.

Conjugate

Although infants and children younger than two years old do not recognize TI immunogens efficiently, they can respond immunologically to T-cell-dependent (TD) immunogens such as proteins. The chemical conjugation of Ps to a carrier protein converts the Ps from a TI to a TD immunogen. As a consequence, Ps-protein conjugate vaccines can elicit protective IgG and immunological memory in infants and young children. This strategy is particularly important for encapsulated bacteria such as Hib and *S. pneumoniae* (pneumococcal; Pn) owing to the preponderance of invasive diseases caused by these bacteria in children younger than two years old, in whom a Ps vaccine is ineffective. There are four different licensed Hib conjugate vaccines,[67] all with different carrier proteins (TT, DT, CRM_{197} and an outer membrane protein complex from *N. meningitidis* Group B) of different sizes and immunological character, distinct Ps chain lengths and distinct conjugation chemistries. Given these differences, the four vaccines display one or more differences in the following immunological properties: response of 2-month-old infants to the first dose of vaccine, responses of four- and six-month-old infants to the second and third doses, response of children older than one year to a booster dose, kinetics of decay of antibody levels, peak of antibody titer and age at which protection from clinical disease first can be shown.

Pn bacteria consist of ca. 90 serotypes, as reflected in distinct capsular Ps structures. For designing a pediatric Pn conjugate vaccine, seven serotypes have been recognized as responsible for 60-75% of the major pediatric Pn diseases (acute otitis media, pneumonia, meningitis). A heptavalent vaccine was recently licensed.[68] Other vaccines being tested in advanced clinical trials consist of a mixture of up to 11 individual Pn Ps conjugates.[69]

Anti-Idiotypic Antibodies

The **idiotype** (Id), that is, idiotypic determinant, represents unique antigenic determinants associated with the hypervariable region of the antibody molecule. An antibody-1 (Ab-1) can be defined as an antibody recognizing a particular antigen, e.g., vaccine candidate. The Id on Ab-1 itself can act as an immunogen; the antibodies that bind to the Id on Ab-1 are referred to as **anti-idiotypic antibodies** (anti-Id) or Ab-2. The paratope on Ab-1 is the binding site for the particular antigen; thus, the binding site of an anti-paratope antibody is a molecular "mimic" of the original antigen. If the paratope and the Id on Ab-1 represent the same or overlapping sites, then the Ab-2 and particular antigen both bind at that site and thus have similar conformations (Ab-1 is the image of both the antigen and Ab-2). By virtue of the antibody-binding site of Ab-2 mimicking the conformation of the particular antigen, Ab-2 molecules themselves can be used as vaccine candidates in which an epitope (the Id) is presented on a carrier molecule (whole Ab-2). It was shown that vaccination of chimpanzees with anti-Id that mimicked HBsAg protected the animals from infection with HBV.[70]

Numerous technologies exist for using an antigen as a vaccine candidate, either directly or by augmenting its immunogenicity as described earlier. Furthermore, an antibody molecule (Ab-2) is not necessarily a desirable immunological carrier for an antigen (anti-Id). Hence, the situations in which the use of anti-Id would be the preferred vaccine strategy are quite limited in number. Certain tumor antigens cannot be recognized immunologically by the host, because these antigens are self-antigens, often being expressed in low levels in the host. Nevertheless, the Ab-2 that is the mimic of the tumor antigen, yet not necessarily identical in structure to the antigen (hence not a self-antigen), can elicit an immune response against the tumor antigen.[71] When the tumor antigen is a defined Ps that cannot be isolated or synthesized in quantities sufficient for vaccine studies, an anti-Id of the mimic of the Ps can be a useful cancer vaccine candidate.[72] The ultimate utility of anti-Id as a vaccine strategy remains to be established. Furthermore, to obtain the highest degree of specificity as a vaccine candidate, one would derive a monoclonal antibody (MAb) as an anti-Id and make it into a recombinant human or humanized MAb.

DNA (Nucleic Acid) Vaccines

It was shown that after cells in vivo take up DNA encoding vaccine antigen(s), the antigens can be secreted or can be associated with the cell surface in a way that would trigger a humoral or cellular immune response. Furthermore, the uptake of DNA can be facilitated by chemical formulation or delivery by a virus or bacteria. The latter approaches fit the definition of a DNA-based vaccine as one that cannot replicate in humans. "Naked", facilitated and virally-delivered DNA vaccines recently have entered clinical studies.

"Naked" DNA

One strategy has been to inject intramuscularly a solution of DNA encoding a vaccine antigen.[73] Cells take up the plasmid DNA, transcribe it and synthesize the antigen, which may be processed in a similar way as in a live viral infection. The advantages of using DNA are the relative technical ease of preparation and the ability to direct the synthesis of multiple copies of mRNA, hence an expected amplification of both antigen synthesis and immune response. Such vaccines have been shown to be effective in many animal models of infection, especially virus models.[74] As a novel route of delivery, naked DNA has been applied to mouse skin from which it is taken up by hair follicles to stimulate an immune response.[75] While naked DNA does elicit the production of specific antibodies, it is particularly proficient at eliciting cellular immune responses.

Formulated DNA

Facilitation can be at the level of cellular uptake, expression or immunological activation. One strategy has been the incorporation of DNA into microprojectiles that then are "fired" into cells, which produce the encoded antigen. This "gene gun" technique has been reported to be potent at eliciting immune responses and has undergone initial clinical use.[76] For improving the efficiency of uptake, DNA has been coated with cationic lipids, lipospermines or other molecules which neutralize their charge and have lipid groups for facilitating membrane transfer.[77] Such formulations also are being researched for alternate routes of administration (e.g., oral, nasal) which may elicit mucosal immunity. The anesthetic bupivacaine given in conjunction with DNA has been shown to enhance DNA uptake and expression.[78] ADP-ribosylating exotoxins given together with DNA and applied to the skin can stimulate transcutaneous immunization.[79] The base composition of the DNA may affect its potency in that unmethylated CpG dinucleotides have been shown to induce B-cell proliferation and immunoglobulin secretion and to adjuvant responses to DNA vaccines.[80]

Viral Delivery

The above nucleic acid-based vaccines all result in the deposition into a cell of a plasmid. For delivery of DNA by fowlpox or canarypox virus, the expression cassette for the recombinant protein is integrated into the viral genome. These avian poxviruses can infect mammalian (human) cells but not produce infectious virus;[81] hence this can be considered a nucleic acid-based approach. This single round of self-limiting infection may be sufficient to elicit broad immunity to a pathogen whose recombinant polypeptide is expressed by these avian poxviruses in infected cells, while reactogenicity should be minimal, given the inability of the virus to spread within the host. A variation on the design of the expression plasmid is to use a virus-based DNA expression system that can amplify the level of RNA and protein expression as occurs in a live virus infection, as developed for Sindbis virus vectors.[82]

Bacterial Delivery

Bacteria that replicate intracellularly can be engineered to deliver plasmid DNA into cells for the expression of recombinant proteins. *S. flexneri* has been attenuated by making a deletion mutant in an essential gene (*asd*). While such a strain can be propagated in vitro in the presence of diaminopimelic acid (DAP) and can invade cells, it cannot replicate in vivo, where DAP is not available. A plasmid harboring a eukaryotic promoter and recombinant gene was transformed into this strain. The resultant recombinant *S. flexneri* strain was shown to be able to invade mammalian cells in vitro and to express the plasmid-encoded protein as a potential vaccine antigen.[83] Since *S. flexneri* replicates in the intestine and stimulates mucosal immunity, this vector may be delivered orally for delivering DNA to cells where mucosal immunity is stimulated. Other attenuated strains of bacterial species, e.g., *Salmonella*,[84] that can invade mammalian cells but not divide also can deliver recombinant plasmids orally for expressing recombinant proteins as vaccine antigens.

Formulation of Antigens

The immunological effectiveness of vaccines (other than live) may be enhanced by their formulation, which refers to the final form of the vaccine to be administered in vivo. In addition to the "active substance" (antigen or DNA), the formulation may contain an adjuvant and/or delivery system in addition to excipients. The adjuvant is a substance that stimulates an increased humoral and/or cellular immune response to a coadministered antigen. The delivery system is a vehicle for assuring the presentation of the vaccine to cells of the immune system or for stabilizing and releasing the antigen over an extended period of time. There may be overlap in structure and function between adjuvants and delivery systems. Many future vaccines are expected to contain new adjuvants and delivery systems. This topic has been addressed extensively in reviews by others.[85,86]

Adjuvants

Aluminum salts, such as hydroxide or phosphate, are currently the only adjuvants widely licensed for human use. This adjuvant has been used for decades in vaccines injected into more than 1 billion people worldwide. The vaccine antigen binds stably to the aluminum salt by ionic interactions and forms a macroscopic suspension in solution.[87] This adjuvant preferentially promotes a Th2-type immune response, i.e., antibody-based, and thus is not useful in applications where inducing a cell-mediated immune response is needed for protection. While aluminum salts have been useful for certain vaccines (e.g., hepatitis B, pertussis), for other vaccine antigens they are not potent enough for inducing antibody responses which are high enough to be optimally effective. Aluminum salts have not been shown to be useful for presentation of vaccines by the oral or intranasal routes. Therefore, many chemicals, biochemicals from natural sources, and proteins with immune-system activity (cytokines[88]) have been researched as potential adjuvants. The adjuvanticity of virtually all known formulations is associated with local or systemic side-effects which may be mechanism-based or nonspecific. The ideal adjuvant needs to achieve a balance between degree of side-effects and immune-enhancement.

Certain bacterial toxins with ADP-ribosylating activity have received considerable attention as mucosal adjuvants in terms of molecular engineering. In particular, CT was shown to be active as a mucosal adjuvant for a coadministered antigen[89] when presented by the oral, nasal, vaginal or rectal routes, as was shown subsequently for the heat-labile toxin (LT) of ETEC. These toxins are composed of a catalytic A subunit and a pentameric B subunit that binds to GM1 ganglioside on many cell types. However, both CT and LT are toxic in humans, especially by the oral route through which they induce diarrhea. To dissociate the toxicity and adjuvanticity of CT and LT, point mutations have been made which result in reduced or eliminated ADP-ribosylating activity, reduced toxicity, and the apparent retention of adjuvanticity in mice.[90] An alternative approach has been to eliminate the B subunit and substitute a synthetic dimeric peptide derived from *Staphylococcus aureus* Protein A (DD) which binds to immunoglobulin (Ig). The fusion protein of the CTA subunit with the DD domain binds to Ig^+ cells, appears devoid of toxicity, retains ADP-ribosylating activity, and is active as an adjuvant in mice.[91] The tolerability and effectiveness of these engineered adjuvants needs to be validated in humans.

Delivery Systems

Besides presenting an antigen or DNA to cells of the immune system, a delivery system may perform other key functions. There may be a depot effect whereby the antigen is maintained in an appropriate in vivo site for continual immune stimulation. There may be an enhancement of vaccine stability in vivo. For mucosally-delivered vaccines, the delivery system may enable efficient presentation and uptake by M cells, followed by transcytosis into Peyer's patches and presentation to lymphocytes for the induction of mucosal immunity.[92] For certain formulations, the vaccine may be maintained in vivo inside a physical structure for a significant period of time, during which it is released slowly or in pulsatile fashion such that it may function as a one dose vaccine. No delivery systems have been widely licensed. Gaining clinical and pharmaceutical experience with new delivery systems and adjuvants remains a key goal in the field.

Conclusion

Technological developments in the past decade have rapidly expanded the number of general strategies for making new vaccines. In the next decade the number of approaches will continue to expand and technical aspects further refined, such that most antigens could be presented in a highly immunogenic form in the context of a live or subunit vaccine. Protein antigens alternatively can be expressed through a nucleic acid-based vaccine. Further understanding of gene function in viral and bacterial pathogens should enable live vaccines to be more stably and predictably attenuated as vaccines and as live vectors for immunizing against other pathogens. Adjuvant technologies should advance to the point where formulations which are more potent than aluminum salts, yet as well tolerated, gain widespread use for subunit/inactivated vaccines and where oral delivery of purified proteins for immunization becomes feasible. Similarly, formulations of DNA may improve the potency of DNA vaccines and its ability to be delivered by routes that elicit mucosal immunity.

As all these technological advances proceed, it is likely that the limiting factor in developing new vaccines for human use will continue to be a more comprehensive understanding of immunology. Some areas in which increased knowledge would have a practical payoff for vaccine development are the immunobiology of pathogens, the type and specificity of immune response required for persistent protection against disease, the attainment of mucosal immunity and the optimal vaccination strategy to achieve this protection. There also should be significant developments in applications to noninfectious diseases, such as cancer and autoimmune diseases.

References

1. Weill-Halle B. Oral vaccination. In: Rosenthal SR, ed. BCG Vaccination Against Tuberculosis. Boston: Little, Brown 1957.
2. Sabin AB, Boulger LR. History of Sabin attenuated poliovirus vaccine. J Biol Stand 1973; 1:115-118.
3. Enders JF, Katz SL, Milovanovic MV et al. Studies on an attenuated measles-virus vaccine I. Development and preparation of the vaccine: Technics for assay of effects of vaccination. N Engl J Med 1960; 263:153-159.
4. Buynak EB, Hilleman MR. Live attenuated mumps virus vaccine. I. Vaccine development. Proc Soc Exp Biol Med 1966; 123:768-775.
5. Plotkin SA, Farquhar JD, Katz M et al. Attenuation of RA27/3 rubella virus in WI-38 human diploid cells. Am J Dis Child 1969; 118:178-185.

6. Takahashi M, Okuno Y, Otsuka T et al. Development of a live attenuated varicella vaccine. Biken J 1975; 18:25-33.
7. Henderson DA. Smallpox eradication. Proc R Soc Lond 1977; 199:83-97.
8. Vesikari T, Kapikian AZ, Delem A et al. A comparative trial of Rhesus monkey (RRV-1) and bovine (RIT 4237) oral rotavirus vaccines in young children. J Infect Dis 1986; 153:832-839.
9. Rennels MB, Glass RI, Dennehy PH et al. Safety and efficacy of high-dose Rhesus-human reassortant rotavirus vaccines—Report of the national multicenter trial. Pediatrics 1996; 97:7-13.
10. Clark HF, Offit PA, Ellis RW et al. The development of multivalent bovine rotavirus (strain WC3) reassortant vaccine for infants. J Infect Dis 1996; 174:S73-80.
11. Maassab HF, DeBorde DC. Development and characterization of cold-adapted viruses for use as live virus vaccines. Vaccine 1995; 3:355-371.
12. Halperin SA, Newtruck AC, Eastwood BJ. Safety and immunogenicity of a new influenza vaccine grown in mammalian cell culture. Vaccine 1998; 16:1331-5.
13. Treanor JJ, Kotloff K, Betts RF et al. Evaluation of trivalent, live, cold-adapted (CAIV-T) and inactivated (TIV) influenza vaccines in prevention of virus infection and illness following challenge of adults with wild-type influenza A (H1N1), A (H3N2), and B viruses. Vaccine 2000; 18:899-906.
14. Germanier R, Furer E. Isolation and characterization of galE mutant Ty21a of *Salmonella typhi*: A candidate strain for a live, oral typhoid vaccine. J Infect Dis 1975; 131:553-558.
15. Levine MM, Black RE, Ferreccio C et al. Clinical Typhoid Committee. Large-scale field trial of Ty21a live oral typhoid vaccine in enteric-coated capsule formulation. Lancet 1987; 2:1049-1052.
16. Farrell H, McLean CS, Harley C et al. Vaccine potential of a herpes virus type 1 mutant with an essential glycoprotein gene deleted. J Virol 1994; 68:927-32.
17. Tacket CO, Losonsky G, Nataro JP et al. Onset and duration of protective immunity in challenged volunteers after vaccination with live oral cholera vaccine CVD 103-HgR. J Infect Dis 1992; 166:837-841.
18. Moss B. Vaccinia virus vectors. In: Ellis R, ed. Vaccines: New Approaches to Immunological Problems. New York: Marcel Dekker. 1992; 345-357.
19. Lee MS, Roos JM, McGuigan LC et al. Molecular attenuation of vaccinia virus: Mutant generation and animal characterization. J Virol 1992; 66:2617-2630.
20. Leong KH, Ramsay AJ, Boyle DB et al. Selective induction of immune responses by cytokines coexpressed in recombinant fowlpox virus. J Virol 1994; 68:8125-8130.
21. Graham FL, Prevec L. Adenovirus-based expression vectors and recombinant vaccines. In: Ellis R, ed. Vaccines: New Approaches to Immunological Problems. New York: Marcel Dekker. 1992; 363-390.
22. Schlesinger S. Alphaviruses—vectors for the expression of heterologous genes. Trends Biotechnol 1993; 11:18-22.
23. Gonzalez C, Hone D, Noriega FR et al. Salmonella typhi vaccine strain CVD 908 expressing the circumsporozoite protein of Plasmodium falciparum: Strain construction and safety and immunogenicity in humans. J Infect Dis 1994; 169:927-931.
24. Butterton JR, Beattie DT, Gardel CL et al. Heterologous antigen expression in *Vibrio cholerae* vector strains. Infect Immunol 1995; 63:2689-2696.
25. Noriega FR, Losonsky G, Wang JY et al. Further characterization of *DaroA DvirG Shigella flexneri* as a mucosal *Shigella* vaccine and a live-vector vaccine for delivering antigens of enterotoxigenic *Escherichia coli*. Infect Immunol 1996; 64:23-27.
26. Cherry JD, Brunell PA, Golden GS et al. Report of the task force on pertussis and pertussis immunization—1988. Pediatrics 188; 81:939.
27. Clemens JD, Sack DA, Harris JR et al. Field trial of oral cholera vaccines in Bangladesh: Results from three-year follow-up. Lancet 1990; 355:270-273.
28. Svennerholm A-M, Holmgren J, Sack DA. Development of oral vaccines against enterotoxigenic *Escherichia coli* diarrhoea. Vaccine 1989; 7:196-198.
29. Clemens JD, Sack DA, Harris JR et al. Impact of B subunit killed whole-cell and killed whole-cell-only oral vaccines against cholera upon treated diarrhoeal illness and mortality in an area endemic for cholera. Lancet 1988; 1:1375-1379.
30. Murdin AD, Barreto L, Plotkin S. Inactivated polio vaccines: Past and present experience. Vaccine 1996; 14:735-746.

31. Crawford CR, Faiza AM, Mukhlis FA et al. Use of zwitterionic detergent for the preparation of an influenza virus vaccine. 1. Preparation and characterization of disrupted virions. Vaccine 1984; 2:193-198.
32. Plotkin SA. Rabies vaccine prepared in human cell cultures: Progress and perspectives. Rev Infect Dis 1980; 2:433-447.
33. Hoke CH, Nisalak A, Sangawhipa N. Protection against Japanese encephalitis by inactivated vaccines. N Engl J Med 1988; 319:608-614.
34. Provost PJ, Hughes JV, Miller WJ et al. An inactivated hepatitis A viral vaccine of cell culture origin. J Med Virol 1986; 19:23-31.
35. Werzberger WA, Mensch B, Kuter B et al. A controlled trial of a formalin-inactivated hepatitis A vaccine in healthy children. New Engl J Med 1992; 327:453-457.
36. Choo QL, Kuo G, Weiner AJ et al. Isolation of a cDNA clone from a blood-borne non-A, non-B viral hepatitis genome. Science 1989; 244:359-362.
37. Tam AW, Smith MM, Guerra ME et al. Hepatitis E virus: molecular cloning and sequencing of the full-length viral genome. Virol 1991; 185:120-131.
38. Pizza M, Scarlato V, Masignani V et al. Identification of vaccine candidates against serogroup B Meningococcus by whole-genome sequencing. Science 287:1816-20.
39. Hilleman MR, Bertland AU, Buynak et al. Clinical and laboratory studies of HBsAg vaccine. In Vyas GN, Cohen SN, Schmid R (eds.). Viral Hepatitis. Philadelphia, Franklin Institute Press, 1978; 525–541.
40. Greco D, Salmaso S, Mastrantonio P et al. A controlled trial of two acellular vaccines and one whole-cell vaccine against pertussis. New Engl J Med 1996; 334:341-348.
41. Gustafson L, Hallander HO, Olin P et al. A controlled trial of a two-component acellular, a five-component acellular, and a whole-cell pertussis vaccine. New Engl J Med 1996; 334:349-355.
42. Schmitt H-J, Wirsing von König, Neiss A et al. Efficacy of acellular pertussis vaccine in early childhood after household exposure. J Am Med Assoc 1996; 275:37-41.
43. Jones FG, Moss JM. Studies on tetanus toxoid. I. The antitoxic titer of human subject following immunization with tetanus toxoid and tetanus alum precipitated toxoid. J Immunol 1936; 30:115-125.
44. Ramon G. Sur le pouvoir floculant et sur les proprietes immunisantes d'une toxin diphterique rendue anatoxique (anatoxine). Compt Rend Acad Sci 1923; 177:1338-1340.
45. Chazono M, Yoshida I, Konobe T. The purification and characterization of an acellular pertussis vaccine. J Biol Stand 1988; 16:83-89.
46. Nencioni L, Pizza MG, Bugnoli M et al. Characterization of genetically inactivated pertussis toxin mutants: Candidates for a new vaccine against whooping cough. Infect Immunol 1990; 58:1308-1315.
47. Giannini G, Rappuoli R, Ratti G. The amino-acid sequence of two non-toxic mutants of diphtheria toxin: CRM_{45} and CRM_{197}. Nucleic Acids Res 1984; 12:4063-4069.
48. Valenzuela P, Medina A, Rutter WJ et al. Synthesis and assembly of hepatitis-B virus surface-antigen particles in yeast. Nature 1982; 298:347-350.
49. Thanavala Y, Yang Y-F, Lyons P et al. Immunogenicity of transgenic plant-derived hepatitis B surface antigen. Proc Natl Acad Sci USA 1995; 92:3358-3361.
50. Van Hoecke C, Comberbach M, De et al. Evaluation of the safety, reactogenicity and immunogenicity of three recombinant outer surface protein (OspA) Lyme vaccines in healthy adults. Vaccine 1996; 14:1620-1626.
51. Corey L, Langenberg AGM, Ashley et al. Recombinant glycoprotein vaccine for the prevention of genital HSV-2 infection. J Am Med Assoc 1999; 282:331-40.
52. Jansen KU, Rosolowsky M, Schultz LD et al. Vaccination with yeast-expressed cottontail rabbit papillomavirus (CRPV) virus-like particles protects rabbits from CRPV-induced papilloma formation. Vaccine 1995; 13:1509-1514.
53. Crawford SE, Labbe M, Cohen J et al. Characterization of virus-like particles produced by the expression of rotavirus capsid proteins in insect cells. J Virol 1994; 68:5945-5952.
54. Miyamura K, Kajigaya S, Momoeda M et al. Parvovirus particles as platforms for protein presentation. Proc Nat Acad Sci USA 1994; 91:8507-11.

55. Kingsman SM, Kingsman AJ. Polyvalent recombinant antigens: A new vaccine strategy. Vaccine 1988; 6:304-306.
56. Zavala F, Cochrane AH, Nardin EH et al. Circumsporozoite proteins of malaria parasites contain a single immunodominant region with two or more identical epitopes. J Exp Med 1983; 157:1947-1957.
57. Cachia PJ, Glasier LM, Hodgins RR et al. The use of synthetic peptides in the design of a consensus sequence vaccine for *Pseudomonas aeruginosa*. J Pept Res 1998; 52:289-99.
58. Vreden SGS JP, Oettinger T, Sauerwein RW et al. Phase I clinical trial of a recombinant malaria vaccine consisting of the circumsporozoite repeat region *of Plasmodium falciparum* coupled to hepatitis B surface antigen. Am J Trop Med Hyg 1991; 45:533-538.
59. Schodel F, Peterson D, Hughes J et al. Hybrid hepatitis B virus core antigen as a vaccine carrier moiety. I. Presentation of foreign epitopes. J Biotechnol 1996; 44:91-96.
60. Herrington DA, Clyde DF, Losonsky G et al. Safety and immunogenicity in man of a synthetic peptide in malaria vaccine against *Plasmodium falciparum* sporozoites. Nature 1987; 328:257-259.
61. Wang CY, Looney DJ, Li et al. Long-term high-titer neutralizing activity induced by octomeric synthetic HIV-1 antigen. Science 1991; 254:285-288.
62. Vitiello A, Ishioka G, Grey HM et al. Development of a lipopeptide-based therapeutic vaccine to treat chronic hepatitis B infection. I. Induction of a primary cytotoxic T-lymphocyte response in humans. J Clin Inv 95:341-9.
63. Nestle FO, Alijagic S, Gilliet M et al. Vaccination of melanoma patients with peptide- or tumor lysate-pulsed dendritic cells. Nat Med 1998; 4:328-32.
64. Rodrigues LP, Schneerson R, Robbins JB. Immunity to *H. influenzae* type b I. The isolation, and some physicochemical, serologic and biologic properties of the capsular polysaccharide of *H. influenzae* type b. J Immunol 1971; 107:1071-1080.
65. Gotschlich EC, Liu TY, Artenstein MS. Human immunity to the meningococcus. III. Preparation and immunochemical properties of the group A, group B and group C meningococcal polysaccharides. J Exp Med 1969; 129:1349-1365.
66. Kass EG. Assessment of the pneumococcal polysaccharide vaccine. Rev Infect Dis 1981; 3:S1 S197.
67. Kniskern PJ, Marburg S, Ellis RW. *Haemophilus influenzae* type b conjugate vaccines. In: M. Powell, M. Newman eds. Vaccine design: The subunit approach. New York: Plenum Publishing Corporation 1995; 673-694.
68. Black S, Shinefeld H, Fireman B et al. Efficacy, safety and immunogenicity of heptavalent pneumococcal conjugate vaccine in children. Ped Infect Dis J 2000; 19:187-95.
69. Klein D, Ellis RW. Pneumococcal conjugate vaccines. In: Levine MM, Woodrow GC, Kaper JB, Cobon GS, eds. New Generation Vaccines. New York: Marcel Dekker 1997; 503-526.
70. Kennedy RC, Adler-Storthz K, Henkel RD et al. Immune response to hepatitis B surface antigen: Enhancement by prior injection of antibodies to the idiotype. Science 1983; 221:853-855.
71. Herlyn D, Wettendorff M, Iliopoulos D et al. Modulation of cancer patients' immune responses by administration of anti-idiotypic antibodies. Viral Immunol 1989; 2:271-276.
72. Diakun KR, Matta KL. Synthetic antigens as immunogens: III. Specificity analysis of an anti-anti-idiotypic antibody to a carbohydrate tumor-associated antigen. J Immunol 1989; 142:2037-2040.
73. Wolff JA, Malone RW, Williams P et al. Direct gene transfer into mouse muscle in vivo. Science 1990; 247:1465-1468.
74. Leitner WW, Ying H, Restifo NP. DNA- and RNA-based vaccines: principles, progress and prospects. Vaccine 2000; 18:765-77.
75. Fan H, Lin Q, Morrissey GR et al. Immunization via hair follicles by topical administration of naked DNA to normal skin. Nature Biotech 1999; 17:870-2.
76. Tacket CO, Roy MJ, Widera G et al. Phase 1 safety and immune response studies of a DNA vaccine encoding hepatitis B surface antigen delivered by a gene delivery device. Vaccine 1999; 17:2826-9.
77. Remy J-S, Sirlin C, Vierling P et al. Gene transfer with a series of lipophilic DNA-binding molecules. Bioconjugate Chem 1994; 5:647-654.

78. MacGregor RR, Boyer JD, Ugen KE et al. First human trial of a DNA-based vaccine for treatment of human immunodeficiency virus type 1 infections: safety and host response. J Infect Dis 1998; 178:92-100.
79. Glenn GM, Scharton-Kersten T, Vassell R et al. Transcutaneous immunization using bacterial ADP-ribosylating exotoxins as antigens and adjuvants. Infect Immun 1999; 67:1100-6.
80. Krieg AM, Yi A-K, Matson S et al. CpG motifs in bacterial DNA trigger direct B-cell activation. Nature 1995; 374:546-549.
81. Fries LF, Tartaglia J, Taylor J et al. Human safety and immunogenicity of a canarypox-rabies glycoprotein recombinant vaccine: an alternative poxvirus vector system. Vaccine 1996; 14:428-434.
82. Dubensky TW, Driver DA, Polo JM et al. Sindbis virus DNA-based expression vectors: Utility for in vitro and in vivo gene transfer. J Virol 1996; 70:508-519.
83. Sizemore DR, Branstrom AA, Sadoff JC. Attenuated *Shigella* as a DNA delivery vehicle for DNA-mediated immunization. Science 1996; 270:299-302.
84. Darji A, Guzman CA, Gerstel B et al. Oral somatic transgene vaccination using attenuated *S. typhimurium*. Cell 1997; 91:765-75.
85. Cox JC, Coulter AR. Adjuvants—a classification and review of modes of action. Vaccine 1997; 15:248-256.
86. Singh M, O'Hagan D. Advances in vaccine adjuvants. Nature Biotech 1999; 17:1075-81.
87. Shirodkar S, Hutchinson RL, White JL et al. Aluminum compounds used as adjuvants in vaccines. Pharm Res 1990; 7:1282-1288.
88. Lin R, Tarr PE, Jones TC. Present status of the use of cytokines as adjuvants with vaccines to protect against infectious diseases. Clin Infect Dis 1995; 21:1438-1449.
89. Elson CD, Falding W. Generalized systemic and mucosal immunity in mice after mucosal stimulation with cholera toxin. J Immunol 1984; 132:2736-2744.
90. Douce G, Turcottee C, Cropley I et al. Mutants of *Escherichia coli* heat-labile toxin lacking ADP-ribosylating activity acts as non-toxic mucosal adjuvants. Proc Natl Acad Sci USA 1995; 92: 1644-1648.
91. Agren LC, Ekman L, Lowenadler B et al. A genetically-engineered nontoxic vaccine adjuvant that combines B-cell targeting with immunomodulation by cholera toxin A1 subunit. J Immunol 1997; 158:3936-3946.
92. Mestecky J, Moldoveanu Z, Michalak SM et al. Current options for vaccine delivery systems by mucosal routes. J Controlled Release 1997; 48:243-57.

CHAPTER 2

Clinical Issues for New Vaccine Technologies

Luc Hessel

Introduction

Vaccination as a means of preventing infectious diseases arguably has had the greatest impact on human health of any medical intervention.[1] Since the pioneer work of Jenner and Pasteur, the development of vaccines has been the consequence of the uninterrupted introduction of a series of new technologies. The discovery of toxoids, purification of polysaccharides, cell culture enabling virus culture, controlled methods for attenuation of viruses, conjugation of polysaccharides to proteins, production of protein vaccines in genetically-engineered cells and reassortment of viruses have been among the basic technologies used so far in the development of vaccines.[1,2] Application of the tools of modern biotechnology has resulted in an array of vaccine candidates coming from many sources and created the promise of prevention or treatment for many more infectious and chronic diseases.[3,4] It has also revolutionized the capability to engineer and produce vaccines that are potentially safer, more effective, easier to produce and less costly.[5] The forthcoming new technologies form a continuum in the innovative process that has always been characteristic of vaccine development. Different new technologies are currently considered with more attention, such as

1. genetically-engineered vaccines,
2. live vectors,
3. nucleic acid vaccines,
4. new delivery systems or
5. new adjuvants.[1,2]

This biotechnology revolution poses a tremendous challenge for traditional vaccine development to provide adequate and timely assessments so that maximal benefits might be reaped from these advances.[6] Indeed, successful development of vaccines is a time-intensive process requiring years of commitment from a network of scientists and a continuum of regulatory and manufacturing entities.[7]

The vaccine development process leading to licensure is pyramidal and selective.[4] It is step-wise and bridges basic research, development, large-scale production in an approved facility, and clinical evaluation to establish safety and protective efficacy.[7] Although not basically different than for conventional vaccines, the clinical contribution to the development of new vaccines is of special importance at many stages of the discovery process and can be schematically defined as follows:

1. definition of the medical needs,
2. choice of the rationale and elaboration of the strategy based on preclinical development,
3. demonstration of proof-of-principle in early clinical research,
4. design and implementation of an integrated clinical development plan and
5. continuous assessment of safety and efficacy.

New Vaccine Technologies, edited by Ronald W. Ellis. ©2001 Eurekah.com.

After a review of the core essence of the work, some specific clinical issues raised by each of the new technologies will be addressed.

Definition of the Medical Needs

The ultimate goal of vaccine research is to develop vaccines that are useful and effective. Vaccines must address public health needs and be a logical means of controlling the disease of interest. In theory, the field of application of vaccines is extremely broad, including at least all diseases caused by microorganisms. Recently the field of vaccines has been expanded to other fields such as cancer, allergy, autoimmune diseases, contraception and drug addictions.[4,8-11] Choices have to be made, guided by epidemiological studies, experiments and observations evaluating the burden of the disease as well as the existence of alternative prophylactic measures and the availability of curative treatments.[12] The exercise is not easy. In addition to ubiquitous diseases of major importance such as pneumococcal infections or HIV, special consideration is given to tropical diseases which have a tremendous impact on health, with hundreds of thousand of deaths annually in children.[4] Thus, a strong scientific base and rationale, a firm quantification of disease burden to be prevented (including mortality, acute morbidity, long-term sequelae and the associated direct and indirect economic costs), and clear identification of target populations are central to the successful development of new vaccines. The review of the development of selected vaccines recently performed by the National Vaccine Advisory Committee in the USA clearly showed that vaccine development moved forward expeditiously when the scientific base was well established, whereas development efforts often stalled when the science was less mature due to a lack of clear direction and endpoints.[7]

Moving from Preclinical to Clinical Development

Choice of the Rationale

Once the new target has been defined, the difficult step is to understand the basis for natural protection against the pathogen, which will enable the identification of the relevant immunological approach for developing a candidate vaccine. In some cases the nature of the immune response needed for protection is well known and the antigen is clearly identified (e.g., polysaccharide vaccines, neutralizing antibodies against viral diseases). In many cases, however, there is no clear evidence of surrogate markers of protection. The identification of candidate immunogens relies on several approaches, including a thorough analysis of the human immune response to disease and a clear understanding of the biology of the causative organism. The techniques used for these approaches are more and more complex, and interpretation of results is not easy. Thus, laboratory animal assays are essential tools at several stages of the research, development and production of improved and novel vaccines, delivery systems or adjuvants.[13] Demonstration of immunogenicity in animals is an absolute prerequisite to clinical studies of a product. Preclinical studies in animals also include extensive toxicology studies in order to demonstrate the absence of major safety issues. In most areas of vaccine research, animal models are developed that contribute to the characterization of protective immunity, the study of the safety and immunogenicity of various formulations and the preclinical evaluation of the protective efficacy.[13] In spite of their limitations, animal studies are still irreplaceable and taken with caution they are of great help.

Choice of a Strategy

Once the antigen has been identified, there often are several ways to produce or express it; these include attenuation of the live microorganism to lose virulence while maintaining immunogenicity, a protein purified from the microorganism itself, a recombinant protein made in bacteria, yeast or mammalian cells, the antigen expressed by a recombinant live attenuated bacterium or virus or a plasmid DNA construct. Every approach has its own merits and limits in terms of the type of elicited immune response, ease of production, ease of controls and risks for the environment. All these aspects will have an impact on the objectives and methodology of the clinical trials to be performed as well as on the overall acceptability of the vaccine by regulatory agencies. Sometimes, as illustrated by the first generation of respiratory syncytial virus (RSV) formalin-inactivated parenteral vaccines, the legacy of safety concerns has a marked impact on subsequent vaccine development.[14]

Thus, before moving to clinical trials in man, the vaccine candidate must have been designed with a sound scientific rationale. Based on either known or likely protective antigens or on a live strain with genetic deletion of known virulence factors, there must be an expectation of efficacy and safety formally demonstrated in an appropriate model using a dose and a route of administration that will be proposed in clinical trials. Animal studies demonstrating the immunogenicity and, if an appropriate model exists, the efficacy of the vaccine candidate against laboratory- or wild-type pathogens represent the ideal approach. It is also highly desirable that the vaccine be prepared in a formulation as close as possible to the final manufacturing process, including antigen preparation, adjuvants, volume, etc. The critical scale-up from bench-scale to pilot lots and then to large-scale production is often a particularly vulnerable point in the development process of new vaccines.[7] Thus, early establishment of the product profile and characteristics will considerably help the regulatory process.

Demonstration of the Proof-of-Principle

Once preclinical development has been completed, it is time to turn to clinical testing. The scope of the challenge may be limited. If the protective antigen or the type of immune response responsible for protection is well known, it will be necessary only to demonstrate that the product developed at the laboratory level achieves expectations. This is the objective of classical phase I studies. If the product is safe and raises the expected immune response, the decision may be taken quickly to bring the product to full development.

In some other cases, the validity of the approach is not ensured even when safety and immunogenicity have been established, and the proof-of-principle will be qualified only after protective efficacy results are known. This especially applies when

1. animal models do not exist or are not relevant,
2. clinical or immunological markers are not available and
3. safety issues are central to the acceptability of the vaccine. In this case human challenge studies may be recommended as long as they are ethical and do not endanger volunteers' health.

Human challenge tests can represent a good marker of the actual efficacy of the vaccine candidate against laboratory or wild pathogens and across serotypes. They will also contribute to identifying immunological correlates of protection, to comparing protection conferred by the vaccine candidate to that of a clinical infection, and to giving some information of the duration of immunity. Human challenge tests represent very useful tools for the early screening of vaccine candidates, but their use remain limited by several issues relating to their reproducibility, to the possible lack of correlation between experimental disease and natural infection and to general and specific ethical concerns.[15]

However, in a few cases when the disease is very common, preliminary efficacy data can be obtained at a very early stage. This is the case for some respiratory and diarrheal diseases such as rotavirus diarrhea or RSV infection where the incidence is so high that studies in small numbers of children may allow to estimate the value of the approach.

Design and Implementation of a Clinical Development Plan

General Design

The goal of prelicensure studies is to show that the vaccine is safely tolerated in terms of local and systemic reactions and to demonstrate that it confers protection against the target disease.[12] The Clinical Development Plan (CDP) describes and justifies in a logical and phased way the clinical studies to be done in order to answer well-defined questions for documenting the safety, immunogenicity and efficacy of the vaccine and delineating its conditions of use in order to obtain the marketing approval in the proper indications. The CDP must take into account numerous aspects of the future vaccine, especially its regulatory and marketing strategy as well as industrialization issues and include all questions to be clinically addressed in the Product License Application (PLA). It also needs to anticipate short- and long-term evaluation to be considered after the vaccine has been licensed, including concomitant administration with other vaccines, antibody persistence and need for booster injections, administration in special populations and development of surveillance systems to address long-term safety issues.[7,12] Phase I trials usually enroll 10-100 adult volunteers to assess initial safety tolerability and acceptable vaccine dosage in humans. Phase II trials (100-1,000 persons) seek to expand knowledge about the safety and immunogenicity of the vaccine, to select the optimal formulation of a candidate product, and to identify the most suitable schedule for vaccine administration (including dosage, route of administration, optimal interval between primary series and boosters, if needed) for subsequent evaluation in phase III efficacy studies.[16] Phase III trials aim to show that the candidate vaccine is efficacious in conferring protection on a targeted, at-risk population under controlled conditions. Safety issues are also examined to the extent that the sample size and study duration permit. As with any clinical trial, issues such as case definition, case finding, trial design and sample size (which can range from 1,000 to 100,000 subjects according to the incidence of the disease) must be considered carefully.[17] Because sample sizes for these pivotal studies are large, they often require several years to complete enrollment and follow-up.[18] Although not basically different from those relating to conventional vaccines, special attention must be paid to some methodological aspects when designing a CDP for vaccines using novel technologies (see below).

Objectives

Many aspects must be considered when preparing a CDP. They include clear definitions of objectives and how they will be addressed. Precise objectives are central to the development of a proper methodology, the establishment of relevant clinical and statistical hypotheses and the determination of the number of subjects to be included. According to the type of technology or delivery system used, considerations include specific regulatory requirements as well as manufacturing issues in terms of stability and consistency of the vaccine lots and expected safety issues, e.g., person-to-person transmission or environmental risk for genetically-engineered live vector vaccines. Similarly, objectives will be different when the vaccine is intended for pre- or postexposure prophylaxis as opposed to a therapeutic vaccine. What is expected from the vaccine candidate should also be defined, e.g., an improved immunogenicity or safety profile compared to an existing vaccine or an extension of the target population such as non- or poor-responders, immunocompromised, neonates, etc. Whereas most vaccines are aimed at controlling the spread of a disease or even eradicating it, future vaccines may control the progression of an existing disease, e.g., cancer.

Regulatory Issues

The development of novel vaccine technologies led to considerable changes in the international regulations in order to address safety issues linked to the use of genetically-modified organisms (GMOs), live vectors and any new means of influencing immunological mechanisms through adjuvants. In addition to the relevant guidance established by the International Conference on Harmonization (ICH), precise guidelines have been defined in the USA and Europe which have to be taken into consideration in the design of the CDP for a new vaccine.[19] In Europe, the recently issued CPMP Note for Guidance on Clinical Evaluation of New Vaccines gives very precise methodological requirements on the assessment of efficacy, immunogenicity and safety during both the pre- and postlicensure periods.[20] The guideline insists also on the fact that nonconventional vaccines require special attention with regard to vectors, immune responses, immunological mechanisms and safety considerations. Indeed, special requirements and laws have been defined regarding certain technologies like GMOs. Directive 90/220/EEC addresses the potential risk to human health and to the environment, including plants and animals, which may be associated with investigational products containing GMOs. An environmental risk assessment (ERA) (which does not include naked DNA) is requested for any biological entity capable of replicating or of transferring genetic material. An updated Note for Guidance has been issued in December 1999 on the quality, preclinical and clinical aspects of gene transfer medicinal products. In addition to international guidelines many national regulations and ad hoc committees have been established which may influence the content of a CDP. For instance, in the USA, the FDA also has issued guidance for industry for the evaluation of combined vaccines,[21] plasmid DNA vaccines,[22] and human somatic cell therapy and gene therapy.[23]

Clinical Endpoints and Surrogate Markers

The ultimate clinical endpoint may be to prevent infection (e.g., HIV), the final disease (e.g., AIDS), severe disease (e.g., pertussis), disease progression (therapeutic vaccines) or survival (rabies). It is not always possible to directly address such primary endpoints for some vaccines due to the epidemiology of the disease, long-term clinical endpoints or multiple endpoints. Should prevention of CMV infection be focused on prevention of congenital infections or of secondary infection among women of childbearing age?[24]

When it is not feasible to obtain clinical endpoints, appropriate immunological data, determined by an adequately validated method, may be used to support evidence of efficacy.[20] Indeed the clinical efficacy of vaccines depends on the elicited immune response, whether humoral, mucosal or cellular. Surrogate markers are parameters for predicting efficacy without knowing the exact level of protection. They are mainly biological markers of humoral and/or cellular immunity. Serological surrogates are mostly used and defined as a predefined antibody concentration correlating with clinical protection.[20] This especially applies when a reference vaccine exists and is routinely used. This may also apply to safety issues that could be related to potential autoimmune mechanisms. While sometimes scientifically well-grounded (e.g., antibody protective levels against *Haemophilus influenzae* type b or hepatitis B surface antigen), it may be very difficult in many instances to establish a precise correlation between an antibody level and clinical protection.[25] This especially applies to pertussis or rotavirus infections. Attempts to establish such correlations might be difficult when the measured antibodies are not specific enough (ELISA), although they could remain relevant for comparison with a reference vaccine. Antibody assays are especially acceptable when the function of the antibody is known. Neutralizing or bactericidal activities or measurement of avidity may represent valid surrogate markers of efficacy as long as it contributes directly to protection. Assessment of cellular immunity is also subject to many limitations, since the assays require very demanding biological monitoring, are not always well standardized, often rely on cytokine induction, are variable and irreproducible and very expensive. Mucosal immunity might be an important parameter to be measured when the vaccine-preventable disease is a local infection, or starts in the mucosa and invades (e.g., upper respiratory infections), or sexually transmitted diseases.[26] Unfortunately, the methods for measuring local immune responses are not well standardized and the collection and storage of such samples is problematic.[25] Other surrogate markers such as the evaluation of bacterial or viral load may also be extremely useful in the development of therapeutic vaccines.

Overall, any information on the characteristics of the immune response according to the known or presumed activities of the vaccine represents a very valuable tool for establishing vaccine efficacy;[20] these include the level, class, subclass and function of specific antibody produced, the induction of mucosal or cell-mediated immunity, formation of neutralizing antibodies, cross-reactive antibodies, etc.

Study Design

As described above, the traditional design for developing a new vaccine consists in establishing the safety and immunogenicity of the product through phase I, II and III studies. This sequential approach may not always be applicable for novel technology vaccines. Phase I studies may not be possible when the candidate vaccine has no relevance for healthy adult volunteer, e.g., cancer vaccines. In other instances (e.g., live vectors), it may need a confined environment and long surveillance. Phase II studies may not be conducted as expected in the target population. This applies to cancer vaccines that are basically indicated at early stages of cancer when the immune system is still able to develop adequate protective responses. However, due to potential safety issues and ethical concerns, inclusion criteria may be restricted to advanced disease where an actual therapeutic effect might not be apparent. Selection of volunteers or patients can be an issue when sero-eligibility is required or when the pharmacological properties of the vaccine relies on HLA specificity or other immune characteristics of the host. Defining the relevant control group is critical when a placebo is not acceptable or when no control vaccine is available, especially for therapeutic or postexposure vaccines. Concomitant administration of other vaccines according to routine recommendations can constitute a confounding factor for safety and immunogenicity evaluation. Overall, the methodological approach for a relevant clinical study design mainly relies on the quality and accuracy of the objectives, the existence of validated clinical or biological endpoints and on a clear statistical hypothesis.

Statistical Issues

The design of clinical trials obviously depends on its objectives but will be influenced by the statistical hypothesis, the tests to be used and the power requested to meet them. This will affect the number of subjects to be included, the choice of the control vaccine and of the evaluation criteria to be chosen as primary or secondary endpoints. The criteria for determining sample size in each trial must be based on methodological and statistical considerations, clinical, immunological and epidemiological issues and tailored to the clinical hypothesis.[20] When immunogenicity is the sole primary endpoint to evaluate efficacy, appropriate randomization of the target population should be ensured. Controlled trials, ideally placebo-controlled double-blind, have been the gold standard in vaccine development for many decades. It provides the ideal means to demonstrate the efficacy of a new vaccine by showing its superiority to a placebo or a passive control vaccine. As more and more effective vaccines become available, the objective of clinical investigation changes. Comparative trials can seek equivalence or noninferiority of the candidate vaccine rather than its superiority to an effective standard reference in active controlled trials.[27] The fundamental assumption of such trials is that showing noninferiority is evidence of efficacy. This especially applies to vaccines combining multiple antigens already available as separate vaccines and to the use of new technologies to improve the safety of the manufacturing process of already existing vaccines without increasing their efficacy. In this design, the choice of the control vaccine is critical. It has to be a well-accepted standard vaccine across different geographical regions and one for which efficacy has been satisfactorily and consistently proven. This methodological approach may have an important impact of the size of the study population as it requires a much larger sample size than a placebo-controlled superiority trial.[27] In noninferiority trials, the purpose it to show that the efficacy of the test treatment is not inferior to that of the reference treatment by more than a minimum clinically relevance difference (δ), which has to be predefined at the trial design stage. Typically, the sample size should be such that, if the treatments are equivalent or the test vaccine is superior, there is a high probability (>80%) that the lower limit of a 95% confidence interval for the difference in protection rate will not fall below the predefined δ.[20,27] One key issue is to define the clinical relevance of differences in serological responses in order to avoid disqualifying a candidate vaccine because it does not meet narrow statistical differences. Indeed, the choice of the δ value depends on the expected seroprotection/conversion rates and will strongly influence the size of the study population. In individual trials, δ can often be set to about 10 percentage points but will need to be smaller for very high protection rates and larger with lower protection rates.[20]

Ethical Issues

Ethical issues represent a major aspect in the design of a clinical development plan for novel technology vaccines. They are obviously critical for testing candidate vaccines in special groups, e.g., for cancer vaccines where survival may be the primary endpoint, as well as for HIV vaccines, and when placebo controls are used or challenge tests requested.[15,20] Genetically-engineered live vectors may evoke long-term safety concerns, generating demanding clinical and biological long-term monitoring. Developing vaccines for infants or neonates also represents an ethical challenge, as recently illustrated with a reassortant rotavirus vaccine.[28] The content of informed consent forms therefore become an important aspect of the protocols. Setting up data safety monitoring committees in charge of reviewing and validating adverse events, independently from the investigators and the sponsor, represent the best way to address such issues.

Continuous Assessment of Safety and Efficacy

In many respects, safety is a critical issue for novel technology vaccines. This applies for both prophylactic and therapeutic vaccines, especially if there is an existing alternative for prevention or cure of the given disease. As the result of safety considerations, certain vaccines (e.g., live virus or bacterial vaccines) may require considerable additional information relating to the vaccine strain, including comparison of the properties of the shed vaccine isolates versus the parent vaccine strain, the pattern of shedding, transmissibility to contacts,[29] and its potential to survive in the environment or to enter the food chain.[30] Similarly special attention on safety has been developed for genetically-engineered live vectors using GMOs. New adjuvants targeted at stimulating a specific immune response will justify paying attention to specific safety issues such as autoimmune diseases or potentially rare and distant adverse events (AEs).[31,32] In the latter case, it may be difficult to link AEs to the vaccine when such diseases are of unknown origin or incidence. Development of new delivery systems, such as jet injectors and mucosal administration, also will lead to defining specific approaches for new potential safety issues. Safety will be addressed both during the pre- and postlicensure phases.

In the prelicensure evaluation phase, potential safety hazards ideally will have been identified in preclinical laboratory tests, e.g., genetic stability, as well as animal testing where applicable. Clinical assessment will remain the key factor, in particular for defining proper control groups, identifying relevant markers for autoimmunity and addressing the risk of person-to-person transmission. When GMOs are used, ERA also will be required. Prelicensure evaluation nevertheless will be limited to a relatively small number of subjects, usually not exceeding 10,000 followed for short periods of time. Ensuring the safety of new vaccines as they become licensed and are given to larger numbers of persons is critical to ensuring the trust of the public and stability of the immunization program.[7]

Therefore, post marketing surveillance (PMS) is essential to assess rare, unexpected, distant AEs.[33] PMS is traditionally based on spontaneous reporting of AEs following immunization by health-care workers as a tool to generate signals to be further investigated in epidemiological studies. In an attempt to overcome the deficiencies of passive reporting systems, "active" surveillance through prospective or retrospective observational studies is becoming more frequently used in the immediate postlicensure period. They allow the assessment of rare AEs that cannot be described in the selected and relatively small population involved in the clinical development of the vaccine. This requires a procedure for identifying all clinically significant events that occur within defined postvaccination periods.[12,33] Most recent regulations even consider these PMS studies as conditional for granting a license.[20] New tools must be developed to allow a more systematic and cheaper way of addressing long-term safety issues linked to the use of novel technology vaccines. This especially includes establishing national or international disease registries that could be linked to vaccination registries for allowing the determination of possible changes in the epidemiology of a disease or a group of disorders and the implementation of the use of a new vaccine.[12,33]

Specific Issues

According to the novel technology used, some specific issues will apply to clinical development of new vaccines, as described in more detail in the relevant chapters of this book.

Live Vectors

By using live vectors multiplying in the organism and expressing foreign genes, one can deliver protective antigens from pathogens that themselves might be considered as unsafe as an attenuated vaccine.[1] Nevertheless, this technology raises different clinical issues that need to be addressed at different times of the development process. The rational choice of the vector is closely dependent on the pathophysiology of the wild organism.[34] Live *Salmonella* recombinants are widely explored because of their potential ability to reach the lymphoid tissue of the gut and thus deliver the antigens to immunocompetent cells. Similarly *Shigella* are considered potential vectors because of the possibility of passing the gut mucosa. BCG or BCG-like organisms are selected as potential vectors because of their capacity to trigger T-cell responses and also because of the adjuvant effect of the bacterial membrane. Many viruses are also explored as live vectors expressing foreign antigens; among the most studied are poxviruses, adenoviruses and more recently flaviviruses. The very first point to address clinically is the attenuation of the live vector that does not allow for any residual effect. Even if the balance between attenuation and antigen expression has been found, the immunogenicity of the vector itself is another frequent issue. It may be previous immunity (this is definitely the case for vaccinia immunity which interferes with pox-recombinants) or it may be that the first contact with the vectored vaccine will elicit a stronger response to the vector than to the foreign gene, thus jeopardizing the response to a further injection of the same live vector. These points represent key evaluation criteria to be addressed very early in phase I and II trials.

Recombinant DNA technology using attenuated viral or bacterial strains falls under regulatory requirements regarding GMOs. They will justify specific extensive safety evaluation linked to the residual virulence of the vector strain, as well as an environmental analysis of virus replication and survivability of the vaccine strain in various environments.[35] Person-to-person transmission will request special attention as well as documentation of the stability and potential reversion of the strain to a virulent one through shedding and immunological markers.

Therefore all possible tests on suitable animal models (including SCID/transgenic animals) as well as in vitro tests will be necessary before moving to clinical trials. Phase I/II studies could need to be conducted in confined environments and have an ERA performed. The clinical evaluation of the consistency of manufacturing also represents an important part of the clinical development.

Cancer Vaccines

Understanding the interactions between the immune system and cancer cells, especially the molecular mechanisms of immune recognition and immune regulations, so that antitumor immunity could be amplified forms the basis for the development of effective means to treat cancer patients. In contrast to prophylactic vaccination against infectious agents, in which the generation of humoral immunity is the most important feature, cancer vaccine development focuses mainly on the generation of antigen-specific T-cell responses.[9] Whatever the fundamental approach is, tumor antigens or cells, cytokines or dendritic cells, the development of cancer vaccines represents a real challenge, as clinical studies in man are often the only way of supporting proof-of-concept for many vaccine candidates. Animal models are not always available or relevant for selecting antigens. Correlation between immunological markers and clinical efficacy is difficult to establish. In addition, phase I study in healthy volunteers are generally not applicable, such that phase II studies often will be the first approach for supporting the concept. But they cannot be easily conducted in the final target population (minimal vs. advanced disease). Clinical endpoints are not easy to define or acceptable (e.g., survival). Control groups are difficult to define due to concomitant therapies or surgery and lack of standardized therapeutic protocols (impact on stratification). Finally, AE reporting could be very demanding in a population at high risk of experiencing many AEs unrelated to vaccination that will have to be followed up for a long period of time.

Adjuvants

It is recognized that impurities had a nonnegligible effect and thus a real impact on the efficacy of old vaccines. A weaker immunogenicity profile generally appears as the price to pay for safer and purer vaccine so that there is now a need for strong adjuvants of immunity. The development of new adjuvants represents a major goal for new vaccine technology. Such new vaccine formulations should allow improved immunogenicity of antigens in the general population or in persons whose immune system is immature (neonates) or suffers from genetically or acquired deficiencies. The clinical development of new adjuvants illustrates a number of difficulties mentioned above. The target population may be poor or nonresponders, justifying prescreening of a large population. Such populations include immunodeficient patients or neonates, thus raising special immunological evaluation and safety issues. Immunological endpoints for adjuvants may be very specific when they are targeted at stimulating cellular or mucosal immunity. Defining the type of immune responses (Thl vs. Th2) will also require specific tests with demanding monitoring processes. Depending on the type of adjuvant, specific short- and long-term, local and systemic safety issues will have to be addressed, including potential mutagenic effects and induction of autoimmunity justifying extensive safety surveillance.[35]

DNA Vaccines

A new approach to vaccination opened up recently with the demonstration that direct inoculation of DNA encoding a foreign antigen can initiate protective immune responses.[36] DNA vaccines containing the genes of foreign proteins provide cells in the vaccinated host with the code for the synthesis of the immunizing antigens. Such vaccines have a number of advantages over more classical vaccines, although their use is limited to protein antigens and so far to skin or muscle delivery. They raise antibody against the natural form of proteins, can elicit both cytolytic T-cell and antibody responses and long-lived immunity.[35] In addition they hold promise for safer, less expensive, easier-to-produce and easier-to-administer than conventional vaccines. Finally DNA vaccines offer a broad range of applications including new antigens (viral, bacterial and parasitic), combination of multiple antigens and biased Th1 or Th2 responses.[36]

Although DNA technology will apply to vaccines for which some clinical and immunological endpoints are well established, the applied preclinical and clinical research on DNA vaccines will have to address very specific additional issues, including a better characterization of the induced immune responses and special consideration for safety. Determining the relative roles of the target site (muscle or skin), better understanding of the T-helper preference and evaluation of vaccine vectors and delivery systems will represent specific clinical and immunological endpoints for phase I and II studies. Regarding safety issues, regulatory agencies considering the application of DNA vaccines for human use have already identified key areas of concern, including mutagenic events linked to the integration of the plasmid DNA into the genome of transfected cells, potential induction of immune tolerance to the vaccine antigen or induction of autoimmunity and/or of antibodies to the plasmid DNA. Based on animal experiments and current knowledge of the role of natural infections in the induction of autoimmunity, these potential safety issues appear very unlikely to be manifest.[36-38] These complicated issues will nevertheless represent a key aspect of the pre- and postlicensure clinical development of DNA vaccines. It will be based not just on the identification of relevant biological markers but also mostly on the recently developed epidemiological surveillance systems mentioned above that provide background data on rare diseases and prospective registries.[12,33]

Conclusion

Designing and performing clinical trials with a novel vaccine technology is a complex process which needs to be carefully analyzed before entering into long and costly clinical trials in order to address all specific issues that may impact on the licensibility of the vaccine. Numerous guidelines and regulations have been issued recently which help designing relevant clinical plans but which must be interpreted with some flexibility. Thus, early discussions with regulatory authorities of the content of the CDP of a novel technology vaccine are key for further acceptability of the registration file.[7] Finally, one must accept that, even at the dawn of the new millennium, the process of vaccine research and development continues to be, at least in part, an empirical and methodic science. In the era of molecular vaccinology and immunology, developing an effective vaccine does not necessarily mean understanding the precise mechanisms by which most vaccines work.

Acknowledgment

The author is grateful to Dr Michel Cadoz for his helpful comments and suggestions during the preparation of the manuscript.

References

1. Liu MA. Vaccine developments. Nature Medicine 1998; 4:515-519.
2. Ellis RW. New technologies for making vaccines. Vaccine 1999; 17:1596-1604.
3. Levine MM. The legacy of Edward Jenner. Br Med J 1996; 312:1177-1183.
4. Division of the Microbiology and Infectious Diseases, National Institute of Allergy and Infectious Diseases, National Institutes of Health, Bethesda, MD. Accelerated development of vaccines, The Jordan Report 2000.
5. Woodrow GC. An overview of biotechnology as applied to vaccine development. In Woodrow GC, Levine MM eds. New generation vaccines. New York, NY: Marcel Dekker, 1990:31-41.
6. Mitchell VS, Philipose NM, Sanford JP eds. The Children's vaccine initiative: Achieving the vision. Washington, DC: National Academic Press, 1993.
7. National Vaccine Advisory Committee. Lessons learned from a review of the development of selected vaccines. Pediatrics 1999; 104:942-950.
8. Linehan DC, Goedegebuure PS, Eberlain TJ. Vaccine therapy for cancer. Ann Surg Oncol 1996; 3:219-228.
9. Pardoll DM. Cancer vaccines. Nature Medicine 1998; 4:525-531.
10. Cookson WOCM, Moffatt MF. Alchemy for asthma. Nature Medicine 1998; 4:500-501.
11. Rose NR, Wick G, Berger P et al. Immunological hazards associated with human immunization with self or self-like antigens. In: GGL Ada, PD Griffin, eds. Vaccines for Fertility Regulation. The assessment of their safety and efficacy. Cambridge: Cambridge University Press 1991:1121-146.
12. Chen RT, Orenstein WA. Epidemiologic methods in immunization programs. Epidemiol Rev 1996; 18:99-117.
13. Makela H et al. Animal models for vaccines to prevent infectious diseases. The European Commission COST/STD-3 initiative on European vaccine research. Vaccine 1996; 14:717-731.
14. Kapikian AZ, Mitchell RH, Chanock RM et al. An epidemiological study of altered clinical reactivity to respiratory syncytial virus (RS) infection in children previously vaccinated with an inactivated RS vaccine. Am J Epidemiol 1969; 89:405-421.
15. Levine MM. Experimental challenge studies in the development of vaccines for infectious diseases. In: Plotkin S, Brown F, Horaud F, eds. Preclinical and Clinical Development of New Vaccines. Dev Biol Stand. Basel: Karger 1998; 95:169-174.
16. Tacket CO, Mattheis M, Rennels MB. Initial clinical evaluation of new candidate vaccines: Phase 1 and 2 clinical trials for safety, immunogenicity and preliminary efficacy. In: Levine MM, Woodrow GC, Kaper JB, Cobon GS, eds. New generation vaccines. New York, NY: Marcel Dekker 1997:35-46.
17. Herrington DA. Initial clinical evaluation of new vaccine candidates. In: Woodrow GC, Levine MM eds. New generation vaccines. New York, NY: Marcel Dekker 1990:43-49.
18. Clemens JD, Naficy A, Rao MR. Long term evaluation of vaccine protection methodological issues for phase 3 trials and phase 4 studies. In: Levine MM, Woodrow GC, Kaper JB et al, eds. New generation vaccines. New York, NY: Marcel Dekker 1997:47-67.
19. Georges AM. Regulatory aspects of clinical trials with vaccines. In: Plotkin S, Brown F, Horaud F, eds. Preclinical and Clinical Development of New Vaccines. Dev Biol Stand Basel: Karger 1998; 95:69-74.
20. Committee for Proprietary Medicinal Products. Note for Guidance on Clinical Evaluation of New Vaccines. CPMP/EWP/463/97.
21. Guidance for industry for evaluation of combination vaccines for preventable diseases: Production, testing and clinical studies. FDA, CEBER April 1997.
22. Points to consider on plasmid DNA vaccines for preventive infectious disease indications. FDA, CBER 1996.
23. Guidance for human somatic cell therapy and gene therapy. FDA, CBER March 1998.
24. Adler SP, Starr SE, Plotkin SA et al. Immunity induced by primary human cytomegalovirus infection protects against secondary infection among women of childbearing age. J Infect Dis 1995; 171:36-32S.

25. Kayhty H. Immunogenicity Assays and surrogate markers to predict vaccine efficacy. In: Plotkin S, Brown F, Horaud F, eds. Preclinical and Clinical Development of New Vaccines. Dev Biol Stand. Basel, Karger, 1998; 95:175-180.
26. Kaul D, Ogra PL. Mucosal responses to parenteral and mucosal vaccines. In: Plotkin S, Brown F, Horaud F eds. Preclinical and Clinical Development of New Vaccines. Dev Biol Stand. Basel, Karger, 1998; 95:141-146.
27. Hwang IK, Morikawa T. Design issues in noninferiority/equivalence trials. Drug Information Journal 1999; 33:12205-1218.
28. Abramson JS, Baker CJ, Fisher MC et al. Possible association of intussusception and rotavirus vaccination. Pediatrics 1999; 104:575.
29. Levine MM, Barry EM. Inadvertent nontarget effects of live vaccines. In: Levine MM, Woodrow GC, Kaper JB, Cobon GS, eds. New generation vaccines. New York, NY: Marcel Dekker, 1997:1179-1184.
30. Simanjuntak CH, O'Hanley P, Punjabi NH et al. The safety, immunogenicity, and transmissibility of single-dose live oral cholera vaccine CVD 103-HgR in 24 to 59 month old Indonesian children. J Infect Dis 1993; 168:1169-1176.
31. Shoenfeld Y, Aron-Maor A. Vaccination and auto-immunity -«Vaccinosis»: A dangerous liaison? J Autoimmunity 2000; 14:1-10.
32. Rose NR. Immunological hazards associated with vaccination of humans. J Autoimmunity 2000; 14:11-13.
33. Miller E, Waight P, Farrington P. Safety assessment postlicensure. In: Plotkin S, Brown F, Horaud F, eds. Preclinical and Clinical Development of New Vaccines. Dev Biol Stand. Basel: Karger 1998; 95:235-243.
34. Cryz S et al. Vaccine Delivery Systems. The European Commission COST/STD-3 initiative on European vaccine research. Vaccine 1996; 14:665-690.
35. Ellis RW, Douglas RG. New vaccine technologies. JAMA 1994; 271:929-931.
36. Robinson HL, Ginsberg HS, Davis HL et al. The scientific future of DNA for immunization. American Academy of Microbiology 1996. http://www.asmusa.org/aca1.htm.
37. Nichols WW, Ledwith BJ, Manam SV et al. Potential DNA vaccine integration into host cell genome. In: Liu M, Hilleman MR, Kurth R eds. DNA vaccines: A new era in vaccinology. Ann New York Academy of Science 1995; 772:30-39.
38. Mor G, Singla M, Steinberg AD et al. Do DNA vaccine induce autoimmune disease? Human Gene Therapy 1997; 8:293-300.

Chapter 3

Vaccine Regulatory Issues

Marion F. Gruber, Paul G. Richman and Julianne C. M. Clifford

Introduction

The fundamental goals in developing new vaccine technologies are to improve current vaccines for existing clinical indications and to develop new immunogens for both pediatric and adult use. Recent progress in the field of recombinant DNA technology and advances in basic immunology have accelerated the development of novel vaccine approaches to modulate the immune response. Examples include non-replicating antigen delivery systems, genetically modified vectors expressing foreign antigens, DNA vaccines, as well as the use of novel adjuvants and different modes of vaccine administration.

The diversity and complexity of these products present new regulatory challenges, because specific standards for the criteria of safety, purity and potency for these products may not exist and current experience largely relies on animal models. The problem is further complicated by the variety of regulatory submissions for prevention of viral, bacterial and parasitic diseases. Therefore, the safety, purity and potency of these products are evaluated using a case-by-case approach that is product-specific and indication-based. Consideration is also given to potential risks versus the benefits of using the product in the target population. It is important to realize that regulatory policy evolves in response to advances in technologies. Thus, the purpose of this chapter is to provide a framework for product characterization from which preclinical and clinical testing strategies will need to be developed according to current knowledge and state-of-the-art technology.

Federal Regulations Pertaining To Vaccines

In the United States, vaccines are regulated as biological products by the Center for Biologics Evaluation and Research (CBER) of the U.S. Food and Drug Administration (FDA). The CBER is currently organized into offices with specific areas of expertise and responsibilities for regulating biologic products made by manufacturers which are or desire to be licensed for marketing in the United States. Preventive vaccines and therapeutic vaccines for infectious diseases are reviewed and regulated by the Office of Vaccines Research and Review (OVRR), whereas most vaccines intended for therapeutic indications other than infectious diseases are reviewed and regulated by the Office of Therapeutic Research and Review (OTRR).

Current authority for the regulation of vaccines along with other biologics resides in Section 351 of the Public Health Service Act as well as specific Sections of the Food, Drug and Cosmetic Act.[1] A single set of basic regulatory approval criteria apply to vaccines, regardless of the technology used in their manufacture. These regulations are contained in Title 21 of the Code of Federal Regulations (21 CFR).[2] They include a description of the production, testing and establishment standards pertaining to the manufacture of a biological product and are intended to assure that the product is safe and meets the quality and purity characteristics that it claims to possess (Table 3.1).

Table 3.1. Regulations applicable to the development, manufacture, licensure and use of vaccines

Title 21, Code of Federal Regulations[1] Chapter 1-FDA, DHHS	Subject
Subchapter A-General	
25	Environmental impact considerations
50	Protection of human subjects
56	Institutional Review Boards
58	Nonclinical laboratory studies, good laboratory practice regulations
Subchapter C-Drugs: General	
201	Labeling
202	Prescription drug advertising
210	Current good manufacturing practice in manufacturing, processing, packing, or holding of drugs, general
211	Current good manufacturing practice (GMPs) for finished pharmaceuticals
Subchapter D-Drugs for Human Use	
312	New drugs for investigational use
314	Applications for FDA approval to market a new drug
Subchapter F-Biologics	
600-680[2]	
600	Biological products, general, definitions
	Establishment standards
	Establishment inspection
	Adverse experience reporting
601	Licensing
610	General biological products standards

[1]Food and Drug Administration, Department of Health and Human Services.
[2]Parts 606, 607, 640, 660 and 680 apply to blood, blood products, diagnostic tests, and allergenics.

These regulations also cover the type and conduct of clinical studies that should be performed during product development. The Food and Drug Administration Modernization Act (FDAMA) of 1997 amends specific parts of these regulations and renews the Prescription Drug User Fee Act with amendments (PDUFA 2).[3,4] A "managed review process" was introduced with the initial enactment of PDUFA of 1992, and CBER implemented for the first time specific timelines for product and establishment license application (PLA and ELA) review.[5] The re-authorized

PDUFA 2 program outlines the five year review performance goals for the following areas: license applications, supplements and re-submissions; meeting management; clinical holds; major dispute resolution; special protocol assessment and agreement; electronic applications and submissions; discipline review; and complete response letters. All of these provisions apply to biologic products including vaccines and new vaccine technologies, and most are described in Guidance Documents written to assist manufacturers. In addition to these administrative Guidance Documents, FDA has developed scientific Guidance Documents for combination vaccines and DNA vaccines which, with applicable documents prepared by the International Conference on Harmonization (ICH), are particularly useful in emerging areas of new vaccine technologies because they specify a degree of detail beyond what is included in the regulations. These and other documents are available through the CBER website http://www.fda.gov/cber/publications.htm. As new Guidance Documents are constantly being issued, the reader is encouraged to refer to the CBER web site for current updates.

To obtain a U.S. license to market a new biologic product, an applicant must show that such a product is safe, effective and potent and that it can be manufactured in a consistent manner. Title 21 of the CFR, Part 600, provides definitions for safety, purity and potency:

1. **Safety** is defined as "the relative freedom from harmful effect to persons affected, directly or indirectly, by a product when prudently administered…" (21 CFR 600.3(p))[6]
2. **Purity** is defined as "relative freedom from extraneous matter in the finished product, whether or not harmful to the recipient or deleterious to the product." (21 CFR 600.3 (r))[6]
3. **Potency** is defined as "specific ability or capacity of the product as indicated by appropriate laboratory tests or by adequately controlled clinical data obtained through the administration of the product in a manner intended, to effect a given result." (21 CFR 600.3(s))[6]

Given the diversity of novel vaccine products, applying these criteria would necessitate a careful consideration of the character of the product, the methods of manufacture and the indication (Table 3.2). Therefore, it is critical that early in product development a dialogue be initiated between the CBER and the vaccine developer to decide whether the existing methods and standards for evaluation of the product with regard to the above criteria are adequate and also to agree on the design of preclinical and clinical programs.

Premarketing Phase

Product development is a continuous process from the pre-IND stage through the various phases of the clinical investigation, the license application and postmarketing, throughout which the interaction with CBER is critical (Figure 3.1). Meetings with industry serve as a forum for the agency to provide guidance to firms during product development and facility design, and to facilitate their compliance with the regulations governing development and postmarketing approval of products. Meetings that the sponsor can request are described in 21 CFR 312.47 and are categorized under PDUFA 2 into three types.[4,7] Type A meetings are necessary for proceeding with product development and need to be scheduled within 30 calendar days from the sponsor request date. Most meetings with FDA are type B meetings, i.e., pre-IND, end-of-Phase 1 or -Phase 2/pre-Phase 3, and pre-BLA meetings that need to be scheduled within 60 calendar days. Type C meetings are all other meetings that need to be scheduled within 75 calendar dates.

Preclinical Testing

Early in pre-clinical development, the primary goal of product testing should be to demonstrate that the investigational product is safe and exhibits activity that justifies testing in humans. Documentation of preclinical safety is one of the prerequisites to move a vaccine from

Table 3.2 Examples of antigen delivery systems/new generation vaccines

Non-replicating antigen delivery systems:	
	biodegradable microspheres
	liposomes
	proteosomes
DNA vaccines	
Subunit antigens	
Live attenuated viruses and bacteria:	
	e.g., *S. typhi* Ty 21a, *V. cholerae* CVD 103 hgR, rotavirus
Genetically modified live bacterial and viral vectors expressing selective virulence antigens from unrelated microorganisms:	
	e.g., *S. typhi, Shigella, V. cholera,* BCG
Mucosal adjuvants:	
	e.g., mutant forms of *E. coli* heat-labile enterotoxin, LT, and *V. cholerae* exotoxin, CT

the laboratory to the clinic. Toxicity studies should be conducted to assess the potential toxic effects of the vaccine antigens to target organs, including the hematopoietic and immune systems, as well as systemic toxicity.

Recent advances in recombinant DNA and protein technology allow the generation of highly purified subunit and recombinant protein vaccines. These are frequently not sufficiently immunogenic per se and require coadministration with an adjuvant. For example, for mucosal vaccines, much attention has focused on detoxified cholera toxin (CT), *E. coli* heat labile toxin (LT) or even cytokines as adjuvants to increase the immunogenicity of the vaccine antigen.[8-11] In these cases, toxicity studies of the specific antigen-adjuvant combination are important. Such studies can be particularly informative if the adjuvant is a novel component for which only limited or no toxicology data are available or if little or no clinical experience is available with this adjuvant in previous antigen-adjuvant formulations.[12]

Where different routes of administration are proposed, i.e., intranasal, oral, rectal or vaginal, multiple safety/toxicity studies may have to be conducted in a suitable animal model addressing specific safety concerns associated with vaccine administration via these routes. Caution should be used when extrapolating safety data obtained using a certain route of administration compared to other routes.

Reproductive toxicity studies are requested if the vaccine is specifically intended for use in a study that enrolls pregnant women and also will be needed if the product is indicated for a target population that includes women of childbearing age. ICH guidance documents should be consulted when designing such studies.[13,14]

While it is not current practice in vaccine testing to conduct pharmacokinetic studies, they may provide useful information, for example, when using controlled-release delivery systems such as liposomes, microspheres and proteosomes, or alternate routes of administration.[15,16] When live attenuated bacterial or viral organisms are used, preclinical safety studies also may include a demonstration of adequate attenuation of the organisms and absence of reversion to wild-type.[17,18] For nucleic acid vaccines, preclinical studies should be designed to determine whether the vaccine may have an immunotoxic effect by either inducing tolerance to the vaccine antigen or by potentially inducing autoimmunity.[19] Integration of the nucleic acid vaccine into the genome of the vaccinated subject may be the most important theoretical risk to consider in preclinical studies for these products.

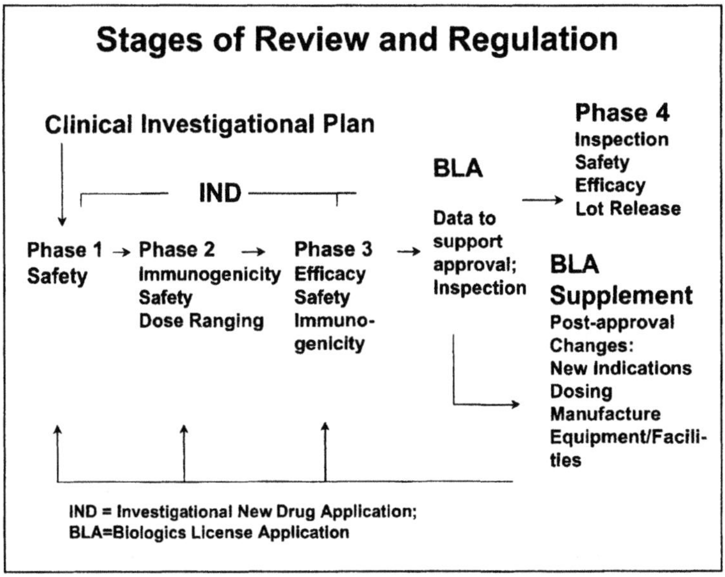

Fig. 3.1. Stages of review and regulation of biological products: Sequence of key events in product development through the premarketing Investigational New Drug (IND) and licensing phases (BLA, PLA), and the postapproval marketing phase. Solid lines indicate additional development submissions when significant changes are made to the product or its indication.

In addition to identifying potential toxicity problems requiring further clinical monitoring, data derived from preclinical studies also should provide the basis for the selection of the dosage levels, schedules and routes of administration to be evaluated in clinical trials. While responses in animals may not predict the human response, immunization of animals may provide valuable proof of concept to support a clinical development plan. For vaccines, preclinical studies should be designed to assess the relevant immune response, e.g., humoral and cell-mediated immune response in vaccinated animals. Where appropriate, challenge/protection studies in animal models with the corresponding infectious agent are encouraged. Of primary concern in interpreting these data should be how closely the animal model resembles the human disease that the vaccine is intended to prevent.

Investigational New Drug Application

Before initiating a clinical investigation for a vaccine, a sponsor submits to the CBER an Investigational New Drug (IND) Application.[20] The information that must be submitted in support of an IND is described in the US Code of Federal Regulations, 21 CFR 312.[21] Briefly, the IND should describe the scientific rationale for the vaccine and general investigational plan; preclinical animal safety data; the composition, source and method of manufacture of the product; control tests for in-process monitoring, final-product release; and stability testing; the proposed Phase 1 clinical protocol and names and qualifications of the clinical investigators.

The chemistry, manufacturing and control (CMC) section should include a detailed description of the manufacturing procedures employed. Bacterial and viral seeds and master and production cell banks will need to be characterized and described in detail. When genetic constructs are used, i.e., live attenuated vector vaccines expressing foreign antigen, the application should describe in detail the vector construction, including the source of plasmids used, and all cloning procedures. For a live attenuated and/or modified bacterial vector, the safety concern is whether the vector is virulent. Thus, assays demonstrating sufficient attenuation and lack of reversion to wild-type are necessary. For example, data regarding the colonization and reversion rates, as well as morbidity and mortality following the administration of the attenuated variant to animals, are useful. If whole cells or subunits are inactivated by mutation or chemical detoxification, absence of reversion to toxicity and completeness of inactivation need to be demonstrated.

The source and quality of the starting material will need to be described. If bovine materials are used in production, information that the country from which the material is derived is free of bovine spongiform encephalopathy (BSE) should be available because of concerns of the possible relationship of BSE with new variant human Creutzfeldt-Jakob disease. In-process testing should be performed to ensure control over the manufacturing process and manufacturing consistency. For the establishment of batch-to-batch consistency, characterization of the vaccine product should be undertaken during product development to establish analytical criteria that are used subsequently in evaluating new batches. These characterization and chemical purity tests may include various chromatographic and electrophoretic techniques, specific ELISA procedures and protein assays. Finally, it should be noted that technical advances have increased the ability to control and analyze the manufacture of biotechnology products such as synthetic peptide and recombinant DNA-derived vaccines allowing these products to be more clearly evaluated by end-product testing.

If changes in manufacturing occur during product development, these changes and concomitant in-process testing modifications should be described in detail. It should be noted that a new product may require additional preclinical studies to support its initial clinical use. Information about the purification process should be included and stability data provided showing that the product is stable for the duration of the clinical trial. Additionally, initial clinical use of the new product may need to occur in small study groups or with dose escalation to establish safety. Also, a direct comparison of the old and new products may need to be undertaken in a randomized clinical trial to establish a "bridge" between the performance of the two products.

Ultimately, the clinical development of a product is affected by its indication or target population and thus, the study population, laboratory and clinical evaluation, trial design and endpoints chosen depend on this indication.[22,23] The clinical immunogenicity, safety and efficacy of a vaccine is evaluated in various phases of IND studies defined in 21 CFR 312.21.[24] The initial phase 1 trial is usually an open-label study designed to evaluate the safety and perhaps the immunogenicity of the product in a small number of subjects, who are usually healthy adult volunteers at low risk of contracting the infection against which the vaccine is indicated. Phase 2 studies are designed to provide more definitive dose, schedule, and route of administration data and provide information about the common side effects associated with its use. Phase 3 studies are typically large-scale trials designed to collect pivotal efficacy data and expanded safety information.

Although most of the novel vaccine technologies are still in preclinical testing, some of these, such as live attenuated bacterial and viral vectors and DNA vaccines, have been studied in clinical trials for years.[17,18] Concerns relating to the safety of live recombinant microorganisms as vector systems have prompted a more cautious approach to their use in humans. In this case, shedding of live organisms would be examined during phase I, and possibly also in further phases. If challenge studies of normal volunteers are justified to screen vaccine candidates, a thorough safety monitoring of the subjects is mandatory and the proposed procedures for containment of the live biological material during clinical studies should be provided.[22] Study exclusion criteria and the consent form should address topics such as immunosuppressed vaccines and immunosuppressed contacts.

It is important that standardized and validated assays are in place that measure and detect the vaccine's elicited response in clinical samples. FDA historically has relied on serologic assay results to assess the immunogenicity of vaccines, for example diphtheria and tetanus toxoids and the *H. influenzae* type b vaccine.[22,25]

Serologic immune responses have been used for "bridging" of data, e.g., comparison of immunogenicity in different populations; comparison of different lots of vaccines, different manufacturing procedures or formulations; or comparison of proposed commercial lots with those used in efficacy studies. Immune correlates of protection are used as endpoints for such studies when possible. However, some vaccines (even licensed ones) do not have an identified immune correlate of protection (examples: BCG and rotavirus). In the absence of an identified immune correlate of protection, the clinical relevance of a decreased immune response for a new product or formulation may be difficult to interpret. Further considerations for clinical studies in these situations can be found in an article by Goldenthal et al.[23]

In summary, product development during the premarketing phase should apply a continuum approach; that is, as product development proceeds the appropriate data should be obtained demonstrating that the vaccine meets requirements for safety, purity and potency. By Phase 3, the product should be sufficiently defined and product specifications and final product formulation established.

Biologics License Application (BLA)

When the relevant IND studies are completed and when the product has been shown to be safe and effective, the sponsor submits a license application to manufacture and distribute the product commercially. Historically, two license applications have been required, one for the product and one for the establishment in which the product is manufactured.[26] However, as a result of FDAMA of 1997, CBER now accepts the submission of one single Biologic License Application (BLA) which will continue to include detailed information about product manufacture and clinical use, but the information about the establishment will be simplified.[27] To assist manufacturers in completing the various parts of the BLA, CBER has published for vaccines a document intended to provide guidance to applicants in completing the CMC section and the establishment description information.[28]

The establishment requirements for biological products continue to include the current Good Manufacturing Practice (cGMP) regulations found in parts 210, 211, 600 and 606. Although the establishment information required to be submitted in a BLA is reduced, FDA intends to use the requirements of 21 CFR part 211 during inspections of facilities, to help assure that biological products have the proper raw material controls, process validation and controls, as well as sensitive and validated test methods and specifications that are necessary to ensure the safety, purity, potency and effectiveness.

The essential elements for vaccine licensure are the availability of clinical data demonstrating that the product is safe and effective, a detailed description of the manufacturing process including data demonstrating control and consistency of the production process as well as validation of all methods, assays, systems and equipment employed; a demonstration that the vaccine product can be manufactured consistently within the defined and agreed upon specifications; data on stability; information about the package insert and labeling and environmental assessment. In general, prior to licensure of the vaccine, CBER's Vaccines and Related Biological Products Advisory Committee is asked to comment on the adequacy of the data to support safety and efficacy in the target population.[29]

Testing of Licensed Vaccines

The regulations for General Biological Standards 21 CFR 610 specify that each lot must be tested for potency, general safety, sterility, purity and identity.[30] For most vaccines a lot release protocol providing an outline of production and a summary of the test results and established specifications is submitted along with samples for CBER review and testing. Release or rejection is based on a review of all test results obtained by the manufacturer and by CBER.

Potency tests [21 CFR 610.10] measure biological activity or correlate to biological activity and are also used as one measure of product stability. The initial concept of potency testing for vaccines was to quantify the active moiety of the vaccine with known bioactivity for which the antigenic component(s) were not well defined (i.e., whole cell pertussis vaccine). More recently, recombinant DNA methodology and modern chemistry techniques have resulted in the manufacture of highly purified products that can be characterized in great detail. However, the ability to measure the "relevant" biological activity for such products may be lacking. For some vaccines, potency testing may rely on physicochemical characterization, such as amount of antigen, size of the antigen or other chemical parameter correlating to in vivo biological activity. For example, for certain antigen-proteosome mixtures potency may be defined by weight based on protein content, as well as characterization of antigen to proteosome ratios. For some products it may be possible to demonstrate potency by immunological methods such as ELISA detection of vaccine epitopes. Yet, the establishment of relevant potency tests for other products may be more difficult. For example, in the case of bacterial- or viral-vectored vaccines expressing inserts encoding heterologous vaccine antigens, determining the "biological activity" of the entire construct as measured by CFU or infectious titer is not the most relevant potency test. Instead, the quantification of expression of the insert in vivo would seem to be a more informative measurement of potency. One approach to quantify expression of the heterologous antigen may be the evaluation of the effective dose (ED_{50}) of the vectored vaccine. The difficulty in developing appropriate potency assays for these vaccines is further magnified with constructs expressing multiple epitopes. Several points to consider are a) the means by which one can demonstrate the contribution of each construct or insert to the overall immunogenicity or bioactivity of the product, b) the need for assays which detect individual immune responses to each epitope expressed, and c) whether antigenic competition occurs when multiple epitopes are expressed.

The general safety test (21 CFR 610.11) on the filled final container material is performed using guinea pigs and mice. This test is required for all vaccines licensed in the U.S. to assure that extraneous toxic substances have not been introduced during production or filling. If the product has intrinsic toxicity, it may be necessary to modify the test from that specified in the regulations. CBER should be consulted regarding any modification and development of the general safety test that will be used. Manufacturers should note that the general safety test is not a substitute for other safety and toxicity testing of the product that may be appropriate.

Sterility testing (21 CFR 610.12) is a U.S. requirement for all vaccines intended for injection. However, for non-parenterally administered products bioburden measurements performed according to USP may be sufficient. These include a determination of bioburden count, an estimation of the bacterial and fungal load and screening on selective media. If organisms grow they should be speciated to ensure the absence of pathogens.

Purity tests (21 CFR 610.13) need to be performed and may include an evaluation of residual moisture, endotoxin content, and quantification of process additives introduced during manufacture. Although not required, some vaccines are also tested for pyrogenic substances by intravenous injection in rabbits. In general, the limulus amoebocyte lysate (LAL) test may be substituted for the rabbit pyrogenicity test when the LAL test has been validated in the presence of product.[31]

Finally each lot must be tested for identity (21 CFR 610.14). For vaccines, this may be a physical or chemical test or an immunologic assay for the included antigens. For example, for DNA vaccines, a restriction enzyme analysis and DNA sequence determination for the plasmid construct would be appropriate.

Other more specific tests designed to provide additional assurance of safety, purity or product consistency may be required. These tests may be described in the regulations for a particular product or described in guidelines that are developed by the agency. Once the product is licensed, testing is conducted according to the exact specifications in the manufacturer's license application. Results of such testing must be within the prescribed limits for the material to be distributed commercially. Depending on the product, confirmatory testing may also be performed by the CBER. New regulations have been developed that may exempt particular products from the requirements for lot release.[32] Ordinarily, this would only be appropriate to consider for a product with an extensive postlicensure experience. To be exempted from lot release, the product must repeatedly meet the standards for potency, purity and stability of bulk and final container material while using a consistent process and maintaining a satisfactory inspectional history. A sponsor needs to apply for this exemption.

Postmarketing Phase

For any new product approval, sponsors may be asked to perform postmarketing "Phase 4" studies. These studies are often important to provide additional data on less common or rare adverse events in large numbers of vaccine recipients. The protocol and proposed timelines for such Phase 4 studies should be provided to FDA prior to initiation.

If a licensed applicant wants to make a change in the product, labeling, production process, quality controls, equipment, facilities, or responsible personnel established in the approved application, such changes need to be reported to FDA.[33] Changes in manufacturing will be reported as either

1. a supplement requiring prior approval,
2. a supplement submitted at least 30 days prior to distribution of the product made using the change,
3. an annual report to the approved license application. These three reporting categories are based on the potential risk of the change having an adverse effect on the product. Significant labeling changes, e.g., revision of the package insert based on data for a new indication, will require submission of a prior approval supplement. However, safety clarifications may be implemented concomitantly with the submission of a supplement and editorial changes may be summarized in an annual report. Guidance for industry documents that assist in determining the category in which a manufacturing or labeling change will be placed are available.[34]

For childhood vaccines, the National Childhood Vaccine Injury Act of 1986 requires explicit, extensive record-keeping, including the identification of any significant problems encountered in the production, testing or handling of the product.[35]

For all licensed vaccines, adverse reactions reports are entered into the Vaccine Adverse Event Reporting System with the reports submitted by manufacturers, physicians, parents and patients.[36] CBER and the CDC monitor this database for incidents of concern and patterns of events that appear to be vaccine related.

In summary, vaccines licensed in the US must meet requirements for safety, purity and potency. Although general requirements exist for these parameters, case-specific standards for each vaccine are agreed upon by the manufacturer and CBER/OVRR. The final standards for any vaccine are relevant to the technology used to produce the vaccine. Finally, by expanding the managed review process to all phases of product development and standardizing the format and content requirements for biologics premarketing applications (BLA), the regulatory timelines for approval of new vaccine technologies are being shortened allowing faster access of the intended target population to these important products.

References

1. Public Health Service Act, July 1, 1944, Chap.373, Title III, Sec. 352, 58 Stat. 702, currently codified at 42 U.S.C., Sec 262.
2. Code of Federal Regulations, Title 21, Food and Drugs, Washington DC, US Government Printing Office, 1998.
3. Food and Drug Administration Modernization Act of 1997, Public Law 105-115.
4. Prescription Drug User Fee Act letter, November 12, 1997.
5. Prescription Drug User Fee Act 1992, Public Law 102-571.
6. Code of Federal Regulations, Title 21, Parts 600, Washington DC, US Government Printing Office, 1998.
7. Code of Federal Regulations, Title 21, Part 312.47, Washington, DC, US Government Printing Office, 1998.
8. Chen, KS, Stroeber W. Cholera holotoxin and its B subunit enhance Peyer's Patch B cell responses induced by orally administered influenza virus:Disproportionate cholera toxin enhancement of the IgA B cell response. Eur J Immunol 1990; 20:433-436.
9. Chong C, Friberg M, Clements JD. LT(R192G), a non-toxic mutant of the heat-labile enterotoxin of *Escherichia coli*, elicits enhanced humoral and cellular immune responses associated with protection against lethal oral challenge with *Salmonella* spp. Vaccine 1998; 16:732-740.
10. Ruby J, Ramsay A, Karuplah G et al. Recombinant vaccine vectors that co-express cytokines—A new vaccine strategy. Vaccine Res 1992; 1:347-356.
11. Abraham E and Shah S. Intranasal immunization with liposomes containing IL-2 enhances bacterial polysaccharide antigen-specific pulmonary secretory antibody response. J Immunol 1992; 149:3719-3726.
12. Goldenthal KL, Cavagnero CR, Alving CR, et al. Safety evaluation of vaccine adjuvants: National cooperative vaccine development meeting working group. AIDS Res Hum Retroviruses 1993; 9:47-51.
13. ICH, Harmonized Tripartite Guideline Detection of toxicity to reproduction for medicinal products FR 1994; 183(59):48746-48752.
14. ICH, Harmonized Tripartite Guideline "Detection of toxicity to reproduction for medicinal products: Addendum on toxicity to male fertility", FR 1996; 67:(61)15359-15361.
15. Kuntz RM and Saltzman WM. Polymeric controlled delivery for immunization Tibtech 1997; 15:364-368.
16. Chaicumpa W, Chongsa-ngunan M, Kalambaheti PW et al. Immunogenicity of liposome-associated and refined antigen oral cholera vaccine in Thai volunteers Vaccine 1998; 16:678-684.
17. Levine MM, Galen J, Barry E et al. Attenuated *Salmonella* as live oral vaccines against typhoid fever and as live vectors. J Biotechn 1996; 44:193-196.
18. Tackett CO, Losonsky G, Nataro JP et al. Onset and duration of protective immunity in challenged volunteers after vaccination with live oral cholera vaccine CVD103-HgR. J Inf Dis 1992; 166:837-841.
19. Smith HA. Regulatory considerations for nucleic acid vaccine. Vaccine 1994; 12:1515-1519.
20. Novak JM, Barrett, J, McVittie LD et al. The Biological IND. In: Mathieu M, ed.: Biologics Development: A regulatory overview. Waltham, MA, USA: PAREXEL 2nd ed 1997:49-81.
21. Code of Federal Regulations, Title 21, Part 312, Washington, DC, US Government Printing Office, 1998.
22. Goldenthal KL, McVittie LD. The clinical testing of preventive vaccines for infectious disease indications. In: Mathieu M, ed. Biologics Development: A regulatory overview. Waltham, MA, USA: PAREXEL 2nd ed. 1997:123-139.
23. Goldenthal KL, Vaillancourt JM, Geber A et al. Preventive HIV Type I vaccine clinical trials: A regulatory perspective. AIDS Res Hum Retroviruses 1998; 14:333-340.
24. Code of Federal Regulations, Title 21, Part 312.21, Washington, DC, US Government Printing Office, 1998.
25. Käyhty H. Immunogenicity assays and surrogate markers to predict vaccine efficacy. In: Plotkin S, ed. Preclinical and clinical development of new vaccines. Karger, Basel, Switzerland, 1998:175-180.

26. Code of Federal Regulations, Title 21, Secs. 601.10, 601.20. Washington, DC, Us Government Printing Office, 1998.
27. Proposed rule: Implementation of Biologics License Elimination of establishment license and product license, FR. 1998; 63(147):40858-40865.
28. Guidance for Industry: Content and Format of Chemistry, manufacturing and controls information and establishment description information for a vaccine or related product, FR. 1999; 64(2):518-519.
29. Freas W, Gertzog J. The FDA's Biological Products Advisory Committees. In: Mathieu M, ed.: Biologics Development: A regulatory overview. Waltham, MA, USA: PAREXEL 2nd ed. 1997:253-268.
30. Code of Federal Regulations, Title 21, Sec 610, Washington, DC, US Government Printing Office, 1998.
31. Guidance for Industry: Human, Biological, and animal drugs and medical devices; availability of guideline for use of the limulus amebocyte lysate (LAL) test, FR. 1988; 53(33):5044-5055.
32. Guidance on Alternatives to lot release for licensed biological products, FR. 1993; 58(137): 38771-38773.
33. Code of Federal Regulations, Title 21, Sec 601.12, Washington, DC, US Government Printing Office, 1998.
34. Guidance for Industry: Changes to an approved application for a biological product, FR. 1997; 62(142):39904-39906.
35. National Childhood Vaccines Act: Requirements for Permanent Vaccination Records and for Reporting of Selected Events After Vaccination, 37 MMWR 1988:197-200.
36. CDC. Vaccine Adverse Event Reporting System—United States, 39 MMWR 1990:730-733.

Chapter 4

In-Licensing Issues and Vaccine Technologies

Dale R. Spriggs

Introduction

The pharmaceutical industry is undergoing profound changes as the advances in biotechnology give rise to new technologies and approaches to discovering and developing new products. Conventional combinatorial chemistry and high-throughput screening combined with functional genomics, pharmacogenomics, proteomics, transgenics and a whole host of variations on these technologies are driving the pace of innovation.

The vaccine industry is also evolving rapidly as a result of these new technologies. Within the last few years we have seen the emergence of nucleic acid vaccines, bacterial genomics, novel vectoring systems, new delivery systems such as food crops and transcutaneous immunization.

In the face of this technology explosion, companies in the pharmaceutical industry are constantly trying to enhance their internal capabilities to keep up and looking for new ideas and inventions outside the company in order to stay competitive. Translating discoveries into useful products requires massive amounts of resources and time and—for scientists outside the pharmaceutical industry or in small biotech companies—requires that the inventor license the invention to a company that has the capability and desire to develop the technology to commercialization. This process (licensing, or in-licensing from the company's perspective) is the focus of this chapter.

I will try to outline some of the general principles and give advice about how licensing can work to the benefit of licensor and licensee. I will discuss these issues mainly from the corporate perspective and focus most of the discussion on vaccines, adjuvants, and delivery systems and how companies might think about these technologies .

Why Do Companies In-License?

Although this may seem like a simple question, the answers are often complex and depend on the company with which you are dealing. The key driver is competition. The vaccine industry is very competitive, and no one company has the right constellation of antigens and technology to successfully compete in all major areas. Each company will develop its own competitive strategy and determine the role of in-licensing in the mix.

The most obvious reason to in-license is to fill in the R&D pipeline or add a specific product to a company's portfolio (Table 4.1). For example, a company might have a DTaP vaccine, but they may not have the next component (IPV, HiB) to add for the next-generation product. If they don't have an internal candidate, they will need to in-license an existing vaccine to keep their development underway. Companies also need to have the right mix of early- and late-stage candidates, and there are constant pressures to balance the portfolio to ensure a steady stream of new products coming to market. It is also important to understand that most vaccines never make it to licensure; indeed, studies have shown that only 20-25% of vaccines

New Vaccine Technologies, edited by Ronald W. Ellis. ©2001 Eurekah.com.

Table 4.1. Why do companies in-license?

To Fill in Their R&D Pipeline
To Enhance Manufacturing Capabilities
To Explore New Technologies
To Block Competitors

Table 4.2. How do companies evaluate licensing opportunities?

Technical Issues
Intellectual Property/Patents
Strategic Fit
Marketing/Commercial Issues

that enter clinical development actually get licensed. A company has to plan for the inevitable failures to keep up its momentum.

Companies also will in-license cutting-edge technology or products as part of an exploratory effort to develop a new platform to discover new antigens. The most recent example of this approach is the surge in the number of companies that have started genomics efforts to identify possible vaccine and therapeutic targets. A company might also look, for example, at new adjuvants that may have a longer time-horizon for development, but which may be useful for a variety of vaccines. These early-stage efforts could have important strategic value to the company, but it is important to realize that these are long-range projects and the returns (scientific and financial) may not show up for a long time.

Another approach centers on in-licensing technology that can improve the manufacturing process and hence make the existing products more competitive and profitable. A major component of product cost is the cost of goods for manufacturing and packaging, and companies continually look for opportunities to keep these under control. These are often referred to as enabling technologies. These technologies could include basic process improvements, enhanced expression systems, or new vectors and cell lines that could improve early development projects. Later-stage development improvements such as sterile-filling, syringe-filling, and packaging improvements also can have important benefits on products in development or already licensed.

Finally, it is important to realize that companies may occasionally license a technology to block competitors. Because business is so competitive, it may be advantageous to pay modest licensing fees to use the intellectual property as a placeholder and to prevent a competitor from acquiring the technology. Alternatively, a company may reflexively license a technology in response to a similar deal by a competitor. In either case, always remember that intellectual property is precisely that—property. As such, it can either be developed or used as part of a deal to help the company carry out its business strategy. This is one reason that the due-diligence provisions in license agreements are so important: they should clearly spell out how the parties define and document progress in the project and spell out how and when the rights revert back to the inventor/institution if the company is not devoting enough energy to development.

How Do Companies Evaluate Licensing Opportunities?

Each company has its own set of rules for evaluating licensing opportunities. A small biotech company will view the risk/opportunity equation much differently than will a larger company.

The following are a few general criteria that a company might use to determine whether to proceed with a deal and how a licensing deal might take shape based on the opportunity. These various deliberations go on in parallel (Table 4.2).

Technical Risk

This is the most obvious level at which most companies will start the classic "due diligence" process of finding out how good the technology really is. If the project is still in a research phase, the company scientists will be interested in conducting a thorough review of the data to see how the studies were done and get a sense of whether the data support the claims. They also will assess the feasibility of repeating some of the studies either in-house or through a third party.

At another level, the company will need to assess the potential development risks when the technology is scaled up from the laboratory to manufacturing. For example, if a new viral vaccine was derived and grown in monkey cells cultured in T-flasks, the development group would want to assess the likelihood of being able to grow the virus in human cells in serum-free growth media in bioreactors. This assessment is, of course, fraught with uncertainty and requires educated guesses based on reviews of the literature and in-house experience with similar problems. A more experienced company can obviously make a better-informed decision on these issues, and their ability to address these concerns should be part of what you look for in a good licensing partner.

Most companies also will include preliminary marketing issues in this technical review. For example, if preliminary market research indicates that the vaccine should be a lyophilized format, the formulation group may be consulted to assess how (or if) this could be done. Again, companies will vary in their attention to these issues.

One of the major areas of technical risk is the clinical development plan. Even at an early stage, a company needs to think through these issues because the clinical trials effort is the longest and most costly part of the whole plan. This assessment will usually start with an overview of the likely sequence of trials that could lead to licensure of the product. In this situation, one would look for similar products developed for the same target population for guidance. If a company is considering an injectable pediatric vaccine, there will be numerous examples of the kinds of safety, immunogenicity, and efficacy data needed for licensure. On the other hand, a new vaccine like Lyme vaccine poses a whole host of questions about potential safety concerns and how one would do efficacy trials. In any assessment, the company will have to weigh the risks of moving into new disease areas or indications in which it has no experience.

Finally, a very important risk that is linked closely with clinical is regulatory risk. The pharmaceutical industry is highly regulated by the Food and Drug Administration (in the U.S.); indeed, the FDA should be viewed as the most important "customer" until the product is on the market (even after this, the FDA plays a major role in the life of the product). The company's goal is to get a product licensed so it can be used, and this will not happen if a company cannot deal with issues raised by the FDA during the development process. Even if you do everything right, there's no guarantee that you will get approved. There are many examples of products, including vaccines, that have been rejected at the final stages of the approval process; indeed, even licensed product such as the recent live, attenuated rotavirus vaccine can be withdrawn because of safety concerns. At every stage, the FDA is critical, and the manner in which a company handles these interactions is vital to its survival—and, by extension, the likelihood that it will be able to turn an in-licensed opportunity into a commercial reality.

Intellectual Property/Patents

One of the earliest issues that arises during discussions is the patent position. With many projects, the intellectual property (IP) is not yet patented, and this inevitably leads to questions about whether this technology actually can be patented and in what form the patent might issue. In addition, there is always the possibility of missing important IP. Aside from wanting clear proprietary protection, a company also wants to avoid getting into a situation in which it will have to pay royalties that eat into profits.

For example, if a company is interested in licensing a recombinant protein vaccine expressed in *E. coli*, it might have to worry about the following: Are there any patents that claim this DNA sequence and that would therefore dominate; will they have to obtain licenses to use specific plasmids/promoters or proprietary strains of *E. coli*? Are there process patents that they may need to access as part of the development (novel purification scheme); will they need to license a patented stabilizer to include in the final formulation? All of these issues can increase the development and manufacturing cost and final cost of the commercial product; these costs substantially affect the overall financial return to the company, hence its potential interest in the licensing opportunity.

Strategic Fit

Most companies have internal processes to evaluate and prioritize ongoing projects, and at some level they have to assess the implications of bringing in a new project. How will it be integrated? If it is a new technology such as a new viral vector, can it be used as a platform to express a variety of antigens that the company is developing? How widely might a new adjuvant be tested with existing vaccines or vaccines in development? Does the company have the expertise and capability to take on a new project, or will they have to dedicate new resources to it and hire additional people? For example, a strategy to in-license and develop DNA vaccine processes would fit better in a company with ongoing prokaryotic R&D activities than in a company that works only with mammalian cell systems.

Marketing/Commercial

The final major area of risk and opportunity that a company explores is the potential commercial opportunity with the new product or technology. This activity is at the heart of the corporate enterprise. Through a series of discussions and financial analyses, a company tries to visualize how the new product could be used and how the company could develop the product in a manner that will yield the highest value.

If a company does not have experience in a certain market, its marketing people may do some preliminary research to understand the market size, segmentation, and needs. They also need to define if there are competing licensed products or products under development. If so, how might this new product be better (e.g., safety, immunogenicity, route of immunization)? If there are no competitors, how does the company plan to create the market? How might these issues be handled in various regions of the world?

Another important preliminary issue to explore is reimbursement. Who is going to pay for the vaccine, and what are their concerns? Are there similar products that could be used to model the possible market issues? In the vaccine field, we have a mix of government, managed care, and private payers, each of which will have to be polled early on. Vaccines, like other pharmaceuticals, are scrutinized for cost-benefit and cost-effectiveness, and this variable must be considered early in the development effort and should be considered carefully during all phases of the process.

Ideally, the marketing group also should have an understanding of the trends that are shaping the field and how they might influence the success of the product. They need to see

Table 4.3. How can an inventor make the technology more attractive?

Focus on the Science
Cultivate a Company Champion
Learn About the Industry

larger issues such as vaccine safety, combination vaccine strategies, shifts in reimbursement policy, or liability issues and try to project how these might influence the likely success of the new vaccine. They need to understand how the competitive landscape could change and how the company would respond.

How Can an Inventor Make the Technology More Attractive?

Because of the time it takes to work through the complex web of due-diligence and decision-making leading up to a decision about whether a company will in-license a new technology or product, an inventor has an opportunity to influence the outcome by paying attention to a few points (Table 4.3).

First and foremost, make sure that the science underlying your technology is of the highest quality and is effectively communicated. Of all the risks outlined above, this is the only one that the inventor can control directly. Early on, you will be dealing primarily with company scientists who later will be asked to comment on the strength of the data. Present a clear picture of how the work evolved, how hypotheses were tested, and how the experimental design of your studies supports the conclusions. If the science is solid, you will inspire confidence in your technology and greatly enhance your chances of getting the deal done. As a corollary, the more data you have that reinforces the intellectual property that the company is interested in and the further you have moved into early development, the more valuable it will be to the company. You will thereby increase your chances of getting the best financial terms possible.

During your discussions with the company, you will interact with various scientists and people from business development, the latter of whom will actually negotiate the license agreement. It is very important to cultivate one or more people from these groups to be the "champion" for your project. The champion is someone in the company who really supports the technology and is willing to fight for it internally if necessary. Early on, you will get a sense of who might be a strong supporter, and when that person asks for information or makes suggestions, you should do whatever you can to comply. You should never assume that the science will speak for itself; you need a champion to ensure the greatest chance for getting the technology licensed and developed.

Another important area to focus on is to gain some understanding of the vaccine industry. During discussions with business development people and the scientists in the company, you should try to get a sense of their perspective on what forces are driving the business. This information can be helpful at a variety of levels. For example, if you appreciate that a company has invested hundreds of millions of dollars developing a pediatric combination vaccine, you will have a better sense of how difficult it will be to convince them to replace one of the antigens of the combination, even if the new antigen is more immunogenic or safer. This can happen rarely—as seen in the case of acellular pertussis vaccines—but it requires a special constellation of public and private interests converging to drive the development.

Likewise, adding an experimental adjuvant to existing vaccines poses interesting problems. From a regulatory standpoint, the combination of vaccine and adjuvant is a distinct product, and the company must therefore get a separate license to sell this combination, even if they have years of experience with the vaccine alone. A company will have to weigh the potential gain in immunogenicity with the possible costs to reformulate the vaccine, ensure that the product maintains an acceptable safety profile, and conduct a series of clinical trials to support the new product. To date, these have been difficult issues to balance.

Framework of a Licensing Agreement

Although each licensing agreement has its own unique features, I will try to outline a few general principles that drive the negotiations. Once both parties agree that they are interested in moving forward the negotiations usually coalesce around three main topics: license exclusivity, money, and IP. Simply put, the perfect deal from the company's perspective would give them worldwide, exclusive rights to all IP (existing or developed later) with very little money at risk. Inventors and their institutions usually disagree with this approach.

If a company wants an exclusive, worldwide license to the intellectual property, they will be willing to pay more than if they are willing to take a nonexclusive license. The perceived value of the IP will be driven by all the factors discussed above, especially the market forecast for the product(s) that might come out of the development effort. IP will also be a major negotiation point when considering who has rights to new IP generated under terms of the agreement; for example, the agreement will specify what new IP would be held by the inventor, the company, or jointly. The company will usually request the right to be the first to review any new IP and discuss possible licensing terms if they are interested (the so-called right-of-first-refusal). These issues can become surprisingly contentious and can consume a good deal of the negotiation time.

Most of the negotiations will focus on the financial terms of the deal. These terms will be driven by several factors, including the following: the nature of the technology, the stage of development, the strength of the IP portfolio, the type of company (small biotech vs. large pharmaceutical), the exclusivity provisions and territories included in the agreement, and most importantly, the perceived commercial value of the technology.

The final agreement will consist of a series of milestones that will trigger payments from the company. These will usually include an upfront payment at the time the deal is signed and payments at key development milestones such as filing the first IND (Investigational New Drug Application) with the FDA; completion of Phase 1, 2, and 3 trials; filing a Product License Application; and marketing approval. Depending on the product, these payments can add up to many millions of dollars.

The agreement also will define the royalties that will be paid to the inventor/institution once the product is on the market. Negotiations from the company's perspective of course will focus on the potential market opportunity, but also they will include consideration of other licensing obligations and the associated costs. These will have an impact on profit projections and overall value to the company (as discussed previously). The final agreement will specify that the company will pay the inventor/institution fees based on percentage of sales of the product; these typically are from 1%-2% up to 20%, and in some cases can add up to many millions of dollars a year. Indeed, several biotechnology companies, including BioChem Pharma, have been built on the royalties received from other companies who have successfully developed and marketed drugs or vaccines discovered by the smaller company.

Although a detailed discussion of the wide variety of issues that can be included in a license agreement is beyond the scope of this chapter, the agreement can contain several other provisions, including research support, equity investment in the biotechnology company, payment of patent application and maintenance costs, and support with potential litigation. Agreements also will describe what sublicensing rights the company has and how money from the sublicensee will be split between the parties. Another very important component is the section specifying what rights the parties have to terminate the agreement and what happens to all the data, IP, and regulatory documents that were generated during the collaboration. This is not a comfortable topic to negotiate when both parties want to focus on the positive and consummate the deal, but it is extremely important.

What Does the Future Hold?

The flood of new technology and IP mentioned in the *Introduction* creates interesting challenges for companies trying to integrate in-licensing efforts with ongoing business activities. While competitive forces will continue to drive in-licensing needs, the sheer volume of new technology will make it more difficult for a company to create what it wants most: a simple proprietary product. In the future, companies will have to cobble together license rights to a larger number of patents and IP to create a defensible portfolio. To do this, they will have to pay a larger percentage of product revenue to cover royalties on the patents; this will, in turn, create additional internal pressures as the industry tries to contain R&D and manufacturing costs in order to stay competitive. This conflict may have the effect of decreasing the value of each patent since few, if any, will dominate in a diverse portfolio of IP. This is, however, a good situation for people interested in business development and technology transfer, who will be busy trying to sort out appropriate licensing and milestone fees for these complex and overlapping IP concerns.

Coda

The in-licensing process can take years, depending on the complexity of the technology and the types of ongoing interactions (e.g., research agreements) that might be negotiated with the license. Every company handles these issues differently, and the strategy can evolve, even during the course of the negotiations. Listen carefully to what your corporate colleagues are saying, and try to gain an understanding of what motivates them. If all goes well, you will have a long-standing relationship that will ultimately yield new and improved products for human health.

Chapter 5

Live Vaccines

Alan R. Shaw

Live attenuated viral vaccines represent the most effective means of inducing a broad immune response against viruses that can be cultivated in vitro. These vaccines mimic a natural infection and thereby induce both cellular and humoral responses required for efficient defense against subsequent exposure to the natural agent. In most cases, this is achieved with a single dose, an increasingly important attribute in an era of ever more complex immunization recommendations.

This chapter will focus primarily on the live attenuated viral vaccines currently (or soon to be) licensed for use. The author hopes to convey a sense of how these vaccines were developed, how well they work, where current concerns lie, and where future development is headed.

Smallpox Vaccine

A review of live virus vaccines must, by definition, begin with a discussion of smallpox vaccines. Smallpox was endemic in most of the world throughout recorded history. Records from Glasgow in the 18th century showed that smallpox was responsible for 40% of childhood deaths, and the death rate for children under ten from any cause was 50% overall. In Europe at that time, smallpox was also the leading cause of blindness. Case fatality rates were on the order of 20%.[1]

For centuries, there had been a practice of inoculating naive individuals with material derived from lesions of persons suffering from active smallpox. The idea behind this practice was that a small dose of the active agent would protect from later disease. This inoculation, or variolation (after the name of the disease, variola), was successful. When it worked, and more often than not it did, the inoculated person developed a local rash which resolved. Success depended upon getting the dose right and having a skilled inoculator. In addition to the risk of developing frank smallpox, there was also the liability of the inoculated person being a source of infectious virus and thus spreading the disease. Another risk was the spread of other maladies from the smallpox "donor," notably syphilis.

The first successful exercise in the induction of immunity to an infectious agent through deliberate pre-exposure to a similar agent is attributed to Edward Jenner in the late 18th century. Jenner, a country physician in England, noticed that milkmaids were seldom scarred by smallpox. He also made the connection between the bovine disease, cowpox, and a mild rash suffered by these milkmaids. Jenner himself was aware of the practice of variolation and had practiced it himself; variolation had been introduced into England by Lady Mary Wortley Montague after seeing it performed during a visit to Turkey.[2] Applying this technique, but using "lymph" from a lesion on a cow with cowpox, Jenner made the first "vaccination" in 1796. The subject, an eight-year-old boy named James Phipps, was challenged by variolation six weeks later, and no rash typical of variolation occurred. This one-person clinical trial was written up for publication in the *Philosophical Transactions* of the Royal Society, but it was

New Vaccine Technologies, edited by Ronald W. Ellis. ©2001 Eurekah.com.

rejected on the grounds that it was "at variance with established knowledge".[1] After a hiatus due to a lack of appropriate cowpox extract (cowpox was only a sporadic disease in England), Jenner resumed his tests in 1798, vaccinating and challenging another 15 people. The results were privately published as a pamphlet in September of 1798 and are referred to as Jenner's *Inquiry*.

Jenner's vaccine was not an instant, universal success. Objections were raised by, amongst others, the traditional variolators for the obvious reason, and by religious leaders for thwarting the will of God by preventing a disease.[1] Reason did prevail, and Jenner's results were reproduced by Pearson and Woodville at the St. Pancras Smallpox Hospital in London in 1799.[1] Jenner's pamphlet was widely translated and distributed, and the practice of "vaccination" (a term attributed to another English doctor, Richard Dunning in 1803[1]) spread rapidly, in spite of practical barriers. A major problem was the relative rarity of cowpox, a disease which was only sporadic in England and quite rare if not unknown elsewhere. Various means of transporting bovine lymph were developed, primarily as dried material on linen fibers or ivory spikes. Vaccine was then produced by serial transfer in calves that were, in some cases, taken from place to place to provide for mass vaccinations. An alternative was to transfer vaccine from a vaccinee with cowpox rash to new subjects. Vaccination did produce mild local pustules, which could be used as a source of fresh virus. This, too, had practical limitations.

It is easy to imagine how the provenance of what we now call vaccinia virus might have been lost over time. Passage of virus from arm to arm in humans, from calf to calf and even through horses, sheep and water buffalo would tend to blur the definition of vaccinia. Jenner himself postulated that cowpox was really a fortuitous infection of bovines by an equine disease known as "grease".[3] Modern analyses of vaccinia, cowpox and smallpox show that they are quite independent viruses, and it is difficult to see how vaccinia could have evolved from cowpox. One possibility is that vaccinia is a derivative of horsepox.[4-7]

The last case of naturally transmitted smallpox occurred in 1977 in Somalia,[5] and in 1980, the WHO declared the world free from smallpox. This declaration has held up so far, and the debate at the time of this writing is whether the existing laboratory stocks of smallpox virus should be destroyed and whether the virus is being developed into a biological warfare agent.

Japanese Encephalitis Vaccines

Japanese encephalitis is a viral illness caused by a member of the Flavivirus family. Japanese encephalitis virus (JEV) is transmitted by mosquitoes of the Culex genus and can infect pigs, horses, humans and wading birds. Humans are a dead-end host leading to low levels of viremia, but the virus can penetrate the blood-brain barrier and cause severe encephalitis. Case fatality rates are in the range of 15-25% while neurological sequelae in survivors are common (reviewed in 8 and 9). Seroprevalence approached 90% by the age of 15 in endemic areas, suggesting that subclinical infections are quite common. The virus is endemic in Japan, Korea, China, Southeast Asia and parts of the Indian subcontinent and is most common in rural areas with standing water (e.g., rice paddies).

JEV was first isolated in Japan in 1935 as the Nakayama strain. A killed virus vaccine was produced for military use by E.R. Squibb in 1944-45,[10] but it was never licensed for general use. An inactivated whole-virus vaccine has been licensed in Japan since 1954. This vaccine is produced in adult mouse brain and has undergone a series of process upgrades over the years (reviewed in 8 and 11) in order to reduce carryover of neuronal material. This vaccine, produced by Biken, is widely used in Japan and elsewhere in the Far East as a five-dose series. In 1992 the vaccine was licensed and recommended in the United States for use by travelers to endemic areas in an accelerated 3-dose regimen. Versions of this vaccine have also been produced by local organizations in Thailand and India.[12,13] Second-generation recombinant DNA-based experimental vaccines have been made using poxvirus vectors.[14]

There is currently one licensed live attenuated JEV vaccine based on the SA-14 isolate taken from a mosquito in 1954. The virus was passaged 11 times in adult mice and 100 times in primary hamster kidney cells at 36-37°C. This yielded strain SA14A, a virus that was not neurovirulent when injected intracerebrally in mice and rhesus monkeys. This virus was plaque-purified 14 times on primary chick embryo cells and primary hamster kidney cells yielding a highly attenuated clone, "5-3". This virus had the appropriate lack of neurovirulence, but it grew poorly in mice. Five further passages in suckling mice followed by two rounds of plaque-purification gave the final virus, SA14-14-2.[15,14] This vaccine was licensed in China in 1988 as a single-dose product. Subsequent studies have shown that performance is substantially improved by a second dose administered one year after the first.[16] This vaccine is propagated in primary hamster kidney cells, a cell substrate not as yet accepted by any of the major regulatory bodies. As a consequence, this vaccine is used only in China.

Mechanism of Attenuation

Not much is known about the molecular basis of attenuation of JEV. Genomic sequencing has yielded hints of mutations in the E protein gene, particularly at amino acid position 138,[17,18] which may be responsible for an attenuated phenotype in mice. The existence of molecular clones that can be modified at will should aid in sorting out attenuation, at least in experimental animals.

Yellow Fever Vaccines

Yellow fever virus (YFV), a mosquito-borne flavivirus, is the cause of a deadly hemorrhagic fever. The virus grows primarily in the liver, but CNS sequelae are common. Yellow Fever is endemic in tropical areas of Africa and South America. YFV being the first of the family to be cultivated, is the prototype of the flaviviruses. Much of the current work on YFV is done as modeling for other flaviviruses, such as hepatitis C virus, which cannot be cultured.

Two yellow fever vaccines were produced in the 1930s. One vaccine, developed in France, came to be known as the French Neurotropic Vaccine. The original wild-type isolate, Dakar, was passaged in mouse brain 128 times to yield the root virus. The resulting vaccine was used largely in French-speaking Africa and was quite successful when used in individuals 14 years of age or older. Use in children was less successful, with an excess of post-vaccination encephalitis. Use of this vaccine was discontinued in 1980. A detailed review of the French Neurotropic Vaccine was written by Barrett in the *Annals of the Pasteur Institute.*

The more widely used vaccines, based on the 17D strain, were derived from the Asibi isolate in an effort begun by Stokes in 1928 and continued through the 1930s by Theiler and Smith.[19] The YFV 17D lineage includes six passages in rhesus monkeys, one passage in mosquito cells and up to 286 passages in chick embryos.[20] Two separate vaccines have been made from the 17D line, the 17D-204 vaccine made in ten different countries, and the 17DD strain produced in Brazil. Both vaccines are made in chick embryos but differ in passage level; the 17D-204 group is produced at passage 233 to 237 while the 17DD vaccine is produced at passage 284 to 286.[21] Vaccines in use today contain 5000-10,000 pfu, but the WHO specification for potency has long been expressed in mouse LD_{50} units, with 1000 LD_{50} being the minimum human dose (Barrett, 1997).

These 17D-based vaccines have been quite safe and effective over a span of more than five decades. Early variability in safety and efficacy, linked to large differences in passage history and practice, led to what we now call the seed lot system of vaccine production which was adopted by the WHO in 1949. Yellow fever vaccines are used in limited areas of the world where the vector and the virus are endemic.

Mechanism of Attenuation

A substantial effort has been made to characterize the antigenic and genetic changes that led from the Asibi isolate to the 17D-204 and 17DD vaccines, beginning with oligonucleotide fingerprinting and monoclonal antibodies[22,23] and ending with genomic sequencing (reviewed by Barrett[24]). Sequences are now available for the original Asibi isolate, the 17D-204 vaccine, the 17DD vaccine and the FNV vaccine. For a genome of 10,882 nucleotides having undergone well over 200 passages in (mostly) chick embryos in different places under different conditions, there are remarkably few differences among these vaccines and their parent. Most changes held in common are clustered in the envelope protein gene and in the NS2a/b gene. As is often the case in molecular biology, having a lot of sequence information is only the first step in understanding a biological phenomenon. The use of molecular clones[25] site-directed mutagenesis and swapping of genes from one strain into another, ultimately, could yield information about attenuation of these viruses, at least in mice. Even with this capability, the story may not be a simple one since at least some vaccines appear to be non-clonal, with populations of viruses of different plaque size and corresponding virulence phenotypes in mice.[26] Furthermore, point mutations in either the NS5 or envelope proteins have been reported to alter the biological characteristics in mice.[27,28] It is likely that the true molecular basis of attenuation in man will be difficult to define.

Future Directions for Yellow Fever Vaccines

The available vaccines are almost invisible success stories. Through their use, yellow fever has been brought down to the level of a public-health problem instead of the nightmare it once was. Nevertheless, use of YFV vaccines is less than it should be in many areas. Cost is not a major issue, since these vaccines are relatively inexpensive. As global vaccination strategies are refined, consideration should be given to yellow fever.

The current vaccines are produced in embryonated chicken eggs. This method has the advantage of simplicity but has the drawback of all that goes with chicken husbandry. There is the potential for changing the management of the cell substrate, but for the production of a vaccine of this type, one can question the economics of this kind of investment of time and regulatory effort. Recent work, (reviewed by Monath[26]) has improved the thermal stability of the lyophilized vaccine so that distribution of an effective product may be better assured.

Poliovirus Vaccines

Poliovirus is a positive-strand RNA virus, a member of the enterovirus family. Infection with poliovirus is often asymptomatic, but in a subset of cases infection results in severe damage of motor neurons which in turn results in muscle wasting and paralysis. Poliovirus has been with man through recorded history, with images of atrophied limbs and assisted walking found in materials from ancient Egypt. Paralytic poliomyelitis experienced an upsurge at the end of the 19th Century and has been postulated to be linked to improvements in sanitation.

Poliomyelitis was first recognized as an infectious disease as a consequence of epidemiological studies in Sweden and Vermont in the 1890s, and its viral etiology was demonstrated in 1908 when Landsteiner and Popper passaged an agent from human neural tissue extracts in primates. An approach to vaccination was made possible in 1949 by Enders, Weller and Robbins[29] when they succeeded in culturing poliovirus in non-neuronal tissue.

The first vaccine against polio was a formalin-inactivated preparation introduced by Jonas Salk in 1955. This was an effective vaccine up to a point, but it soon became clear that a more potent vaccine would be required to achieve optimal control of poliomyelitis. There were several competing approaches to a live attenuated vaccine in the 1950s. One, described by Cox,[30] involved passage of poliovirus in suckling hamster brain and subsequent adaptation to the yolk sac of embryonated eggs.

The strategy that won out was the passage of virus in primate tissue culture and selection of virus variants with reduced neurotropic properties and was championed by Sabin, Koprowski and Melnick.[31] Much of the early data on vaccine development was presented in meetings and is not readily accessible in the traditional literature. The development and implementation of poliovirus vaccines, both live and inactivated, is one of the triumphs of medicine and public health.

Attenuation

As mentioned above, the strains used in the Sabin vaccine were selected on the basis of reduced growth in the neurons of monkeys and chimpanzees. In the last 20 years, molecular biological methods have permitted the unraveling (at least partially) of the basis of attenuation. From the perspective of late 1999, it looks like a pretty simple story. The key piece of information that forms the point of departure is the rare but consistent reversion of the vaccine strains and the appearance of vaccine-associated paralytic poliomyelitis (VAPP, see below). This has afforded the opportunity to determine what genetic composition correlates with safety and what changes correlate with neurovirulence.

The poliovirus genome is about 7.5 kilobases long and encodes 12 proteins. The genome is translated as one long polyprotein that is then cleaved by a virus-encoded site-specific protease. At the 5' end, or left end as the genome is generally drawn, there is a long untranslated sequence of about 750 nucleotides. This sequence forms substantial secondary structures, stems and loops, which interact with proteins required for translation and replication.[32-34] Comparing the nucleotide sequences of the vaccine strains, their wild-type progenitors, and the multiple isolates of virus from VAPP cases, one can line up the mutations which are common to viruses that have displayed virulence in man.[35-37] Distilling an enormous body of work into the space allowed here is impossible, but the simple view is that a single mutation in the 5'-untranslated sequence at positions 480 (Type 1 virus), 481 (Type 2 virus) and 472 (Type 3 virus) is necessary for reversion to neurovirulence. Type 3 vaccine strain reverts the most often of the three, and reversion is also typified by a change in the VP1 sequence and, perhaps, changes in the polymerase sequence as well.[38] Type 2 vaccine reverts less frequently, and a similar set of minimal mutations has been described.[39-42] Type 1 vaccine virus appears to be inherently more stable than types 2 and 3 since it reverts and causes VAPP the least often. When it does revert, a large number of mutations are found.[41,43-45] The genetic stability of the type 1 vaccine strain has led two groups to propose that, if one wanted to make new strains of type 2 and type 3 vaccines, one good way to do this would be to start with a molecular clone of type 1 and substitute the surface antigen genes from type 2 and type 3 at the gene level.[46,47]

One aspect of poliovirus vaccine production is the requirement for each bulk batch of each strain to undergo testing for neurovirulence in monkeys. In this test, the candidate vaccine bulk is tested by intracranial and intraspinal injection in a series of monkeys in parallel with a reference strain. The reference strain is chosen to represent the safety limit for vaccine use. This test is reviewed by Furesz.[48] In an effort to develop tests that do not rely upon primates, several alternatives have reached the stage of validation where they are now under consideration for implementation. One is a set of transgenic mouse models wherein the human gene for the poliovirus receptor has been substituted for the analogous gene in a mouse. The resulting mouse supports the growth of human poliovirus, and upon intraspinal injection of a neurovirulent strain it develops clinical symptoms of paralysis. One mouse, the TgPVR21 line, is appropriate for testing types 2 and 3; a second mouse, TgPVR1 may be used for testing of type 148.[49-51] A second set of tests based on molecular methods measures the presence and proportion of mutations at the critical point in the 5'-untranslated region. The most advanced test, known as the MAPREC test,[45,52,53] has a good correlation with the transgenic mouse test and also tracks well with the monkey neurovirulence test. Analogous tests based on chemiluminescence[54] or translation efficiency[55] have been developed as well.

Any discussion of attenuation, neurovirulence and reversion would be incomplete without touching upon cell substrates. The vast majority of live oral poliovirus vaccine (OPV) has been made in primary monkey kidney cells. Recently, Pasteur-Merieux-Connaught has licensed in Europe a Sabin vaccine produced on Vero cells.[56,57] No notable differences have been found in the clinic. Some years ago, early small clinical trials were conducted with a live Sabin-type vaccine grown on the human diploid fibroblast line MRC-5 with acceptable results,[58] but this vaccine has not made it to the marketplace.

Current Concerns and Future Developments

The immunological performance of the Sabin vaccines has been very good, some might say miraculous given the adverse conditions under which they have been used. Wild-type poliovirus has been eliminated from most areas of the world. The index for eradication is the disappearance of poliomyelitis due to wild-type virus. In the United States, the last case of naturally acquired poliomyelitis occurred in 1979, after which time all cases were either imported or due to vaccine reversion. Similar results have been obtained in the developed world and most of the developing world as well. As this is written, naturally acquired poliomyelitis continues to occur in parts of sub-Saharan Africa and in certain parts of south Asia. As a consequence, vaccine-associated paralytic poliomyelitis (VAPP) is the primary if not the sole source of poliomyelitis in much of the world. In light of this, vaccination recommendations have begun to shift. For example, in the United States, the Advisory Committee on Immunization Practices (ACIP) recommended in early 1997 that the first two doses of polio vaccine be given as inactivated polio vaccine (IPV), followed by two doses of OPV.[59] Straight 4-dose OPV or IPV regimens were deemed acceptable under certain conditions, but the stated intent was to phase out the use of OPV. An excellent discussion of informed physician and parental choice[60] covers this subject. In 1999 the ACIP revised the recommendations in favor of four doses of IPV.[61] Although IPV is more expensive than OPV, the trend towards IPV usage will be enhanced by the introduction of combined vaccine formulations containing IPV, DTaP, Hib and hepatitis B antigens.

As a given infectious disease is conquered through vaccination, attention turns from the success to the side-effects. The major concern about OPV is the above-mentioned rare reversion to neurovirulence and subsequent VAPP.[62,63] Reversion is most common with type 3 vaccine virus, less so with type 2 and even less with type 1. Mutations coinciding with reversion lie primarily in the 5' untranslated region of the genome as discussed above in the section on *Attenuation*. There is an extensive literature on the viruses isolated from VAPP cases, too much to cover in detail here. The most interesting findings cover potential recombination between vaccine strains[64,65] or perhaps between vaccine strains and wild-type virus,[66] as well as the variety of genetically different strains isolated from different sites in a given patient with VAPP.[67] It must be kept in mind that VAPP is a rather rare event, around 8-10 cases per year in the United States, or one per 2.5 million doses.[68,62] This rate will no doubt decline with the increased use of IPV.

One sidebar to the OPV-IPV debate is a theory regarding the role of infectious agents in the development of insulin-dependent diabetes mellitus (IDDM). Enteroviruses such as Coxsackie B virus have been proposed as etiological agents in IDDM. It has been suggested that Coxsackie B virus shares with poliovirus epitopes important for cellular immunity, and that vaccination with OPV provides better cross-protection against Coxsackie B than does IPV and thus may reduce the risk of IDDM.[69] This is an intriguing idea that merits further study.

The other major polio-related subject in the literature is the global effort to eradicate polio. In 1988 the WHO set a goal of eradicating polio from the globe by the year 2000. In late 1999, there are two areas where poliovirus circulates at a low level, parts of sub-Saharan Africa and south Asia. Reported cases are dropping to a few hundred per year, and regular updates are published in the CDC's *MMWR* and the WHO's *Weekly Epidemiological Record.* This effort has been a collaboration among the WHO, CDC, UNICEF, vaccine manufacturers and private organizations, perhaps most notably Rotary International. Although the stated goal of elimination by the year 2000 may be missed by a year or two, the end is in sight. The present debate is how to play out the end-game. There are two opposing points of view. On the one side, the WHO[70] contends that once the last case of poliomyelitis is recorded, vaccination with OPV should be stopped, period. Intense surveillance would identify any late-breaking cases, and these would be contained by intensive vaccination in the surrounding area. The other view is that OPV should be phased out in favor of IPV for several years.[71] The United States has embraced this latter approach in the latest ACIP recommendations. The concerns that support this argument are 1), the ability of poliovirus to survive in the environment, 2) disposition of laboratory and manufacturing stocks of wild-type poliovirus, and 3), circulation of reverted vaccine virus. As mentioned earlier, the advent of IPV-containing DT_aP vaccines will help this shift in vaccination practice. At issue here is the cost of IPV, which is greater than that of OPV, and the capacity to provide this vaccine for the developing world. In the end, we may have one policy for one part of the world, and another for the rest.

Measles Vaccines

Measles virus, member of the paramyxovirus family, is the most contagious viral pathogen known to man. Before the advent of effective vaccines, essentially the entire birth cohort was infected by a very early age. Although measles is normally a self-limiting disease in healthy subjects, the rash, cough and high fever make it an extremely unpleasant experience. The author remembers vividly having measles at the age of five and being quite miserable. Measles is a particularly nasty disease for two reasons. First, measles virus is immunosuppressive, and this results in a variety of opportunistic infections. Natural measles infection also has a group of relatively rare but serious long-term sequelae including measles inclusion body encephalitis and subacute sclerosing panencephalitis. The pathogenesis of measles infection and its complications is the subject of a recent review by Schneider-Schaulies and ter Meulen.[72]

Early vaccine development began in the 1950s with an inactivated virus strategy which was abandoned due to short-lived immunogenicity and the appearance of atypical measles disease in vaccinees exposed to wild-type virus.[73-76] This phenomenon is attributed to inappropriate priming of CD4 T-cell responses typical of an Arthus reaction.[77]

Attenuation of Measles Viruses

Current live attenuated measles vaccines can be traced to the isolation of measles virus by Enders and Peebles in 1954.[78] This virus was attenuated by Enders and co-workers,[79] by passage in human kidney and human amnion cells to yield the Edmonston-Enders strain (Edmonston was the surname of the child from whom the virus was isolated). This strain went on to become the parent of several live attenuated measles vaccines developed in the United States, Europe, Russia, Czechoslovakia and Japan. The family tree is shown in Figure 5.1.[80] A recent analysis of the nucleic acid sequences of the major viral antigen genes shows remarkably small differences between strains despite the varied methods of virus passage used to generate these strains.[81] Four other vaccines developed independently in Japan (CAM-70), China (Changchun-47, Shanghai-191) and Russia (Leningrad-16), show greater differences as one might expect given the geographic diversity of the original isolates. It is not possible to attribute any of the known

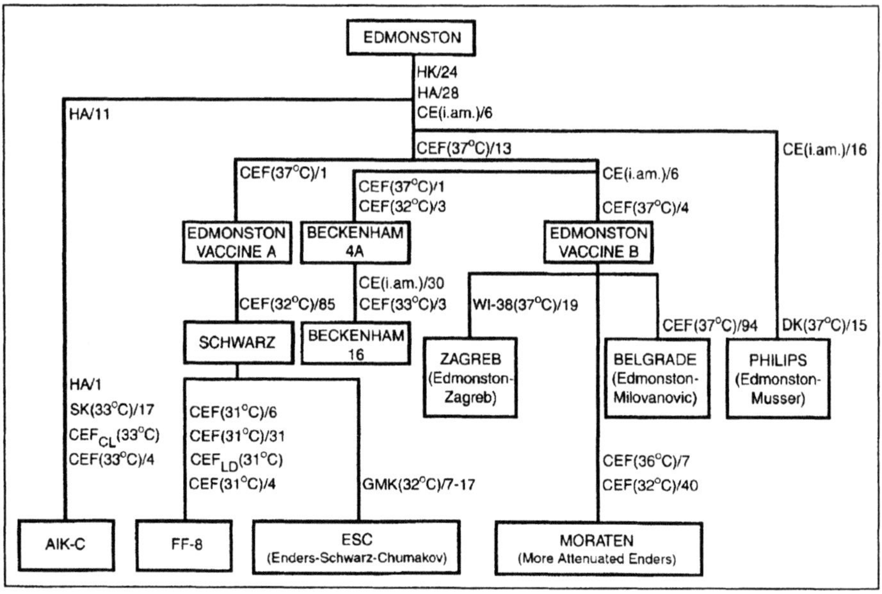

Fig. 5.1. The derivation of strains of live attenuated measles vaccine.

or perceived differences in clinical performance to any of the known mutations in these antigens. Studies of non-transcribed regions of the genome have contributed to phylogenetic knowledge.[82]

As noted elsewhere in this review, attenuation of viral vaccine strains is a highly empirical process. An excellent example of this is the history of the Merck "more attenuated Edmonston-Enders" vaccine. The first live measles vaccine licensed by Merck, Sharp & Dohme in 1963 was based on the Edmonston B strain.[83,84] This strain was reactogenic in children to the extent that it was administered concomitantly with a dose of standardized anti-measles immune globulin at a separate site. Obviously, this was only a short-term solution due to complexity and inconvenience. The second-generation solution was a further attenuated virus derived from the Edmonston B strain, grown for another eight passages in chick fibroblasts at 36°C and 40 passages at 32°C. This vaccine had appropriate immunogenicity and tolerability without added immune globulin. The new strain was also propagated in cells derived from chicken flocks free of infectious avian leukosis virus, a novel move at that time. The new strain, called "Moraten", was marketed as Attenuvax® from 1968 onwards. As an aside, it is worth noting here that although the term "Moraten" was originally applied to this particular derivative of the Edmonston B strain, the same word is now used as a trademark and applied by Swiss Serum and Vaccine Institute to their measles vaccine, the Edmonston-Zagreb strain. This has led to some confusion in the field. The Schwarz vaccine strain,[85] widely used throughout the world, was attenuated by a similar strategy but without the detour through an intermediate vaccine requiring immune globulin.

Current Concerns and Future Directions

The current live attenuated measles vaccines are clearly effective, especially when used in a diligently applied program including one dose around one year of age followed by a second dose several years later. This is the norm in most parts of the developed world, Japan being one exception (see below). In the developing world, things are a bit different. The Expanded Program for Immunization includes a dose of measles vaccine early in life. In areas of the world where measles remains endemic there is a desire to protect infants by vaccinating earlier than the commonly recommended age of 12-18 months. The problem encountered is maternal antibody that persists in the infant for several months, the duration being largely a function of the mother's antibody level. This reduces the "take" of the live vaccine and results in waning immunity. One common strategy is to vaccinate around 8-9 months of age with a follow-up dose around 18 months. Obviously, this entails the cost of two visits, two doses and it carries the risk of missing that second dose. This scheme is effective when it can be implemented. As vaccinated girls become mothers, this problem appears to go away in that vaccine-induced maternal antibody is of a lower titer.[86] In a general sense, however, coverage for measles vaccination remains a public health issue in some parts of the developing world where measles is still the cause of around one million deaths per year.

Another approach to this problem was to raise the potency of the first dose of measles vaccine. This was attempted in several settings including the Gambia, Mexico, Peru, Guinea Bissau, Senegal and Haiti. Vaccines with a titer of $10^{4.7}$-$10^{5.0}$ infectious units of virus were administered at the age of 6-9 months in these studies, about ten times the average dose of fresh vaccine at release and two logs greater than the recommended minimum dose.[87] In settings of lower socio-economic status an unexplained excess mortality, primarily in girls, was noted two to three years after vaccination.[88,87] There was no association with vaccine strain or vaccine manufacturer, as both Schwarz and Edmonston-Zagreb vaccines were supplied from two different sources. As a result, the high-dose strategy was abandoned in favor of the two-dose scheme.

The other major current concern, which also impacts on vaccine coverage, is a series of articles by A. Wakefield and coworkers in the medical literature which purport to link the onset of chronic inflammatory bowel disease, Crohn's Disease and autism to measles (or measles/mumps/rubella) vaccination.[89,90] This allegation, which appears to lack any truly supporting evidence, is discussed below under "MMR" vaccines.

Indigenous measles transmission has been interrupted in the United States, in much of the Americas and in parts of Europe. Since measles virus has no known animal reservoir and since measles vaccines are widely available at affordable prices, a discussion has begun within public health circles of the possibility of eradicating measles.[91] This is an attractive idea, but there are several obstacles that must, and can, be overcome. In parts of the developed world, measles is not a public health priority and outbreaks do occur. Thus, a renewed effort must be made to increase vaccine coverage and to encourage the two-dose policies that have been shown to be successful. There is also the perception that funding for such an effort is an insurmountable problem. While it is true that funding for vaccine programs is a perennial issue, the current management of the WHO appears to be taking a keen interest in vaccination along with the World Bank and private donors such as the Gates Foundation. Given the progress made against smallpox and polio, why not measles next?

Rubella Vaccines

Rubella virus is the single member of the rubivirus genus within the togaviridae family. Rubella virus has a small icosahedral capsid enveloped in a lipid-protein membrane. Rubella, also known as German measles, is normally a benign self-limiting disease of infants and children characterized by fever, rash and arthralgia or arthritis of a transient nature. Rubella infection of pregnant women is a different matter. The virus is readily transmitted to the fetus and can cause fetal wasting and severe birth defects. This congenital rubella syndrome is the primary pubic health driver for development and application of rubella vaccines. An excellent review of rubella disease[92] has been published by the late A. Galazka of the WHO.

Rubella virus was isolated by two groups in 1962 in the United States, and attenuation of the virus in African Green Monkey (AGMK) (strain HPV-77) was reported in 1966.[93,94]

In contrast to most other viral pathogens discussed in this section, a variety of independent isolates of rubella virus have been attenuated and developed as vaccine strains. In the United States and Europe, three strains of rubella vaccines have been marketed; the HPV-77 strain grown in duck embryos by Merck Sharp & Dohme and in dog kidney cells by Parke-Davis, the Cendehill strain grown in rabbit kidney cells by the Research Institute for Technology (RIT, now SB Biologicals), and the RA27/3 strain developed by Stanley Plotkin of the Wistar Institute and propagated in WI-38 human diploid fibroblasts. In addition, the Benoit strain grown in duck embryos was developed by Merck, but it was never marketed.[10] Five other strains have been developed and used in Japan; To-336, TCRB19, Matsuura, Matsuba and Takahashi. In Japan, local isolates were developed in favor of American or European strains due to a perception that rubella viruses circulating in Japan were less teratogenic. The origins of these vaccines are reviewed in more detail by Shishido and Ohtawara.[95]

The HPV-77 strain was the first to be licensed for use in the United States. Attenuation of the virus was achieved by 77 passages of the M33 strain in AGMK cells; the vaccine was produced at the fifth subsequent passage in duck embryos to yield the HPV-77 DE5 product sold by Merck Sharp & Dohme as Meruvax® from 1968 to 1978. HPV-77 DK vaccine was produced by passage of the same seed in dog kidney cells by Parke-Davis. This product had a limited lifespan as it was judged to be excessively reactogenic. The Cendehill strain developed in Belgium, grown in rabbit kidney cells and marketed as Cendevax®, was widely used in Europe through the 1970s.

A third vaccine strain, RA27/3, developed at the Wistar Institute in Philadelphia, was grown in WI-38 human diploid fibroblasts and was originally marketed in Europe by Wellcome as Almevax®.[96,97] Multiple comparative clinical studies showed that overall RA27/3 was superior to both the Cendehill and HPV-77 DE5 vaccines in terms of reactogenicity (arthritis/arthralgia) and immunogenicity at low doses.[98-106] As a consequence, RA27/3 is used almost exclusively in most of the world with the notable exception of Japan where the Japanese strains are used. RA27/3 also has an advantage over the HPV-77 DE5 vaccine in ease of manufacture. As mentioned earlier, RA27/3 is produced in WI-38 cells and was one of the first, if not the first, attenuated human vaccine to be produced in human cells. This was a rather radical idea at the time (1964) due to concerns expressed by Sabin at one point during the previous decade about the potential transmission of leukemia virus by such cells. A debate which appears in Laurence[107] and which must have taken place multiple times concludes that concerns about such transmission were without foundation, and decades of experience have borne this out. Cells such as the WI-38 line can be, and have been, extensively characterized, and they can be conveniently frozen away for future use. On the other hand, one can only imagine the challenge of producing embryonated waterfowl eggs on demand at commercial scale under GMP conditions!

Attenuation

As with most other attenuated vaccines, little is understood about the molecular mechanism of attenuation of rubella viruses. A substantial body of sequence information is available including the entire sequence of the RA27/3 strain[108] and comparative sequences of the E1 genes of Japanese and western vaccines and wild-type isolates.[109,110] Rubella is relatively homogeneous at the nucleotide level with strain-to-strain variation about 2%. It would appear that many different roads lead from wild-type to the attenuated phenotype. Attenuation has been achieved by passage in human cells (RA27/3), rabbit kidney cells (Cendehill), AGMK plus duck embryo cells (HPV-77 DE5) and different combinations of AGMK, bovine kidney, swine kidney, chick embryo and rabbit kidney cells (Japanese vaccines, reviewed by Shishido and Ohtawara).[95] Current sequencing technology, infectious clone schemes and site-directed mutagenesis could, in principle, be used to generate vaccine strains that logically should embody the minimal genotype required for attenuation starting from a wild-type isolate. However, the ultimate test in humans would be difficult to justify, so we must be satisfied with what we have.

Current Concerns and Future Directions

Rubella vaccines as mature products have relatively well-known safety and efficacy profiles. Two topics constitute the bulk of clinical discussion in the literature in the 1990s. First is the question of arthritis and arthralgia related to these vaccines. As mentioned earlier, natural rubella disease is frequently accompanied by acute joint symptoms that generally resolve, especially in younger individuals. The frequency and severity of joint symptoms is greater in adults and especially in women.[111] The HPV-77 DK12 vaccine was withdrawn from use due to an unacceptable frequency of arthritis and arthralgia in children.[112] There has been some debate about the association of acute or chronic arthralgia or arthritis in women. The Institute of Medicine conducted and extensive study of the available information[113,114] and concluded that there is an association between rubella vaccination and acute joint complaints in women. The Institute also stated that there may be an association with chronic or recurrent arthritis or arthralgia, but that a better study would be required to answer this question. Such a study was carried out at Kaiser Permanente, a large health maintenance organization in California, and this group determined that there was no association.[115]

The second concern revolves around vaccination strategy for areas just now implementing rubella vaccine; who do you vaccinate and when so as to have the greatest impact on congenital rubella syndrome (CRS), and how do you vaccinate safely women of childbearing age? In an ideal world, one might vaccinate all girls at puberty. This should render them immune during their childbearing years and should eliminate CRS. Such a strategy was used in the United Kingdom with mixed results,[116] and rubella vaccine is now applied along with measles and mumps vaccines during the second year of life (with a booster in gynecological settings where serologically indicated). Another strategy, tested in Brazil with good results, is a "pulse vaccination" of all individuals between 1 and 10 years of age followed by universal vaccination of the birth cohort at 1 year of age.[117] In the United States, rubella vaccination is part of the standard pediatric regimen delivered along with measles and mumps vaccines at 12-18 months and again before school entry. Experience has shown that herd immunity is the most important metric for success, as in any vaccination program against an endemic virus.

The American College of Obstetrics and Gynecology has a detailed algorithm for deciding who should be vaccinated later in life. Women not clearly seropositive (and not pregnant) should be vaccinated either before attempting to conceive or post partum.[118,119] This raises the question of what to do if a pregnant woman is inadvertently vaccinated against rubella. Although this is clearly contraindicated, it does happen in the real world. One strategy may be to test for anti-rubella IgM as an indicator of a primary response in a seronegative individual.[120] A number of studies of fetal tissues recovered following termination of pregnancies post vaccination have shown that rubella virus can be cultured and thus show that the vaccine strain RA27/3 can pass from the mother to the fetus.[121-123] However in the vast majority of these cases there was no evidence of fetal abnormality. Therefore, termination of pregnancy following inadvertent vaccination is not automatically indicated.[124]

Rubella vaccination has achieved the goal of reducing the number of cases of CRS in areas where vaccine coverage is high. When coverage falls, as it did during the late 1980s in the United States, CRS cases can increase, only to decrease upon renewed vigor of vaccination efforts. Future work may be limited to attempts to define in molecular terms the mechanism of rubella-related arthritis.

Mumps Vaccines

Mumps virus is a member of the Paramyxovirus family, along with measles, respiratory syncytial virus and parainfluenza virus. This group accounts for a substantial fraction of pediatric disease and has been the target of vaccine development efforts for decades. The paramyxoviruses are negative-strand RNA viruses encoding 6 or 7 genes that are transcribed separately. Mumps virus causes parotitis, which in some cases can be visually dramatic. Other frequent consequences are mild meningitis and orchitis. Severe meningitis and deafness are less common occurrences, but before vaccines were available, mumps was the leading cause of deafness. As with many "pediatric" viral illnesses, the same infection in naive adults is more serious and can result in sterility in males or mastitis in females. Mumps was also a significant cause of myocarditis in children.

Mumps Vaccine Development

Several independent vaccine development programs have yielded licensed products.

Jeryl Lynn Strain[125,126]

Mumps virus was originally isolated, cultured, and shown to be attenuated in monkeys through passage in embryonated eggs by Enders and coworkers in 1946. Early attempts to create a vaccine at Merck in the late 1950s based on the Enders strain failed to yield a satisfactory virus. In 1963, a culture taken from the throat of a 5-year-old girl was cultured in embryonated hen's eggs, and this became the Jeryl Lynn strain of mumps. The virus was passaged 12,17 and 27 times in embryonated eggs and was tested in volunteers in the Philadelphia area. The 17 times passaged virus showed the appropriate combination of immunogenicity and lack of pathogenicity, while the 12 and 27 passage vaccines were either reactogenic, causing parotitis, or poorly immunogenic.[127] Subsequent trials demonstrated the safety and efficacy of this vaccine.[128] The Jeryl Lynn strain was licensed as Mumpsvax® in the United States in 1968 and was included as the mumps component of M-M-R®, the first trivalent measles, mumps and rubella vaccine in 1971.

Urabe AM9 Strain

The Urabe strain began as an isolate taken from a sick child in Japan in 1967. The virus was passaged in embryonated eggs and, as in the case of the Jeryl Lynn strain, it was tested at several successive passage levels to find the best compromise of safety and immunogenicity. The Urabe AM9 strain was licensed in Japan in 1980 and subsequently in Europe and elsewhere around the world. The majority of the Urabe AM9 virus distributed was sold as a component of SB's Pluserix and Pasteur Merieux's Trimovax. During the late 1980s and early 1990s reports from Canada and the United Kingdom raised concerns about vaccine-related aseptic meningitis (reviewed by Brown and Wright).[129] This ultimately led to the withdrawal of Urabe-containing vaccines from Europe and Canada, although these vaccines remain in distribution in other parts of the world.

Rubini Strain

The Rubini strain of mumps was isolated from the urine of a boy with rubella and propagated in human diploid cell, passaged in embryonated eggs and then adapted to MRC-5 human diploid fibroblasts.[130,131] This vaccine was licensed in Switzerland in 1985 and has been most widely used as a component of the Swiss Serum and Vaccine Institute's Triviraten measles, mumps and rubella vaccine. Epidemiological studies in Switzerland called into question the efficacy of the Rubini strain[132-135] with calculated efficacy as low as 6%. Given the lengthy passage history of the Rubini strain and the steep gradient between reactogenicity and overattenuation demonstrated during the development of the Jeryl Lynn and Urabe AM9 strains, it is reasonable to conclude that the Rubini strain may be overattenuated.

Strains Used in Japan

As with the rubella vaccines, a group of 5 independent strains have been used in Japan. First is the Urabe AM9 strain described above. Others include the Hoshino L32 strain propagated in embryonated eggs.

Leningrad-3 Strain

The Leningrad 3 strain was developed from a mixture of five isolates of wild-type virus propagated in guinea pig embryo cells and Japanese Quail embryos.[136] Vaccines containing the Leningrad-3 strain of mumps have been used in the former Soviet Union and eastern Europe since 1974. This strain was shown to be highly immunogenic and efficacious,[137] but it was also shown to be associated with aseptic meningitis at a rate of one case per thousand vaccinees, comparable to the higher estimates for the Urabe AM9 strain.

Leningrad-Zagreb Strain

The L-Zagreb strain of mumps virus was developed by further attenuation of the Leningrad-3 strain in chick embryo fibroblast culture.[138] This vaccine has been used in Croatia and Slovenia. The L-Zagreb strain is also produced locally in India.

Sofia Strain

The Sofia strain was developed in Bulgaria from a local isolate propagated and attenuated in primary guinea pig kidney cell culture. This vaccine was used in Bulgaria from 1980 to 1982. At this point, concerns about neurovirulence caused it to be withdrawn and replaced with the Leningrad-3 strain imported from Russia.[139]

RIT 4385

The RIT 4385 strain was isolated by SB from a commercial vial of Mumpsvax® (Merck, Sharp & Dohme). In an effort to develop a unique strain, several single plaques were picked and the SH gene (a particularly variable region of the genome) was sequenced. A plaque with a sequence unique compared to the available information[140] was selected for further propagation in chick embryo cells and was called JL-1 (SB patent). This strain is marketed by SB as a component of a trivalent vaccine, Priorix®.

Current Concerns

The principal concern in the 1990s has been the association of certain mumps vaccine strains with a low, but measurable, incidence of aseptic meningitis[131] and parotitis.[141] Parotitis has been reported to be primarily related to the Hoshino strain used in Japan. Aseptic meningitis has been shown to be associated with the Urabe AM9 strain, as mentioned earlier, as well as with the Sofia and Leningrad-3 strains. Conflicting data are reported for the L-Zagreb strain.[142,143]

Although the problem has been side-stepped in Europe and Canada by the withdrawal of vaccines containing the Urabe AM9 strain, a substantial effort has been invested in sorting out the molecular basis of meningitis. Much attention has been directed to the fact that the Urabe AM9 strain is a mixture of at least two viruses (as are the Jeryl Lynn and Leningrad-3 strains). One species has an A residue at nucleotide 1081, the first position in codon 335 of the hemagglutinin-neuraminidase (HN) gene while the other species has a G. These RNA sequence differences lead to amino acid differences; the A-1081 virus has a lysine while the G-1081 has a glutamic acid residue at position 335. The A-1081 virus is routinely grown from cerebrospinal fluid samples from cases of aseptic meningitis post vaccination. How the two species arose from a stock that was twice plaque-purified is unclear, and the fact that growth of the A-1081 virus is favored in Vero cells, the cells routinely used for laboratory culture of mumps virus, further clouds the issue. The change from a glutamic acid to a lysine at position 335 not only changes the net charge of the HN protein by two but it also affects an important epitope on the surface of the virus. For a detailed discussion of this subject, see Brown and Wright[144] and Afzal, Yates and Minor.[145]

Summary

Mumps virus seems to have a uniquely steep gradient between pathogenicity and overattenuation. Examples of both have been discussed above. It is worth noting that all of these vaccine strains have excellent performance at the level of seroconversion. They all had safety profiles that were deemed acceptable by their respective developers in their respective contexts. Concerns were raised only upon gathering significant bodies of clinical data and field experience, combined with the opportunity to compare efficacy and adverse events among vaccine strains.

Trivalent Measles-Mumps-Rubella Vaccines

As the live attenuated measles, mumps and rubella vaccines were in development during the 1960s, the idea of ultimately combining these products into a single package evolved in parallel.[146] Although control of measles was the initial goal due to measles' high level of morbidity and mortality, mumps and rubella were important public health targets as well. If one could administer all three viruses at the same time, achieve the same level of protective immunity, and maintain the good tolerability profiles of the single components, one would have a truly convenient and effective public health tool. That challenge has been met, and the trivalent "MMR" vaccines have played a critical role in the control of these three diseases in many parts of the world.

Vaccine Development

The first trivalent vaccine, M-M-R®, was developed in the United States by Merck Sharp and Dohme[146-148] and was licensed for use in 1971. Parallel efforts in Japan,[147] the Soviet Union,[137,149-153] Europe and the United States[154-156] yielded comparable clinical results.[157,158] The biggest barrier to success in this effort was the potential for immunological interference between the three viruses. Since the minimum dose required for seroconversion had been shown to vary over two orders of magnitude for the three viruses, new clinical trials were required to establish the appropriate relationship between the potencies of the three components in this new context.[159,160,148] As newer trivalent vaccines are developed, this issue must be readdressed.[161] In the end, a general set of lower limits of potency have been established; 1000 $TCID_{50}$ for measles, 5000 $TCID_{50}$ for mumps, and 1000 $TCID_{50}$ for rubella.

There are currently five major trivalent vaccines on the market:

M-M-R®II, (Merck) containing the more attenuated Enders Edmonston measles strain, the Jeryl Lynn mumps strain, and the RA27/3 rubella strain. This vaccine is licensed in North America, Europe, and many other places worldwide. M-M-R®II was licensed in 1978. Its predecessor, M-M-R®, was licensed in 1971. M-M-R® contained the HPV-77 duck embryo passaged rubella strain and had a different stabilizer formulation.[162] M-M-R®, being the first vaccine of its type, had a name that was easily converted into a generic, "MMR", so that there can be considerable confusion in the field. This has led to a tendency in promotion to link the data from one vaccine producer to the product of another, perhaps without justification since these vaccines are not generic products of identical composition.

Pluserix®, (SmithKline Beecham Biologicals) containing the Schwarz measles strain, the Urabe AM-9 mumps strain, and the RA27/3 rubella strain. This vaccine is marketed in some areas outside Europe, the United States and Canada.

Trimovax®, (Pasteur Merieux Connaught) containing the Schwarz measles strain, the Urabe AM-9 mumps strain and the RA27/3 rubella strain. This vaccine is also marketed in some areas outside Europe and the United States.

Triviraten®, (Swiss Serum and Vaccine Institute) containing the Edmonston-Zagreb measles strain, the Rubini mumps strain, and the RA27/3 rubella strain. This vaccine is marketed in some European countries and in various countries in the Far East and the Pacific.

Priorix®, (SmithKline Beecham Biologicals) containing the Schwarz measles strain, the RIT 4385 mumps strain, and the RA27/3 rubella strain. This is a relatively new vaccine, a replacement for Pluserix® which was withdrawn from many markets due to the frequency of aseptic meningitis related to the Urabe mumps component (see earlier mumps section for more details). Priorix® is marketed in Europe and other selected areas of the world.

In the late 1960s and through the 1970s, live measles, mumps and rubella vaccines were considered to be single-dose immunogens capable of inducing life-long immunity. Serological studies showed that antibody persisted for prolonged periods.[128,163-165] In the absence of natural boosting, antibody levels do tend to decline over time. In addition, there is a small minority of subjects who, for some reason, fail to seroconvert after one dose of vaccine. Furthermore, not all children who should be vaccinated during the second year of life are vaccinated. This led to the recommendation of a second dose of MMR vaccine in many parts of the world during the 1980s,[164,166,167] and inspection of CDC (*MMWR*) data on the incidence of all three diseases shows the positive impact of this decision.

Although administering the second dose of vaccine is clearly policy, there remains the question of "when is this second dose best administered?" In the United States, the public (ACIP) and private (AAP) recommendations for timing of the second dose were at odds,[168] with the ACIP recommending the second dose upon school entry (4-5 years of age) and the AAP recommending administration of the second dose at 10-12 years of age. There are good and valid arguments to be made for both positions. Earlier administration is favored by convenience (it is more difficult to get adolescents to the clinic) and lower reaction rates. There is an approximately 50% increase in post-vaccination complaints, notably joint pain which is most likely linked with the rubella component in older subjects.[168,169] This is fully consistent with the known reactogenicity profile of rubella vaccine and the much more severe consequences of rubella disease beyond early childhood discussed earlier. Second-dose administration on school entry also reduces the build-up of susceptibles, and this rationale has led Australia to shift from a late revaccination policy to a school-entry schedule.[170] Later administration is favored by the notion of boosting or catching the unvaccinated with these immunogens, rubella in particular, as individuals approach their reproductive years[169] and to reduce the likelihood of congenital rubella syndrome. This has been shown to work well from a serological perspective.[166] A third point of view discussed in the United Kingdom[171] is that a second dose should not be given automatically to all children but rather those who are identifiably seronegative or not previously vaccinated. This is a reasonable notion in an ideal world, but the logistics of serological analysis and record-keeping make it impractical. Overall, the weight of opinion seems to favor earlier rather than later administration of the second dose, due to convenience and the growing realization that declining immunity is less of a problem than rare primary vaccine failure and more common failure to vaccinate.[168] In the United States, the ACIP has recommended lowering the age for primary immunization from 18 months to 12-15 months so as to accommodate the rest of the vaccination schedule, and they also continue to stress the importance of a second dose plus an adolescent catch-up program for those who were missed earlier.[172] Setting aside the issue of timing, the most important parameter, as with any vaccine, is coverage of the population. Serological studies[173,164] and the outbreaks of measles in the early 90s show that vigorous implementation of vaccination policies to ensure coverage and herd immunity is the key to control of disease.[174]

The MMR vaccines share at least two properties that impact the precautions and contraindications in the label. First, the measles and mumps components are produced in some sort of embryonated egg system (except for Triviraten® where these viruses are grown in MRC-5 human diploid fibroblasts). For example, the measles and mumps viruses in M-M-R®II are grown in cultured cells taken from embryonated eggs,[175] while others may be grown directly in the eggs. Regardless, the contraindications in the labels included vaccination of persons known to be allergic to eggs. Numerous studies have shown that vaccination of egg-allergic individuals can be performed safely[176-180] although rare reactions may occur.[181] With this data, various policy-making groups have recommended that egg-allergy no longer be considered a contraindication for vaccination with MMR, e.g., although as with any vaccine, the ability to treat any anaphylactic reaction should be at hand. The second contraindication common to live virus vaccines is vaccination of immunocompromised individuals. HIV-positive subjects had been a particular concern. After reviewing the available data and assessing the risk-benefit parameters of severe measles in HIV-positive persons, the ACIP now recommends that vaccination with MMR should be carried out routinely in asymptomatic HIV-positives and should be considered for symptomatic subjects. A third consideration was vaccination of children with mild illnesses (colds, coughs, etc.). Originally, it was recommended that vaccination should be deferred until the child had recovered. Recently, based on accumulated data, the ACIP changed this recommendation[172] to allow vaccination of children with mild illnesses.

Current Concerns

The public health track record of M-M-R® and M-M-R®II has been described repeatedly in the CDC *Weekly Morbidity and Mortality Report.* The data described therein are derived exclusively from the use of M-M-R® and M-M-R®II, and their monovalent components in the earlier years, since these have been to date the only vaccines used in the United States.[146] Similarly, Finland has used M-M-R®II exclusively in a national public health campaign since 1982, with the result of elimination of all three diseases.[163] In other settings, these vaccines may perform better than much of the epidemiological data on breakthrough disease would suggest[182] since about half of the measles-mumps-rubella-like illnesses in vaccinees can be attributed to other pathogens when appropriate diagnostic procedures are applied.

As mentioned at the outset, when a vaccine succeeds in reducing the incidence of a disease to near invisibility, attention turns to potential side-effects of the vaccine. Since a successful vaccine will have been given to essentially all individuals within a population, it is nearly impossible to assess the cause-effect relationship between vaccination and a rare event that occurs some time afterwards. Some information can be derived from carefully conducted case-controlled epidemiological studies, and one can attempt to answer the question of biological plausibility. However, when the public press weighs in, and since bad news sells more newspapers and airtime than does good news, the public health may suffer.

A case in point is the allegation made by a group of researchers in the United Kingdom that Crohn's disease, other inflammatory bowel disorders, and autism are triggered by MMR vaccines.[183] In this article the authors describe a syndrome of inflammatory bowel disease linked with a pervasive developmental disorder in a group of 12 children. This would appear to be a new and interesting relationship in and of itself, but in this same article the authors suggest that vaccination with MMR might be the triggering event. The same group had published earlier findings of persistent measles virus in intestinal samples from patients with Crohn's disease,[184-187] although numerous other laboratories have attempted to reproduce these results without success.[188,189] An excellent review of the evolution of this argument was written by

Akobeng and Thomas.[190] A series of independent follow-up studies have shown no association between vaccination and either autism or inflammatory bowel disease,[191-193] and the Chief Medical Officer of the United Kingdom has issued an official press release summing up the deliberations of over 30 experts on this question and concluding that there is no link between vaccination and either of these conditions. In parallel, there has been a running commentary in the "Letters" section of *The Lancet* on the merits of the argument and on the larger question of publication of such work. This story has also played largely in the British press with the result of falling confidence in vaccination and falling vaccine coverage. One article mentioned the resurgence of "measles parties" in the United Kingdom. In conclusion, the weight of the data so far are not in favor of an association between autism or inflammatory bowel disease and vaccination with MMR, but the consequences of even the suggestion of such an association can have an unfortunate impact on public health.

One other concern, which pales in significance compared to the situation above, is allergy to gelatin, largely confined to Japan. Many pediatric vaccines, particularly DTaP produced in Japan, contain some level of gelatin as a stabilizer,[194] and these are administered in a three-or-four-dose regimen prior to vaccination with measles, mumps and rubella vaccines. This may induce some level of anti-gelatin immunity in a very small minority of subjects, which results in an immediate-type hypersensitivity reaction upon subsequent administration of measles or mumps vaccine that in Japan contained high-molecular-weight gelatin.[194] In Japan there is now a move away from high-molecular-weight gelatin in vaccines. M-M-R®II contains highly hydrolyzed gelatin as a stabilizer and is less likely to raise antibody than is high molecular weight gelatin. Reactions of this type to gelatin are relatively rare in the United States with one well-documented case reported by Kelso.[195] The labels for M-M-R®II as well as for other trivalent vaccines states that persons with allergies to any of the vaccine components should be vaccinated with caution.

Quadrivalent Measles-Mumps-Rubella-Varicella Vaccines

The one obvious enhancement to MMR vaccines would be the addition of varicella vaccine to the cocktail. This has been an elusive goal for over ten years. Early studies with M-M-R®II and VARIVAX® at the standard doses gave excellent seroconversion rates for all four components, but the geometric mean titer (GMT) of antibody against varicella was only half of what is obtained with monovalent VARIVAX®, and breakthrough varicella disease is more common in vaccinees with low anti-varicella antibody titers.[196-198] Similar results have been reported[199-202] in studies using a quadrivalent vaccine candidate prepared by SB. The unstated assumption in these studies is that measles vaccine virus interferes with the varicella response by a mechanism specific to local immune response, since co-administration of M-M-R®II and VARIVAX® at standard doses gives the expected GMT of antibody for all four components.[203] One approach to the interference problem, taken by Berger and Just, showed that by making appropriate co-dilutions of SB's Varilrix® and Pluserix®,[204] one could obtain good seroconversion and anti-varicella titers. For some reason this idea has not been carried forward. Recent studies by Kuter at Merck (unpublished) have shown that by raising the varicella titer above a certain threshold good overall performance of all four components may be obtained. Thus, after over a decade and a half of "cut-and-try" clinical studies, the goal is in sight.

Varicella Vaccines

Chickenpox Disease

Since the varicella vaccine is a recent development, most of us are personally familiar with this disease (the author had a severe case in the fifth grade). Varicella zoster virus (VZV) is a highly contagious agent spread primarily by the respiratory route. The incubation time is between 7-21 days and is followed by an extremely itchy maculo-papular rash that lasts about 7 days. Most cases of chickenpox in normal healthy individuals follow this unpleasant but relatively benign course. The most common medically important sequelae of VZV infection are ataxia, pneumonia, encephalitis, and invasive bacterial infections of the skin due to scratching of the lesions. In immunocompromised individuals, however, chickenpox can be a devastating disease with a high case-fatality rate. Particularly at risk are leukemic patients undergoing chemotherapy and bone-marrow transplant recipients. Patients receiving steroids or other immunosuppressing agents are also subject to greater risk of complications and a higher case fatality rate. Finally, age is a risk factor. The disease is more severe, and the case-fatality rate in adults is reported to be several times the rate in children.

Chickenpox Vaccine Development

The current crop of live attenuated chickenpox vaccines is derived from the Oka attenuated strain developed by Takahashi, Asano and coworkers at the University of Osaka. An excellent review by M. Takahashi covers the history of this effort in detail.[205] The primary motivation for developing this vaccine in Japan was the rapid spread of chickenpox through pediatric wards in hospitals. The first clinical trials were, in fact, post-exposure prophylaxis studies around index cases in pediatric wards. Success in this challenging environment led to true prophylaxis trials and ultimately to licensure of the Oka vaccine in Japan in 1987.

The development of the Oka strain is a classic story of attenuation by passage in an alternate host cell. VZV is a difficult virus to grow in culture since infected cells do not release virus into the culture medium but rather spread from cell to cell by cell contact. It also has a limited host range in vitro. One cell type that could support VZV growth turned out to be the guinea pig embryo cell.[205] Takahashi isolated his primary virus from a three-year-old boy named Oka and propagated this virus through human embryonic lung cells for 11 passages. This virus pool was then applied to guinea pig embryo cells for 12 more passages. The resulting virus pool was plaque-purified once and then grown in WI-38 human diploid lung cells for early vaccine production.

The attenuated phenotype of the Oka strain is attributed to the passage of the virus in guinea pig cells. Several independent Japanese isolates were subjected to similar guinea pig cell passages, and these strains exhibited similar in vitro properties compared to the Oka strain.[205] Indirect confirmation of the role of guinea pig passage in attenuation comes from earlier experience at Merck. A local isolate, the KMcC strain, was passaged up to 60 times in human diploid fibroblasts, and clinical trials were carried out at various intermediate stages.[206] A satisfactory balance of tolerability and immunogenicity was not obtained.

Early Oka-based vaccines were made in WI-38 cells, but at a subsequent point the human diploid lung MRC-5 line was substituted. There are four current producers of Oka-based chickenpox vaccine, Biken (closely associated with Osaka University), Merck and Co. Inc., SmithKline Beecham Biologicals, and Pasteur Merieux Connaught. Although all four vaccines are derived from the same root strain, manufacturers have received their starting materials from Japan at slightly different passage levels, and subsequent handling by each producer differs in some subtle, but perhaps important, ways. As a consequence of these differences, each vaccine is distinguished in the trade by a company-specific suffix (Oka/Biken, Oka/Merck, Oka/RIT, Oka/PMC). Since live attenuated viruses are the products of their passage histories, these vaccines should not necessarily be considered as generic products but should stand on their own merits of safety and efficacy.

Vaccine Attenuation

VZV is a double-stranded DNA virus, and this general type of virus is relatively stable in genetic terms compared to most RNA viruses. Much of this stability may be attributed to the editing, or (error-correction), functions of DNA polymerases relative to RNA polymerases, which usually lack this activity. Whatever the mechanism behind this genetic stability may be, the result is that there are relatively few known sequence differences among varicella strains and between the Oka vaccine strains and wild-type virus. The few differences reported to date include a particular change in single Bgl1 and Pst1 restriction sites between Oka vaccine and North American or European wild type viruses.[207] This has been exploited to "type" virus isolates and discriminate between vaccine virus and wild-type.[208,209] Differences in the number of repeated sequences in the gene 5 sequence have also been reported.[210-212] Recently, a series of mutations in the IE62 (transactivation protein) gene were described (M. Takahashi, unpublished, D.DiStefano, unpublished) as differences between Oka vaccine virus and the wild-type Oka parent. How these changes relate to the attenuated phenotype remains unclear.

Two pathological features of varicella zoster virus have clearly changed as a result of the Oka passage history. VZV has the ability to enter a state of latency in dorsal root and trigeminal ganglia and to emerge at a later date to cause zoster. Studies suggest that the Oka vaccine may cause zoster in children at a rate lower than that seen after normal chickenpox.[213] Levin and colleagues showed that the Oka vaccine strain grows less well than does wild-type virus in human embryonic neuronal cultures.[214] The hallmark of chickenpox is the pronounced rash. In a recent study, Arvin and coworkers suggest that the vaccine strain has altered its tissue tropism,[215] retaining the ability to grow well in lymphocytes but growing less well in skin. The molecular basis behind these changes may, or may not, emerge from extensive DNA sequencing efforts underway at the USFDA, the CDC, at Osaka University and at Merck.

Current Concerns

The varicella vaccines are the first live virus vaccines to be licensed in recent years and the first-ever herpesvirus vaccines. As such, there are a number of questions that are asked repeatedly during any discussion of these vaccines. With, at this writing, a total of 19 million doses delivered over five years for VARIVAX® alone, we have a significant body of experience upon which to base our answers (P. Coplan and R. Sharrar, Merck, personal communications). The most frequently asked questions, and answers, are:

"What is the duration of immunity?" A cohort from earlier clinical trials has been followed for seropositivity for over 12 years in the United States, and so far we have seen little drop in seropositivity or antibody titer. Of course, boosting due to exposure to wild-type disease can play a role here, but in addition we have not seen an increase in the breakthrough rate or severity or time.

"Can the vaccine virus be transmitted to healthy contacts?" During our postmarketing surveillance we have documented three such cases. In all three cases, the vaccinees had developed a (mild) varicella-like rash, and the contacts were either siblings or parents with intimate and frequent interactions with the vaccinees. Thus, it can happen, but it is extremely rare.

"Can the vaccine strain cause zoster?" During our clinical studies and post-marketing surveillance we have identified cases of zoster in vaccinees. The rate appears to be some fraction, between one quarter to one eighth, of what one would see in children who had suffered natural chickenpox.

"Is the vaccine effective after exposure to wild-type virus?" Takahashi's original studies[205] included post-exposure trials that showed protection when vaccine was administered within three days of exposure. Several more recent US studies confirm this result.[216-218]

"What happens if a pregnant woman gets vaccinated?" VARIVAX® is contraindicated during pregnancy, but nevertheless several hundred instances of inadvertant vaccination have been brought to Merck's attention; these have been actively sought by the company. So far, no untoward effects have been noted.

"Can one vaccinate immunocompromised individuals?" The current label for the vaccine used in the United States contains a contraindication for vaccination of individuals with immunocompromising conditions such as HIV, steroid therapy, chemotherapy for neoplasias, transplantation, etc. Clinical studies are underway to address carefully many of these situations. In the specific case of leukemic subjects, VARIVAX® is available under the strict terms of an open Investigational New Drug application at no charge.

Future Development

Varicella vaccines are recommended for administration at approximately 12-18 months of age. This coincides with the time of administration of combined measles, mumps and rubella vaccines in most part of the world where varicella vaccines are likely to be used. Data developed during the clinical trials of VARIVAX® show that this vaccine can be delivered at a separate site at the same time as M-M-R®II with no deleterious effects with regard to safety, seroconversion or titer of antibody for any of the four viruses;[217,218] similar results have been obtained with other varicella vaccines and MMR vaccines. When mixed together and delivered at the same site, however, the result is different. Several studies with different varicella vaccines from different producers have shown that a mixed "MMRV" vaccine yields good seroconversion rates for all four components and good antibody titers against measles, mumps and rubella. Geometric mean titers of anti-varicella antibody are diminished by half or more.[204] Although this has not been studied systematically, the measles vaccine component is thought to be responsible for the apparent suppression of the varicella. There is some hope that this interference may be overcome by increasing the relative varicella content of the MMRV cocktail while retaining the safety profile of VARIVAX® (B. Kuter, Merck, personal communication).[219, 220] The other development effort is at the other end of the age spectrum. One of the long-term sequelae of chickenpox is shingles or zoster. VZV establishes a state of latency in the dorsal root and trigeminal ganglia. The mechanism of latency is not well understood, but there is an epidemiological correlation between advancing age, declining cellular (but not humoral) immunity to VZV, and increasing incidence of zoster.[221-225] The obvious hypothesis that follows is that boosting cellular immunity by vaccination should impact the incidence and/or severity of zoster. Preliminary studies[226-229] tend to support this notion, but much larger studies will be required to answer this question in a definitive manner. To this end, the United States Veterans' Administration in, collaboration with Merck, has initiated a long term double-blinded, placebo-controlled study of the effects of vaccination on zoster in persons over 60 years of age.

Rotavirus Vaccines

Introduction

Rotavirus, a member of the reovirus family, is the single identifiable cause of up to 50% of severe gastroenteritis in children under three years of age. Rotavirus infects the villus epithelial cells of the gut, and under normal circumstances it is confined to this cell type in this compartment. The disease is characterized by effusive "rice-water" diarrhea, vomiting, fever and listlessness. The major worry in cases of rotavirus diarrhea is dehydration. Aggressive oral rehydration is the treatment of choice and is highly successful when applied in a timely manner. The mechanism of diarrhea due to rotavirus was long thought to depend on destruction of the gut epithelium and the osmotic barrier it provides. Recently, however, a specific mechanism has been elucidated. A non-structural protein, NSP4, encoded by gene 10 of rotavirus has been shown to be an activator of chloride channels in gut epithelial cells. Mutation of the channel-specific sequence of NSP4 has been shown to abrogate the diarrhea-triggering activity of NSP4.[230]

Immunity to rotavirus, even now that effective vaccines have been developed, is not well understood. Severe rotaviral disease is uncommon in infants under 6 months of age, and it is thought that this apparent protection is due to maternal antibody. There are multiple serotypes of human rotaviruses that circulate in the human population. In an elegant epidemiological study,[231] it has been shown that approximately three exposures and infections with rotavirus are required to induce immunity that prevents further clinically apparent rotavirus disease. Each individual exposure seems to induce additive protection since disease becomes successively less pronounced. So far, no laboratory measurement of immunity[232,233] has jumped out and claimed its place as the reliable index of protection. This, coupled with the multiplicity of serotypes and shifting endemicity, presents the developer of a rotavirus vaccine with a formidable challenge.

Rotavirology

Human rotaviruses carry their genetic information as a set of 11 double-stranded RNAs each encoding a single protein. This genomic arrangement allows the efficient creation and selection of "reassortant" or recombinant, in the classical sense, rotaviruses by simple co-infection of a cell with two types of rotavirus. This has been crucial to the development of two lines of rotavirus vaccines as will be discussed below. The rotaviruses are non-enveloped viruses carrying two types of proteins on their outer surface. Nomenclature of these proteins and their corresponding serotyping is complicated by multiple systems and revisions of these systems over time.[234-236] Human and animal rotaviruses share the same typing systems. Further complication comes from close relationships between human and animal rotaviruses and their ability to reassort in nature. The basic nomenclature has the glycoprotein product of gene segment 9, called virus protein 7 (VP7), as the primary antigen for serotyping. Serotypes are named by glycoprotein type, e.g., G1, G2, G3, etc. The second surface protein is the product of gene 4 and is called VP4. These VP4s have their own serotyping system and are called P1, P2, P3, etc. In addition, the different examples of VP4 are now best distinguished by sequence differences and thus are assigned a genotype as well. VP4 is the protein that interacts with the cell surface during attachment, and it generally requires cleavage with a trypsin-like enzyme for "activation". Thus, rotavirus cultures require trypsin as a component of the medium, and it is easy to imagine how this might complicate a culture system! Rotaviruses are rather finicky about what cell type they will grow in. To date, monkey cells (MA104, FRhL, CV1, Vero, primary AGMK) have been used almost exclusively for propagation of rotaviruses.

Rotaviruses are democratic viruses in the sense that the incidence of rotavirus disease is similar in Bangladesh and Beverly Hills; everyone gets it. Prevalence of serotypes does vary by locale to some degree. For example, in the United States and in Europe the G1 serotype dominates with G3 usually in second place and G4 plus G2 trailing behind. In India, G1 dominates, but G5 is also common. In South America, Brazil especially, G1 dominates and G9 is commonly found. Another complication for vaccine developers immunity is that is believed to be type-specific. Rotavirus disease (and presumably infection) is highly seasonal in temperate climates with a peak in the winter to early spring months. In the United States, the seasonal wave of disease spreads from the southwest to the northeast. The reasons for this can only be speculated upon.

Rotavirus Vaccine Development

The first attempts at making vaccines against rotavirus were pure Jennerian efforts. Three such attempts have been made beginning in the early 1980s. One attempt used a bovine rotavirus called NCDV (newborn calf diarrhea virus, also known as RIT4237 or RIT4256), a virus which was delivered orally and was well-tolerated in infants.[237-241] Development of this vaccine was dropped by SB due to variable protection from disease in early clinical trials.[237-241] A second attempt with a bovine virus (WC3 for Wistar Calf 3) was made a few years later by the Wistar Institute and Pasteur Merieux serums and vaccines.[240,242,243] This, too, was abandoned due to variable protection results in clinical trials.[244,245] The third and more extensive attempt was by workers at the NIH (and later involving Wyeth) using a rhesus monkey rotavirus (RRV) also known as MMU 18006. The RRV had the theoretical advantage of being more like the human rotavirus in its capacity to replicate in the human gut and in being a close mimic of human G3 type rotavirus.[240,246-248] Ultimately, this vaccine was also set aside in favor of a reassortant approach due to variability of clinical performance. A detailed review of the history of the pure animal virus vaccines has been made by Vesikari.[249]

Of course, research and development does not occur in isolation, and while the above animal virus vaccines were being tested, the importance of serotypic immunity was being elaborated.[240,248,250-252] This led Kapikian and co-workers at the NIH and Clark and coworkers at the Wistar Institute and Children's Hospital of Philadelphia (CHOP) to explore the properties of rhesus RRV and bovine WC3 viruses, respectively, carrying the human VP7 and VP4 proteins on their surfaces.[253-256] These reassortant viruses are created by co-infecting cells with the relevant animal virus plus a wild-type human rotavirus of the desired serotype. Progeny viruses carrying 10 animal virus genes plus gene 9 (or gene 4) from the human virus can be selected by using neutralizing sera against the parent animal virus VP7 (or VP4).

Both RRV and WC3 reassortants have proved to be substantial improvements over their pure animal virus predecessors in terms of uniformity of protection against disease. The majority of data published to date pertains to the RRV reassortants, since this effort preceded the bovine reassortant trials by several years. A large number of studies have been carried out by the NIH and Wyeth-Lederle, and these have been reviewed in detail.[249,257,258] A large subset of these trials were done in Finland, and this allows one to explore parameters of dose, composition, regimen, etc., in a relatively uniform environment.[259] The overall conclusion of this massive effort is that the reassortant RRV vaccine works best when given as three doses at nominally two, four, and six months of age, at a dose of 10^5 pfu per reassortant in a cocktail containing RRV-G1, RRV-G2, RRV-G4 and RRV which is itself a mimic of human G3 virus. The licensed vaccine, Rotashield®, protects against 60-70% of all rotavirus disease and at least 90% of severe disease when used in developed country settings.

A limited series of trials has been carried out in South America using variants (dose, regimen, composition) of Rotashield® prototypes.[260,261] The original reports of these studies were disappointing, in that protection from disease was significantly lower than that reported in the United States and Finland.[262] More recent reanalyses of the data from these trials, taking into account methods of scoring disease severity and trial design differences, shows a more encouraging picture with protection on the order of 50%.[261] This is a significant issue, given the need for an effective rotavirus vaccine in developing countries.

Data from the clinical trials of Rotashield® and with the RRV predecessor indicate that there is some room for improvement in tolerability. In normal infants, the vaccine causes a low-grade fever in ~20% of subjects, mainly after the first dose, and significant fever in 1-2%.[263, 264] First dose fevers are more frequent in older infants, and as a consequence, the first dose of vaccine is not to be given beyond the age of six months.[265] Several "second-generation" vaccines are in development with the hope of coming up with something better tolerated and with perhaps better protective efficacy. The bovine WC3-based reassortant vaccine developed by F. Clark and P. Offit of CHOP and Merck is the next likely candidate for licensure.[266] In its present form this vaccine contains five reassortant WC3-based viruses carrying human G1/2/3/4 VP7 proteins plus human VP4 type P1a 8, the dominant human VP4. Final dose has not been established, but the 2-4-6 month regimen is in common with Rotashield®. The WC3 reassortants appear to have less capacity to cause fevers and have been delivered to infants older than six months with no apparent reactions. Protection from disease in small trials appears to be at least equivalent to that of Rotashield®, although direct comparative studies would be necessary to resolve this.

The next rotavirus vaccine in the queue is strain 89-12, a human G1 attenuated "nursery strain" isolated by R. Ward and D. Bernstein of the Children's Hospital of Cincinnati. Early clinical studies show good protection against G1 rotavirus disease along with low-grade fever in 19% of vaccinees.[267] This vaccine is being developed by SB. A second human rotavirus strain, "RV3" has also been tested in the clinic.[268]

The original developers of the RRV strain, A. Kapikian and colleagues at the NIH, have also created a bovine reassortant vaccine. This has been in early clinical trials[269] and has shown promising performance.

Current Concerns

In July of 1999, the CDC issued a recommendation to withhold Rotashield® until November of 1999 due to a potential concern about intussusception. A number of cases of intussusception have been reported following the first dose of Rotashield® in infants 2-3 months of age. This is somewhat below the expected age for this condition. This was not judged to be associated with either natural or vaccinal rotavirus infection[270] at the time of licensure. The CDC conducted a series of epidemiological studies during the months of July to October of 1999. On October 22, 1999, the ACIP withdrew the recommendation of immunization with Rotashield® due to a strong causal link with intussusception within the first 3-10 days following administration of the first and second doses. At the time this review is being written, the pathophysiological link between Rotashield and intussusception remains unclear. What is clear is that this association will complicate the development of future rotavirus vaccines by raising dramatically the scope of safety trials.

A major question that applies to all rotavirus vaccines is barriers to more widespread use of these vaccines. Rotashield® was licensed for use in the United States and was under review in Europe. The reported mortality rates for rotavirus disease in these regions are low, on the order of 100-200 deaths per year, but the medical and economic burden is substantial.[271] In the developing world, some 800,000 annual deaths are attributed to rotavirus-related diarrhea. The vaccines available in the foreseeable future are cell-culture-based products with multiple components, produced in the United States or in Europe under strict GMP guidelines and meeting a high standard of quality. Cost of production is high compared to the vaccines comprising the EPI portfolio. The global public-health community has a formidable task ahead in implementing these vaccines in worldwide immunization programs.

Live Attenuated Influenza Vaccines

The concept of a live, attenuated, nasally-administered influenza virus vaccine has been around since the mid-1960s. One has been in widespread use in Russia or the former Soviet Union for almost ten years. A second was licensed in the UK for a brief period in the '70s. Another should be available in the United States in the year 2002. Still others have been tested in the clinic and have shown some promise, but have never been licensed.

Influenza is a classic short-incubation respiratory disease. The virus is transmitted primarily by aerosols and is taken up in the respiratory tract. Incubation time is about 2 days, and the disease runs its course in healthy individuals within a week. We are all familiar with the symptoms of sneezing, runny nose, cough, congested airways, fever and myalgia. Influenza is a seasonal disease with its peak in the winter.

Influenza virus is an orthomyxovirus with an eight-segmented, negative-stranded RNA genome. The eight segments encode 10 protein products; two segments each encode two polypeptides. The virus carries two surface proteins, a hemagglutinin (HA) and a neuraminidase (NA), and these are the primary targets of the humoral immune response. There are two medically important types of influenza virus, A and B. The segmented nature of the influenza virus genome, like that of the rotaviruses, lends itself to recombination. Two viruses infecting one cell can reassort their genes to create progeny viruses of new genetic composition. This happens in nature and is the cause of major antigenic "shift" in the circulating virus population. The circulating human virus can pick up a new hemagglutinin gene, a new neuraminidase gene or both from another virus, frequently an avian or a swine influenza virus;[272] then, man is suddenly exposed to an influenza virus his immune system has not seen before. Major pandemic influenza infections, with substantial morbidity and mortality in healthy individuals, are the result. Influenza virus, as an RNA virus, undergoes mutation and selection by the human immune system and these more subtle changes due to mutation of the HA or NA genes result in antigenic "drift". Influenza nomenclature includes as a number the HA type, e.g., H3, and the NA type, e.g., N2, as well as a geographic name plus year of isolation. Thus, the virus A/Dallas/72 (H2N1) would be an A-type virus, isolated in Dallas in 1972 and having HA of the type 2 and NA of type 1. The year and place designations identify, in a sense, the uniqueness of the HA and NA subtypes. This nomenclature is not Y2K compliant. Additional numbers give an indication of passage level.

This shifting and drifting antigen profile makes influenza vaccine production a moving target. A surveillance network operated by the WHO and the CDC scans the world for new influenza virus serotypes and subtypes. Trends are tracked, and early each year a group of influenza experts issues a recommendation for the composition of the coming autumn's flu vaccine. Since the standard inactivated vaccines contain three different sets of virus antigens, generally two A types and one B type, it can happen that three new vaccine strains have to be developed, grown up, tested and released in a matter of months. Furthermore, this year's vaccine might not be good for next year, so the whole process starts over again every year.

The cyclic process of vaccine production makes live attenuated influenza vaccines, made by traditional, empirical attenuation on a strain-by-strain basis, almost impossible. There is not enough time allowed by the logistics of the system to attenuate and show safety and efficacy, not to mention produce and release vaccine batches. This obstacle has been overcome through the development of attenuated "master strains" with well-established phenotypes linked to mutations in specific gene segments. These master strains can be reassorted with new virulent isolates so as to generate customized vaccine strains carrying the desired HA and NA genes along with the other genes from the attenuated master strain.

Influenza Virus Attenuation

Several independent approaches have yielded candidate attenuated influenza vaccine strains. These were initially considered as vaccines themselves, but as technology for genetic analysis advanced, the notion of deliberately reassorted strains became feasible.[273,274]

One of the early approaches to attenuating influenza viruses was repetitive passage in embryonated eggs, following the same strategy that had proved successful for YFV. In Japan, an attenuated vaccine, the A/Okuda strain, was developed after 280 cycles of growth in eggs.[275] In the United Kingdom at Wellcome Laboratories the A/Okuda strain was reassorted with an H3N2 strain (A/Finland/4/74) to create the WRL 105 vaccine.[275-279]

A second strategy was to select for resistance to guinea pig and horse serum in the growth medium. This was used in the generation of the attenuated A/PR8/34 strain developed by SmithKline and French/RIT. The PR8 attenuated strain was reassorted with several different A-type virus HA and NA donors to create four independent vaccine strains, Alice, RIT 4025,

RIT 4050, and RIT 4199.[280] Analysis of gene segments tracking with attenuated phenotype suggested that the genes for P3 (now called PB2), P1 (now called PB1), M and NS were all implicated in the attenuated phenotype. These viruses were tested in the clinic where they showed promising immunological performance in selected volunteers.[281-289] This culminated in the licensure in the U.K., South Africa and a handful of other countries of a live influenza vaccine, NASOFLU, produced by SB in 1976. Concerns about possible over-attenuation and an uncertain early pedigree,[274] and more importantly a regulatory requirement for a clinical trial of each new reassortant, led to the eventual discontinuation of this line of work (W. Vandersmissen, SB, personal communication).

The third strategy was to select influenza viruses that were either temperature-sensitive, cold-adapted, or both. The theory behind this strategy is that a virus with a limited ability to replicate only in the cooler upper part of the respiratory tract should be appropriately attenuated in humans. Much of the early work in this area occurred in the Soviet Union in the 1950s and 1960s and is not readily accessible in English.[290,291] Three groups have had major roles in this effort.

During the 1970s and 1980s, the laboratory of Brian Murphy and Robert Chanock at the NIH developed a series of temperature-sensitive strains of influenza virus, most notably the *ts*-1E strains of A/Hong Kong/68 and the *ts*- 1A2 variant of A/Udorn/72.[292-297] The ts-1E strain were shown to be attenuated through mutations in the PB1 and NS genes[298-300] while the ts-1A2 strains had mutations in the PB2 and PA genes of the polymerase.[299,301] Clinical studies of these two strains and reassortants based upon them were promising, with some advantage in favor of the double mutation in the polymerase genes of the ts-1A2 series[299-302] and the occurrence of revertants of the ts-1E strains.[303,304]

In the 1960s, Maassab and co-workers at the University of Michigan developed a strain of influenza virus that was both temperature-sensitive and cold-adapted, the A/Ann Arbor/6/60 strain. This virus was attenuated by successive passages in monkey kidney cells followed by a series of passages in embryonated eggs at decreasing temperatures.[305-307] This virus, along with strain B/Maryland/1/59 (succeeded by B/Ann Arbor 1/66), was grown at 25°C in embryonated eggs and thus is known as a "cold-adapted" strain; wild-type influenza virus does not grow at this temperature. The A/Ann Arbor/6/60 cold-adapted strain is the subject of an extensive literature. This strain is attenuated in ferrets and appeared to be genetically stable when recovered from humans.[307,308] Comparison clinical studies of ts-1A2 vaccine and the cold-adapted Ann Arbor vaccines favored further development of the latter strain.[302] Single-gene reassortant studies showed that four genes (PA, PB1, PB2 and M) from the cold-adapted strain each contribute to the attenuated phenotype of the virus.[309] Sequence analysis of the complete gene complement of the cold-adapted Ann Arbor strain showed clearly that all six of the non-HA/NA segments carried mutations with respect to the wild-type parent virus.[310] Clinical studies of monovalent and trivalent vaccines based on the A/Ann Arbor/6/60 and B/Ann Arbor 1/66 in institutionalized patients showed good tolerability and protection from disease.[311] An excellent review of the laboratory and clinical development of the "Maassab Strains" has been published by Murphy.[312]

As mentioned earlier, a group at the Research Institute for Experimental Medicine in Leningrad/St. Petersburg began a program of attenuating influenza viruses in the 1950s. This culminated in the development and local distribution of live attenuated, cold-adapted influenza A and B vaccines with properties similar to the Maassab strains.[313,314] These vaccines have been tested in very large field studies in Russia[313] with very good results.

In addition to the changes in the polymerase, membrane, nucleoprotein and non-structural protein sequences, it has been shown recently that mutations in the NA sequence may also be temperature-sensitive.[315] Mutating a specific cysteine pair in the NA gene yielded a stable ts phenotype.

Future Development

Inactivated influenza vaccines dominate field in most parts of the world, but they suffer from a perceived (if not real) deficit in efficacy and tolerability.[273] Based on the general success of live polio, measles and smallpox vaccines, it has been postulated that one might do better with a live influenza vaccine.[273] This has been the driving force behind four decades of laboratory and clinical development. As this is written, two vaccines appear to have met the criteria for implementation into public health programs. One is the Russian vaccine, a cold-adapted, temperature sensitive set of A and B strains. This vaccine is produced and distributed within Russia in excess of five million doses per year for use in children and adults (L. Rudenko, personal communication). This vaccine has been reported to have good tolerability and field efficacy against respiratory disease in Russia.

A combination live attenuated, cold-adapted, temperature-sensitive vaccine based on the Maassab strains from Ann Arbor has been taken through commercial development and large-scale clinical trials by Aviron and has been licensed to Wyeth-Lederle for distribution. This vaccine is currently under consideration by the regulatory authorities in the United States.

Large-scale double-blinded, placebo-controlled studies with this vaccine in children[316] showed 89-94% protective efficacy against laboratory-confirmed cases of influenza when administered as a nasal spray in a one or two-dose regimen. Perhaps more interesting is the finding in this study that the rate of otitis media in vaccinated children was 30% lower than that of the control group. The role of influenza virus in otitis media has been debated for some time. At one point there was a speculation that a live influenza vaccine administered nasally might cause otitis media.[317] Although otitis media is considered to be a bacterial disease, the initiating event is often a viral infection of the respiratory tract. Preventing the viral disease should, logically, have an impact on the bacterial infection. Studies in older adults using this type of vaccine showed roughly equivalent performance when compared with currently available inactive vaccine, but the coadministered combination of live and killed vaccines showed a significant advantage.[318]

Current energies are directed towards finalizing licensure of the Wyeth-Lederle/Aviron vaccine in the United States. The availability of this vaccine should trigger a reconsideration of current influenza vaccination practices, especially for the pediatric population.

References

1. Henderson DA. Edward Jenner's vaccine. Public Health Rep 1997; 112(2):116-21.
2. Rathbone J. Lady Mary Wortley Montague's contribution to the eradication of smallpox letter. Lancet 1996; 347(9014):1566.
3. Baxby D. Edward Jenner's Inquiry; A bicentenary analysis. Vaccine 1999; 17(4):301-7.
4. Baxby D. Vaccines for smallpox letter. Lancet 1999; 354(9176):422-3.
5. Fenner F. Portraits of viruses: The poxviruses. Intervirology 1979; 11(3):137-57.
6. Baxby D. In: Collier and Mahy, eds. Poxviruses. London: Arnold 1998.
7. Alcami A, Smith GL. Receptors for gamma-interferon encoded by poxviruses: Implications for the unknown origin of vaccinia virus. Trends Microbiol 1996; 4(8):321-6.
8. Barrett AD. Japanese encephalitis and dengue vaccines. Biologicals 1997; 25(1):27-34.
9. Jelinek T, Nothdurft HD. Japanese encephalitis vaccine in travellers. Is wider use prudent? Drug Saf 1997; 16(3):153-6.
10. Hilleman MR. Personal historical chronical of six decades of basic and applied research in virology, immunology, and vaccinology. Immunological Reviews 1999; 170:7-27.
11. Bhau LN, Saxena SN. Vaccines against Japanese encephalitis. J Commun Dis 1989; 21(1):75-80.
12. Rojanasuphot S, Charoensuk O, Kitprayura D et al. A field trial of Japanese encephalitis vaccine produced in Thailand. Southeast Asian J Trop Med Public Health 1989; 20(4):653-4.
13. Manabe S. Development of second generation JE vaccine with genetic engineering techniques. Southeast Asian J Trop Med Public Health 1989; 20(4):643-6.

14. Konishi E, Kurane I, Mason PW et al. Poxvirus-based Japanese encephalitis vaccine candidates induce JE virus- specific CD8+ cytotoxic T lymphocytes in mice. Virology 1997; 227(2):353-60.
15. Oya A. Japanese encephalitis vaccine. Acta Paediatr Jpn 1988; 30(2):175-84.
16. Shen-Gunther J, Walker JL, Johnson GA et al. Hepatic venoocclusive disease as a complication of whole abdominopelvic irradiation and treatment with the transjuglar intrahepatic portosystemic shunt: Case report and literature review. Gynecol Oncol 1996; 61(2):282-6.
17. Sumiyoshi H, Tignor GH, Shope RE. Characterization of a highly attenuated Japanese encephalitis virus generated from molecularly cloned cDNA. J Infect Dis 1995; 171(5):1144-51.
18. Lin YL, Liao CL, Yeh CT et al. A highly attenuated strain of Japanese encephalitis virus induces a protective immune response in mice. Virus Res 1996; 44(1):45-56.
19. Theiler M, Smith H. The use of yellow fever virus modified by in vitro cultivation for human immunization. Jour Exp Med 1937; 65(6):787-800.
20. Post PR, Santos CN, Carvalho R et al. Molecular analysis of yellow fever virus 17DD vaccine strain. Mem Inst Oswaldo Cruz 1991; 86(2):239-46.
21. Monath TP, Kinney RM, Schlesinger JJ et al. Ontogeny of yellow fever 17D vaccine: RNA oligonucleotide fingerprint and monoclonal antibody analyses of vaccines produced world-wide. J Gen Virol 1983; 64 Pt 3:627-37.
22. Cane PA, Gould EA. Immunoblotting reveals differences in the accumulation of envelope protein by wild-type and vaccine strains of yellow fever virus. J Gen Virol 1989; 70(Pt 3):557-64.
23. dos Santos CN, Post PR, Carvalho R et al. Complete nucleotide sequence of yellow fever virus vaccine strains 17DD and 17D-213. Virus Res 1995; 35(1):35-41.
24. Barrett AD. Yellow fever vaccines. Biologicals 1997; 25(1):17-25.
25. Li GP, Rice CM. Mutagenesis of the in-frame opal termination codon preceding nsP4 of Sindbis virus: Studies of translational readthrough and its effect on virus replication. J Virol 1989; 63(3):1326-37.
26. Liprandi F. Isolation of plaque variants differing in virulence from the 17D strain of yellow fever virus. J Gen Virol 1981; 56(Pt 2):363-70.
27. Xie H, Ryman KD, Campbell GA et al. Mutation in NS5 protein attenuates mouse neurovirulence of yellow fever 17D vaccine virus. J Gen Virol 1998; 79(Pt 8):1895-9.
28. Ryman KD, Ledger TN, Campbell GA et al. Mutation in a 17D-204 vaccine substrain-specific envelope protein epitope alters the pathogenesis of yellow fever virus in mice. Virology 1998; 244(1):59-65.
29. Enders JF, Weller TH, Robbins FC. Cultivation of the Lansing srain of poliomyelitis virus in cultures of various human embryonic tissues. Science 1949; 49:234.
30. Cox HR. Active immunization against poliomyelitis. April 2, 1953 classical article see comments. J Urban Health 1998; 75(3):584-99.
31. Melnick JL. Thermostability of poliovirus and measles vaccines. Dev Biol Stand 1996; 87:155-60.
32. Mosimann SC, Cherney MM, Sia S et al. Refined X-ray crystallographic structure of the poliovirus 3C gene product. J Mol Biol 1997; 273(5):1032-47.
33. O'Reilly EK, Kao CC. Analysis of RNA-dependent RNA polymerase structure and function as guided by known polymerase structures and computer predictions of secondary structure. Virology 1998; 252(2):287-303.
34. Stewart SR, Semler BL. RNA structure adjacent to the attenuation determinant in the 5'-noncoding region influences poliovirus viability. Nucleic Acids Res 1998; 26(23):5318-26.
35. Friedrich F. Genomic modifications in Sabin vaccine strains isolated from vaccination-associated cases, healthy contacts and healthy vaccinees. Acta Virol 1996; 40(3):157-70.
36. Chumakov KM, Dragunsky EM, Norwood LP et al. Consistent selection of mutations in the 5'-untranslated region of oral poliovirus vaccine upon passaging in vitro. J Med Virol 1994; 42(1):79-85.
37. Minor PD. Attenuation and reversion of the Sabin vaccine strains of poliovirus. Dev Biol Stand 1993; 78:17-26.
38. Baker S, Richards OC, Ehrenfeld E. Elongation activity of poliovirus RNA polymerase derived from Sabin type 1 sequence is not temperature sensitive. J Gen Virol 1995; 76(Pt 8):2081-4.
39. Chumakov K, Norwood L, Parker M et al. Assessment of the viral RNA sequence heterogeneity for control of OPV neurovirulence. Dev Biol Stand 1993; 78:79-89; discussion 88-9.
40. Macadam AJ, Pollard SR, Ferguson G et al. Genetic basis of attenuation of the Sabin type 2 vaccine strain of poliovirus in primates. Virology 1993; 192(1):18-26.

41. Guillot S, Otelea D, Delpeyroux F, Crainic R. Point mutations involved in the attenuation/neurovirulence alternation in type 1 and 2 oral polio vaccine strains detected by site-specific polymerase chain reaction. Vaccine 1994; 12(6):503-7.
42. Rezapkin GV, Fan L, Asher DM, Fibi MR, Dragunsky EM, Chumakov KM. Mutations in Sabin 2 strain of poliovirus and stability of attenuation phenotype. Virology 1999; 258(1):152-60.
43. Mulders MN, Reimerink JH, Stenvik M et al. A Sabin vaccine-derived field isolate of poliovirus type 1 displaying aberrant phenotypic and genetic features, including a deletion in antigenic site 1. J Gen Virol 1999; 80(Pt 4):907-16.
44. Georgescu MM, Balanant J, Macadam A et al. Evolution of the Sabin type 1 poliovirus in humans: Characterization of strains isolated from patients with vaccine-associated paralytic poliomyelitis. J Virol 1997; 71(10):7758-68.
45. Rezapkin GV, Chumakov KM, Lu Z et al. Microevolution of Sabin 1 strain in vitro and genetic stability of oral poliovirus vaccine. Virology 1994; 202(1):370-8.
46. Almond JW, Stone D, Burke K et al. Approaches to the construction of new candidate poliovirus type 3 vaccine strains. Dev Biol Stand 1993; 78:161-9; discussion 169-70.
47. Kohara M, Abe S, Yoshioka I et al. Development of candidates for new type 2 and type 3 oral poliovirus vaccines. Dev Biol Stand 1993; 78:141-8.
48. Furesz J, Contreras G. Some aspects of the monkey neurovirulence test used for the assessment of oral poliovirus vaccines. Dev Biol Stand 1993; 78:61-70.
49. Levenbook I, Nomura T. Development of a neurovirulent testing system for oral poliovirus vaccine with transgenic mice. Lab Anim Sci 1997; 47(2):118-20.
50. Stephanou G, Andrianopoulos C, Vlastos D et al. Induction of micronuclei and sister chromatid exchange in mouse splenocytes after exposure to the butadiene metabolite 3,4-epoxy-1- butene. Mutagenesis 1997; 12(6):425-9.
51. Racaniello VR. Infectious cDNA, cell receptors and transgenic mice in the study of Sabin's poliovirus vaccines. Biologicals 1993; 21(4):365-9.
52. Furesz J, Levenbook I. New assays for the quality control of live oral poliovirus vaccine. Acta Microbiol Immunol Hung 1998; 45(3-4):391-9.
53. Wood DJ, Macadam AJ. Laboratory tests for live attenuated poliovirus vaccines see comments. Biologicals 1997; 25(1):3-15.
54. Old MO, Logan LH, Maldonado YA. Development of chemiluminescent probe hybridization, RT-PCR and nucleic acid cycle sequencing assays of Sabin type 3 isolates to identify base pair 472 Sabin type 3 mutants associated with vaccine associated paralytic poliomyelitis. J Virol Methods 1997; 68(2):109-18.
55. Svitkin YV, Alpatova GA, Lipskaya GA et al. Towards development of an in vitro translation test for poliovirus neurovirulence. Dev Biol Stand 1993; 78:27-32.
56. Mallet L, Pelloquin F, Brigaud M et al. Comparative study of poliovirus excretion after vaccination of infants with two oral polio vaccines prepared on Vero cells or on primary monkey kidney cells. J Med Virol 1997; 52(1):50-60.
57. Chezzi C, Dommann CJ, Blackburn NK et al. Genetic stability of oral polio vaccine prepared on primary monkey kidney cells or Vero cells—effects of passage in cell culture and the human gastrointestinal tract. Vaccine 1998; 16(20):2031-8.
58. Shepherd WM, Langford DT, Kelly A. A single-blind, placebo-controlled comparison of the reactivity and antigenicity of trivalent oral poliomyelitis vaccine prepared in monkey kidney or human diploid cell substrates. J Biol Stand 1983; 11(1):29-33.
59. Modlin JF, Halsey NA, Thoms ML et al. Humoral and mucosal immunity in infants induced by three sequential inactivated poliovirus vaccine-live attenuated oral poliovirus vaccine immunization schedules. Baltimore Area Polio Vaccine Study Group. J Infect Dis 1997; 175 Suppl 1:S228-34.
60. Brooks HW, Gleadhill A, Wagstaff AJ et al. Fallibility of plasma urea and creatinine as indices of renal function in diarrhoeic calves treated with conventional or nutritional oral rehydration solutions. Vet J 1997; 154(1):35-9.
61. From the Centers for Disease Control and Prevention. Revised recommendations for routine poliomyelitis vaccination. Jama 1999; 282(6):522.
62. Nkowane BM, Wassilak SG, Orenstein WA et al. Vaccine-associated paralytic poliomyelitis. United States: 1973 through 1984. Jama 1987; 257(10):1335-40.
63. Friedrich F. Neurologic complications associated with oral poliovirus vaccine and genomic variability of the vaccine strains after multiplication in humans. Acta Virol 1998; 42(3):187-94.

64. Driesel G, Diedrich S, Kunkel U et al. Vaccine-associated cases of poliomyelitis over a 30 year period in East Germany. Eur J Epidemiol 1995; 11(6):647-54.
65. Friedrich F, Da-Silva EF, Schatzmayr HG. Type 2 poliovirus recombinants isolated from vaccine-associated cases and from healthy contacts in Brazil. Acta Virol 1996; 40(1):27-33.
66. Li J, Zhang LB, Yoneyama T et al. Genetic basis of the neurovirulence of type 1 polioviruses isolated from vaccine-associated paralytic patients. Arch Virol 1996; 141(6):1047-54.
67. Georgescu MM, Delpeyroux F, Tardy-Panit M et al. High diversity of poliovirus strains isolated from the central nervous system from patients with vaccine-associated paralytic poliomyelitis. J Virol 1994; 68(12):8089-101.
68. Prevots DR, Sutter RW, Strebel PM et al. Completeness of reporting for paralytic poliomyelitis, United States, 1980 through 1991. Implications for estimating the risk of vaccine- associated disease. Arch Pediatr Adolesc Med 1994; 148(5):479-85.
69. Juhela S, Hyoty H, Uibo R et al. Comparison of enterovirus-specific cellular immunity in two populations of young children vaccinated with inactivated or live poliovirus vaccines. Clin Exp Immunol 1999; 117(1):100-5.
70. Hull HF, Aylward RB. Ending polio immunization see comments. Science 1997; 277(5327):780.
71. Kimman TG, Koopmans MP, van der Avoort HG. Ending polio immunization: When and how are we sure that the needle is out of the haystack? Vaccine 1999; 17(7-8):624-7.
72. Schneider-Schaulies S, ter Meulen V. Pathogenic aspects of measles virus infections. Arch Virol Suppl 1999; 15:139-58.
73. Carter, Conway TJ, Cornfield D et al. Serologic response of children to inactivated measles. JAMA 1979, 1962:848-53.
74. Nader PR, Warren RJ. Reported neurologic disorders following live measles vaccine. Pediatrics 1968; 41(5):997-1001.
75. Guinee VF, Henderson DA, Casey HL et al. Cooperative measles vaccine field trial. I. Clinical efficacy. Pediatrics 1966; 37(4):649-65.
76. Fulginiti VA, Eller JJ, Downie AW et al. Altered reactivity to measles virus. Atypical measles in children previously immunized with inactivated measles virus vaccines. Jama 1967; 202(12):1075-80.
77. Polack FP, Auwaerter PG, Lee SH et al. Production of atypical measles in rhesus macaques: Evidence for disease mediated by immune complex formation and eosinophils in the presence of fusion-inhibiting antibody. Nat Med 1999; 5(6):629-34.
78. Enders JF, Peebles TC. Propagation in tissue cultures of cytopathogenic agents from patients with measles. Proc. Soc. Exp. Biol., N.Y. 1954; 86:2:277-86.
79. J.F. E, Katz SL, Milovananovic MV et al. Studies on attenuated measles-virus vaccine. I. Development and preparations of the vaccine: Technics for assay of effects of vaccination. New Engl J Med 1960; 263:153-9.
80. Hirayama M. Measles vaccines used in Japan. Rev Infect Dis 1983; 5(3):495-503.
81. Rota JS, Wang ZD, Rota PA et al. Comparison of sequences of the H, F, and N coding genes of measles virus vaccine strains. Virus Res 1994; 31(3):317-30.
82. Heider A, Santibanez S, Tischer A et al. Comparative investigation of the long non-coding M-F genome region of wild-type and vaccine measles viruses. Arch Virol 1997; 142(12):2521-8.
83. Enders JF, Katz SL, Milovanovic V et al. Studies on attenuated measles-virus vaccine. I. Development and preparations of the vaccine: Technics for assay of effects of vaccination. New Engl J Med 1960; 263:153-9.
84. Katz, Enders. Studies on an attenuated measles-virus vaccine. . Development and preparations of the vaccine technics for assay of effects of vaccination. Am J Dis Child 1959; 98:605-607.
85. Schwarz AJ. Immnization against measles: Development and evaluation of a highly attenuated live masles vaccine. Ann Paediat (Basel) 1964; 202:241-52.
86. Carson MM, Spady DW, Albrecht P et al. Measles vaccination of infants in a well-vaccinated population. Pediatr Infect Dis J 1995; 14(1):17-22.
87. Diaz-Ortega JL, Forsey T, Clements CJ et al. The relationship between dose and response of standard measles vaccines. Biologicals 1994; 22(1):35-44.
88. Halsey NA. Increased mortality after high titer measles vaccines: Too much of a good thing. Pediatr Infect Dis J 1993; 12(6):462-5.
89. Thompson NP, Fleming DM, Pounder RE et al. Crohn's disease, measles, and measles vaccination: A case-control failure letter. Lancet 1996; 347(8996):263.
90. Thompson NP, Montgomery SM, Pounder RE et al. Is measles vaccination a risk factor for inflammatory bowel disease? see comments. Lancet 1995; 345(8957):1071-4.

91. Olive JM. Measles immunization policies and control in Europe. Pediatr Pulmonol Suppl 1997; 16:284-5.
92. Galazka A. Rubella in Europe. Epidemiol Infect 1991; 107(1):43-54.
93. Parkman PD, Meyer HM, Jr., Kirschstein RL et al. Attenuated rubella virus. I. Development and laboratory characterization. N Engl J Med 1966; 275(11):569-74.
94. Meyer HM, Jr., Parkman PD, Panos TC. Attenuated rubella virus. II. Production of an experimental live-virus vaccine and clinical trial. N Engl J Med 1966; 275(11):575-80.
95. Shishido A, Ohtawara M. Development of attenuated rubella virus vaccines in Japan. Jpn J Med Sci Biol 1976; 29(5):227-53.
96. Laurence GD. The production and testing of vaccines in human diploid cells, with special reference to rubella. Postgrad Med J 1972; 48(Suppl):24-9.
97. Plotkin SA, Beale AJ. Production of RA27/3 rubella vaccine and clinical results with the vaccine. Dev Biol Stand 1976; 37:291-6.
98. Zealley H, Morrison AM, Freestone DS. Dose response studies with Wistar RA 27/3 strain live attenuated rubella vaccine. J Biol Stand 1974; 2(2):111-9.
99. Freestone DS, Reynolds GM, McKinnon JA et al. Vaccination of schoolgirls against rubella. Assessment of serological status and a comparative trial of Wistar RA 27/3 and Cendehill strain live attenuated rubella vaccines in 13-year-old schoolgirls in Dudley. Br J Prev Soc Med 1975; 29(4):258-61.
100. Plotkin SA, Farquhar JD, Ogra PL. Immunologic properties of RA27-3 rubella virus vaccine. A comparison with strains presently licensed in the United States. Jama 1973; 225(6):585-90.
101. Grillner L. Neutralizing antibodies after rubella vaccination of newly delivered women: A comparison between three vaccines. Scand J Infect Dis 1975; 7(3):169-72.
102. Macdonald H, Tobin JO, Cradock-Watson JE et al. Antibody titres in women six to eight years after the administration of RA2713 and Cendehill rubella vaccines. J Hyg (Lond) 1978; 80(3):337-47.
103. Fogel A, Gerichter CB, Barnea B et al. Response to experimental challenge in persons immunized with different rubella vaccines. J Pediatr 1978; 92(1):26-9.
104. Schluederberg A, Horstmann DM, Andiman WA et al. Neutralizing and hemagglutination-inhibiting antibodies to rubella virus as indicators of protective immunity in vaccinees and naturally immune individuals. J Infect Dis 1978; 138(6):877-83.
105. Gershon AA, Frey HM, Borkowsky W et al. Live attenuated rubella virus vaccine: Comparison of responses to HPV- 77-DE5 and RA 27/3 strains. Am J Med Sci 1980; 279(2):95-7.
106. Brandling-Bennett AD, Jackson RS, Halstead SB et al. Serologic response to revaccination with two rubella vaccines. Am J Dis Child 1976; 130(10):1081-4.
107. Laurence GD. The production and testing of vaccines in human diploid cells, with special reference to rubella. Postgrad Med J 1972; 48:Suppl 3:24-9.
108. Pugachev KV, Abernathy ES, Frey TK. Genomic sequence of the RA27/3 vaccine strain of rubella virus. Arch Virol 1997; 142(6):1165-80.
109. Katow S, Minahara H, Ota T et al. Identification of strain-specific nucleotide sequences in E1 and NS4 genes of rubella virus vaccine strains in Japan. Vaccine 1997; 15(14):1579-85.
110. Frey TK, Abernathy ES. Identification of strain-specific nucleotide sequences in the RA 27/3 rubella virus vaccine. J Infect Dis 1993; 168(4):854-64.
111. Bayer AS. Arthritis related to rubella: A complication of natural rubella and rubella immunization. Postgrad Med 1980; 67(5):131-4.
112. Spruance SL, Metcalf R, Smith CB et al. Chronic arthropathy associated with rubella vaccination. Arthritis Rheum 1977; 20(2):741-7.
113. Howson CP, Katz M, Johnston RB, Jr. et al. Chronic arthritis after rubella vaccination. Clin Infect Dis 1992; 15(2):307-12.
114. Howson CP, Fineberg HV. Adverse events following pertussis and rubella vaccines. Summary of a report of the Institute of Medicine. Jama 1992; 267(3):392-6.
115. Black S, Shinefield H, Ray P et al. Risk of hospitalization because of aseptic meningitis after measles- mumps-rubella vaccination in one- to two-year-old children: An analysis of the Vaccine Safety Datalink (VSD) Project. Pediatr Infect Dis J 1997; 16(5):500-3.
116. Freestone DS. Vaccination against rubella in Britain: Benefits and risks. Dev Biol Stand 1979; 43:339-48.
117. Massad E, Azevedo-Neto RS, Burattini MN et al. Assessing the efficacy of a mixed vaccination strategy against rubella in Sao Paulo, Brazil. Int J Epidemiol 1995; 24(4):842-50.

118. Villarejos VM, Arguedas JA, Hernandez GC et al. Clinical laboratory evaluation of rubella virus vaccine given to postpartum women without pregnancy preventive. Obstet Gynecol 1973; 42(5):689-95.
119. Katzenstein M. Rubella vaccine—drug or menace. Adv Exp Med Biol 1972; 27:413-7.
120. Banatvala JE, Druce A, Best J et al. Specific IgM responses after rubella vaccination; potential application following inadvertent vaccination during pregnancy. Br Med J 1977; 2(6097):1263-4.
121. Fleet WF, Jr., Benz EW, Jr., Karzon DT et al. Fetal consequences of maternal rubella immunization. Jama 1974; 227(6):621-7.
122. Hayden GF, Herrmann KL, Buimovici-Klein E et al. Subclinical congenital rubella infection associated with maternal rubella vaccination in early pregnancy. J Pediatr 1980; 96(5):869-72.
123. Stuart-Harris C. Rubella immunization in the USA and in Great Britain. Midwives Chron 1980; 93(1111):278-9.
124. Tookey PA, Jones G, Miller BH et al. Rubella vaccination in pregnancy. CDR (Lond Engl Rev) 1991; 1(8):R86-8.
125. Hilleman MR, Weibel RE, Buynak EB et al. Live attenuated mumps-virus vaccine. IV. Protective efficacy as measured in a field evaluation. N Engl J Med 1967; 276(5):252-8.
126. Weibel RE, Stokes J, Jr., Buynak EB et al. Live attenuated mumps-virus vaccine. 3. Clinical and serologic aspects in a field evaluation. N Engl J Med 1967; 276(5):245-51.
127. Hilleman MR, Buynak EB, Weibel RE et al. Live, attenuated mumps-virus vaccine. N Engl J Med 1968; 278(5):227-32.
128. Weibel RE, Buynak EB, McLean AA et al. Long-term follow-up for immunity after monovalent or combined live measles, mumps, and rubella virus vaccines. Pediatrics 1975; 56(3):380-7.
129. Brown EG, Wright KE. Genetic studies on a mumps vaccine strain associated with meningitis. Rev Med Virol 1998; 8(3):129-142.
130. Gluck R, Hoskins JM, Wegmann A et al. Rubini, a new live attenuated mumps vaccine virus strain for human diploid cells. Dev Biol Stand 1986; 65:29-35.
131. Carroll L, Hoenigmann-Stovall N, Whitehead GI, 3rd. Interpersonal consequences of narcissism. Psychol Rep 1996; 79(3 Pt 2):1267-72.
132. Chamot E, Toscani L, Egger P et al. Estimation of the efficacy of three strains of mumps vaccines during an epidemic of mumps in the Geneva canton (Switzerland). Rev Epidemiol Sante Publique 1998; 46(2):100-7.
133. Afzal MA, Minor PD. Immune response and vaccine efficacy letter. Vaccine 1999; 17(15-16):1813.
134. Schwarzer S, Reibel S, Lang AB et al. Safety and characterization of the immune response engendered by two combined measles, mumps and rubella vaccines see comments. Vaccine 1998; 16(2-3):298-304.
135. Schlegel M, Vernazza PL. Immune response and vaccine efficiency letter; comment. Vaccine 1998; 16(13):1256.
136. Boczon K, Winiecka J, Kociecka W et al. The diagnostic value of enzymatic and immunological tests in human trichinellosis. Tropenmed Parasitol 1981; 32(2):109-14.
137. Smorodintsev AA, Nasibov MN, Jakovleva NV. Experience with live rubella virus vaccine combined with live vaccines against measles and mumps. Bull World Health Organ 1970; 42(2):283-9.
138. Beck M, Welsz-Malecek R, Mesko-Prejac M et al. Mumps vaccine L-Zagreb, prepared in chick fibroblasts. I. Production and field trials. J Biol Stand 1989; 17(1):85-90.
139. McTaggart SJ. The two-dose policy for measles vaccination letter. Med J Aust 1994; 161(9):573.
140. Afzal MA, Pickford AR, Forsey T et al. The Jeryl Lynn vaccine strain of mumps virus is a mixture of two distinct isolates published erratum appears in J Gen Virol 1994 Aug;75(Pt 8):2139. J Gen Virol 1993; 74(Pt 5):917-20.
141. Nakayama T, Oka S, Komase K et al. The relationship between the mumps vaccine strain and parotitis after vaccination letter. J Infect Dis 1992; 165(1):186-7.
142. Galazka AM, Robertson SE, Kraigher A. Mumps and mumps vaccine: A global review. Bull World Health Organ 1999; 77(1):3-14.
143. Odisseev H, Gacheva N. Vaccinoprophylaxis of mumps using mumps vaccine, strain Sofia 6, in Bulgaria. Vaccine 1994; 12(14):1251-4.
144. Brown EG, Dimock K, Wright KE. The Urabe AM9 mumps vaccine is a mixture of viruses differing at amino acid 335 of the hemagglutinin-neuraminidase gene with one form associated with disease see comments. J Infect Dis 1996; 174(3):619-22.

145. Afzal MA, Yates PJ, Minor PD. Nucleotide sequence at position 1081 of the hemagglutinin-neuraminidase gene in the mumps Urabe vaccine strain letter; comment. J Infect Dis 1998; 177(1):265-6.
146. Gogel WC, Tietz JD. Absence of compensation and reasoning-like processes in the perception of orientation in depth. Percept Psychophys 1992; 51(4):309-18.
147. Swartz TA, Klingberg W, Klingberg MA. Combined trivalent and bivalent measles, mumps and rubella virus vaccination. A controlled trial. Infection 1974; 2(3):115-7.
148. Buynak EB, Weibel RE, Whitman JE, Jr et al. Combined live measles, mumps, and rubella virus vaccines. Jama 1969; 207(12):2259-62.
149. Beck M, Smerdel S, Dedic I et al. Immune response to Edmonston-Zagreb measles virus strain in monovalent and combined MMR vaccine. Dev Biol Stand 1986; 65:95-100.
150. Wegmann A, Gluck R, Just M et al. Comparative study and evaluation of further attenuated, live measles vaccines alone and in combination with mumps and rubella vaccines. Dev Biol Stand 1986; 65:69-74.
151. Nakayama T, Urano T, Osano M et al. Evaluation of live trivalent vaccine of measles AIK-C strain, mumps Hoshino strain and rubella Takahashi strain, by virus-specific interferon-gamma production and antibody response. Microbiol Immunol 1990; 34(6):497-508.
152. Boulianne N, De Serres G, Ratnam S et al. Measles, mumps, and rubella antibodies in children 5-6 years after immunization: Effect of vaccine type and age at vaccination. Vaccine 1995; 13(16):1611-6.
153. Just M, Berger R, Gluck R et al. Evaluation of a combined vaccine against measles-mumps-rubella produced on human diploid cells. Dev Biol Stand 1986; 65:25-7.
154. Schell K, Kenny MT, Jackson JE. Serological evaluation of measles virus (Schwarz strain), mumps virus (Jeryl Lynn strain) and rubella virus (Cendehill strain) combination vaccines. J Biol Stand 1975; 3(2):231-9.
155. Walters VW, Miller SA, Jackson JE et al. A field with a liver measles-mumps-rubella vaccine. Clin Pediatr (Phila) 1975; 14(10):928-33.
156. Bloom JL, Schiff GM, Graubarth H et al. Evaluation of a trivalent measles, mumps, rubella vaccine in children. J Pediatr 1975; 87(1):85-7.
157. Robertson CM, Bennett VJ, Jefferson N, Mayon-White RT. Serological evaluation of a measles, mumps, and rubella vaccine. Arch Dis Child 1988; 63(6):612-6.
158. Ozaki K, Kuno-Sakai H, Kimura M. Evaluation of mumps antibodies after measles-mumps-rubella (MMR) vaccines. Tokai J Exp Clin Med 1987; 12(5-6):305-11.
159. Landrigan PJ, Murphy KB, Meyer HM, Jr et al. Combined measles-rubella vaccines. Virus dose and serologic response. Am J Dis Child 1973; 125(1):65-7.
160. Minekawa Y, Takahashi M, Ueda S et al. Studies on live rubella vaccine. VI. Seroepidemiological surveillance and immunization of adolescents in high school. Biken J 1975; 18(3):149-54.
161. Reich M. A prescription for taking medicine properly. J Am Osteopath Assoc 1988; 88(11):1421-4.
162. Weibel RE, Carlson AJ, Jr., Villarejos VM et al. Clinical and laboratory studies of combined live measles, mumps, and rubella vaccines using the RA 27/3 rubella virus. Proc Soc Exp Biol Med 1980; 165(2):323-6.
163. Davidkin I, Valle M. Vaccine-induced measles virus antibodies after two doses of combined measles, mumps and rubella vaccine: A 12-year follow-up in two cohorts. Vaccine 1998; 16(20):2052-7.
164. Miller E, Hill A, Morgan-Capner P et al. Antibodies to measles, mumps and rubella in UK children 4 years after vaccination with different MMR vaccines. Vaccine 1995; 13(9):799-802.
165. Weibel RE, Buynak EB, McLean AA et al. Persistence of antibody in human subjects for 7 to 10 years following administration of combined live attenuated measles, mumps, and rubella virus vaccines. Proc Soc Exp Biol Med 1980; 165(2):260-3.
166. Goncalves G, Tavares F, de Andrade HR. A seroconversion study of the measles component of the MMR vaccine in adolescents of the town of Sabrosa. Acta Med Port 1998; 11(12):1079-84.
167. Nishiwaki H, Nishimori T, Teramura M et al. Results of long follow-up study for peptic ulcer patients treated with gastrectomy. Nippon Geka Gakkai Zasshi 1988; 89(11):1789-95.
168. Katz SL. Harmony on the second dose of measles vaccine comment. Pediatrics 1997; 100(5):891-2.
169. Davis RL, Marcuse E, Black S et al. MMR2 immunization at 4 to 5 years and 10 to 12 years of age: A comparison of adverse clinical events after immunization in the Vaccine Safety Datalink project. The Vaccine Safety Datalink Team see comments. Pediatrics 1997; 100(5):767-71.
170. Heath TC, Burgess MA, Forrest JM. Moving the second dose of measles-mumps-rubella vaccine to school entry: Implications for control of rubella. Commun Dis Intell 1998; 22(8):157-8.

171. Miller E, Nokes DJ, Anderson RM. Measles, mumps, and rubella vaccination letter. Bmj 1992; 304(6839):1440-1.
172. Watson JC, Hadler SC, Dykewicz CA et al. Measles, mumps, and rubella—vaccine use and strategies for elimination of measles, rubella, and congenital rubella syndrome and control of mumps: Recommendations of the Advisory Committee on Immunization Practices (ACIP). MMWR Morb Mortal Wkly Rep 1998; 47(RR-8):1-57.
173. Zach K, Nicoara C, Germann D et al. Age-related seroprevalence of measles, mumps and rubella antibodies in 1996. Schweiz Med Wochenschr 1998; 128(17):649-57.
174. Schlenker TL, Bain C, Baughman AL et al. Measles herd immunity. The association of attack rates with immunization rates in preschool children see comments. Jama 1992; 267(6):823-6.
175. McAleer WJ, Buynak EB, Weibel RE et al. Measles, mumps and rubella virus vaccines prepared from virus produced by unit process. J Biol Stand 1975; 3(4):381-4.
176. Kemp A, Van Asperen P, Mukhi A. Measles immunization in children with clinical reactions to egg protein see comments. Am J Dis Child 1990; 144(1):33-5.
177. Beck SA, Williams LW, Shirrell MA et al. Egg hypersensitivity and measles-mumps-rubella vaccine administration. Pediatrics 1991; 88(5):913-7.
178. Fasano MB, Wood RA, Cooke SK et al. Egg hypersensitivity and adverse reactions to measles, mumps, and rubella vaccine. J Pediatr 1992; 120(6):878-81.
179. Bruno G, Grandolfo M, Lucenti P et al. Measles vaccine in egg allergic children: Poor immunogenicity of the Edmoston-Zagreb strain. Pediatr Allergy Immunol 1997; 8(1):17-20.
180. Bruno G, Giampietro PG, Grandolfo ME et al. Safety of measles immunisation in children with IgE-mediated egg allergy letter. Lancet 1990; 335(8691):739.
181. Trotter AC, Stone BD, Laszlo DJ et al. Measles, mumps, rubella vaccine administration in egg-sensitive children: Systemic reactions during vaccine desensitization see comments. Ann Allergy 1994; 72(1):25-8.
182. Davidkin I, Valle M, Peltola H et al. Etiology of measles- and rubella-like illnesses in measles, mumps, and rubella-vaccinated children. J Infect Dis 1998; 178(6):1567-70.
183. Wakefield AJ, Murch SH, Anthony A et al. Ileal-lymphoid-nodular hyperplasia, non-specific colitis, and pervasive developmental disorder in children see comments. Lancet 1998; 351(9103):637-41.
184. Lewin J, Dhillon AP, Sim R et al. Persistent measles virus infection of the intestine: Confirmation by immunogold electron microscopy. Gut 1995; 36(4):564-9.
185. Chadwick N, Bruce IJ, Schepelmann S et al. Measles virus RNA is not detected in inflammatory bowel disease using hybrid capture and reverse transcription followed by the polymerase chain reaction. J Med Virol 1998; 55(4):305-11.
186. Chen RT, DeStefano F. Vaccine adverse events: Casual or coincidental? comment see comments. Lancet 1998; 351(9103):611-2.
187. Wakefield AJ, Pittilo RM, Sim R et al. Evidence of persistent measles virus infection in Crohn's disease. J Med Virol 1993; 39(4):345-53.
188. Haga Y, Funakoshi O, Kuroe K et al. Absence of measles viral genomic sequence in intestinal tissues from Crohn's disease by nested polymerase chain reaction see comments. Gut 1996; 38(2):211-5.
189. Bower H. New research demolishes link between MMR vaccine and autism news. Bmj 1999; 318(7199):1643.
190. Akobeng AK, Thomas AG. Inflammatory bowel disease, autism, and the measles, mumps, and rubella vaccine. J Pediatr Gastroenterol Nutr 1999; 28(3):351-2.
191. Stokkers PC, van den Berg M, Rings E. Measles vaccination and inflammatory bowel disease. J Pediatr Gastroenterol Nutr 1998; 27(2):239-40.
192. DeStefano F, Chen RT. Negative association between MMR and autism comment. Lancet 1999; 353(9169):1987-8.
193. Peltola H, Patja A, Leinikki P et al. No evidence for measles, mumps, and rubella vaccine-associated inflammatory bowel disease or autism in a 14-year prospective study letter. Lancet 1998; 351(9112):1327-8.
194. Nakayama T, Aizawa C, Kuno-Sakai H. A clinical analysis of gelatin allergy and determination of its causal relationship to the previous administration of gelatin-containing acellular pertussis vaccine combined with diphtheria and tetanus toxoids see comments. J Allergy Clin Immunol 1999; 103(2 Pt 1):321-5.
195. Kelso JM. The gelatin story editorial; comment. J Allergy Clin Immunol 1999; 103(2 Pt 1):200-2.

196. Brunell PA, Novelli VM, Lipton SV et al. Combined vaccine against measles, mumps, rubella, and varicella. Pediatrics 1988; 81(6):779-84.
197. Reuman PD, Sawyer MH, Kuter BJ et al. Safety and immunogenicity of concurrent administration of measles-mumps- rubella-varicella vaccine and PedvaxHIB vaccines in healthy children twelve to eighteen months old. The MMRV Study Group. Pediatr Infect Dis J 1997; 16(7):662-7.
198. Mori C, Tooriyama T, Imagawa T et al. Nucleotide sequence at position 1081 of the hemagglutinin-neuraminidase gene in the mumps virus Urabe vaccine strain letter; comment see comments. J Infect Dis 1997; 175(6):1548-9.
199. Vesikari T, Ohrling A, Baer M et al. Evaluation of live attenuated varicella vaccine (Oka-RIT strain) and combined varicella and MMR vaccination in 13-17-month-old children. Acta Paediatr Scand 1991; 80(11):1051-7.
200. Arbeter AM, Baker L, Starr SE et al. Combination measles, mumps, rubella and varicella vaccine. Pediatrics 1986; 78(4 Pt 2):742-7.
201. Arbeter AM, Baker L, Starr SE et al. The combination measles, mumps, rubella and varicella vaccine in healthy children. Dev Biol Stand 1986; 65:89-93.
202. Just M, Berger R, Just V. Evaluation of a combined measles-mumps-rubella-chickenpox vaccine. Dev Biol Stand 1986; 65:85-8.
203. Shinefield HR, Black SB, Staehle BO et al. Safety, tolerability and immunogenicity of concomitant injections in separate locations of M-M-R II, VARIVAX and TETRAMUNE in healthy children vs. concomitant injections of M-M-R II and TETRAMUNE followed six weeks later by VARIVAX. Pediatr Infect Dis J 1998; 17(11):980-5.
204. Berger R, Just M. Interference between strains in live virus vaccines. II: Combined vaccination with varicella and measles-mumps-rubella vaccine. J Biol Stand 1988; 16(4):275-9.
205. Takahashi S, Machikawa F, Noda A et al. Detection of immunoglobulin G and A antibodies to rubella virus in urine and antibody responses to vaccine-induced infection. Clin Diagn Lab Immunol 1998; 5(1):24-7.
206. Neff BJ, Weibel RE, Villarejos VM et al. Clinical and laboratory studies of KMcC strain live attenuated varicella virus. Proc Soc Exp Biol Med 1981; 166(3):339-47.
207. LaRussa P, Steinberg S, Arvin A et al. Polymerase chain reaction and restriction fragment length polymorphism analysis of varicella-zoster virus isolates from the United States and other parts of the world. J Infect Dis 1998; 178 Suppl 1:S64-6.
208. LaRussa P, Steinberg S, Meurice F et al. Transmission of vaccine strain varicella-zoster virus from a healthy adult with vaccine-associated rash to susceptible household contacts. J Infect Dis 1997; 176(4):1072-5.
209. Salzman MB, Sharrar RG, Steinberg S et al. Transmission of varicella-vaccine virus from a healthy 12-month-old child to his pregnant mother see comments. J Pediatr 1997; 131(1 Pt 1):151-4.
210. Kinchington PR, Ling P, Pensiero M et al. The glycoprotein products of varicella-zoster virus gene 14 and their defective accumulation in a vaccine strain (Oka). J Virol 1990; 64(9):4540-8.
211. Ling P, Kinchington PR, Ruyechan WT et al. A detailed analysis of transcripts mapping to varicella zoster virus gene 14 (glycoprotein V). Virology 1991; 184(2):625-35.
212. Kinchington PR, Ling P, Pensiero M et al. A possible role for glycoprotein gpV in the pathogenesis of varicella- zoster virus. Adv Exp Med Biol 1990; 278:83-91.
213. White CJ. Clinical trials of varicella vaccine in healthy children. Infect Dis Clin North Am 1996; 10(3):595-608.
214. Assouline JG, Levin MJ, Major EO et al. Varicella-zoster virus infection of human astrocytes, Schwann cells, and neurons. Virology 1990; 179(2):834-44.
215. Moffat JF, Zerboni L, Sommer MH et al. The ORF47 and ORF66 putative protein kinases of varicella-zoster virus determine tropism for human T cells and skin in the SCID-hu mouse. Proc Natl Acad Sci U S A 1998; 95(20):11969-74.
216. Hall S, Galil K, Watson B et al. The use of school-based vaccination clinics to control varicella outbreaks in two schools. Pediatrics 2000; 105(1):e17.
217. White CJ, Stinson D, Staehle B et al. Measles, mumps, rubella, and varicella combination vaccine: Safety and immunogenicity alone and in combination with other vaccines given to children. Measles, Mumps, Rubella, Varicella Vaccine Study Group. Clin Infect Dis 1997; 24(5):925-31.
218. Watson BM, Laufer DS, Kuter BJ et al. Safety and immunogenicity of a combined live attenuated measles, mumps, rubella, and varicella vaccine (MMR(II)V) in healthy children published erratum appears in J Infect Dis 1996 Jun;173(6):1529. J Infect Dis 1996; 173(3):731-4.

219. Miller AE. Selective decline in cellular immune response to varicella-zoster in the elderly. Neurology 1980; 30(6):582-7.
220. Russell AS, Maini RA, Bailey M et al. Cell-mediated immunity to Varicella-Zoster antigen in acute Herpes zoster (shingles). Clin Exp Immunol 1973; 14(2):181-5.
221. Levin MJ, Barber D, Goldblatt E et al. Use of a live attenuated varicella vaccine to boost varicella-specific immune responses in seropositive people 55 years of age and older: Duration of booster effect. J Infect Dis 1998; 178 Suppl 1:S109-12.
222. Levin MJ, Hayward AR. The varicella vaccine. Prevention of herpes zoster. Infect Dis Clin North Am 1996; 10(3):657-75.
223. Levin MJ, Murray M, Rotbart HA et al. Immune response of elderly individuals to a live attenuated varicella vaccine. J Infect Dis 1992; 166(2):253-9.
224. Hayward A, Levin M, Wolf W et al. Varicella-zoster virus-specific immunity after herpes zoster. J Infect Dis 1991; 163(4):873-5.
225. Gershon A, Silverstein S. Live attenuated varicella vaccine for prevention of herpes zoster. Biologicals 1997; 25(2):227-30.
226. Zhang M, Zeng CQ, Dong Y et al. Mutations in rotavirus nonstructural glycoprotein NSP4 are associated with altered virus virulence. J Virol 1998; 72(5):3666-72.
227. Dong Y, Zeng CQ, Ball JM et al. The rotavirus enterotoxin NSP4 mobilizes intracellular calcium in human intestinal cells by stimulating phospholipase C-mediated inositol 1,4,5- trisphosphate production. Proc Natl Acad Sci U S A 1997; 94(8):3960-5.
228. Tian P, Hu Y, Schilling WP et al. The nonstructural glycoprotein of rotavirus affects intracellular calcium levels. J Virol 1994; 68(1):251-7.
229. Ball JM, Tian P, Zeng CQ et al. Age-dependent diarrhea induced by a rotaviral nonstructural glycoprotein see comments. Science 1996; 272(5258):101-4.
230. Matson DO. Protective immunity against group A rotavirus infection and illness in infants. Arch Virol Suppl 1996; 12:129-39.
231. Ward RL, Bernstein DI. Lack of correlation between serum rotavirus antibody titers and protection following vaccination with reassortant RRV vaccines. US Rotavirus Vaccine Efficacy Group. Vaccine 1995; 13(13):1226-32.
232. Hoshino Y, Kapikian AZ. Classification of rotavirus VP4 and VP7 serotypes. Arch Virol Suppl 1996; 12:99-111.
233. Nomenclature of human rotaviruses: Designation of subgroups and serotypes. Bull World Health Organ 1984; 62(3):501-7.
234. Vesikari T, Isolauri E, D'Hondt E et al. Protection of infants against rotavirus diarrhoea by RIT 4237 attenuated bovine rotavirus strain vaccine. Lancet 1984; 1(8384):977-81.
235. Vesikari T, Isolauri E, Delem A et al. Clinical efficacy of the RIT 4237 live attenuated bovine rotavirus vaccine in infants vaccinated before a rotavirus epidemic. J Pediatr 1985; 107(2):189-94.
236. Maldonado Y, Hestvik L, Wilson M et al. Safety and immunogenicity of bovine rotavirus vaccine RIT 4237 in 3- month-old infants. J Pediatr 1986; 109(6):931-5.
237. Vesikari T, Ruuska T, Delem A et al. Neonatal rotavirus vaccination with RIT 4237 bovine rotavirus vaccine: A preliminary report. Pediatr Infect Dis J 1987; 6(2):164-9.
238. Hanlon P, Hanlon L, Marsh V et al. Trial of an attenuated bovine rotavirus vaccine (RIT 4237) in Gambian infants. Lancet 1987; 1(8546):1342-5.
239. Christy C, Madore HP, Pichichero ME et al. Field trial of rhesus rotavirus vaccine in infants. Pediatr Infect Dis J 1988; 7(9):645-50.
240. Vesikari T. Clinical and immunological studies of rotavirus vaccines. Southeast Asian J Trop Med Public Health 1988; 19(3):437-47.
241. Piperno G, Mead K. Transport of a novel complex in the cytoplasmic matrix of Chlamydomonas flagella. Proc Natl Acad Sci U S A 1997; 94(9):4457-62.
242. Clark HF, Furukawa T, Bell LM et al. Immune response of infants and children to low-passage bovine rotavirus (strain WC3). Am J Dis Child 1986; 140(4):350-6.
243. Clark HF, Borian FE, Bell LM et al. Protective effect of WC3 vaccine against rotavirus diarrhea in infants during a predominantly serotype 1 rotavirus season. J Infect Dis 1988; 158(3):570-87.
244. Bernstein DI, Smith VE, Sander DS et al. Evaluation of WC3 rotavirus vaccine and correlates of protection in healthy infants. J Infect Dis 1990; 162(5):1055-62.
245. Georges-Courbot MC, Monges J, Siopathis MR et al. Evaluation of the efficacy of a low-passage bovine rotavirus (strain WC3) vaccine in children in Central Africa. Res Virol 1991; 142(5):405-11.

246. Vesikari T, Kapikian AZ, Delem A et al. A comparative trial of rhesus monkey (RRV-1) and bovine (RIT 4237) oral rotavirus vaccines in young children. J Infect Dis 1986; 153(5):832-9.
247. Flores J, Perez-Schael I, Gonzalez M et al. Protection against severe rotavirus diarrhoea by rhesus rotavirus vaccine in Venezuelan infants. Lancet 1987; 1(8538):882-4.
248. Losonsky GA, Rennels MB, Lim Y et al. Systemic and mucosal immune responses to rhesus rotavirus vaccine MMU 18006. Pediatr Infect Dis J 1988; 7(6):388-93.
249. Vesikari T. Rotavirus vaccines against diarrhoeal disease. Lancet 1997; 350(9090):1538-41.
250. Ward RL, Sander DS, Schiff GM et al. Effect of vaccination on serotype-specific antibody responses in infants administered WC3 bovine rotavirus before or after a natural rotavirus infection. J Infect Dis 1990; 162(6):1298-303.
251. Green KY, Taniguchi K, Mackow ER et al. Homotypic and heterotypic epitope-specific antibody responses in adult and infant rotavirus vaccinees: Implications for vaccine development. J Infect Dis 1990; 161(4):667-79.
252. Bernstein DI, Sander DS, Smith VE et al. Protection from rotavirus reinfection: 2-year prospective study. J Infect Dis 1991; 164(2):277-83.
253. Clark HF, Borian FE, Plotkin SA. Immune protection of infants against rotavirus gastroenteritis by a serotype 1 reassortant of bovine rotavirus WC3. J Infect Dis 1990; 161(6):1099-104.
254. Madore HP, Christy C, Pichichero M et al. Field trial of rhesus rotavirus or human-rhesus rotavirus reassortant vaccine of VP7 serotype 3 or 1 specificity in infants. The Elmwood, Panorama, and Westfall Pediatric Groups. J Infect Dis 1992; 166(2):235-43.
255. Pichichero ME, Marsocci SM, Francis AB et al. A comparative evaluation of the safety and immunogenicity of a single dose of unbuffered oral rhesus rotavirus serotype 3, rhesus/human reassortant serotypes 1, 2 and 4 and combined (tetravalent) vaccines in healthy infants. Vaccine 1993; 11(7):747-53.
256. Treanor JJ, Clark HF, Pichichero M et al. Evaluation of the protective efficacy of a serotype 1 bovine-human rotavirus reassortant vaccine in infants. Pediatr Infect Dis J 1995; 14(4):301-7.
257. Kapikian AZ, Hoshino Y, Chanock RM, Perez-Schael I. Efficacy of a quadrivalent rhesus rotavirus-based human rotavirus vaccine aimed at preventing severe rotavirus diarrhea in infants and young children. J Infect Dis 1996; 174 Suppl 1:S65-72.
258. Vesikari T. Rotavirus vaccine studies in Europe. Acta Paediatr Suppl 1999; 88(426):9-13.
259. Vesikari T, Joensuu J. Review of rotavirus vaccine trials in Finland. J Infect Dis 1996; 174 Suppl 1:S81-7.
260. Perez-Schael I, Guntinas MJ, Perez M et al. Efficacy of the rhesus rotavirus-based quadrivalent vaccine in infants and young children in Venezuela see comments published erratum appears in N Engl J Med 1998 Apr 2;338(14):1002. N Engl J Med 1997; 337(17):1181-7.
261. Lanata CF, Black RE, Flores J et al. Immunogenicity, safety and protective efficacy of one dose of the rhesus rotavirus vaccine and serotype 1 and 2 human-rhesus rotavirus reassortants in children from Lima, Peru. Vaccine 1996; 14(3):237-43.
262. Tarpey I, Stacey SN, McIndoe A, Davies DH. Priming in vivo and quantification in vitro of class I MHC-restricted cytotoxic T cells to human papilloma virus type 11 early proteins (E6 and E7) using immunostimulating complexes (ISCOMs). Vaccine 1996; 14(3):230-6.
263. Rennels MB, Losonsky GA, Young AE et al. An efficacy trial of the rhesus rotavirus vaccine in Maryland. The Clinical Study Group. Am J Dis Child 1990; 144(5):601-4.
264. Joensuu J, Koskenniemi E, Vesikari T. Symptoms associated with rhesus-human reassortant rotavirus vaccine in infants. Pediatr Infect Dis J 1998; 17(4):334-40.
265. Rennels MB, Losonsky GA, Shindledecker CL et al. Immunogenicity and reactogenicity of lowered doses of rhesus rotavirus vaccine strain MMU 18006 in young children. Pediatr Infect Dis J 1987; 6(3):260-4.
266. Clark HF, Offit PA, Ellis RW et al. The development of multivalent bovine rotavirus (strain WC3) reassortant vaccine for infants. J Infect Dis 1996; 174 Suppl 1:S73-80.
267. Edmonds ME. Progress in care of the diabetic foot. Lancet 1999; 354(9175):270-2.
268. Barnes GL, Lund JS, Adams L et al. Phase 1 trial of a candidate rotavirus vaccine (RV3) derived from a human neonate. J Paediatr Child Health 1997; 33(4):300-4.
269. Clements-Mann ML, Makhene MK, Mrukowicz J et al. Safety and immunogenicity of live attenuated human-bovine (UK) reassortant rotavirus vaccines with VP7-specificity for serotypes 1, 2, 3 or 4 in adults, children and infants. Vaccine 1999; 17(20-21):2715-25.

270. Rennels MB, Parashar UD, Holman RC et al. Lack of an apparent association between intussusception and wild or vaccine rotavirus infection. Pediatr Infect Dis J 1998; 17(10):924-5.
271. Tucker AW, Haddix AC, Bresee JS et al. Cost-effectiveness analysis of a rotavirus immunization program for the United States. Jama 1998; 279(17):1371-6.
272. Zhou NN, Senne DA, Landgraf JS et al. Genetic reassortment of avian, swine, and human influenza A viruses in American pigs. J Virol 1999; 73(10):8851-6.
273. Richman DD. Use of temperature-sensitive mutants for live, attenuated influenza- virus vaccines editorial. N Engl J Med 1979; 300(3):137-8.
274. Tyrrell DA. Using the genetics of influenza virus to make live attenuated vaccines. Lancet 1978; 1(8057):196-7.
275. Morris CA, Freestone DS, Stealey VM et al. Recombinant WRL 105 strain live attenuated influenza vaccine. Immunogenicity, reactivity, and transmissibility. Lancet 1975; 2(7927):196-9.
276. Evans AE, Letley E, Ferris RD et al. WRL 105 strain live attenuated influenza vaccine; comparison of one and two dose schedules. J Hyg (Lond) 1976; 77(3):327-32.
277. Nicholson KG, Tyrrell DA, Freestone DS. WRL 105 strain (H3N2) live attenuated influenza vaccine: Acceptability, reactivity, and antibody response in normal, bronchitic, and geriatric volunteers. Lancet 1976; 1(7973):1309-11.
278. Moffat MA, Stealey VM, Freestone DS et al. Assessment of elicited antibody responses, clinical reactions and transmissibility of WRL 105 live influenza vaccine. J Biol Stand 1976; 4(2):91-5.
279. White WG, Freestone DS, Bowker CH et al. A clinical trial of WRL 105 strain live attenuated influenza vaccine comparing four methods of intranasal vaccination. Dev Biol Stand 1976; 33:202-6.
280. Florent G. Gene constellation of live influenza A vaccines. Arch Virol 1980; 64(2):171-3.
281. Beare AS, Schild GC, Hall TS et al. Antigenic characteristics of swine influenza virus closely related to human Hong Kong strain and results of experimental infection in volunteers. Lancet 1971; 1(7694):305-7.
282. Beare AS, Hall TS. Recombinant influenza-A viruses as live vaccines for man. Report to the Medical Research Council's Committee on Influenza and other Respiratory Virus Vaccines. Lancet 1971; 2(7737):1271-3.
283. Spencer MJ, Cherry JD, Powell KR et al. Clinical trials with Alice strain, live, attenuated, serum inhibitor- resistant intranasal influenza A vaccine. J Infect Dis 1975; 132(4):415-20.
284. Spencer MJ, Cherry JD, Powell KR. Clinical trial with "R-75" strain live, attenuated, serum inhibitor- resistant intranasal influenza B vaccine. J Clin Microbiol 1977; 5(6):584-7.
285. Miller LW, Hume EB, O'Brien FR et al. Alice strain live attenuated influenza (H3N2) vaccin in an elderly population. Am J Epidemiol 1975; 101(4):340-6.
286. Minor TE, Dick EC, Dick CR et al. Attenuated influenza A vaccine (Alice) in an adult population: Vaccine- related illness, serum and nasal antibody production, and intrafamily transmission. J Clin Microbiol 1975; 2(5):403-9.
287. Hall CB, Douglas G, Jr., Fralonardo SA. Live attenuated influenza virus vaccine trial in children. Pediatrics 1975; 56(6):991-8.
288. Delem A. Protective efficacy of RIT 4025, a live attenuated influenza vaccine strain, and evaluation of heterotypic immunity to influenza A viruses in ferrets. J Hyg (Lond) 1977; 79(2):203-8.
289. Douglas RG, Jr., Betts RF. Studies with live attenuated influenza vaccines. Am J Clin Pathol 1978; 70(1 Suppl):153-8.
290. Smorodintsev. J. Micro. Epidemiol 1954; 9:22.
291. Zhdanov VM. First International Conference on Vacines Against Viral and Rickettsial Diseases of Man. Book 1967.
292. Mills J, Chanock V, Chanock RM. Temperature-sensitive mutants of influenza virus. I. Behavior in tissue culture and in experimental animals. J Infect Dis 1971; 123(2):145-57.
293. Murphy BR, Wood FT, Massicot JG et al. Temperature-sensitive mutants of influenza virus. XVI. Transfer of the two ts lesions present in the Udorn/72-ts-1A2 donor virus to the Victoria/3/75 wild-type virus. Virology 1978; 88(2):244-51.
294. Murphy BR, Wood FT, Massicot JG et al. Temperature-sensitive mutants of influenza virus. XV. The genetic and biological characterization of a recombinant influenza virus containing two ts lesions produced by mating two complementing, single lesion ts mutants. Virology 1978; 88(2):231-43.
295. Murphy BR, Chalhub EG, Nusinoff SR et al. Temperature-sensitive mutants of influenza virus. 3. Further characterization of the ts-1(E) influenza A recombinant (H3N2) virus in man. J Infect Dis 1973; 128(4):479-87.

296. Murphy BR, Chalhub EG, Nusinoff SR et al. Temperature-sensitive mutants of influenza virus. II. Attenuation of ts recombinants for man. J Infect Dis 1972; 126(2):170-8.
297. Palese P, Ritchey MB. Live attenuated influenza virus vaccines. Strains with temperature- sensitive defects in P3 protein and nucleoprotein. Virology 1977; 78(1):183-91.
298. Murphy BR, Holley HP, Jr., Berquist EJ et al. Cold-adapted variants of influenza A virus: Evaluation in adult seronegative volunteers of A/Scotland/840/74 and A/Victoria/3/75 cold-adapted recombinants derived from the cold-adapted A/Ann Arbor/6/60 strain. Infect Immun 1979; 23(2):253-9.
299. Murphy BR, Chanock RM, Levine MM et al. Temperature-sensitive mutants of influenza A virus: Evaluation of the A/Victoria/75-ts-1A2 temperature-sensitive recombinant virus in seronegative adult volunteers. Infect Immun 1979; 23(2):249-52.
300. Markoff LJ, Thierry F, Murphy BR et al. Relationship of genotype of recombinants of influenza A/Hong Kong/68-ts- 1Evirus used as live virus vaccines to virulence in humans. Infect Immun 1979; 26(1):280-6.
301. Massicot JG, Murphy BR, van Wyke K et al. ts P1 and P3 genes are responsible for satisfactory level of attenuation of ts-1A2 recombinants bearing H1N1 or H3N2 surface antigens of influenza A virus. Virology 1980; 106(1):187-90.
302. Murphy BR, Rennels MB, Douglas RG, Jr. et al. Evaluation of influenza A/Hong Kong/123/77 (H1N1) ts-1A2 and cold- adapted recombinant viruses in seronegative adult volunteers. Infect Immun 1980; 29(2):348-55.
303. Mostow SR, Hopkins JA, Wright PF. Behavior of vaccine revertants of temperature-sensitive mutants of influenza virus in ferret tracheal organ culture. Infect Immun 1979; 26(1):193-6.
304. Richman DD, Murphy BR, Chanock RM. Demonstration of a non-temperature-sensitive growth-restricting mutation in a ts mutant of influenza A virus: Implications for live virus vaccine development. Virology 1977; 83(2):356-64.
305. Maassab HF. Adaptation and growth characteristics of influenza virus at 25 degrees C. Nature 1967; 213(76):612-4.
306. Maassab HF. Plaque formation of influenza virus at 25 degrees C. Nature 1968; 219(154):645-6.
307. Maassab HF. Biologic and immunologic characteristics of cold-adapted influenza virus. J Immunol 1969; 102(3):728-32.
308. Davenport FM, Hennessy AV, Maassab HF et al. Pilot studies on recombinant cold-adapted live type A and B influenza virus vaccines. J Infect Dis 1977; 136(1):17-25.
309. Subbarao EK, Perkins M, Treanor JJ et al. The attenuation phenotype conferred by the M gene of the influenza A/Ann Arbor/6/60 cold-adapted virus (H2N2) on the A/Korea/82 (H3N2) reassortant virus results from a gene constellation effect. Virus Res 1992; 25(1-2):37-50.
310. Cox NJ, Kitame F, Kendal AP et al. Identification of sequence changes in the cold-adapted, live attenuated influenza vaccine strain, A/Ann Arbor/6/60 (H2N2). Virology 1988; 167(2):554-67.
311. Miyazaki C, Nakayama M, Tanaka Y et al. Immunization of institutionalized asthmatic children and patients with psychomotor retardation using live attenuated cold-adapted reassortment influenza A H1N1, H3N2 and B vaccines. Vaccine 1993; 11(8):853-8.
312. Murphy, 1993, Infectious Diseases in Clinical Practice, 2;174-181
313. Rudenko LG, Slepushkin AN, Monto AS et al. Efficacy of live attenuated and inactivated influenza vaccines in schoolchildren and their unvaccinated contacts in Novgorod, Russia. J Infect Dis 1993; 168(4):881-7.
314. Nicholson KG, Tyrrell DA, Oxford JS et al. Infectivity and reactogenicity of reassortant cold-adapted influenza A/Korea/1/82 vaccines obtained from the USA and USSR. Bull World Health Organ 1987; 65(3):295-301.
315. Basler CF, Garcia-Sastre A, Palese P. Mutation of neuraminidase cysteine residues yields temperature- sensitive influenza viruses In Process Citation. J Virol 1999; 73(10):8095-103.
316. Belshe RB, Mendelman PM, Treanor J et al. The efficacy of live attenuated, cold-adapted, trivalent, intranasal influenzavirus vaccine in children see comments. N Engl J Med 1998; 338(20):1405-12.
317. Kessler H. Live attenuated influenza virus vaccine letter. Lancet 1980; 2(8191):431-2.
318. Nichol KL, Mendelman PM, Mallon KP et al. Effectiveness of live, attenuated intranasal influenza virus vaccine in healthy, working adults: A randomized controlled trial, see comments. Jama 1999; 282(2):137-44.

Chapter 6

Recombinant Live Attenuated Viral Vaccines

Richard R. Spaete

Advantages and Concerns Associated with the Use of Live Attenuated Vaccines

Vaccination with live attenuated viruses in general offers a number of advantages as a strategy to evoke an effective and long lasting immune response. Foremost among these is the possibility of a nearly complete presentation of the antigenic repertoire of the pathogen to the immune system. Both structural and non-structural antigens are available to stimulate humoral and cell-mediated immune responses in association with either class I or class II HLA molecules. Antigenic structures will be appropriately presented in authentic states of conformation and oligomerization. Taking advantage of the infectious nature of the virus obviates the need for an adjuvant/antigen delivery system. Obviously, the ability of the vaccine virus to amplify will also increase the antigenic dose and influence the magnitude of the immune response. The ability to amplify depends on the state of attenuation or the nature of the block to replication (as with DISC viruses discussed below).

Delivery of vaccines via the pathogen's natural portal of entry including mucosal and oral routes, will be increasingly used in the future. By mimicking the route or portal of infection, this mode of administration: 1) stimulates immunity at the portal of entry, and 2) stimulates multiple components of the immune system. Intranasal administration offers the possibility of achieving protection of the upper and lower respiratory tract. Intranasal delivery can also stimulate immunity in the urogenital tract. In common with vaccine delivery by parenteral routes, these vaccines also present multiple protective antigens in their native conformation. Another attractive feature of this mode of delivery is that needles are not necessary for administration. This feature is especially popular in pediatric populations!

In comparison with other vaccine approaches, live vaccines also have the advantage of relatively low cost of manufacture. The ability of the vaccines to replicate is the primary reason for this advantage. In many instances, one eucaryotic cell can yield several doses of vaccine. Costs associated with product characterization and general safety testing are common to any vaccine approach and live vaccines do not a priori incur any additional cost burden. Necessarily, the nature of the tests will be unique to the product depending on whether the vaccine is a subunit, a killed virus or a polynucleotide vaccine. In some cases, the benefit to the consumer may be economic as well as improved compliance resulting from the need for fewer doses with live attenuated vaccines.

Finally, and most importantly on the advantage side of the ledger, is the fact that they work. With a few exceptions, the public health benefits that have been gained in preventing viral infectious diseases have been won by use of live vaccine approaches. Smallpox has been eliminated. Measles, mumps, rubella and poliovirus have been effectively controlled with live attenuated virus vaccines. The success of three herpesvirus vaccines is outlined below.

New Vaccine Technologies, edited by Ronald W. Ellis. ©2001 Eurekah.com.

Concerns associated with the use of live vaccines usually revolve around the issue of genetic stability. Reversions to virulence are one potential manifestation of genetic instability. These reversions can result from repair of mutations responsible for attenuation or from compensatory mutations elsewhere in the genome that restore some measure of virulence. Another source of instability is the potential for recombination with wild type viruses. Using recombinant vaccine candidates with deletions or mutations at widely separated loci will guard against both of these concerns. Deletions will prevent repair, and when widely separated in the genome, will ensure that any recombinant will be attenuated. RNA viruses with segmented genomes offer a special case where reassortment of segments can lead to recombinants with mixed parentage. Again, attenuated vaccines will carry multiple mutations on disparate segments so that it is highly unlikely that the process of reassortment in a mixed infection will lead to a virulent virus.

A possible example of inadequate attenuation is that of vaccination-associated mumps meningitis following vaccination with preparations containing the Urabe strain of the mumps virus. Although not definitively causally linked to this sequelae, combined measles/mumps/rubella (MMR) vaccines containing the Urabe strain have been withdrawn from use in several countries. Judgements differ as to the risk-benefit ratio of including this strain in MMR vaccines for the prevention of mumps as opposed to the inclusion of other vaccine strains of the mumps virus for this vaccination preventable disease.

The concept of creating a virus more virulent than the original wild type assumes that the vaccine will introduce genes or alterations not present in the wild type population which confer some selective advantage in overcoming the host response. For example, one can imagine vaccine viruses carrying foreign genes, genetically engineered duplications of endogenous targets of immunity, added regulatory genes or with other changes altering tissue tropism. Concern about engineering a more virulent virus is a valid concern which may only be resolved by preclinical studies (where possible) and clinical trials. Depending on the clinical endpoint(s) being measured, only very large numbers of individuals in clinical trials (in some cases >10,000), will truly address this issue. It should be kept in mind that many vaccines take deliberate advantage of routes of delivery to achieve attenuation, presumably through altered tissue tropism. Investigational vaccines with genetic insertions have thus far been attenuated in clinical trials.

Viruses which are capable of a state of latency or persistence and which can periodically reactivate pose special concerns for the use of live attenuated vaccine approaches. The concern most commonly voiced is fear of a new recombinant arising in a vaccinated individual as the result of recombination following an infection by a wild type virus. Again, deletions at widely separated loci in the vaccine virus should result in recombinants which carry at least one of the attenuating mutations. All of these recombination scenarios presume that the vaccine virus and wild type virus gain access to the same cell, at the same time, and cellular recombination mechanisms exist to mediate the exchange of genetic information. Finally, these scenarios presuppose that the vaccinated individual's immunity would not control even the wild type virus.

Vaccine Efforts for Herpesviruses

The herpesvirus family comprise a group of large double stranded DNA viruses that share the property of establishment of a lifelong persistent or latent state in their host following the initial infection. Members of the family of human viruses include herpes simplex virus (HSV), human cytomegalovirus (HCMV), and Epstein Barr virus (EBV), which are the prototypes for the Alphaherpesvirinae, the Betaherpesvirinae and the Gammaherpesvirinae subfamilies, respectively.[1] These viruses periodically reactivate with different consequences for the host usually dependent on the immune status at the time of reactivation. For example, HCMV often reactivates without consequence in an immunologically competent individual. Reactivation of HCMV or Varicella-zoster virus (VZV), an alphaherpesvirus, can result in a clinically apparent reactivation in individuals with compromised or waning cellular immune responses. The challenge presented by the herpesviruses to developing an effective vaccine is to prevent the disease consequences associated with the initial infection and deliver a vaccine which if it does reactivate, will do so without causing disease or result in minimal shedding episodes that reduce the chances for transmission to others (or which if it does occur is of no consequence to the new recipient).

Vaccine induced immunity to the herpesviruses has been possible to achieve. Effective live attenuated veterinary vaccines exist for Aujeszky's disease and Marek's disease, caused by members of the alphaherpesviruses in pigs and chickens, respectively.[2,3] In particular, the Aujeszky's disease vaccine was developed as a genetically engineered pseudorabies virus (PRV), a herpesvirus which causes a severe and often fatal infection in piglets. The vaccine, named Tolvid®, was generated from a wild-type PRV-strain Aujeszky with a deletion in thymidine kinase (*tk*) which is responsible for its attenuated phenotype, and a deletion of glycoprotein X (a gG homolog), to provide a serological marker to distinguish vaccine virus from field isolates.[4] Tolvid® is approved by the USDA for sale. In addition to providing an example of an efficacious vaccine for a herpesvirus, it stands as the prototype live attenuated genetically engineered viral vaccine.

Varivax® is a live attenuated vaccine licensed by the FDA in 1995 and manufactured by Merck & Company for use in preventing varicella (chicken pox) in humans.[5] It is a classically derived vaccine developed by Dr. Takahashi and colleagues using vesicular fluid from a child named Oka. The Oka strain was attenuated by repeated passage in guinea pig embryo and human fibroblasts and first used in 1974 to prevent the spread of varicella in hospitalized children in Japan.[6] The exact molecular basis of the attenuation is not currently understood. The properties of this vaccine have been reviewed in the previous chapter by Shaw.

HSV

Infections with HSV-2 are a major public health problem which has continued despite effective antiviral therapy.[7] According to a recent epidemiological report by Fleming et al,[8] one in five Americans now harbor HSV-2 infections. This finding represents an age-adjusted increase of seroprevalence of 30 percent in the 13 years between the second and third National Health and Nutrition Examination surveys taken between 1976 and 1994, with the greatest increase among young Caucasian individuals. The majority of infected individuals have asymptomatic infections and do not experience the ulcerative lesions that are commonly recognized as genital herpes.[9] This aspect of the epidemiology of the disease coupled with the ability of HSV to establish latent infection, spontaneously reactivate, and shed asymptomatically has exacerbated this problem to epidemic proportions. A vaccine will be the most cost effective way to reduce the morbidity associated with this unmet medical need.[10] No licensed vaccine is currently available to address the problems posed by genital herpes infection.

The experiences using a live attenuated recombinant HSV vaccine targeted for humans have been limited. To date, none of the candidates have been advanced into Phase 3 clinical trial development. The earliest use of a recombinant HSV vaccine in human clinical trials was reported in 1992 as a Phase 1 trial sponsored by Pasteur Mérieux Connaught.[11] The vaccine candidate, R7020, is a HSV-1 strain F-based vaccine candidate in which a region near and including the internal repeats is replaced with a HSV-2 strain G-based fragment encoding glycoprotein gG2.[12] In addition, the thymidine kinase (*tk*) gene fused to the HSV-1 $\alpha 4$ promoter was inserted in a new location between the HSV-2 fragment and the residual inverted repeat sequences. R7020 lacks UL56, UL24 (as a result of removing *tk* from its natural locus), one copy each of genes $\alpha 0$, $\gamma_1 34.5$, ORF P, ORF O, and one complete copy of the sequences which give rise to the LAT transcripts. The function of each of these genes is only partly understood, but the constellation effect of the insertion/deletions has resulted in an attenuated virus. This construct was shown to be safe in all animal models tested.[12-14] In a double blind, randomized, placebo controlled Phase 1 dose escalation trial in humans, R7020 was well tolerated in HSV-1 / HSV-2 seronegative volunteers when given subcutaneously in doses ranging from $10^{2.2}$ to $10^{5.2}$ median tissue culture infectious doses ($TCID_{50}$). However, reactogenicity was observed in HSV-2 seronegative volunteers who were asymptomatically positive to HSV-1 after doses of $10^{4.2}TCID_{50}$. Two doses at $10^{5.2}TCID_{50}$ appeared to be required to elicit a strong ELISA or neutralizing antibody response.[11] Continued development of R7020 as a vaccine candidate has not been reported.

The most recent human vaccine clinical trial has utilized a glycoprotein H (gH)-deleted HSV-2 strain HG52 in a Phase 1 trial to evaluate safety. This type of virus is termed a DISC (disabled infectious single cycle) virus because it can only complete a single round of replication in a normal cell since the gH gene, essential for viral replication, has been deleted. DISC viruses can be grown to titers suitable for manufacturing in a complementing cell line expressing the gH glycoprotein.[15-17] In principle, DISC viruses represent an elegant strategy for attenuation. Because they cannot spread within the vaccinated host, they should be safe. The strategy, however, limits the ability to amplify the antigenic signal delivered on immunization. Theoretically, DISC viruses should be able to present the entire antigenic repertoire of the virus except gH to both the humoral and cellular compartments of the immune response. The DISC virus will allow presentation to the cellular response in the context of class I molecules and will be able to elicit a CD8+ cytotoxic T lymphocyte response. Preclinical workups of HSV-1 and HSV-2 prototypes have shown protection against HSV disease in both a prophylactic and therapeutic mode in the guinea pig model.[18,19]

Recently, immunogenicity and safety of a DISC HSV-2 vaccine candidate was evaluated in a Phase 1 clinical trial in HSV-2 seropositive and seronegative volunteers sponsored by Cantab Pharmaceuticals.[20,21] Seventy-two healthy adult male and female volunteers were enrolled and divided into three groups: 24 $HSV2^+$ individuals not experiencing frequent reactivations; 24 $HSV1^+/HSV2^-$, and 24 $HSV1^-/2^-$. Subjects were randomized to receive one of three formulations ($2x10^3$, $3x10^4$, and $3x10^5$ pfu/dose) or placebo. All subjects received two 0.5 ml doses of vaccine by SC injection, each given on day 0 and 56. No serious adverse events were reported. Humoral and cell mediated responses were evaluated and the results will be reported elsewhere. HSV could not be recovered from the inoculation site or distant sites after immunization.

Trials at higher doses of the vaccine are now under way. Until recently this vaccine was being developed in collaboration with Glaxo.

Other strategies employing live attenuated recombinant approaches to engineering HSV have been reported in preclinical development.[22-25] At Aviron, the strategy has focused on genes that are required for efficient replication in the central nervous system (CNS) tissue of mice and guinea pigs. The $\gamma_1 34.5$ gene has been the starting point for these efforts since viruses harboring deletions in this gene are unable to replicate in the CNS of mice and are impaired for reactivation.[26-29] One candidate, RAV9395, is a HSV-2 strain G-based recombinant which carries deletions in UL55 and UL56 in addition to both copies of the $\gamma_1 34.5$ gene. The function and contribution of UL55 and UL56 to an attenuated phenotype on this genetic background is unclear. Nevertheless, this candidate has shown promise for the development of a live attenuated HSV-2 vaccine in mucocutaneous challenge studies in the guinea pig model.[30] RAV9395 appeared to be unable to reactivate efficiently from latency in this model and protected guinea pigs from significant HSV disease following challenge with wild type HSV-2(G). Exploration of the optimal genetic constellation in conjunction with the $\gamma_1 34.5$ gene deletions is the focus of the continuing effort at Aviron. Specific gene deletions are being introduced using overlapping cosmid transfection technology (see HCMV below).

HCMV

Infection with human cytomegalovirus (HCMV) is nearly universal in humans. Twenty to eighty percent of the adult population has serological evidence of prior infection with this virus by the age of 21. HCMV is narrowly restricted in its host range to the human host. Modes of transmission include maternal to fetal (congenital), maternal cervical secretion or by breast milk (perinatal), urine, respiratory secretions and sexual transmission. Once infected, the individual harbors a lifelong latent or persistent infection, that may frequently reactivate with no apparent clinical symptoms. In the fetus and immunocompromised host, however, HCMV may cause significant morbidity and mortality.[31]

Every year in the U.S. 4,000 to 8,000 children are born who will have sequelae from an intrauterine HCMV infection. Most of these children are born to women who have primary HCMV infection during pregnancy. Approximately 10% of the neonates born with congenital HCMV disease will die. The sequelae of infection in the survivors can profoundly affect the central nervous system and include deafness and mental retardation. In immunocompromised individuals, HCMV infection can lead to retinitis in AIDS patients, and hepatitis, pneumonia and/or graft rejection in transplant recipients. Significant healthcare costs are associated with HCMV infections. An estimated health care savings of $2.5 million dollars per year accrue when 24 cases of congenital infection are prevented.[32] The disease burden associated with HCMV in the U.S. is estimated at $1.86 billion dollars annually.[33] Recently, the Institute of Medicine has given HCMV a Level I status (Most Favorable), in terms of rating vaccine prospects with respect to cost of development and potential for improvement of quality of life (http://bob.nap.edu/readingroom/books/vacc21/).

Naturally acquired immunity to HCMV usually has protective benefits against a second infection.[34] This observation provides a foundation for pursuing a live, attenuated vaccine to prevent HCMV disease. Thirty to 50% of women who have a primary HCMV infection during pregnancy will transmit the infection to the fetus. In contrast, less than 1% of immunocompetent women who have preconceptual immunity to HCMV transmit the infection. Long term follow up of infected newborns also demonstrated that preconceptual maternal immunity reduces the incidence and severity of neurological sequelae.[35,36] In addition, solid organ transplant recipients who have immunity to HCMV prior to immunosupression and transplantation have reduced risk of severe HCMV disease compared to seronegative recipients.[37] Currently, no vaccine against HCMV is commercially available for humans.

HCMV is a herpesvirus with a double stranded DNA genome of 230 kilobase pairs (kbp) encoding at least 226 open reading frames.[38] We have found that clinical isolates and the Toledo strain carry at least 19 additional open reading frames which are not found in the laboratory isolate AD169 and 17 of these are not found in some Towne strain variants.[39] Although this region of difference was localized to the so-called UL/b' region, other differences in the Towne and Toledo genomes have been catalogued (unpublished data). We hypothesize that these changes occurred during long-term passage in cell culture.

At Aviron, the goal of the HCMV vaccine program is to generate a safe and efficacious live, attenuated HCMV strain targeted to the prevention of congenital CMV disease. Starting with the Towne and Toledo strains which have been tested in controlled human trials,[40,41] we have created a set of four recombinant vaccine candidates. In each candidate, defined regions of the Towne genome were replaced with the collinear regions of Toledo by utilizing overlapping cosmids derived from the two parental strains. Taken together this set of four chimeras survey the entire genetic content of the Toledo genome (Fig. 6.1.). This strategy is intended to increase the level of immunogenicity of the prospective vaccine candidates to that achieved in a naturally infected individual without incurring potential concomitant symptoms. Using this approach, we have initiated a Phase 1 clinical trial to assess the safety and immunogenicity of each chimera in an effort to identify a vaccine candidate.

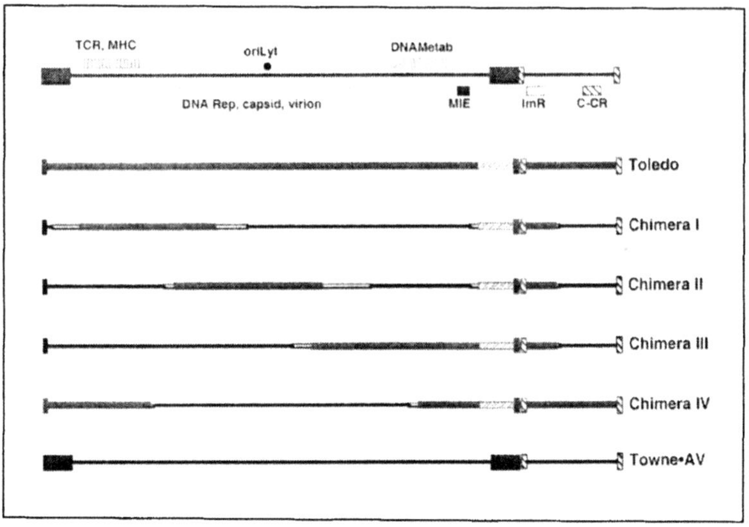

Fig. 6.1. HCMV chimeric vaccine candidates. Chimeras I through IV were generated as recombinant viruses between the Towne parent genetic backbone (thin line), and discrete portions of the Toledo genome (thick line). The UL/b' region which encodes additional HCMV specific genes is shown as a cross-hatched region. Some of the HCMV genes encoded in the prototype virus are shown at the top. As a set, the chimeras survey the genetic content of the entire Toledo genome. TCR=T cell receptor homolog; MHC=MHC class I homolog; MIE=Major Immediate Early gene; IMR=Immune response regulatory genes; C-C R=ß chemokine receptor. Figure courtesy of Dr. George Kemble, Aviron.

Recombinant Vaccine Efforts for RNA Viruses

Yellow fever virus (YFV) is the prototype of the Flaviviridae family which includes many viruses pathogenic for humans such as YFV, Japanese encephalitis virus, tick-borne encephalitis and dengue viruses.[42] The development of the attenuated YF 17D vaccine strain by extensive in vitro and in vivo passage has enabled control of this major public health problem in vaccinated populations.[43] The efficacy of the vaccine is greater than 95% and immunity is good for at least 10 years and is possibly lifelong. OraVax has capitalized on the excellent vaccine profile of YF 17D to use it as a genetic backbone in the creation of chimeric vaccine viruses for other members of the flavivirus and pestivirus families as well as hepatitis C virus. This concept is embodied in the ChimeriVax™ technology (described below).

YFV is a positive sense single stranded RNA genome of 10.9 kbp whose full length cDNA can yield infectious virus.[44,45] The gene order is 5'-C-prM(M)-E- followed by the nonstructural genes. The genome is translated as a single polyprotein which is processed by a number of viral and cellular proteases.[46] Mature virions contain two transmembrane proteins, M, which are produced by proteolytic processing of a glycosylated precursor called prM, and E, the envelope glycoprotein. The C protein forms the structural component of the nucleocapsid. The E protein is the principal target for viral neutralizing antibodies.

The ChimeriVax™ technology uses the YF 17D genome cloned as cDNA and substitutes the analogous "prM-E" regions from a particular target virus, e.g., immunotypes of Japanese encephalitis virus or dengue viruses. The chimeric cDNA is used to generate mRNA which is transfected into cells and, using the YFV 17D replicative machinery, generates vaccine viruses which display the heterologous virus antigens on the envelope (chimeras). These chimeras should be capable of eliciting strong neutralizing antibody responses. The company expects the first vaccine, directed to Japanese encephalitis, to be available early in the next decade.[47]

Vaccine Efforts for Respiratory Viruses

Recent technological developments have enabled engineering of negative-strand RNA viruses through an approach termed "reverse genetics."[48] Using this approach, influenza virus of the Orthomyxovirus family, whose members have segmented RNA genomes, has yielded to genetic manipulation by recombinant methods.[49,50] In addition, other negative-strand RNA viruses including infectious clones of rabies, vesicular stomatitis virus, Sendai virus, measles virus, respiratory syncytial virus and members of the parainfluenza viruses have been rescued and manipulated in ways that have enabled alteration of their growth properties in cell culture.[51]

Influenza

Recently, Aviron has been developing a live attenuated vaccine approach to the prevention of disease caused by types A and B influenza in humans. The vaccine named, FluMist™, is based on the use of an attenuated cold adapted master donor virus. Cold adapted influenza virus was first developed by Dr. H. F. Maassab at the University of Michigan in 1967. Dr. Maassab created weakened influenza strains by propagating the virus at progressively lowered temperatures in primary chick kidney cells until a virus was derived that grew efficiently at 25°C. Two master donor strains were derived in this manner, the cold adapted (*ca*) A/Ann Arbor/6/60 virus and *ca* B/Ann Arbor/1/66 virus. The master virus strains carry *ca* and *ts* phenotypes as well as an attenuated (*att*) phenotype contributed by mutations in four gene segments including the polymerase genes (reviewed by Maassab et al).[52]

Using the master donor virus strains, the vaccine strains are updated each year according to the recommendations of the CDC and the FDA. The virus strains used in the vaccine are prepared as 6:2 reassortants using the appropriate *wt* isolates which contribute two gene segments, those encoding the hemagglutinin (HA) and neuraminidase (NA) surface antigens, crossed with the attenuating master strains (*ca* A/Ann Arbor/6/60 (H2N2) and *ca* B/Ann Arbor/1/66 virus).

Since 1976, at least 68 clinical trials have been performed using the cold adapted influenza virus vaccine involving over 8,000 vaccinees. Most recently, as part of the cooperative research and development agreement between NIAID and Aviron to enable licensure of the trivalent cold adapted influenza vaccine (FluMist™), six of the Division of Microbiology and Infectious Diseases' Vaccine and Treatment Evaluation Units as well as other sites have enrolled 6614 children and 4935 adults in clinical studies designed to assess the safety, immunogenicity, and efficacy of FluMist™. Because FluMist™ is delivered as a nasal spray, it should provide the first practical way to immunize children on an annual basis and interrupt an important link in the spread of influenza in the population. The results of year one of a pediatric efficacy trial in 1602 children, 15-71 months of age, demonstrated greater than 93% protection rate against culture confirmed influenza in children who received two doses.[53] Previously, Aviron completed a challenge efficacy study in healthy seronegative adults. Following vaccination with FluMist™, protection from laboratory-confirmed influenza due to H1N1, H3N2 and B strains in the challenge study was 80%, 78% and 100%, respectively. Aviron intends to continue clinical studies on the effectiveness of FluMist™ in children, healthy adults and high risk adults (to be coadministered with the inactivated influenza vaccine for enhanced protection) and apply for product licensure.

Aviron is also using "reverse genetics" technology to engineer future generations of influenza vaccines which are tailored to the needs of various age groups in the population. Another goal is to be able to rapidly engineer influenza strains in response to human health requirements in the event of an influenza pandemic. In this approach, the HA and NA gene segments from the circulating wild type virus are cloned as cDNAs and used to make the 6:2 reassortants by the reverse genetics technology as illustrated in Figure 6.2. Investigational vaccines generated in this manner have been used in human clinical trials conducted by Aviron and have been shown

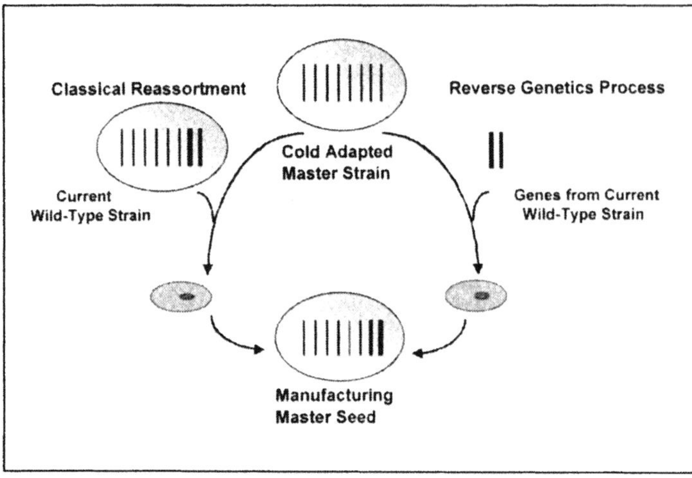

Fig. 6.2. Influenza virus annual update by two different methods. Shown on the left is the "classical method" of preparing the annually updated 6:2 reassortant by coinfection with the appropriate wild type isolate which contributes two gene segments (thick vertical lines). Shown on the right is the "reverse genetics" approach where the HA and NA gene segments from the circulating wild type virus are cloned and introduced into the virus by transfection using the reverse genetics process.

to be safe and well tolerated (unpublished data). A most recent technological development will enable the generation of influenza viruses by cloning only eight cDNA plasmids.[54] This plasmid-based expression system will allow rapid engineering without the need for selection and will reduce the time needed for generation of potential vaccine candidates.

Prospects for the Future

The accumulation of fundamental knowledge concerning the factors which promote virulence and pathogenesis of disease causing viruses will facilitate rational approaches to engineering vaccine candidates. Similarly, when knowledge of immune response mechanisms become specific enough to allow prediction as to what compartments of the immune response should be evoked to elicit disease protection, then accessory gene products such as cytokines or cytokine receptors may be engineered into the vaccines to optimize the acquired immunity that is desirable. For the herpesviruses as a group, it is imaginable that if vaccines are valued that are not capable of reactivation, then it will be possible to effect the required genetic changes as the gene products that control these properties become known. Increased understanding of the biology of vaccines currently in use will inform such a decision since it may turn out that periodic subclinical reactivation is desirable to promote protective long term immunity. For the respiratory viruses, combining genetic manipulation with Jennerian approaches using nonhuman animal viruses will enable engineered viruses carrying human surface antigens for vaccination.[55,56] This approach will take advantage of the attenuated character of the genetic backbone as a delivery vehicle for the appropriate immunogen(s). Engineering in characteristics that enable growth in one body compartment but restrict growth in others will allow exploitation of routes of delivery. As shown for cold adapted influenza viruses, nasal administration takes advantages of the natural route of infection to stimulate appropriate local immunity as well as the lower temperature of the nasal passages to restrict viral replication to the upper respiratory tract. Now it is possible to engineer such properties into viruses to obtain vaccines which will elicit desired virus-host interactions.

Whether such genetically engineered vaccines will be developed will depend on a complex matrix of government and company resources, public policy decisions and political will, regulatory bodies' comfort levels with these approaches, and knowledge of the biology of the human immune response at a fundamental and detailed level. Although it is now generally accepted that prevention strategies such as vaccination are cost effective and cost beneficial, other factors such as market forces and HMO reimbursement policies will also be determinants in decisions to develop genetically engineered vaccines.

Acknowledgment

I thank Drs. Kathleen Coelingh, George Kemble, and Stephen Inglis for critical review of the manuscript, Julie Cordova for helpful discussions of clinical data and Emily Luthra for help in its preparation.

References

1. Roizman B. Herpesviridae. In: Fields BN, Knipe DM, Howley PM, eds. Fields Virology. 3rd ed. Philadelphia, New York: Lippincott-Raven 1996:2221-2230.
2. Wardley RC, Thomsen DR, Berlinski PJ et al. Immune response in pigs to Aujeszky's disease viruses defective in glycoprotein gl or gX. Research in Veterinary Science 1991; 50:178-184.
3. Calneck BW, Witter RL. Marek's Disease. In: Calneck BW, Barnes HJ, Beard CW et al, eds. Diseases of Poultry. Ames, Iowa: Iowa State University Press 1997:369-413.
4. Post LE, Thomsen DR, Petrovskis EA et al. Genetic engineering of the pseudorabies virus genome to construct live vaccines. J Reprod Fert Suppl 1990; 41:97-104.
5. Arvin AM, Gershon AA. Live attenuated varicella vaccine. Annu Rev Microbiol 1996; 50:59-100.
6. Takahashi M, Otsuka T, Okuno Y et al. Live vaccine used to prevent the spread of varicella in children in hospital. Lancet 1974; 2:1288-1290.
7. Whitley RJ. Herpes simplex viruses. In: Fields BN, Howley PM, eds. Fields Virology. 3rd ed. Philadelphia, New York: Lippincott-Raven 1996:2297-2342.
8. Fleming DT, McQuillan GM, Johnson RE et al. Herpes simplex virus type 2 in the United States, 1976 to 1994. N Engl J Med 1997; 337:1105-1111.
9. Wald A, Zeh J, Selke S et al. Virologic characteristics of subclinical and symptomatic genital herpes infections. N Engl J Med 1995; 333:770-775.
10. Arvin AM, Prober CG. Herpes simplex virus type 2–a persistent problem. N Engl J Med 1997; 337:1158-1159.
11. Cadoz M, Micoud M, Seigneurim JM et al. Phase 1 trial of R7020: A live attenuated recombinant herpes simplex virus (HSV) candidate vaccine [abstract 341]. In: Program and abstracts of the 32nd Interscience Conference on Antimicrobial Agents and Chemotherapy (Anaheim, CA). Washington DC: American Society for Microbiology, 1992.
12. Meignier B, Longnecker R, Roizman B. In vivo behavior of genetically engineered herpes simplex viruses R7017 and R7020: construction and evaluation in rodents. J Infect Dis 1988; 158:602-614.
13. Meignier B, Longnecker R, Roizman B. R7017 and R7020 herpes simplex virus recombinant prototype vaccine strains: Animal studies. Vaccines 1998; 88:193-196.
14. Meignier B, Martin B, Whitley RT et al. In vivo behavior of genetically engineered herpes simplex viruses R7017 and R7020. II. Studies in immunocompetent and immunosuppressed owl monkeys (*Aotus trivirgatus*). J Infect Dis 1990; 162:313-321.
15. Farrell H, McLean CS, Harley C et al. Vaccine potential of a herpes virus type 1 mutant with an essential glycoprotein deleted. J Virol 1994; 68:927-932.
16. Forrester A, Farrell H, Wilkinson G et al. Construction and properties of a mutant herpes simplex virus type 1 with glycoprotein gH coding sequences deleted. J Virol 1992; 66:341-348.
17. Speck PG, Efstathiou S, Minson AC. In vivo complementation studies of a glycoprotein H—deleted herpes simplex virus—based vector. J Gen Virol 1996; 77:2563-2568.
18. McLean CS, Erkturk M, Jennings R et al. Protective vaccination against primary and recurrent disease caused by herpes simplex virus (HSV) type 2 using genetically disabled HSV-1. J Infect Dis 1994; 170:1100-1109.

19. Boursnell MEG, Entwisle C, Blakeley D et al. A genetically inactivated herpes simplex virus type 2 (HSV-2) vaccine provides effective protection against primary and recurrent HSV-2 disease. J Infect Dis 1997; 175:16-25.
20. Hickling JK, Chisholm SE, Duncan IA et al. Immunogenicity and safety of a disabled infectious single cycle (DISC) HSV-2 vaccine in phase 1 clinical trials in HSV-2 seropositive and seronegative volunteers [abstract 322]. In: Program and abstracts of the 23rd international herpes virus workshop (York, U.K.). York, U.K.: 23rd International Herpesvirus workshop, 1998.
21. St. Clair Roberts J, Uttridge JA, Hickling JK et al. Safety, tolerability and viral containment of a DISC HSV-2 vaccine evaluated in phase 1 clinical trials. In: Cantab Pharmaceuticals Research Ltd. Publication, 1998.
22. Nguyen LH, Knipe DM, Finberg RW. Replication defective mutants of herpes simplex virus (HSV) induce cellular immunity and protect against lethal HSV infection. J Virol 1992; 66:7067-7072.
23. Morrison L, Knipe DM. Immunization with replication-defective mutants of herpes simplex virus type 1: Sites of immune intervention in the pathogenesis of challenge virus infection. J Virol 1994; 68:689-696.
24. Aurelian L, Kokuba H, Smith CC. Vaccine potential of a herpes simplex virus type 2 mutant deleted in the PK domain of the large subunit of ribonucleotide reductase. Vaccine 1999; 17 (15-16):1951-1963.
25. Da Costa XJ, Bourne N, Stanberry LR et al. Construction and characterization of a replication-defective herpes simplex virus 2 ICP8 mutant strain and its use in immunization studies in a guinea pig model of genital disease. Virology 1997; 232:1-12.
26. Chou J, Kern ER, Whitley RJ et al. Mapping of herpes simplex virus-1 neurovirulence to $\gamma_1$34.5, a gene nonessential to growth in culture. Science 1990; 250:1262-1266.
27. Spivak JG, Fareed MU, Valyi-Nagy T et al. Replication, establishment of latent infection, expression of the latency-associated transcripts and explant reactivation of herpes simplex virus type 1 gamma 34.5 mutants in a mouse eye model. J Gen Virol 1995; 76:321-332.
28. Robertson LM, MacLean AR, Brown SM. Peripheral replication and latency reactivation kinetics of the nonneurovirulent herpes simplex virus 1 variant 1716. J Gen Virol 1992; 73:967-970.
29. Whitley RJ, Kern ER, Chatterjee S et al. Replication, establishment of latency and induced reactivation of herpes simplex virus γ_1 34.5 deletion mutants in rodent models. J Clin Invest 1993; 91: 2837-2843.
30. Spector FC, Kern ER, Palmer J et al. Evaluation of a live attenuated recombinant virus RAV 9395 as a herpes simplex virus type 2 vaccine in guinea pigs. J Infect Dis 1998; 177:1143-1154.
31. Britt WJ, Alford CA. Cytomegalovirus. In Fields DM, Knipe DM, Howley PM, eds. Fields Virology. 3rd ed. Philadelphia, New York: Lippincott-Raven, 1996:2493-2523.
32. Porath A, McNutt RA, Smiley LM et al. Effectiveness and cost benefit of a proposed live cytomegalovirus vaccine in the prevention of congenital disease. Rev Infect Dis 1990; 12(1):31-40.
33. Dobbins JG, Stewart JA, Demmler GJ et al. Surveillance of congenital cytomegalovirus disease, 1990-1991. MMWR Surveillance Summaries 1992; 41:35-43.
34. Adler SP, Starr SE, Plotkin SA et al. Immunity induced by primary human cytomegalovirus infection protects against secondary infection among women of childbearing age. J Infect Dis 1995; 171:26-32.
35. Demmler GJ. Summary of a workshop on surveillance for congenital cytomegalovirus disease. RID 1991; 13:315-29.
36. Alford CA, Stagno S, Pass RF et al. Congenital and perinatal cytomegalovirus infections. RID 1990; 12:S745-S753.
37. Marshall GS, Plotkin SA. Cytomegalovirus vaccines. In: Roizman B, Whitley RJ, Lopez C, eds. The Human Herpesviruses. 1st ed. New York: Raven Press, Ltdl 1993:381-395.
38. Chee MS, Bankier AT, Beck S et al. Analysis of the Protein-Coding Content of the Sequence of Human Cytomegalovirus Strain AD169. In: McDougall JK, Cytomegaloviruses. 154. Berlin: 1990:125-169.
39. Cha T, Tom E, Kemble W et al. Human cytomegalovirus clinical isolates carry at least 19 genes not found in laboratory strains. J Virol 1996; 70:78-83.
40. Plotkin SA, Starr SE, Friedman HM et al. Protective effects of Towne cytomegalovirus vaccine against low-passage cytomegalovirus administered as a challenge. J Infect Dis 1989; 159:860-865.

41. Quinnan GV Jr., Delery M, Rook AH et al. Comparative virulence and immunogenicity of the Towne strain and a nonattenuated strain of cytomegalovirus. Ann Intern Med 1984; 101:478-483.
42. Monath TP, Heinz FX. Flaviviruses. In: Fields BN, Knipe DM, Howley PM, eds. Fields Virology. 3rd ed. Philadelphia, New York: Lippincott-Raven 1996:961-1034.
43. Theiler M, Smith HH. Use of yellow fever virus modified by in vitro cultivation for human immunization. J Exp Med 1937; 65:787-800.
44. Rice CM, Lenches EM, Eddy SR et al. Nucleotide sequence of yellow fever virus: Implications for flavivirus gene expression and evolution. Science 1985; 229:726-733.
45. Rice CM, Grakoui A, Galler R et al. Transcription of infectious yellow fever RNA from full-length cDNA templates produced by in vitro ligation. The New Biologist 1989; 1(3):285-296.
46. Rice CM. Flaviviridae: their viruses and their replication. In: Fields BN, Knipe DM, Howley PM, eds. Fields Virology. 3rd ed. Philadelphia, New York: Lippincott-Raven, 1996:931-959.
47. Peptide Therapeutic's OraVax deal. SCRIP 1999; 2438:14.
48. Palese P, Zheng H, Engelhardt OG et al. Negative-strand RNA viruses: genetic engineering and applications. Proc Natl Acad Sci 1996; 93(21): 1354-11358.
49. Enami M, Luytjes W, Krystal M et al. Introduction to site-specific mutations into the genome of influenza virus. Proc Natl Acad Sci 1990; 87:3802-3805.
50. Enami M, Palese P. High-efficiency formation of influenza virus transfectants. J Virol 1991; 65:2711-2713.
51. Roberts A, Rose JK. Recovery of negative-strand RNA viruses from plasmid DNAs: A positive approach revitalizes a negative field. Virology 1998; 247(1):1-6.
52. Maassab HF, Shaw MW, Heilman CA. Live influenza virus vaccine. In Plotkin SA, Mortimer EA, eds. Vaccines. 2nd ed. Philadelphia: WB Saunders Co., 1994;781-801.
53. Belshe RB, Mendelman PM, Treanor J et al. The efficacy of live attenuated, cold-adapted, trivalent intranasal influenza virus vaccine in children. N Eng J Med 1998; 338:1405-1412.
54. Hoffmann E, Neumann G, Kawaoka Y et al. A DNA transfection system for generation of influenza A virus from eight plasmids. Proc Natl Acad Sci 2000; 97(11):6108-6113.
55. Schmidt AC, McAuliffe JM, Huang A et al. Bovine parainfluenza virus type 3 (BPIV3) fusion and hemagglutinin-neuraminidase glycoproteins make an important contribution to the restricted replication of BPIV3 in primates. J Virol 2000; 74:8922-8929.
56. Haller AA, Miller T, Mitiku M, et al. Expression of the surface glycoproteins of human parainfluenza virus type 3 by bovine parainfluenza virus type 3, a novel attenuated virus vaccine vector. J Virol 2000 74:11626-11635.

CHAPTER 7

Live Viral Vectors

Elizabeth B. Kauffman, Michel Bublot, Russell R. Gettig, Keith J. Limbach, Steven E Pincus, and Jill Taylor

Live viral vector vaccines derived by the insertion of genes encoding sequences from disease organisms offer a number of advantages over live attenuated vaccines, inactivated vaccines, subunit or DNA approaches. The evolution of any successful vaccine must address issues such as safety, efficacy, host range, genetic stability, vector immunity, ease of use, and manufacturing costs. One of the most important safety features of live viral vectors is that the recipients may be immunized against specific antigens from pathogenic organisms without exposure to the disease agent itself. Safety is further regulated by the selection of a viral vector that is either attenuated for the host or unable to replicate in the host although still able to express the heterologous antigen of interest. A vaccine strain that has a history of safety in the target species offers an additional safety feature. However, the possibility of systemic infection in immunocompromised individuals must be considered in the development of all live viral vaccines. Several systems have been developed in which the vector is deleted of essential genes and preparation of the vaccine is carried out in cell systems that provide the missing function. With attenuated vectors, the possibility of reversion to a more virulent strain may be a concern; the use of a vector that has extensive engineered deletions can reduce the chance of reversion by recombination with wild type viruses. Other safety issues with live viral vectors include integration into the host genome and the possibility that insertion of a foreign gene may cause a change of pathogenicity or host range.

The balance between safety and efficacy is important since a viral vector that is too attenuated may not provide sufficient efficacy. Live viral vectors have several basic characteristics that promote strong immune responses. First, when a foreign antigen is delivered by a viral vector it is expressed in its natural host with attendant post translational modifications, resulting in conformationally authentic products that can induce broad and efficacious immune responses. A drawback to the system is that antigens that are not encoded by nucleic acid, such as lipopolysaccharides, cannot be delivered. Second, infection by the vector with accompanying tissue destruction and inflammation produces costimulatory signals that activate immune responses. Third, foreign antigens that do not normally elicit effective immune responses can also be manipulated to enhance immunogenicity or to direct the immune response to a CTL rather than antibody response. Possible modifications include increasing the level of expression of the foreign antigen by changing promoter or insertion site, changing the targeting of the vector or the expressed antigen, or coexpressing immunomodulatory proteins. To broaden immune responses, recent immunization protocols have utilized a combination approach whereby the primary inoculation with one vaccine type such as DNA is followed by one or more boosts with other vaccine types such as a live viral vector or recombinant protein. This approach can potentially ameliorate the vector immunity problem while utilizing the best features of each vaccine system.

New Vaccine Technologies, edited by Ronald W. Ellis. ©2001 Eurekah.com.

Another potential advantage of viral vectored vaccines is the ability to insert multiple genes and epitopes into the same vector, thereby creating a multivalent vaccine that also will provide protection against the vector itself. An important feature of such a vaccine is reduced production costs. One of the drawbacks is vector immunity, a serious problem whereby pre-existing immunity prevents an initial immune response, and repeated immunizations elicit diminishing responses. It can sometimes be overcome by boosting via a different route, boosting with a vector of a different serotype, or by using a combination approach that employs an alternate viral vector or vaccine type. Vaccination with live viral vectors in the presence of maternal antibodies is also a problem, which needs to be addressed when the vaccine is targeted for early infancy. Many live viral vectors offer the advantage of convenient modes of delivery (such as oral, nasal or rectal) rather than needle injection. The ability of most viral vector vaccines to be delivered without adjuvant is also advantageous. The necessity of producing the vaccine in cell culture can offer advantages and disadvantages. Viruses that can be grown to high titer and are designed to be multivalent may be inexpensive to produce. Since the exogenous antigens are expressed within the host after vaccination, elaborate purification protocols are not necessary. However, separation of virus from cell substrate material is desirable, and can introduce another step in the manufacturing process. For many viral vectors the shelf life of the virus is limited and the need for a cold chain creates an additional cost.

Since the first demonstration of the expression of an antigen from a disease organism in a heterologous vector,[1] viral vectors have been developed from a wide spectrum of both DNA and RNA viruses. Construction of a live virus for vaccine delivery starts with the vector itself, with safety as the primary focus. Deletion of non-essential genes often provides a suitable insertion site for the heterologous antigen. Alternatively, deletion of essential genes may be desired in order to provide a safer vector with the essential function provided in trans. Initially viral vector vaccines were designed for the prophylactic treatment of infectious diseases in man and animals. Now the uses of this technology have expanded to include immunotherapy of cancer and HIV, as well as gene therapy. This chapter will not be a comprehensive review, but will attempt to cover the important advances in the field and report on newly developed vectors and novel approaches.

DNA Viruses

Poxvirus Vectors

Poxviruses are large dsDNA viruses belonging to the viral family *Poxviridae*. They have been isolated from a wide variety of animal species including mammals, birds and insects. The pleomorphic, brick-shaped virus particles are enveloped and contain a core and two lateral bodies. The linear DNA, ranging in size from 130-375 kb, is not infectious since it requires essential viral proteins located in the core to initiate replication and gene expression. Poxviruses replicate exclusively in the cytoplasm of infected cells and therefore encode genes that enable viral transcription and translation with minimal requirements from the nucleus.[2]

The most commonly used strategy for construction of poxvirus recombinants involves the insertion of foreign genes by homologous DNA recombination in infected cells, a process that occurs normally during the replication of the virus. Because of the large number of nonessential genes, there is a wide choice of insertion sites. Since poxviruses replicate in the cytoplasm and use their own transcription system, poxvirus promoters are required for the successful expression of foreign genes. The level and time of expression may be finely tuned by choice of promoter, and abbreviated or synthetic promoters have been used effectively. Due to the ability to delete a large number of nonessential genes and the fact that the virus can tolerate the addition of a large amount of foreign DNA (at least 25000 bp for vaccinia), the capacity of this viral vector is enormous. In fact, a vaccinia-based recombinant stably expressing seven antigens from the human malaria parasite *P. falciparum* has been tested in Phase I clinical trials.[3] Replication of poxvirus vectors in the cytoplasm rather than the nucleus offers the advantages that unwanted splicing events in foreign genes and integration into the host genome do not occur.

Since the pioneering work on vaccinia virus as a cloning vector,[4,5] the use of poxviruses in the development of potential vaccine candidates in both human and veterinary medicine has been extensive. Poxvirus vectors have been used to express a wide variety of viral and bacterial antigens, parasite antigens, immunomodulators and tumor-associated antigens (see refs. 6-12 and Table 7.1).

From the early use of vaccinia laboratory strains such as WR, work has progressed in the direction of safer poxvirus vectors. NYVAC, a highly attenuated vaccinia virus derived from the Copenhagen vaccine strain by deletion of specific genes implicated in virulence[41,42] and ALVAC, derived from a canarypox vaccine strain,[41,43] have been widely used as vaccine vectors in humans, companion animals and livestock animals.[11] The ALVAC vector, being host-restricted to avian species, does not replicate in mammals or mammalian cell lines. Likewise, NYVAC is debilitated and does not productively replicate in most human cell lines. Animal studies with NYVAC showed the vector to be highly attenuated in vivo,[41] which led to a reduction of its biological containment level from BL-2 to BL-1.[44] However, both of these nonreplicating vectors are capable of expressing foreign genes and eliciting a protective immune response when inoculated into a mammalian host.[8] ALVAC-based recombinants have been shown to induce both humoral and cellular (CTL) responses in humans.[28-30]

MVA (modified virus Ankara) is another nonreplicating vaccinia vector that is being utilized. This virus was originally derived as a safe smallpox vaccine by over 570 passages in primary chick embryo fibroblast cells, which caused multiple genomic deletions.[45] MVA has been administered without apparent incident to about 120,000 humans.[46] The MVA virus is replication-defective or has limited replication in most mammalian cells yet still retains the ability to express foreign genes under the control of late promoters since it is blocked at a late stage of morphogenesis.[45,47,48] Significantly, this non-replicating vector appears to induce better responses than replicating vectors,[17] which may be due to the fact that unlike other vaccinia strains, MVA lacks some of the immunomodulatory proteins that interfere with the host response to infection. These characteristics may enhance both the safety and immunogenicity of the vector due to host antiviral and pro-inflammatory activity.[48] MVA recombinants expressing influenza,[13] parainfluenza,[14,15] JEV,[16] and SIV[17] antigens have proved to be immunogenic and effective against virus infection in mouse and monkey models.

Another approach to constructing safer vaccinia vectors involves deletion of essential genes. A replication-defective vaccinia was engineered by deletion of the essential uracil DNA glycosylase (UDG) gene from the WR strain.[24,49] This deletion blocks viral DNA replication and late viral gene expression. Using a complementing cell line that expresses the vaccinia UDG gene, recombinants were generated that express the tick-borne encephalitis virus (TBE) preM and E structural proteins under an early promoter.[24] Mice vaccinated with as little as 10^4 pfu were fully protected from challenge with TBE while a replicating WR recombinant (tk -) expressing the same antigens required at least a 10-fold higher dose to achieve the same level of protection. Both viruses had similar particle/pfu ratios. The enhanced effectiveness of the nonreplicating vector may be due to the absence of expression of immunomodulatory genes because of the block in late gene expression.

Although the safety of the ALVAC vector has been established, recent attempts have been made to improve this vector as an immunization vehicle by increasing the level of foreign gene expression.[50] Previous studies had shown that the vaccinia E3L and K3L genes play a role in conferring resistance to the antiviral effects of IFN$\alpha\beta$.[51-53] It was also known that the E3L gene product functions as a dsRNA binding protein and that the K3L open reading frame has significant homology to the N-terminus of the translation factor eIF2α. Based on this

Table 7.1. Examples of live DNA viral vector vaccine candidates

Virus	Vector[a]	Foreign gene[b]	Replication[c]	Efficacy		References
				Species[d]	Protection[e]	
Poxvirus	Vaccinia (MVA)	*Influenza* HA & N	-	Mouse	+	13
	Vaccinia (MVA)	PIV3 F & HA	-	Cotton Rat	±	14,15
	Vaccinia (MVA)	JEV PrM & E	-	Mouse	+	16
	Vaccinia (MVA)	SIV gag/pol/env	-	Macaque	±	17
	Vaccinia (MVA) (DNA prime/MVA boost)	SIV gag (multiepitope)	-	Macaque	±	18
	Vaccinia (MVA) (DNA prime/MVA boost)	*P. berghei* PbTRAP & PbCs	-	Mouse	+	19
	Vaccinia (NYVAC)	CDV HA & F genes	-	Ferret	+	20,21
	Vaccinia (NYVAC)	*P.falciparum* CSP, PfSSP2, LSA1, MSP1, SERA, AMA1, Pfs25	-	Human	-	3
	Vaccinia (NYVAC)	SIV gag/pol/env – IL12 p35, p40	-	Macaque	±	22
	Vaccinia (WR)	VEE structural proteins	+	Mouse	+	23
	Vaccinia (WR) D4L del	TBE prM/E	-	Mouse	+	24
	Canarypox (ALVAC)	CDV HA & F	-	Ferret	+	20,21
	Canarypox (ALVAC)	RHDV capsid	-	Rabbit	+	25
	Canarypox (ALVAC) (part of combo vaccine)	CDV HA &F	-	Dog	+	26
	Canarypox (ALVAC)	P53 -Human and Murine	-	Mouse	+	27
	Canarypox (ALVAC)	Rabies Glycoprotein G	-	Humans	ab	28

continued on next page

Table 7.1. Examples of live DNA viral vector vaccine candidates (continued)

Virus	Vector[a]	Foreign gene[b]	Replication[c]	Efficacy		References
				Species[d]	Protection[e]	
	Canarypox (ALVAC)	JEV PrM/E/NS1	-	Human	CTL	29
	Canarypox (ALVAC)	HIV1-1 gp120	-	Human	ab/CTL	30
	prime-boost	(boost with subunit HIV-1 gp 120)				
	Canarypox (ALVAC) prime-boost	HCMV gB (boost with live HCMV)	-	Human	ab	31
	Fowlpox	MDV gB, gC, gD	+	Chicken	±	32
	Fowlpox (TROVAC)	NDV F & HA	+	Chicken	+	33
	Fowlpox (TROVAC)	AIV H5 HA gene	+	Chicken	+	34
	Fowlpox (TROVAC)	HEV Hexon & 100K	+	Turkey	+	35
	Raccoonpox	FPV VP2	+	Cats	+ (FPV challenge)	36
		Rabies G				
	Swinepox	PRV antigens	+	Pigs	+	37
	Capripox	BTV VP7	+	Sheep	+	38
	Capripox	Rinderpest F & HA	+	Cattle	+	38
					+	
	Myxoma	RHDV capsid	+	Rabbit	+	39
Adenovirus	Ad5	Rabies G	+	Skunk/Fox	+	83
	Ad5	PRCV S	+	Pig	±	84
	Ad5	FMDVP1	+	Cow	±	85
	Ad5	PRV gD	-	Pig	+	86
	Ad5	FIV env	-	Cat	ab	87

continued on next page

Table 7.1. Examples of live DNA viral vector vaccine candidates (continued)

Virus	Vector[a]	Foreign gene[b]	Replication[c]	Efficacy		References
				Species[d]	Protection[e]	
	FAV-10	IBDV VP2	+	Chicken	+	88
	BAV-3	BHV1gD	+	Calf	+	89
	Ovine adenovirus (prime-boost)	*Taenia ovis* 45W	+	Sheep	+	90
	Ad7	HBV sAg	+	Human	nd	91
	Ad5 (prime-boost)	SIV env	+	Macaque	-/ab/CTL	92
	Ad5	PRV gD	±	Cotton rat	+	93
	Ad5	BHV1 gD	±	Cotton rat	+	94
	Ad4(Ad5,Ad7) (prime-boost)	HIV1 env	+	Chimp	+	95
	Ad5 (prime-boost)	Rabies G	+	Mouse	ab	96
	Ad5	hu MART-1	-	Mouse	+	97
	Ad5	hu p53	-	Human	±	98
Herpesvirus	PRV	E2 from CSFV	+	Pig	+	138
	PRV	CSFV E2	±	Pig	+	139
	BHV-1	BRSV G	+	Bovine	+	140,141
	FHV	FIV env	+	Cat	-	142
	FHV	FeLV env	+	Cat	±	143

continued on next page

Table 7.1. Examples of live DNA viral vector vaccine candidates (continued)

Virus	Vector[a]	Foreign gene[b]	Replication[c]	Efficacy		References
				Species[d]	Protection[e]	
	FHV (prime-boost)	FeLV env & *gag*	+	Cat	±	144
	FHV	FCV capsid precursor	+	Cat	±	145,146
	HVT	NDV F	+	Chicken	+	136,137
	HVT	NDV F & HN	+	Chicken	+	147
	HVT	IBDV VP2	+	Chicken	+	148
	HVT	*Eimeria acervulina* Ea1a	+	Chicken	±	149
	MDV	NDV F	+	Chicken	+	150
	MDV	IBDV VP2	+	Chicken	±	151

a MVA Modified vaccine Ankara; WR Western Reserve strain; Ad5 adenovirus Type 5; FAV fowl adenovirus; BAV bovine adenovirus; Ad7 adenovirus type 7; Ad4 adenovirus type 4; PRV pseudorabies virus; BHV bovine herpes virus; FHV feline herpesvirus; HVT herpesvirus of turkey; MDV Marek's disease virus

b PIV parainfluenza virus; JEV Japanese encephalitis virus; SIV simian immunodeficiency virus; CDV canine distemper virus; VEE Venezuela equine encephalitis; TBE tick-born encephalitis virus; RHDV rabbit hemorrhagic disease virus; HIV human immunodeficiency virus; HCMV human cytomegalovirus; MDV Marek's disease virus; HEV hemorrhagic enteritis virus; AIV avian influenza virus; NDV Newcastle disease virus; FPV feline panleukopenia virus; PRV pseudorabies virus; FIV feline immunodeficiency virus; BTV bluetongue virus; PRCV porcine respiratory coronavirus; FMDV foot-and-mouth disease virus; IBDV infectious bursal disease virus; BHV bovine herpes virus; HBV hepatitis B virus; CSFV classical swine fever virus; BRSV bovine respiratory syncytial virus; FeLV feline leukemia virus; FCV feline calicivirus

c Replication in target species: + productive replication;—no replication; ± limited replication

d Species in which the test was performed

e Protection against challenge: + complete; ± partial;—negative; ab = antibody response; CTL = CTL response information, it was speculated that the vaccinia E3L and K3L genes might play a significant role in the down regulation of PKR activity and that expression of these genes in ALVAC, which does not encode any genes with functional homology to E3L or K3L, might enhance viral gene expression. To test this hypothesis, vaccinia E3L and K3L genes were inserted into an ALVAC-HIV recombinant expressing env and gag genes. In vitro results using several different human cell lines showed not only an enhanced level of expression of env and gag but also an increased persistence of expression over a 24 hr period.[50] Similar results were obtained with an ALVAC recombinant expressing the feline herpesvirus glycoprotein B gene (Gettig, unpublished results). In vivo trials with these second-generation ALVAC recombinants are in progress to assess any improvement in immunological response to the expressed foreign antigen.

Several other poxviruses have been tested for their efficacy as vaccine vectors. For veterinary applications the use of strains that are host-restricted to the target species has the advantage of preventing infection of human contacts. Raccoonpox,[36] swinepox,[37] capripox,[38,39] and myxoma[40] have all been used successfully to express heterologous antigens and to protect against challenge.

The successful testing and licensure of RABORAL V-RG, a vaccinia (Copenhagen strain) recombinant expressing the rabies glycoprotein G, exemplifies the success of vaccinia as a vaccine vector. This product has been used successfully in Europe for the vaccination of red foxes in the wild.[54] The recombinant virus is incorporated into live vaccine baits that are then distributed and eventually consumed by the foxes, thus vaccinating them against rabies. As the result of a widespread vaccination campaign, sylvatic rabies has been eliminated from large areas of Western Europe.[55] Raboral is also licensed in the US for vaccination of raccoons in the wild,[56,57] and has been used experimentally for vaccination of coyotes[58] and grey foxes.[59]

The TROVAC vector, derived from a fowlpox vaccine strain, has been successfully used as a vaccine vector in broiler chickens against Newcastle disease virus (NDV) and avian influenza virus.[34,60] This vector, although replicating in its target species, has been proven to be safe, efficacious and economically feasible at typical doses of $1x10^4$pfu given at day-of-age. Presently, two TROVAC recombinants have been licensed by the USDA: TROVAC-NDV and TROVAC-Avian Influenza-H5.

The success of ALVAC as a vaccine vector is exemplified by the recent licensure of two ALVAC recombinants for use in the veterinary field. ALVAC expressing the HA and F genes from canine distemper virus (included in the Recombitek™ line) has been licensed by the USDA for use in dogs[26] and ALVAC expressing rabies G (Purevax™) has been licensed for use in cats. The Purevax™ product is unique in the veterinary market since it is the first non-adjuvented rabies vaccine for cats. Adjuvented vaccines have been implicated as a potential risk factor for inducing fibrosarcomas at the site of injection.[61]

In addition to their utility as vaccines against infectious disease agents, poxvirus recombinants have also been utilized to generate immune responses against tumor associated antigens (TAAs). A number of nonmutated melanocyte differentiation antigens have been identified in melanoma patients. Since melanoma patients who respond to IL2 treatment develop autoimmune depigmentation (vitiligo),[62] murine models have been developed to examine ways to induce vitiligo by administration of melanoma antigens. Vaccinia recombinants (WR strain) expressing murine homologs of gp100, MART-1, tyrosinase, TRP-1 and TRP-2 were administered to mice. The TRP-1 immunized mice developed vitiligo, and these animals were also protected against challenge with B16 melanoma.[63]

The human carcinoembryonic antigen (CEA) is expressed in a high percentage of adenocarcinomas including breast, colon, pancreas and lung. Vaccinia and ALVAC recombinants expressing CEA were shown to elicit CEA-specific humoral and cellular immune responses in mice and to inhibit the growth of tumors expressing human CEA.[64,65] A prime/boost approach where mice were immunized with vaccinia-CEA followed by multiple boosts with

ALVAC- CEA significantly potentiated anti-tumor effects and CEA-specific T-cell responses.[64] Similarly, responses were enhanced with a combination vaccine consisting of a mixture of vaccinia-CEA and a vaccinia recombinant expressing the murine costimulatory molecule B7.1, or a dual gene vaccine co-expressing CEA and B7.1 in one vaccinia virus.[65] Both vaccinia-CEA and ALVAC-CEA recombinants have been evaluated in Phase I clinical trials in patients with carcinoma. These studies have established the safety of these recombinants and demonstrated they are able to stimulate CEA-specific CTL responses in humans.[66-68]

Human papilloma virus (HPV) has been implicated in cervical cancer, and in animal models CTL responses to HPV regulatory proteins E6 and E7 can protect against relevant tumor challenge. A vaccinia recombinant (Wyeth strain) expressing HPV16 and 18 E6/E7 fusion proteins was given as a single immunization to late stage cervical cancer patients. Vaccination elicited HPV-specific antibody and CTL responses, with no significant clinical side effects.[69]

To broaden responses to single vaccine products that elicit suboptimal humoral or cellular immune responses prime-boost protocols are being used. In this immunization scheme the first (priming) immunization is performed with one method of delivery of the immunogen and subsequent immunizations (boosts) are performed with another method of delivering the immunogen. This strategy also addresses the observation that host responses to vaccinia gene products can limit viral replication and immune responses to subsequent inoculations. In a number of Phase I clinical trials, ALVAC recombinants expressing HIV envelope genes were used to prime immune responses, followed by boosts with recombinant HIV-env protein.[30,70-73] This regimen was shown to induce stronger and broader env-specific immune responses than immunization with the ALVAC recombinants or env protein alone. In a recent Phase I clinical trial, volunteers were immunized with an ALVAC recombinant expressing the gB envelope glycoprotein from HCMV and then were boosted with a single dose of an attenuated strain of HCMV. HCMV-specific neutralizing antibody was detected sooner, reached high titers and persisted longer in subjects primed with ALVAC-gB than in subjects primed with an ALVAC recombinant expressing rabies glycoprotein G.[31] Other successful prime-boost results were achieved by initial priming inoculations with DNA-based vaccines expressing malaria or HIV antigens, followed by boosts with a poxvirus recombinant (vaccinia or fowlpox) expressing the same antigen. This scheme elicited protection against immunodeficiency virus or *Plasmodium berghei* challenge.[19,74-76] In one of the protocols an enhanced response was achieved when a plasmid expressing GM-CSF was included in the priming step of the protocol.[75]

In the past decade, advances in the use of poxvirus vectors have resulted in the licensure of commercial veterinary products and in the development of new vectors and protocols for use in human trials. The next decade may bring the first successful licensure of a poxvirus-based recombinant for either prophylactic or therapeutic use in humans.

Adenovirus Vectors

Adenoviruses are a family of relatively large (30-40 kb) non-enveloped double-stranded DNA viruses. Individual members have been isolated from numerous mammalian and avian species. Clinical symptoms associated with adenovirus infection are usually mild and rarely life threatening. Live attenuated human adenovirus type 4 (Ad4) and type 7 (Ad7) have been used as an oral vaccine by the U.S. military for the past 30 years to prevent respiratory disease in recruits.[77] Live attenuated adenovirus vaccines are also used to protect domesticated animals against adenovirus-associated disease.[78] The success of these vaccines has encouraged the development of adenoviruses as recombinant vectors.

As vaccine vectors, adenoviruses have many advantageous properties. Recombinants can be easily generated, grown to high titers, administered intramuscularly, intranasally or orally, and used to infect a broad spectrum of both dividing and nondividing cells. Three types of recombinants have been developed:

1. replication-competent,
2. helper-free, replication-defective and
3. helper-dependent, replication-defective.

Since an adenovirus capsid can package only about 5% more DNA (1.2 kb) than the wild type genome, genes must be removed to accommodate most exogenous sequences.[79] Insertion of foreign genes into the non-essential E3 and E4 regions (up to 5-6 kb) results in the generation of replication-competent recombinants. On the other hand, insertion into the essential E1 region results in the generation of helper-free, replication-defective recombinants that must be propagated in a cell line that expresses the E1 gene products.[80] Deletion of larger amounts of adenovirus DNA results in the generation of helper-dependent, replication-defective recombinants that must be propagated in a helper-virus system that supplies the missing adenovirus structural and enzymatic functions. A major problem in the amplification of E1-deleted recombinants in E1 expressing cells is the generation of replication-competent adenovirus (RCA) through recombination.[81] However, the recent development of a cell line that lacks overlapping adenovirus sequences suggests that the generation of RCAs can be prevented in the future.[82]

Both replication-competent and replication-defective adenovirus vectors have been shown to be efficacious in generating immune responses to a variety of infectious diseases (Table 7.1). Examples of replication-competent recombinants that have successfully protected target species include a human Ad5-rabies virus glycoprotein recombinant that protected skunks and foxes against rabies challenge,[83] an Ad5-porcine respiratory coronavirus (PRCV) spike recombinant that partially protected pigs against PRCV challenge,[84] and an Ad5-foot-and-mouth disease virus (FMDV) P1 recombinant that partially protected cattle against FMDVchallenge.[85]

Although replication-competent adenovirus vaccines have been used as vaccines for over 30 years, their ability to cause disseminated disease in immunocompromised individuals[99] and their tendency to be excreted by oro-fecal and respiratory routes for months after inoculation are serious safety concerns. In fact, horizontal transmission of the Ad4 vaccine has been observed between married couples[100] and between vaccinated children and family members.[101]Helper-free, replication-defective adenoviruses, which have a lower rate of shedding, would be safer vectors. However, they are not always as efficacious as replication-competent recombinants. A direct comparison between replication-competent and incompetent Ad5 vectors expressing PRV gD was carried out in a cotton rat/PRV challenge model. The protective dose of the replication-defective recombinant was at least 250 times higher than the protective dose of the replicating recombinant.[93] Another comparison in a cotton rat/bovine herpesvirus type 1 (BHV1) challenge model with Ad5 recombinants expressing BHV1 gD also demonstrated that the replication-defective was not as protective as the replication-competent recombinant.[94] Furthermore, a replication-defective Ad5-feline immunodeficiency virus (FIV) env recombinant was not able to protect cats against an FIV challenge.[87]

To avoid the safety issues associated with the shedding of human adenovirus vectors, species-specific adenovirus vectors are being evaluated for veterinary applications. Several replication-competent species-specific adenovirus recombinants have been shown to protect target species against infectious challenge. For example, a bovine adenovirus (BAV-3)-BHV1 gD recombinant protected calves against BHV1 challenge[89] and a fowl adenovirus (FAV-10)-infectious bursal disease virus (IBDV) VP2 recombinant protected chickens against IBDV challenge.[88] Porcine adenovirus-3 and ovine adenovirus vectors have also been evaluated.[102,90] In addition to veterinary applications, nonhuman adenoviruses are also being considered for human use, since they are replication-defective in humans but retain the ability to infect and express recombinant antigens. Moreover, these vectors would not be affected by pre-existing immunity against human adenoviruses.[103,104]

As with poxviruses, prime-boost regimens with multiple immunization products are being evaluated to broaden the immune responses elicited by adenovirus vectors. For example, a prime-boost regimen was carried out with a DNA plasmid expressing the protective antigen (45W) from the cestode parasite, *Taenia ovis*, and an ovine adenovirus recombinant expressing the same antigen. The combination procedure elicited stronger humoral immune responses than either vaccine alone and was able to protect sheep against a *T. ovis* challenge.[90] A prime-boost regimen with adenovirus-HIV-1 gp160 recombinants and an HIV-1 gp120 subunit protein was able to elicit humoral and cellular immune responses and protect chimpanzees against an HIV-1 challenge.[95] In a similar prime-boost regimen, with an Ad5-simian immunodeficiency virus (SIV) env recombinant and an SIV gp120 subunit protein, humoral and cellular immune responses were elicited in macaques, but the animals were not protected against an SIV challenge.[92] Another prime-boost protocol was able to boost the response to a replication-defective Ad5 recombinant expressing rabies G.[96]

In addition to generating immune responses against infectious agents, adenovirus recombinants are also being utilized to elicit immune responses against tumor associated antigens (TAAs). Adenovirus-TAA recombinants have protected mice against a variety of tumor challenges. For example, specific CTL responses were induced by replication-defective Ad2 recombinants expressing human MART-1 or gp100.[105] The gp100 recombinant also protected mice against a B16 challenge. Unfortunately, these adenovirus recombinants elicited neither a clinical response, nor MART-1- or gp100-specific immune responses in metastatic melanoma patients, indicating that they are unable to break tolerance or cure patients with established tumors.[106] Better responses have been obtained with p53, a tumor suppressor protein involved in cell cycle regulation and apoptosis. Repeated intratumoral inoculation of an Ad5-human p53 recombinant was able to mediate antitumor activity and elicit clinical responses in a subset of advanced non-small-cell lung cancer patients.[98]

The systemic administration of cytokines such as IL-2 and IL-12 has proven efficacious in a variety of preclinical tumor models, presumably by boosting immune responses to TAAs. Unfortunately, systemic administration of cytokines to humans is limited due to dose-related toxicities. These toxicities could be avoided by the localized production of cytokines from an intratumorally-inoculated vector. In a mouse model intratumoral administration of a replication-defective Ad5-murine IL-12 recombinant resulted in the regression of established murine tumors and protected mice against subsequent tumor challenge.[107,108] Intratumoral or intravenous inoculation of Ad5-murine IL-12 also suppressed metastases.[108,109] Adenovirus-IL-2 recombinants have also been used successfully in murine tumor models.[110,111] In all of these studies, although some cytokine-related toxicities have been observed,[112] protection has been elicited without the toxicity associated with systemic cytokine administration.

Another approach to enhance the immunogenicity of TAAs involves immunization with antigen-loaded dendritic cells (DCs). DCs are potent antigen-presenting cells and adenovirus vectors, because of their ability to infect non-dividing cells, are uniquely suited to deliver genes to DCs, which are terminally differentiated. In fact, adenovirus can deliver recombinant genes to DCs more efficiently than lipofection, electroporation or $CaPO_4$ precipitation.[113] Moreover, replication-defective adenoviruses do not perturb the maturation or function of DCs.

Immunization with adenovirus recombinant-transduced DCs has been shown to be efficacious in murine tumor models.[114-116] For example, DCs transduced with a replication-defective Ad5-human MART-1 recombinant protected mice against challenge with a murine fibrosarcoma cell line (NFSA) engineered to express human MART-1.[97] Mice carrying established NFSA(MART-1) tumors were also protected. DCs transduced with adenovirus-TAA recombinants also protected mice against a subsequent tumor challenge. Mice immunized with murine DCs transduced with a replication-defective Ad2-murine tyrosinase-related protein-2 (TRP-2) recombinant were protected against a B16 challenge,[116] although a gp100 recombinant did not elicit protection.

In conclusion, adenoviruses are highly attractive vaccine vectors with multiple uses and applications in both infectious disease and cancer programs. To address the regulatory and safety issues associated with licensing an adenovirus-based vaccine, second-generation adenovirus recombinants, such as replication-defective and species-specific vectors, are being developed. To address immunogenicity issues, alternative immunization regimens, such as prime-boost and infected DCs, are being evaluated. The success of these and previous studies may eventually lead to the licensure and marketing of adenovirus-based vaccines.

Herpesvirus vectors

Herpesviruses are large (120-300 nm) enveloped viruses which replicate in the nucleus of infected cells. Their genome consists of a linear dsDNA molecule ranging from 120-230 kbp. The naked DNA of herpesviruses is infectious when transfected into a sensitive cell culture. Herpesvirus genomes contain at least 70 genes, some of which are non-essential for in vitro growth. After primary infection, they persist in specific cells of the infected host in a latent stage in which only a limited number of viral genes are expressed. After different stimuli, the latent virus reactivates and may induce secondary infection.[117]

As vaccine vectors herpesviruses offer several advantages.

1. The large genome size and the presence of numerous nonessential genes facilitates the insertion of large fragments of foreign DNA (>20 kbp).
2. The nuclear replication of herpesviruses and the use of cellular transcription machinery allow the use of strong eukaryotic transcription signals and eventually the expression of spliced genes.
3. The use of an attenuated vaccine strain of herpesvirus as the vector allows the generation of bivalent vaccines that protect against the herpesvirus disease as well as against the disease caused by the agent whose gene is expressed.
4. Most herpesviruses can be administered using a mucosal route, as well as a parenteral route.
5. Herpesviruses can usually be grown to high titers.

The persistence of the vector in a latent state in which the viral genome is not integrated into the cellular genome may be an advantage for gene therapy but may be a safety concern for vaccination. The possible change of the biological properties of the vector due to expression of particular foreign genes (see below) may be a disadvantage of such vectors. The procedure for the manipulation of the large size genome is relatively complex, although a new way to generate herpesvirus recombinant in E. coli has recently been described.[118]

Three strategies that have been developed for utilizing herpesviruses as vectors are:

1. replicative herpesviruses in which the foreign gene is inserted into a non-essential gene,[119]
2. nonreplicative herpesviruses in which an essential gene has been inactivated, the product of this essential gene being provided in trans for the production of the virus,[120-122] and
3. a nonreplicative amplicon produced in presence of a helper virus.[123,124]

A fourth strategy has also been recently described for pseudorabies virus (PRV). It is based on the fact that the PRV glycoprotein D (gD) gene is non-essential for the cell-to-cell spread of the virus but is essential for the release of free infectious virions from infected cells. PRV gD-deleted recombinants replicate and spread into the vaccinated animal by means of direct cell-to-cell spread, but cannot be transmitted to contact animals due to the lack of formation of free infectious virions.[125,126]

The use of human herpesviruses as vectors for human infectious disease has been proposed,[119,127] but most of the current work done with these viruses is devoted to the development of vectors for gene or cancer therapy, especially for gene transfer into the brain. These applications have been reviewed elsewhere[128-131] and will not be developed in this review. In contrast, a number of animal herpesviruses such as PRV,[132] Bovine herpesvirus 1 (BHV-1)[133-135] and herpesvirus of turkey (HVT),[136,137] have been tested as replicative vectors for expression of genes from infectious pathogens. Results of protection induced by animal herpesvirus vectors in the target species are reported in Table 7.1. No animal herpesvirus used as a vector is licensed yet, but several candidates are in development.

Excellent levels of protection induced by a herpesvirus vector have been obtained in chickens. Recombinant HVT or Marek's disease virus (MDV) expressing gene(s) from Newcastle disease virus (NDV) fully protected SPF birds after one administration.[136,147,150] The onset of immunity was rapid (14 days), and the interference with maternal antibodies minimal.

A recent study compared the efficacy of two PRV vectors expressing the E2 gene of classical swine fever virus (CSFV), one being able to produce free infectious particles, and the other being nontransmissible due to the deletion of the gD gene (see above). Both viruses induced excellent levels of protection against CSFV and PRV challenges although protection induced by the transmissible vector was slightly better.[139]

The safety of herpesvirus vectors depends on different factors including the pathogenicity of the parental strain, the route of administration, and the properties of the foreign gene. If the parental strain is virulent, genes involved in virulence will have to be inactivated by deletion and/or insertion of the foreign gene.[152] Herpesviruses may be perfectly safe when administered by the parenteral route, but still virulent when administered by the oro-nasal route.[152-154] The presence of some expressed foreign proteins in the recombinant herpesvirus virion can change the biological properties (pathogenicity, tissue-tropism and host-specificity) of the vector.[155] For instance, the class II membrane glycoprotein G gene from bovine respiratory syncytial virus (BRSV) expressed by a BHV-1 vector was incorporated into virions and the recombinant virus could be partially neutralized by anti-G sera.[156] Inoculation into calves resulted in a more severe clinical disease, and higher titers of the recombinant virus were isolated from the lungs when compared to the parental virus.[141] The engineering of a recombinant herpes simplex 1 (HSV-1) targeting binding to a non-HSV-1 cell surface receptor has also been reported.[157]

Factors influencing the efficacy induced by a herpesvirus vector are numerous. The strength of the promoter that drives expression of the foreign gene has been shown to influence the level of protection.[148,158,146] The ability of the recombinant virus to replicate in the host is another important factor for efficacy. Unfortunately, higher level of in vivo replication of the vector correlates often with lower safety indicating that a delicate balance exists between achieving attenuation to provide safety and maintaining immunogenicity for the efficacy of a vaccine. However, the deletion of a gene of the feline herpesvirus vector has been shown to decrease virulence without having any effect on replication in cats.[153] The level of expression of the foreign gene may sometimes be very low or undetectable. For instance, a bovine herpesvirus 1 (BHV-1) recombinant, into which the original gene coding for bovine respiratory syncytial virus (BRSV) glycoprotein G was inserted, was not expressing detectable amounts of G. In contrast, a good level of G expression was obtained when a synthetic G gene was inserted into the BHV-1 vector.[159,156] The original G gene coming from an RNA virus (which replicates in the cytoplasm) may contain signals (including splicing sites) that result in degradation of the transcripts in the nucleus.

Expression of immunomodulators by herpesvirus vectors[160] has been shown to modify the immune response in tumor[161,162] and chronic inflammation[163,164] models. The mode and protocol of administration of the vector are also important factors for efficacy. A prime-boost strategy has been used successfully with the feline herpesvirus used as a vector for feline leukemia.[144]

Overall, herpesviruses are promising candidate vectors for expression of foreign genes, especially for vaccines in animals and for gene and cancer therapy in humans.

Parvovirus Vectors

Adeno-associated virus (AAV) is a defective human parvovirus. It contains a small, single-stranded DNA genome and can infect a broad spectrum of dividing and nondividing cells. Productive replication requires coinfection with a helper virus, such as adenovirus or herpes simplex virus (HSV). In the absence of helper virus, AAV integrates into the host genome, where it persists in a latent state until rescued by an adenovirus or HSV infection. The ability of the AAV genome to integrate into the genome of a wide variety of dividing and nondividing cells and its lack of association with any human disease make AAV an attractive eucaryotic expression vector, especially suitable for gene therapy applications. Although the present packaging systems for generating AAV recombinants are inefficient and cumbersome, a handful of studies have indicated that AAV could potentially become a useful vaccine vector. For example, mice immunized with AAV recombinants expressing HSV2 gB or gD generated HSV2 gB- and gD-specific immune responses.[165] In another example, intratumoral injection of 293 cells transduced with an AAV-IFN recombinant resulted in the regression of an established Eskol tumor derived from a human hairy cell leukemia cell line.[166]

RNA Viruses

Retrovirus Vectors

Retroviruses are important human and animal pathogens that have been remarkably valuable as experimental tools. The first phase of the 2-step retrovirus life cycle consists of viral entry, reverse transcription of the positive strand RNA genome into double strand DNA, transfer to the nucleus of the cell, and stable integration as a provirus into the host cell genome at random sites. The second phase involves synthesis of viral RNA and viral proteins and assembly of the virion. Retroviruses are unique among the RNA viruses because their replication involves a double stranded DNA phase, which facilitates genetic engineering. Transfection of full-length cDNA molecules leads to integration into the host cell genome and production of replicating virus particles. Retrovirus constructs are among the most widely used vectors for gene transfer and gene therapy.

Since the first use of retroviruses as vectors in the early 1980s,[167] the technology has focused principally on gene therapy for treatment of genetic diseases, cancer, and AIDS. This enormous field has been well reviewed, and will not be developed here.[168-173] However, a limited number of retrovirus recombinants have been constructed for use as immunizing agents against infectious agents. A Schmidt-Ruppin Rous sarcoma virus-derived vector expressing influenza hemagglutinin H7 was used to immunize one-month-old chickens. Challenge with a highly pathogenic H7N7 influenza virus revealed protection, although antibody responses were low.[174] Replication-competent avian leukosis virus was used as a vector for the expression of NDV hemagglutinin-neuraminidase (HN) and phosphoprotein (P) genes. Four-week old chickens immunized with the HN recombinant were protected from disease, although viral infection and replication were observed. However, birds immunized with the P construct developed more severe disease than controls receiving no immunization.[175] Townsend et al[176] expressed different forms of hepatitis B virus (HBV) core and e antigens in retrovirus vectors. After immunization of rhesus monkeys, specific HBV core and e antigen antibody and $CD8^+$ cytotoxic T-lymphocyte responses were elicited. The responses lasted 16 weeks and could be boosted, suggesting a use for this recombinant in treating chronic HBV infection.[177] Retroviral vectors expressing the Gag/Pr or SU protein of the lentivirus equine infectious anemia virus (EIAV) were constructed and used to evaluate EIAV-specific CTL responses in horses. Although equine kidney cells transfected with the recombinants were effective targets for CTL from EAIV-infected horses, only one of five horses immunized with a mixture of these constructs responded with EIAV-specific CTL, indicating a need to boost the response, perhaps by the inclusion of cytokines.[178]

The most advantageous feature of retrovirus vectors for use in gene therapy, the irreversible integration of viral DNA into the host genome, is one of the major drawbacks for immunization against infectious diseases. However, long-term expression of pertinent antigens for immunotherapy of cancer or persistent infection caused by agents such as HIV or HBV would be desirable. Since most retroviruses infect only dividing cells, lentivirus vectors such as EAIV and HIV, which will infect non-dividing cell, are being developed to broaden their range. Future work in the targeting of retroviruses to specific cells will offer additional advantages. Information from the large number of clinical trials that have taken place with retrovirus vectors will be an asset in the development of future vaccines.

Positive-Sense RNA Virus

Alphavirus Vectors

Alphaviruses, members of the Togavirus family, have a positive strand RNA genome that is infectious. Members of this genus that have been proposed as viral vectors include Semliki Forest virus (SFV), Sindbis virus and Venezuelan equine encephalilitis virus (VEE). Alphaviruses offer several attractive features:

1. a broad host range including cells of avian, insect and mammalian origin,
2. high levels of protein expression and
3. the ability to manipulate the genome utitlizing cDNA clones to derive infectious RNA transcripts.

The first step in alphavirus replication is translation of the 5' two thirds of the RNA genome to produce the replicase. The positive strand is copied by replicase into intermediate negative strands, which then serve as template for the synthesis of new positive strands. The replicase utilizes an internal promoter on the negative strand to produce a subgenomic (26S) RNA corresponding to the last third of the viral genome. This subgenomic RNA encodes the structural proteins as a polyprotein precursor, which is cotranslationally cleaved to produce the capsid and envelope glycoproteins. Strategies that have been developed for utilizing alphaviruses as vectors include

1. insertion of small antigenic epitopes into one of the capsid proteins,
2. "double promoter" vectors in which a second copy of the 26S promoter is inserted into the genome and is engineered to express a foreign gene and
3. replicon vectors in which the structural protein genes are deleted and replaced by foreign gene sequences under the control of the 26S promoter.

For replicon vectors the structural genes are provided in trans from a helper construct that is missing all or part of the genes encoding for the nonstructural proteins. The first two types of vector have a limited capacity for heterologous genes and produce replication-competent virus particles, which raises concerns for their stability and safety. The replicon system offers the advantage that by utilizing two helper RNAs to supply the capsid proteins, one can reduce the generation of infectious particles to below detectable limits, and increase the capacity for foreign DNA.[179,180]

VEE, Sindbis and SFV recombinants expressing foreign genes from infectious agents have been evaluated for efficacy in mice, guinea pigs and macaques (see Table 7.2 and the discussion that follows). VEE, SFV and sindbis recombinants expressing influenza HA , NP or a CTL epitope from NP have been shown to protect mice upon intranasal challenge with influenza.[179,183,185,189] These studies have also shown that previous immunity to the alphavirus vector did not reduce the immune response to the influenza proteins nor interfere with generation of a protective immune response.[179,190,183,185] Sindbis recombinants expressing a neutralizing epitope from Rift Valley Fever G2 glycoprotein or a CTL epitope from *Plasmodium yoelii* circumsporozoite protein were able to partially protect mice upon subsequent challenge with the corresponding pathogen.[187,189] SFV and Sindbis recombinants expressing Louping ill virus or Japanese encephalitis structural or nonstructural proteins were able to induce immune responses that protected mice from lethal challenge.[182,188]

VEE replicons expressing Marburg virus (MBGV) GP and NP were shown to protect guinea pigs from viremia and death after MBGV challenge.[186] After administration to cynomolgus macaques, animals receiving the GP-replicon or a combination of the GP and NP-replicons survived MBGV challenge without any signs of illness, while animals receiving the NP-replicon were partially protected.[186] An SFV recombinant expressing SIV-PBj14 Env gp160 protein was used to immunize pigtail macaques. After challenge with SIV-PBj14-bcl3, immunized animals were protected against lethal disease but not viral infection.[181] Cynomolgus monkeys that were immunized with an SFV recombinant encoding the HIV-1 envelope protein were challenged with the chimeric simian-human immunodeficiency virus SHIV-4. Anamnestic HIV-1 specific humoral and T-cell proliferative responses were observed in the immunized animals and, although not protected from viral infection, viral loads were significantly reduced compared to unvaccinated controls.[184]

One potential problem with the two-helper RNA system to generate replicon particles is the need to produce and transfect cells with three separate RNAs. For small scale animal studies this has not presented a problem in producing sufficient encapsidated replicons. The development of stable packaging cell lines which can produce high levels of encapsidated replicons with undetectable levels of replication competent virus has recently been reported for Sindbis and SFV.[191] This system may be useful for large-scale replicon production, which will be necessary for human clinical trials.

Picornavirus Vectors

The Picornavirus family consists of small non-enveloped virions that replicate within the cytoplasm of the infected cell. The single-stranded plus sense RNA genome is translated as a single large polyprotein that is subsequently processed by viral encoded proteinases. The genome has been subdivided into three regions: P1 which encodes the capsid proteins, and the P2 and P3 regions which encode the non-structural proteins required for viral replication. The complete nucleic acid sequence of the three serotypes of poliovirus as well as that of Mengo virus is known and infectious cDNAs are available that provide the opportunity to utilize molecular biology approaches to investigate various regions of these genomes for the utility in the expression of foreign antigens.

Strategies that have been employed for engineering poliovirus vectors include:

1. insertion of small antigenic epitopes into one of the capsid proteins,
2. fusion of foreign ORFs in frame to the N terminus of the poliovirus with a artificial viral proteinase cleavage site in between,
3. insertion of foreign ORFs flanked by artificial viral proteinase cleavage sites into the P1-P2 junction,
4. duplication of 5' noncoding region (internal ribosomal entry site [IRES]) so that foreign proteins are expressed using one IRES and the viral polyprotein is expressed using the other IRES (dicistronic poliovirus), and
5. construction of minireplicons in which poliovirus structural protein genes are replaced by foreign sequences, thus requiring a source of P1 proteins to produce infectious particles. Mengo virus recombinants have been engineered by inserting a small antigenic epitope in frame within the L protein coding sequence or as a fusion protein with L.

Polio and Mengo virus recombinants expressing antigens from HIV, SIV and HBV have been shown to elicit humoral and cellular immune responses in mice or cynomolgus monkeys when administered by systemic or mucosal routes.[192-197] A limited number of studies have described the ability of polio and Mengo virus recombinants to afford protection in murine models (Table 7.2). An attenuated Mengo virus recombinant encoding a 14 amino acid CTL epitope of the NP protein from lymphocytic choriomeningitis virus (LCMV) was able to elicit long lasting protective immunity against lethal intranasal LCMV challenge.[198] To explore the utility of poliovirus vectors for cancer immunotherapy, a poliovirus recombinant containing a 600-nt sequence encoding the C-terminal portion of chicken ovalbumin was used to immunize poliovirus receptor transgenic mice (PVR^+). Ovalbumin-specific CTL was induced and mice were protected from local tumor growth and death after challenge with a murine tumor cell line stably expressing ovalbumin.[199] In another model, PVR^+ mice immunized with a poliovirus replicon expressing the C-fragment of tetanus toxin developed an increase in tetanus and polio-specific antibodies and were partially protected from challenge with a lethal dose of tetanus toxin.[200] To examine whether pre-existing immunity to poliovirus might diminish the immune response to the foreign antigen expressed by the replicon, PVR^+ mice were immunized with a 1/10 human dose of inactivated poliovirus vaccine before administration of the replicon. All mice receiving the inactivated poliovirus vaccine developed poliovirus-specific antibody. After boost with replicon the response against tetanus toxin was comparable to that obtained after two doses of replicon and also resulted in a boost in poliovirus specific titers. Upon challenge with tetanus toxin, all mice receiving the replicon boost after polio vaccine survived without any sign of disease, demonstrating that pre-existing immunity to poliovirus did not prevent elicitation of a protective immune response.[200]

The results reviewed here suggest that vaccines based on attenuated picornavirus vectors may provide an effective system to induce both humoral and cellular immune responses. Because of their ability to replicate in the gastrointestinal tract they have the potential to stimulate the mucosal immune system. Since the capacity for heterologous sequences is limited one could envision administration of a cocktail of recombinants allowing simultaneous immunization against multiple pathogens. In fact a cocktail of six poliovirus recombinants expressing five different SIV proteins was able to elicit antibody responses to all of the SIV antigens after a single immunization of mice.[192] A current limitation with this technology is the observation that most recombinants examined are impaired in viral replication and genetically unstable.[192,201] Stability seems to correlate with the size of the insert, smaller inserts being more stable. Although the minireplicon strategy seems to overcome the stability issue, the production of encapsidated replicon particles requires a source of capsid proteins, which can be supplied in trans by infection with a vaccinia recombinant expressing P1. This strategy has limitations in terms of the cost to produce vaccine candidates since all lots of replicons would require a vaccinia recombinant for generation. Furthermore, the replicons need to be purified since residual vaccinia in the replicon preparation may cause adverse complications upon vaccination of immunocompromised individuals. Development of packaging cell lines producing the appropriate picornavirus structural proteins would allow the production of the replicon particles without the need for a helper virus.

Flavivirus Vectors

The family Flaviviridae, encompassing the Pesti and the Flaviviruses, represents small, enveloped viruses with positive-stranded RNA genomes that contain continuous long open reading frames (ORFs). The genome-length ORF is translated into a polyprotein that is co- and post-translationally processed by viral and host cellular proteases (ref. 202 for review). As previously described for Picornaviruses, the fact that the single-stranded RNA genome is infectious and able to function as an mRNA directing production of all viral proteins has made their genetic manipulation, in theory at least, relatively straightforward. In practice, some instability problems have been encountered in the generation of bacterial vectors containing long

Table 7.2. Examples of live RNA viral vector vaccine candidates

Virus	Vector[a]	Foreign gene[b]	Efficacy Species[c]	Efficacy Protection[d]	References
Alphavirus					
	SFV	SIV-PBj14 env	pigtail macaques	±	181
	SFV	Louping ill virus prME or NS1	mice	+	182
	SFV	Influenza-A NP or HA	mice	+	183
	SFV	HIV-I IIIB env	cynomolgus macaques	-	184
	VEE	Influenza-A HA	mice	+	179,185
	VEE	Marburg virus GP or NP	guinea pigs cynomolgus macaques	+ +	186
	Sindbis	Rift Valley Fever G2	mice	±	187
	Sindbis	JEV prM through NS2A	mice	±	188
	Sindbis	Malaria CTL epitope or Influenza-A NP	mice	±	189
Picornavirus					
	Mengo virus	LCMV CTL epitope	mice	±	198
	Poliovirus	ovalbumin CTL epitope (as a target tumor antigen)	mice	±	199
	Poliovirus	Tetanus toxin C-fragment	mice	±	200
Flavivirus					
	Dengue (type 4)	Langat virus	mice	+	215
Orthomyxovirus					
	Influenza	*Plasmodium yoelii* CS (T and B cell epitopes)	mice	+	241,242
	Influenza	LCMV N (CTL epitope)	mice	+	243
	Influenza	*Pseudomonas aeruginosa* outer membrane protein epitope	mice	+	244
Rhabdovirus					
	VSV	Influenza HA	mice	+	245

a SFV Semliki Forest virus; VEE Venezuelan equine encephalitis; VSV Vesicular stomatitis
b SIV simian immunodeficiency virus; HIV human immunodeficiency virus; JEV Japanese encephalitis virus LCMV lymphochoriomeningitis virus
c Species in which the test was performed
d Protection against challenge: + complete; ± partial; – negative

flavivirus-specific inserts. Several flaviviruses have now been developed as infectious clones including dengue virus type 4,[203] Japanese encephalitis virus,[204] dengue type 2,[205] Kunjin virus,[206] classical swine fever virus,[207,208] tick-borne encephalitis virus,[209] yellow fever virus,[201] hepatitis C virus,[211] and BVDV.[212-214]

The major focus of the engineering of flavivirus genomes has been the creation of attenuated vaccines either by the generation of viral chimeras in which genes associated with virulence have been inserted into a more attenuated recipient virus or by specific mutagenesis of virulent strains. For instance, to reduce the neurovirulence of tick-borne Langat virus the genes encoding the structural proteins pre-membrane (preM) and envelope (E) were substituted for dengue virus type 4 preM and E genes. The resulting chimeric virus was more attenuated in mice than the parental Langat virus but induced full protection against virulent Langat virus challenge.[215] Similarly, the replacement of the preM and E genes of the attenuated 17D strain of yellow fever virus with the same genes from a vaccine strain of Japanese encephalitis virus resulted in a YF/JE chimera. When tested in non-human primates, the recombinant was safe, immunogenic and protective against Japanese encephalitis virus challenge.[216,217]

Negative-Sense RNA Virus Vectors

The major block to the application of genetic engineering to negative-stranded RNA viruses was the fact that the template for the polymerase in these viruses is not free nucleic acid. Instead, the template consists of a ribonucleoprotein complex in which the genomic RNA is tightly encapsulated by the nucleoprotein (N or NP). The naked genomic RNA is therefore non-infectious. Procedures have now been developed to overcome this block and have been applied to both the segmented *Orthomyxoviridae* and *Bunyaviridae* families and the non-segmented order *Mononegavirales*, which includes the families *Filoviridae*, *Paramyxoviridae* and *Rhabdoviridae*. Since excellent reviews are now available describing this productive area of research,[218-223] only a brief summary will be provided here

The first demonstration of the rescue of infectious virus from a cDNA copy of a negative-strand genome was made with influenza virus.[224,225] The procedure required the in vitro production of a biologically active RNP complex, consisting of a cDNA-derived RNA copy of one of the eight segments, which was incubated with purified NP and viral RNA polymerase proteins. The RNP complex was transfected into cells that were infected with a helper influenza virus, which could replicate and transcribe the RNP as well as supply the remaining seven viral RNA segments. Since the resulting virus contained an RNA segment derived from cDNA, it was now possible to introduce mutations or foreign sequences into the influenza genome.[220]

This approach did not prove feasible for the nonsegmented RNA viruses, but a breakthrough came with the rescue of infectious rabies virus from cloned cDNA.[226] A plasmid with a full-length antigenome sense cDNA copy of the rabies virus genome linked to the T7 RNA polymerase promoter was transfected into cells that were co-infected with a vaccinia virus recombinant expressing the T7 polymerase. Plasmids containing the T7 promoter linked to cDNA copies of the N protein and polymerase proteins were also cotransfected into the cells. Infectious rabies virus was recovered.

With some modifications, this reverse genetics technique has now been applied to the production of infectious virus from cDNA copies of both segmented and nonsegmented negative-stranded RNA viruses.[222] The list of rescued viruses includes vesicular stomatitis virus (VSV),[227,228] measles virus,[229] respiratory syncytial virus (RSV),[230] Sendai virus,[231,232] Bunyamwera virus,[233] rinderpest virus,[234] human parainfluenza virus,[235,236] simian virus 5,[237] and Newcastle disease virus.[238]

Another breakthrough occurred with the development of a convenient system allowing the rescue of influenza virus entirely from cloned cDNA.[239,240] The system involved the cotransfection of plasmids encoding each of the 8 RNA segments (under control of the pol I promoter and terminator) and plasmids encoding the three proteins that make up the polymerase complex (under control of the pol II promoter). With this scheme influenza A/WSN/33 was rescued from human embryonic kidney cells. Addition of plasmids encoding the other viral proteins, HA, NA, M1, M2 and NS2 enhanced the efficiency of virus rescue at least ten-fold. This work is a major step forward which should speed the development of new influenza vaccines.

The ability to derive infectious virus from cDNA clones opened the way for site-specific mutagenesis studies on specific genes as well as an investigation of the potential for incorporation of foreign sequences and development of the viruses as vaccine vectors. Examples of expression of relevant foreign proteins, which have resulted in the induction of protective immunity, are shown in Table 7.2 and summarized below.

Several groups have demonstrated protective efficacy induced by recombinant influenza virus vectors against viral, bacterial and parasitic diseases. Influenza recombinants expressing T or B cell epitopes from the circumsporozoite protein of *Plasmodium yoelii*, a rodent malaria parasite, induced protection against malaria infection in a prime boost strategy in combination with vaccinia recombinants expressing circumsporozoite immunogens.[241,242] Castrucci et al[243] demonstrated protection against lymphocytic choriomeningitis virus (LCMV) infection in mice. A CTL epitope from the nucleoprotein of LCMV expressed in the influenza virus neuraminidase stalk induced full protection against LCMV-induced intracerebral challenge. An epitope of the outer membrane protein F of *Pseudomonas aeruginosa* expressed in influenza virus provided significant protection against *P. aeruginosa* intratracheal challenge.[244] Expression of the ectodomain of the E2 glycoprotein of classical swine fever virus[246] and a neutralizing epitope from HIV-1 gp41[247] have also been described, although no protection was recorded in these studies..

Considerable work has now also been performed with VSV. This virus may present particular advantages as a vaccine vector because of its ability to grow to high titers in tissue culture, its broad host range and good safety profile. In 1996, Schnell et al[248] demonstrated the production of four VSV recombinants expressing either the cellular CD4 protein, a CD4-G hybrid containing the ectodomain of CD4 and the transmembrane and cytoplasmic tail of VSV-G or the measles virus hemagglutinin or fusion proteins. Genes were inserted between the G and L genes in the VSV genome. All proteins were transported to the infected cell surface and incorporated into the virion envelope. Further work demonstrated that the production of viruses that lack the G protein and are capable of only one round of infection was possible by inclusion of a plasmid expressing the G protein in the original transfection mixture.[249] The authors then demonstrated that insertion of a functional HIV-1 receptor complex composed of CD4- and CXCR4-derived proteins into the envelope of a G-deficient VSV resulted in a virus which selectively infected HIV-1 envelope expressing cells.[249] This method of targeting HIV-infected cells was proposed as a method of reducing viral load and controlling HIV infection. Recombinant VSV have also been used to examine host range tropism and receptor specificity of Ebola virus[250] and to determine the feasibility of expressing the attachment and fusion glycoproteins of RSV as a potential vaccine.[251] There is only one report of protection induced by a VSV recombinant. Kretzschmar et al[252] described the development of a VSV recombinant expressing the hemagglutinin (HA) glycoprotein from influenza A virus. Inoculation of mice with this recombinant by the intranasal route induced full protection against influenza-induced pathogenesis and death.[245]

The ability to rescue infectious virus from cloned cDNA allows the possibility of forming virus chimeras in which surface glycoproteins from virulent strains can be expressed in an attenuated backbone with the idea that the resulting viruses may be useful as live attenuated vaccines. This approach has been taken for RSV, subgroups A and B[253] and human parainfluenza virus type 1 (hPIV1) and type 3 (hPIV3).[254,255] A recombinant bearing the major protective antigens, F and HN, of hPIV-1 and the remainder of the viral proteins of hPIV3 induced complete protection against hPIV-1 challenge virus replication in hamsters and partial protection against hPIV-3 challenge.[255] Spielhofer et al[256] also used this approach to replace the measles virus F and H surface glycoproteins with the VSV G protein. Mice vaccinated with the chimeric virus were protected against lethal VSV challenge. The ability of these viruses to package additional RNA may allow them to be used as vectors to deliver heterologous genes.

It is clear from these examples that the modification of nonsegmented RNA virus genomes is now an efficient process. The impact that these technologies will have on increasing our understanding of virus biology will be immense. Their effect on the development of new vaccines is harder to predict and has been well summarized.[221] It seems that the major limitation may be in establishing long-term stable expression of the non-essential foreign sequences because of the lack of a proofreading function in the RNA polymerase.

Summary

Live viral vectors hold promise as vaccines for the treatment of infectious diseases, including HIV, and cancer. The successful licensing of poxvirus vaccines for prevention of rabies in the wild and for immunization against diseases of poultry and domestic animals proves that the technology can work. The question is whether the issues of safety, efficacy, host range, genetic stability, vector immunity, ease of use, and manufacturing costs can be addressed satisfactorily. Compared to other vaccine modes viral vectors have some distinct advantages. Induction of both humoral and cellular immune responses are consistently more vigorous for live viral vectors than for inactivated viruses or subunit vaccines. The ability of the foreign protein to be synthesized, processed, and modified in situ have advantages in terms of presentation to the immune system. The immunological help supplied by viral vectors make them attractive compared to subunit and DNA vaccines. The ability of some viral vectors to stably express more than one exogenous antigen could lead to multivalent vaccines which are beneficial in terms of manufacturing cost and ease of use. The largest drawbacks to this vaccine technology are the issues of safety and vector immunity. Vector immunity is being addressed by combination protocols where heterogeneous vectors or vaccine modes prevent the development of strong immunity to the vector. The issues of safety, particularly in immunocompromised or neonatal individuals, are being addressed by the development of genetically defined attenuating mutations (such as NYVAC), non-replicating vectors (such as herpes, adenovirus and alphavirus replicons), and species-specific vectors that do not replicate in the target host (such as ALVAC). Non-replicating vectors belonging to almost all virus families have been utilized. Many of these need to be produced in cells that provide the missing virus function.

The success of combination protocols in broadening immune responses suggests that these types of protocols will be utilized more in the future. The ability of combination approaches to improve immune responses and break tolerance to poor immunogens such as TAAs hold great promise for the future in cancer treatment and well as treatment and prevention of recalcitrant infectious diseases such as HIV. Improvements in each type of immunization vehicle will serve to enhance combination strategies to bring them to desired levels. Other areas that will be developed in the next 10 years include:

1. delivery of vaccine vectors to appropriate cells,
2. regulation of expression of the foreign and vector genes,
3. modification of the foreign gene increasing its presentation to the immune system,
4. modulation of the immune responses by co-expression of immunomodulators and
5. regulatory issues associated with qualification of cell lines carrying complementing virus genes for nonreplicating vectors.

References

1. Gething MJ, Sambrook J. Cell-surface expression of influenza haemagglutinin from a cloned DNA copy of the RNA gene. Nature 1981; 293:620-625.
2. Moss B. Poxviridae: The viruses and their replication. In: Fields BN, Knipe BN, Howley PM, eds. Virology, 3rd ed. Philadelphia: Fields Virology Lippincott-Williams and Wilkins 1996:2637-2671.
3. Ockenhouse CF, Sun P, Lanar DE, Wellde BT et al. PhaseI/IIa safety, immunogenicity and efficacy trial of NYVAC-Pf7, a pox-vectored, multiantigen, multistage vaccine candidate for *Plasmodium falciparum* malaria. J Infect Dis 1998; 177:1664-1673.
4. Panicali D, Paoletti E. Construction of poxviruses as cloning vectors: insertion of the thymidine kinase gene from herpes simplex virus into the DNA of infectious vaccinia virus. Proc Natl Acad Sci USA 1982; 79:4927-4931.
5. Mackett M, Smith GL, Moss B. Vaccinia virus: A selectable eukaryotic cloning and expression vector. Proc Natl Acad Sci USA 1982; 79:7415-7419.
6. Cox WI, Gettig RR, Paoletti E. Poxviruses as genetic vectors. In: Vos J-MH, ed. Viruses in Human Gene Therapy. New York: Chapman and Hall, 1992:141-178.
7. Perkus ME, Tartaglia J, Paoletti E. Poxvirus-based vaccine candidates for cancer, AIDS and other infectious diseases. J Leukocyte Biol 1995; 58:1-13.
8. Perkus ME, Taylor J, Pincus S et al. Live attenuated vaccinia and other poxviruses as delivery systems: public health issues. Annals of the New York Academy of Sciences 1995; 754:222-233.
9. Pincus S, Tartaglia J, Paoletti E. Poxvirus-based vectors as vaccine candidates. Biologicals 1995; 23:159-164.
10. Moss B. Genetically engineered poxviruses for recombinant gene expression, vaccination and safety. Proc Natl Acad Sci 1996; 93:11341-11348.
11. Paoletti E. Applications of poxvirus vectors to vaccination: an update. Proc Natl Acad Sci USA 1996; 93:11349-11353.
12. Perkus ME, Paoletti E. Recombinant virus as vaccination carrier of heterologous antigens. In: Kaufmann SHE (ed). Concepts in Vaccine Development. Berlin: Walter de Gruyter 1996: 379-421.
13. Sutter G, Wyatt LS, Foley PL et al. A recombinant vector derived from the host range-restricted and highly attenuated MVA strain of vaccinia virus stimulates protective immunity in mice to influenza virus Vaccine 1994; 12:1032-1040.
14. Wyatt LS, Shors ST, Murphy BR et al. Development of a replication-deficient recombinant vaccinia virus vaccine effective against parainfluenza virus 3 infection in an animal model. Vaccine 1996; 14(15):1451-1458.
15. Durbin AP, Wyatt LS, Siew J et al. The immunogenicity and efficacy of intranasally or parenterally administered replication-deficient vaccinia-parainfluenza virus type 3 recombinants in rhesus monkeys. Vaccine 1998; 16:1324-1330.
16. Nam JH, Wyatt LS, Chae SL et al. Protection against lethal Japanese encephalitis virus infection of mice by immunization with the highly attenuated MVA strain of vaccinia virus expressing JEV prM and E genes. Vaccine 1999; 17:261-268.
17. Hirsch VM, Fuerst TR, Sutter G et al. Patterns of viral replication correlate with outcome in simian immunodeficiency virus (SIV)-infected macaques: effect of prior immunization with a trivalent SIV vaccine in modified vaccinia virus Ankara. J Virol 1996; 70:3741-3752.
18. Hanke T, Samuel RV, Blanchard TJ et al. Effective induction of simian immunodeficiency virus-specific cytotoxic T lymphocytes in macaques by using a multiepitope gene and DNA prime-modified vaccinia virus ankara boost vaccination regimen. J Virol 1999; 73:7524-7532.
19. Schneider J, Gilbert SC, Blanchard TJ et al. Enhanced immunogenicity for CD8+ T cell induction and complete protective efficacy of malaria DNA vaccination by boosting with modified vaccinia virus Ankara. Nat Med 1998; 4:397-402.
20. Stephensen CB, Welter J, Thaker SR et al. Canine distemper virus (CDV) infection of ferrets as a model for testing morbillivirus vaccine strategies: NYVAC- and ALVAC- based CDV recombinants protect against symptomatic infection. J Virol 1997; 71:1506-1513.
21. Welter J, Taylor J, Tartaglia J et al. Mucosal vaccination with recombinant poxvirus vaccines protects ferrets against symptomatic CDV infection. Vaccine 1999; 17:308-318.

22. Benson J, Chougnet C, Robert-Guroff M et al. Recombinant vaccine-induced protection against the highly pathogenic simian immunodeficiency virus SIV(mac251): Dependence on route of challenge exposure. J Virol 1998; 72:4170-4182.
23. Bennett AM, Lescott T, Phillpotts RJ. Improved protection against Venezuelan equine encephalitis by genetic engineering of a recombinant vaccinia virus. Viral Immunol 1998; 11:109-117.
24. Holzer GW, Remp G, Antoine G et al. Highly efficient induction by a vaccinia virus vector defective in late gene expression. J Virol 1999; 73:4536-4542.
25. Fischer L, Le Gros F-X, Mason PW et al. A recombinant canarypox virus protects rabbits against a lethal rabbit hemorrhagic disease virus (RHDV) challenge. Vaccine 1997; 15:90-96.
26. Pardo MC, Bauman JE, Mackowiak M. Protection of dogs against canine distemper by vaccination with a canarypox virus recombinant expressing canine distemper virus fusion and hemagglutinin glycoproteins. Am J Vet Res 1997; 58:833-836.
27. Roth J, Dittmer D, Rea D, Tartaglia, J et al. P53 as a target for cancer vaccines: Recombinant canarypox virus vectors expressing p53 protect mice against lethal tumor cell challenge. Proc Natl Acad Sci USA 1996; 93:4781-4786.
28. Cadoz M, Strady A, Meignier B et al. Immunization with canarypox virus expressing rabies glycoproteins. Lancet 1992; 339:1429-1432.
29. Konishi E, Kurane I, Mason PW et al. Induction of Japanese encephalitis virus-specific cytotoxic T lymphocytes in humans by poxvirus-based JE vaccine candidates. Vaccine 1998; 16:842-849.
30. Clements-Mann ML, Weinhold K, Matthews TJ et al. Immune responses to human immunodeficiency virus (HIV) type 1 induced by canarypox expressing HIV-1_{MN} gp120, HIV-1_{SF2} recombinant gp120, or both vaccines in seronegative adults. J Infectious Dis 1998; 177:1230-1246.
31. Adler SP, Plotkin SA, Gonczol E et al. A canarypox vector expressing cytomegalovirus (CMV) glycoprotein B primes for antibody responses to a live attenuated CMV vaccine (Towne). J Infect Dis 1999; 180:843-846.
32. Nazerian K, Witter RL, Lee LF et al. Protection and synergism by recombinant fowl pox vaccines expressing genes from Marek's disease virus. Avian Diseases 1996; 40:368-376.
33. Taylor J, Christensen L, Gettig R et al. Efficacy of a recombinant fowl pox-based newcastle disease virus vaccine candidate against velogenic and respiratory challenge. Avian Diseases 1996; 40:173-180.
34. Swayne DE, Beck JR, Mickle TR. Efficacy of recombinant fowl poxvirus vaccine in protecting chickens against a highly pathogenic Mexican-origin H5N2 avian influenza virus. Avian Diseases 1997; 41:910-922.
35. Cardona CJ, Reed WM, Witter RL et al. Protection of turkeys from hemorrhagic enteritis with a recombinant fowl poxvirus expressing the native hexon of hemorrhagic enteritis virus. Avian Diseases 1999; 43:234-244.
36. Hu L, Ngichabe C, Trimarchi CV et al. Raccoon poxvirus live recombinant feline panleukopenia virus VP2 and rabies virus glycoprotein bivalent vaccine. Vaccine 1997; 15:1466-72.
37. Tripathy DN. Swinepox virus as a vaccine vector for swine pathogens. Adv Vet Med 1999; 41:463-480.
38. Wade-Evans AM, Romero CH, Mellor P et al. Expression of the major core structural protein (VP7) of bluetongue virus, by a recombinant capripox virus, provides partial protection of sheep against a virulent heterotypic bluetongue virus challenge. Virology 1996; 220:227-231.
39. Ngichabe CK, Wamwayi HM, Barrett T et al. Trial of a capripoxvirus-rinderpest recombinant vaccine in African cattle. Epidemiol Infect 1997; 118:63-70.
40. Bertagnoli S, Gelfi J, Le Gall G et al. Protection against myxomatosis and rabbit viral hemorrhagic disease with recombinant myxoma viruses expressing rabbit hemorrhagic disease virus capsid protein. J Virol 1996; 70:5061-5066.
41. Tartaglia J, Perkus ME, Taylor J et al. NYVAC: A highly attenuated strain of vaccinia virus. Virology 1992; 188:217-232.
42. Paoletti E, Taylor J, Meignier B et al. Highly attenuated poxvirus vectors: NYVAC, ALVAC and TROVAC. Dev Biol Stand 1995; 84:159-163.
43. Fries LF, Tartaglia J, Taylor J et al. Human safety and immunogenicity of a canarypox-rabies glycoprotein recombinant vaccine: an alternative poxvirus vector system. Vaccine 1996; 14:428-434.
44. Federal Register, Addition of Appendix D-LVI to the NIH Guidelines (13 September 1993); 58:1750.

45. Mayr A, Hochstein-Mintzel V, Stickl H. Abstammung, eigenschaften and verwendung des attenuierten vacciniastammes MVA. Infection 1975; 3:6-14.
46. Hochstein-Mintzel V, Hanichen T, Huber HC et al. [An attenuated strain of vaccinia virus (MVA). Successful intramuscular immunization against vaccinia and variola]. Zentralbl Bakteriol [Orig A] 1975; 230:283-297. German.
47. Sutter G, Moss B. Nonreplicating vaccinia vector efficiently expresses recombinant genes. Proc Natl Acad Sci 1992; 89:10847-10851.
48. Blanchard TJ, Alcami A, Andrea P et al. Modified vaccinia virus Ankara undergoes limited replication in human cells and lacks several immunomodulatory proteins: implications for use as a human vaccine. J Gen Virol 1998; 79:1159-1167.
49. Holzer GW, Falkner FG. Construction of a vaccinia virus deficient in the essential DNA repair enzyme uracil DNA glycosylase by a complementing cell line. J Virol 1997; 71:4997-5002.
50. Tartaglia J. Recombinant poxvirus vaccine candidates: update and perspectives. Res Immunol 1998; 149:79-82.
51. Beattie E, Tartaglia J, Paoletti E. Vaccinia virus-encoded eIF-2a homolog abrogates the antiviral effect of interferon. Virology 1991; 183:419-422.
52. Beattie E, Denzler KL, Tartaglia J et al. Reversal of the interferon-sensitive phenotype of a vaccinia virus lacking E3L by expression of the reovirus S4 gene. J Virol 1995; 69:499-505.
53. Beattie E, Paoletti E, Tartaglia J. Distinct patterns of IFN sensitivity observed in cells infected with vaccinia K3L- and E3L- mutant viruses. Virology 1995; 210:254-263.
54. Brochier B, Kieny MP, Costy F et al. Large scale eradication of rabies using recombinant vaccinia-rabies vaccine. Nature 1991; 354:520-522.
55. Pastoret PP, Brochier B. Epidemiology and elimination of rabies in western Europe . Vet J 1998; 156:83-90.
56. Hanlon CA, Niezgoda M, Hamir AN et al. First North American field release of a vaccinia-rabies glycoprotein recombinant virus. J Wildl Dis 1998; 34:228-239.
57. Roscoe DE, Holste WC, Sorhage FE et al. Efficacy of an oral vaccinia-rabies glycoprotein recombinant vaccine in controlling epidemic raccoon rabies in New Jersey. J Wildl Dis 1998; 34:752-763.
58. Fearneyhough MG, Wilson PJ, Clark KA et al. Results of an oral rabies vaccination program for coyotes. J Am Vet Med Assoc 1998; 212:498-502.
59. Mackowiak M, Maki J, Motes-Kreimeyer L et al. Vaccination of wildlife against rabies: successful use of a vectored vaccine obtained by recombinant technology. Adv Vet Med 1999; 41:571-583.
60. Taylor J, Weinberg R, Kawaoka Y et al. Protective immunity against avian influenza induced by a fowlpox virus recombinant. Vaccine 1988; 6:504-508.
61. Hendrick MJ. Historical review and current knowledge of risk factors involved in feline vaccine-associated sarcomas. J Am Vet Med Assoc 1998; 213:1422-1423.
62. Rosenberg SA, White DE. Vitiligo in patients with melanoma:normal tissue antigens can be targets for cancer immunotherapy. J Immunother Emphasis Tumor Immunol 1996; 19:81-84.
63. Overwijk WW, Lee DS, Surman DR et al. Vaccination with a recombinant vaccinia virus encoding a "self" antigen induces autoimmune vitiligo and tumor cell destruction in mice:requirement for CD4+ T lymphocytes. Proc Natl Acad Sci USA 1999; 96:2982-2987.
64. Hodge JW, McLaughlin JP, Kantor JA et al. Diversified prime and boost protocols using recombinant vaccinia virus and recombinant non-replicating avian pox virus to enhance T-cell immunity and antitumor responses. Vaccine 1997; 15:759-768.
65. Kalus RM, Kantor JA, Gritz L et al. The use of combination vaccinia vaccines and dual-gene vaccinia vaccines to enhance antigen-specific T-cell immunity via T-cell costimulation. Vaccine 1999; 17:893-903.
66. Tsang KY, Zaremba S, Nieroda CA et al. Generation of human cytotoxic T cells specific for human carcinoembryonic antigen epitopes from patients immunized with recombinant vaccinia-CEA vaccine. J Natl Cancer Inst 1995; 87:982-990.
67. Schlom J, Kantor J, Abrams S et al. Strategies for the development of recombinant vaccines for the immunotherapy of breast cancer. Breast Cancer Res Treat 1996; 38:27-39.
68. Marshall JL, Hawkins MJ, Tsang KY et al. Phase I study in cancer patients of a replication-defective avipox recombinant vaccine that expresses human carcinoembryonic antigen. J Clin Oncol 1999; 17:332-337.

69. Borysiewicz LK, Flander A, Numako M et al. A recombinant vaccinia virus encoding human papillomavirus types 16 and 18, E6 and E7 proteins as immunotherapy for cervical cancer. Lancet 1996; 347:1523-1527.
70. Pialoux G, Excler JL, Riviere Y et al. A prime-boost approach to HIV preventive vaccine using a recombinant canarypox virus expressing glycoprotein 160 (MN) followed by a recombinant glycoprotein 160 (MN/LAI). The AGIS Group, and l'Agence Nationale de Recherche sur le SIDA AIDS Res Hum Retroviruses 1995; 11:373-381.
71. Evans TG, Keefer MC, Weinhold KJ et al. A canarypox vaccine expressing multiple human immunodeficiency virus type 1 genes given alone or with rgp120 elicits broad and durable CD8+ cytotoxic T lymphocyte responses in seronegative volunteers. J Infect Dis 1999; 180:290-298.
72. Belshe RB, Gorse GJ, Mulligan MJ et al. Induction of immune responses to HIV-1 by canarypox virus (ALVAC)HIV-1 and gp120 SF-2 recombinant vaccines in uninfected volunteers. NIAID AIDS Vaccine Evaluation Group. AIDS 1998; 12:2407-2415.
73. Salmon-Ceron D, Excler JL, Finkielsztejn L et al. Safety and immunogenicity of a live recombinant canarypox virus expressing HIV type 1 gp120 MN MN tm/gag/protease LAI (ALVAC-HIV, vCP205) followed by a p24E-V3 MN synthetic peptide (CLTB-36) administered in healthy volunteers at low risk for HIV infection. AGIS Group and L'Agence Nationale de Recherches sur Le Sida. AIDS Res Hum Retroviruses 1999; 15:633-645.
74. Sedegah M, Jones TR, Kaur M et al. Boosting with recombinant vaccinia increases immunogenicity and protective efficacy of malaria DNA vaccine. Proc Natl Acad Sci U S A 1998; 95:7648-7653.
75. Sedegah M, Weiss W, Sacci JB Jr et al. Improving protective immunity induced by DNA-based immunization: priming with antigen and GM-CSF encoding plasmid DNA and boosting with antigen expressing recombinant poxvirus. J Immunol 2000; 164:5905-5912.
76. Robinson HL, Montefiori DC, Johnson RP et al. Neutralizing antibody-independent containment of immunodeficiency virus challenges by DNA priming and recombinant pox virus booster immunizations. Nat Med 1999; 5:526-534.
77. Top F. Control of adenovirus acute respiratory disease in U.S. Army trainees. Yale J Biol Med 1975; 48:185-195.
78. Carmichael LE, Joubert JC, Pollock RV. A modified live canine parvovirus vaccine. II. Immune response. Cornell Vet 1983; 73:13-29.
79. Bett AJ, Prevec L, Graham FL. Packaging capacity and stability of human adenovirus type 5 vectors. J Virol 1993; 67:5911-5921.
80. Bett AJ, Haddara W Prevec L et al. An efficient and flexible system for construction of adenovirus vectors with insertions or deletions in early regions 1 and 3. Proc Nat Acad Sci USA 1994; 91:8802-8806.
81. Zhu J, Grace M, Casale J. Characterization of replication-competent adenovirus isolates from large-scale production of a recombinant adenoviral vector. Hum Gene Ther 1999; Jan 1;10(1):113-21.
82. Fallaux FJ, Bout A, van der Velde I. New helper cells and matched early region 1-deleted adenovirus vectors prevent generation of replication-competent adenoviruses. Hum Gene Ther 1998 Sep 1;9(13):1909-17.
83. Charlton K, Artois M, Prevec L et al. Oral rabies vaccination of skunks and foxes with a recombinant human adenovirus vaccine. Arch Virol 1992; 123:169-179.
84. Callebaut P, Enjuanes L, Pensaert M. Adenovirus recombinant expressing the spike glycoprotein of porcine respiratory coronavirus is immunogenic in swine. J Gen Virol 1996; 77:309-313.
85. Sanz-Parra A, Vazquez B, Sobrino F et al. Evidence of partial protection against foot-and-mouth disease in cattle immunized with a recombinant adenovirus vector expressing the precursor polypeptide (P1) of foot-and-mouth disease virus capsid proteins. J Gen Virol 1999; 80:671-679.
86. Adam M, Lepottier M, Eloit M. Vaccination of pigs with replication-defective adenovirus vectored vaccines: the example of pseudorabies. Vet Microbiol 1994; 42:205-215.
87. Gonin P, Fournier A, Oualikene W et al. Immunization trial of cats with a replication-defective adenovirus type 5 expressing the env gene of feline immunodeficiency virus. Vet Microbiol 1995; 45:393-401.
88. Sheppard M, Werner W, Tsatas E et al. Fowl adenovirus recombinant expressing VP2 of infectious bursal disease virus induces protective immunity against bursal disease. Arch Virol 1998; 143:915-930.

89. Zakhartchouk A, Pyne C, Mutwiri G et al. Mucosal immunization of calves with recombinant bovine adenovirus-3: Induction of protective immunity to bovine herpesvirus-1. J Gen Virol 1999; 80:1263-1269.
90. Rothel JS, Boyle DB, Both GW et al. Sequential nucleic acid and recombinant adenovirus vaccination induces host-protective immune responses against Taenia ovis infection in sheep. Parasite Immunol 1997; 19:221-227.
91. Tacket C, Losonsky G, Lubeck M et al. Initial safety and immunogenicity studies of an oral recombinant adenohepatitis B vaccine. Vaccine 1992; 10:673-676.
92. Buge S, Richardson E, Alipanah S et al. An adenovirus-simian immunodeficiency virus env vaccine elicits humoral, cellular, and mucosal immune responses in rhesus macaques and decreases viral burden following vaginal challenge. J Virol 1997; 71:8531-8541.
93. Eloit M, Adam M. Isogenic adenoviruses type 5 expressing or not expressing the E1A gene: efficiency as virus vectors in the vaccination of permissive and non-permissive species. J Gen Virol 1995; 76:1583-1589.
94. Papp Z, Middleton D, Mittal S et al. Mucosal immunization with recombinant adenoviruses: Induction of immunity and protection of cotton rats against respiratory bovine herpesvirus type 1 infection. J Gen Virol 1997; 78:2933-2943.
95. Lubeck M, Natuk R, Myagkikh, M et al. Long-term protection of chimpanzees against high-dose HIV-1 challenge induced by immunization. Nat Med 1997; 6:651-658.
96. Xiang Z, Pasquini S, Ertl H. Induction of genital immunity by DNA priming and intranasal booster immunization with a replication-defective adenoviral recombinant. J Immun 1999, 162:6716-6723.
97. Ribas A, Butterfield L, McBride W et al. Genetic immunization for the melanoma antigen MART-1/ melan-A using recombinant adenovirus-transduced murine dendritic cells. Can Res 1997; 57:2865-2869.
98. Swisher SG, Roth JA, Nemunaitis J et al. Adenovirus-mediated p53 gene transfer in advanced non-small-cell lung cancer. J Natl Cancer Inst 1999; 91:763-771.
99. Krilov L, Rubin L, Frogel M et al. Disseminated adenovirus infection with hepatic necrosis in patients with human immunodeficiency virus infection and other immunodeficiency states. Rev Infect Dis 1990; 12:303-307.
100. Stanley E, Jackson G. Spread of enteric live adenovirus type 4 vaccine in married couples. J Infec Dis 1969; 119:51-59.
101. Mueller R, Muldoon R, Jackson G. Communicability of enteric live adenovirus type 4 vaccine in families. J Infec Dis 1969; 119:60-66.
102. Reddy P, Idamakanti N, Hyun B-H et al. Development of porcine adenovirus-3 as an expression vector. J Gen Virol 1999; 80:563-570.
103. Papp Z, Babiuk L, Baca-Estrada M. The effect of pre-existing adenovirus-specific immunity on immune responses induced by recombinant adenovirus expressing glycoprotein D of bovine herpesvirus type 1. Vaccine 1999; 17:933-943.
104. Hofmann C, Loser P, Cichon G et al. Ovine adenovirus vectors overcome preexisting humoral immunity against human adenoviruses in vivo. J Virol 1999; 73:6930-6936.
105. Zhai Y, Yang J, Kawakami Y et al. Antigen-specific tumor vaccines: Development and characterization of recombinant adenoviruses encoding MART1 or gp100 for cancer therapy. J Immunol 1996; 156:700-710.
106. Rosenberg S, Zhai Y, Yang J et al. Immunizing patients with metastatic melanoma using recombinant adenoviruses encoding MART-1 or gp1000 melanoma antigens. J Natl Can Inst 1998; 90:1894-1900.
107. Gambotto A, Tuting T, McVey D et al. Induction of antitumor immunity by direct intratumoral injection of a recombinant adenovirus vector expressing interleukin-12. Cancer Gene Ther 1999; 6:45-53.
108. Chen L, Chen D, Block E et al. Eradication of murine bladder carcinoma by intratumor injection of a bicistronic adenoviral vector carrying cDNAs for the IL-12 heterodimer and its inhibition by the IL-12 p40 subunit homodimer. J Immunol 1997; 159:351-359.
109. Siders W, Wright P, Hixon J et al. T cell- and NK cell-independent inhibition of hepatic metastases by systemic administration of an IL-12-expressing recombinant adenovirus. J Immunol 1998; 160:5465-5674.

110. Cordier L, Duffour M, Sabourin J et al. Complete recovery of mice from a pre-established tumor by direct intratumoral delivery of an adenovirus vector harboring the murine IL-2 gene. Gene Ther 1995; 2:16-21.
111. Addison C, Braciak T, Ralston R et al. Intratumoral injection of an adenovirus expressing interleukin 2 induces regression and immunity in a murine breast cancer model. Proc Natl Acad Sci USA 1995; 92:8522-8526.
112. Toloza E, Hunt K, Swisher S et al. In vivo cancer gene therapy with a recombinant interleukin-2 adenovirus vector. Can Gene Ther 1996; 3:11-17.
113. Arthur J, Butterfield L, Roth M et al. A comparison of gene transfer methods in human dendritic cells. Can Gene Ther 1997; 4:17-25.
114. Wan Y, Bramson J, Carter R et al. Dendritic cells transduced with an adenoviral vector encoding a model tumor-associated antigen for tumor vaccination. Hum Gene Ther 1997; 8:1355-1363.
115. Gong J, Chen L, Chen D et al. Induction of antigen-specific antitumor immunity with adenovirus-transduced dendritic cells. Gene Ther 1997; 4:1023-1028.
116. Kaplan J, Yu Q, Piraino S, et al. Induction of antitumor immunity with dendritic cells transduced with adenovirus vector-encoding endogenous tumor-associated antigens. J Immun 1999; 163:699-707.
117. Roizman B. Herpesviridae. In: Fields BN, Knipe DM, Howley PM et al, eds. Fields Virology. 3rd ed. Lippincott: Williams and Wilkins 1996:2221-2230.
118. Messerle M, Crnkovic I, Hammerschmidt W et al. Cloning and mutagenesis of a herpesvirus genome as an infectious bacterial artificial chromosome. Proc Natl Acad Sci USA 1997; 94:14759-14763.
119. Shih M.-F., Arsenakis M, Thiollais P et al. Expression of hepatitis B virus S gene by herpes simplex virus type I vectors carrying α- and β-regulated gene chimeras. Proc Natl Acad Sci USA 1984; 81:5867-5870.
120. Nguyen LH, Knipe DM, Finberg RW. Replication-defective mutants of herpes simplex viru (HSV) induce cellular immunity and protect against lethal HSV infection. J Virol 1992; 66:7067-7072.
121. Farrell H, McLean CS, Harley C et al. Vaccine potential of a herpes simplex virus type 1 mutant with an essential glycoprotein deleted. J Virol 1994; 68:927-932.
122. Johnson PA, Friedman T. Replication-defective recombinant herpes simplex virus vectors. Methods in Cell Biology 1994; 43:211-230.
123. Kwong AD, Frenkel N. Herpesvirus amplicon: effect on size on replication of constructed defective genomes containing eucaryotic DNA sequences. J Virol 1984; 51:595-603.
124. Frenkel N, Singer O, Kwong AD. Minireview: the herpes simplex virus amplicon—a versatile defective virus vector. Gene Ther 1994; 1 Suppl 1:S40-46.
125. Heffner S, Kovacs F, Klupp B et al. Glycoprotein gp50-negative pseudorabies virus: A novel approach to nonspreading live herpesvirus vaccine. J Virol 1993; 67:1529-1537.
126. Peeters B, Bouma A, de Bruin T et al. Non-transmissible pseudorabies virus gp50 mutants: a new generation of safe live vaccines. Vaccine 1994; 12:375-380.
127. Lowe RS, Keller PM, Keech BJ et al. Varicella-zoster virus as a live vector for the expression of foreign genes. Proc Natl Acad Sci USA 1987; 84:3896-3900.
128. Marconi P, Krisky D, Oligino T et al. Replication-defective herpes simplex virus vectors for gene transfer in vivo. Proc Natl Acad Sci USA 1996; 93:11319-11320.
129. Robertson ES, Ooka T, Kieff ED. Epstein-Barr virus vectors for gene delivery to B lymphocytes. Proc Natl Acad Sci USA 1996; 93:11384-11340.
130. Andreansky SS, He B, Gillespies GY et al. The application of genetically engineered herpes simplex viruses to the treatment of experimental brain tumors. Proc Natl Acad Sci USA 1996; 93:11313-11318.
131. Boursnell ME, Entwisle C, Ali SA et al. Disabled infectious single cycle (DISC) herpes simplex virus as a vector for immunotherapy of cancer. Adv Exp Med Biol 1998; 451:379-384.
132. Thomsen DR, Marotti KR, Palermo DP et al. Pseudorabies virus as a live virus vector for expression of foreign genes. Gene 1987; 57:261-265.
133. Kit M, Kit S, Little SP et al. Bovine herpesvirus-1 (infectious bovine rhinotracheitis virus)-based viral vector which expresses foot-and-mouth disease epitopes. Vaccine 1991a; 9:564-572.

134. Kit S, Kit M, Di Marchi RD et al. Modified-live infectious bovine rhinotracheitis virus vaccine expressing monomer and dimer forms of foot-and-mouth disease capsid protein epitopes on surface of hybrid virus particles. Arch Virol 1991b; 120:1-17.
135. Bello LJ, Whitbeck JC, Lawrence WC. Bovine herpesvirus 1 as a live virus vector for expression of foreign genes. Virology 1992; 190:666-673.
136. Morgan RW, Gelb J Jr, Schreurs CS et al. Protection of chickens from Newcastle and Marek's diseases with a recombinant herpesvirus of turkeys vaccine expressing the Newcastle disease virus fusion protein. Avian Dis 1992; 36:858-870.
137. Morgan RW, Gelb J Jr, Pope CR et al. Efficacy in chickens of a herpesvirus of turkeys recombinant vaccine containing the fusion gene of Newcastle disease virus: onset of protection and effect of maternal antibodies. Avian Dis 1993; 37:1032-1040.
138. van Zijl M, Wensvoort G, de Kluyver E et al. Live attenuated pseudorabies virus expressing envelope glycoprotein E1 of hog cholera virus protects swine against both pseudorabies and hog cholera. J Virol 1991; 65:2761-2765.
139. Peeters B, Bienkowska-Szewczyk K, Hulst M et al. Biologically safe, non-transmissible pseudorabies virus vector vaccine protects pigs against both Aujeszky's disease and classical swine fever. J Gen Virol 1997; 78:3311-3315.
140. Schrijver RS, Langedijk JPM, Keil GM et al. Immunization of cattle with a BHV1 vector vaccine or a DNA vaccine both coding for the G protein of BRSV. Vaccine 1997; 15:1908-1916.
141. Taylor G, Rijsewijk FA, Thomas LH et al. Resistance to bovine respiratory syncytial virus (BRSV) induced in calves by a recombinant bovine herpesvirus-1 expressing the attachment glycoprotein of BRSV. J Gen Virol 1998; 79:1759-1767.
142. Verschoor EJ, Willemse MJ, Stam JG et al. Evaluation of subunit vaccines against feline immunodeficiency virus infection. Vaccine 1996;14:285-289.
143. Willemse MJ, van Schooneveld SH, Chalmers WS et al. Vaccination against feline leukaemia using a new feline herpesvirus type 1 vector. Vaccine 1996; 14:1511-1516.
144. Wardley RC, Berlinski PJ, Thomsen DR et al. The use of feline herpesvirus and baculovirus as vaccine vectors for the gag and env genes of feline leukaemia virus. J Gen Virol 1992; 73:1811-1818.
145. Yokoyama N, Maeda K, Fujita K et al. Vaccine efficacy of recombinant feline herpesvirus type 1 expressing immunogenic proteins of feline calicivirus in cats. Arch Virol 1996; 141:2339-2351.
146. Yokoyama N, Fujita K, Damiani A et al. Further development of a recombinant feline herpesvirus type 1 vector expressing feline calicivirus immunogenic antigen. J Vet Med Sci 1998; 60:717-723.
147. Heckert RA, Riva J, Cook S et al. Onset of protective immunity in chicks after vaccination with a recombinant herpesvirus of turkeys vaccine expressing Newcastle disease virus fusion and hemagglutinin-neuraminidase antigens. Avian Dis 1996; 40:770-777.
148. Darteil R, Bublot M, Laplace E et al. Herpesvirus of turkey recombinant viruses expressing infectious bursal disease virus (IBDV) VP2 immunogen induce protection against an IBDV virulent challenge in chickens. Virology 1995; 211:481-190.
149. Croneneberg AM, Van Geffen CEH, Dorrestein J et al. Vaccination of broilers with HVT expressing an *Eimeria acervulina* antigen improves performance after challenge with *Eimeria.* Acta Virologica 1999; 43:192-197.
150. Sakagushi M, Nakamura H, Sonoda K et al. Protection of chickens with or without maternal antibodies against both Marek's and Newcastle diseases by one-time vaccination with recombinant vaccine of Marek's disease virus type 1. Vaccine 1998; 16:472-479.
151. Tsukamoto K, Kojima C, Komori Y. Protection of chickens against very virulent infectious bursal disease virus (IBDV) and Marek's disease virus (MDV) with a recombinant MDV expressing IBDV VP2. Virology 1999; 257:352-362.
152. Willemse MJ, Chalmers WS, Sondermeijer PJA. In vivo properties of a feline herpesvirus type 1 mutant carrying a lacZ insertion at the gI locus of the unique short segment. Vaccine 1996; 14:1-5.
153. Willemse MJ, Chalmers WS, Cronenberg AM et al. The gene downstream of the gC homologue in feline herpes virus type 1 is involved in the expression of virulence. J Gen Virol 1994; 75:3107-3116.
154. Kruger JM, Sussman MD, Maes RK. Glycoproteins gI and gE of feline herpesvirus-1 are virulence genes: safety and efficacy of a gI-gE- deletion mutant in the natural host. Virology 1996; 220:299-308.

155. Mulder WA, Priem J, Glazenburg K et al. Virulence and pathogenesis of non-virulent and virulent strains of pseudorabies virus expressing envelope glycoprotein E1 of hog cholera virus. J Gen Virol 1994; 75:117-124.
156. Kuhnle G, Heinze A, Schmitt J et al. The class II membrane glycoprotein G of bovine respiratory syncytial virus, expressed from a synthetic open reading frame, is incorporated into virions of recombinant bovine herpesvirus 1. J Virol 1998; 72:3804-3811.
157. Laquerre S, Anderson DB, Stolz DB et al. Recombinant herpes simplex virus type 1 engineered for targeted binding to erythropoietin receptor-bearing cells. J Virol 1998; 72:9683-9697.
158. Hooft van Iddekinge BJL, de Wind N, Wensvoort G et al. Comparison of the protective efficacy of recombinant pseudorabies viruses against pseudorabies and classical swine fever in pigs; influence of different promoters on gene expression and on protection. Vaccine 1996; 14:6-12.
159. Shiu JS, Liu ST, Chang TJ et al. The presence of RNA splicing signals in the cDNA constuct of the E2 gene of classical swine fever virus affected its expression. J Virol Methods 1997; 69:223-230.
160. Kuhnle G, Collins RA, Scott JE et al. Bovine interleukins 2 and 4 expressed in recombinant bovine herpesvirus 1 are biologically active secreted glycoproteins. J Gen Virol 1996;77:2231-2240.
161. Toda M, Rabkin SD, Martuza RL. Treatment of human breast cancer in a brain metastatic model by G207, a replication-competent multimutated herpes simplex virus 1. Hum Gene Ther 1998; 9:2177-2185.
162. Andreansky SS, He B, van Cott J et al. Treatment of intracranial gliomas in immunocompetent mice using herpes simplex viruses that express murine interleukins. Gene Ther 1998; 5:121-130.
163. Furlan R, Poliani PL, Galbiati F et al. Central nervous system delivery of interleukin 4 by a nonreplicative herpes simplex type 1 viral vector ameliorates autoimmune demyelination. Hum Gene Ther 1998; 9:2605-2617.
164. Chun S, Daheshia M, Lee S et al. Immune modulation by IL10 gene transfer via viral vector and plasmid DNA: implication for gene therapy. Cellular Immunology 1999; 194:194-204.
165. Manning W, Paliard X, Zhou S et al. Genetic immunization with adeno-associated virus vectors expressing herpes simplex virus type 2 glycoproteins B and D. J Virol 1997; 71:7960-7962.
166. Zhang J, Hu C, Geng Y et al. Gene therapy with an adeno-associated virus carrying an interferon gene results in tumor growth suppression and regression. Cancer Gene Ther 1996; 3:31-38.
167. Wei CM, Gibson M, Spear PG et al. Construction and isolation of a transmissible retrovirus containing the src gene of Harvey murine sarcoma virus and the thymidine kinase gene of herpes simplex virus type 1. J Virol 1981; 39:935-944.
168. Gordon EM, Anderson WF. Gene therapy using retroviral vectors. Curr Opin Biotechnol 1994; 5:611-616.
169. Morgan RA. Retroviral vectors in human gene therapy. In: Vos J-MH, ed. Viruses in human gene therapy. Durham: Chapman & Hall Carolina: Academic Press, 1995: 77-107.
170. Bunnell BA, Morgan RA. Gene therapy for infectious diseases. Clin Microbiol Rev 1998; 11:42-26.
171. Jane SM, Cunningham JM, Vanin EF. Vector development: A major obstacle in human gene therapy. Ann Med 1998; 30:413-415.
172. Robbins PD, Ghivizzani SC. Viral Vectors for gene therapy. Pharmacol Ther 1998; 80:30-47.
173. Klimatcheva E, Rosenblatt JD, Planelles V. Lentiviral vectors and gene therapy. Front Biosci 1999; 4:D481-496.
174. Hunt LA, Brown DW, Robinson HL et al. Retrovirus-expressed hemagglutinin protects against lethal influenza virus infections. J Virol 1988; 62:3014-3019.
175. Morrison T, Hinshaw VS, Sheerar M et al. Retroviral expressed hemagglutinin-neuraminidase protein protects chickens from Newcastle disease virus induced disease. Microb Pathog 1990; 9:387-396.
176. Townsend K, Sallberg M, O'Dea J et al. Characterization of CD8+ cytotoxic T-lymphocyte responses after genetic immunization with retrovirus vectors expressing different forms of the hepatitis B virus core and e antigens. J Virol 1997; 71:3365-3374.
177. Sallberg M, Townsend K, Chen M et al. Characterization of humoral and CD4+ cellular responses after genetic immunization with retrovirus vectors expressing different forms of the hepatitis B virus core and e antigens. J Virol 1997; 71:5295-5303.
178. Lonning Sm, Zhang W, Leib SR et al. Detection and induction of equine infectious anemia virus-specific cytotoxic T-lymphocyte responses by use of recombinant retroviral vectors. J Virol 1999; 73:2762-2769.

179. Pushko P, Parker M, Ludwig GV et al. Replicon-helper systems from attenuated Venezuelan Equine Encephalitis Virus: Expression of heterologous genes in vitro and immunization against heterologous pathogens in vivo.Virology 1997; 239:389-401.
180. Smerdou,C, Liljestrom,P. Two-helper RNA system for production of rcombinant Semliki forest virus particles. J Virol 1999; 73:1092-1098.
181. Mossman SP, Bex F, Berglund P et al. Protection against lethal simian immunodeficiency virus(siv)$_{smm}$PBj14 disease by a recombinant Semliki Forest virus gp160 vaccine and by a gp120 subunit vaccine. J Virol 1996; 70:1953-1960.
182. Fleeton MN, Sheahan BJ, Gould EA et al. Recombinant Semliki Forest virus particles encoding the prME or NS1 proteins of louping ill virus protect mice from lethal challenge. J Gen Virol 1999; 80:1189-1198.
183. Berguland P, Fleeton MN, Smerdou C et al.. Immunization with recombinant Semliki Forest virus induces protection against influenza challenge in mice. Vaccine 1999; 17:497-507.
184. Berguland P, Quesada-Rolander M, Putkonen P et al. Outcome of immunization of cynomolgus monkeys with recombinant Semliki Forest virus encoding human immunodeficiency virus type 1 envelope protein and challenge with a high dose of SHIV-4 virus. AIDS Res Hum Retroviruses 1997; 13:1487-1485.
185. Davis NL, Brown KW, Johnson RE. A viral vaccine vector that expresses foreign genes in lymph nodes and protects against mucosal challenge. J Virol 1996; 70:3781-3787.
186. Hevey M, Negley D, Pushko P et al. Marburg virus vaccines based upon alphavirus replicons protect guinea pigs and nonhuman primates. Virology 1998; 251:28-37.
187. London SD, Schmaljohn AL, Dalrymple et al. Infectious enveloped RNA virus chimeras. Proc Natl Acad Sci USA 1992; 89:207-211.
188. Pugachev KV, Mason PW, Shope RE et al. Double-subgenomic Sindbis virus recombinants expressing immunogenic proteins of Japanese encephalitis virus induce significant protection in mice against lethal JEV infection. Virology 1995; 212:587-594.
189. Tsuji M, Bergman CC, Takita-Sonoda Y et al. Recombinant Sindbis viruses expressing a cytotoxic T-lymphocyte epitope pf a malaria parasite or of Influenza virus elicit protection against the corresponding pathogen in mice. J Virol 1998; 72:6907-6910.
190. Caley, IJ, Betts, MR, Irlbeck et al. Humoral, mucosal, and cellular immunity in response to a Human immunodeficiency Virus type 1 immunogen expressed by a Venezuelan equine encephalitis virus vaccine vector. J Virol 1997; 71:3031-3038.
191. Polo JM, Belli BA, Driver DA et al. Stable alphavirus packaging cell lines for sindbis virus and semliki forest virus-derived vectors. Proc Natl Acad Sci USA 1999; 96:4598-4603.
192. Tang S, van Rij R, Silvera D et al. Toward a poliovirus-based simian immunodeficiency virus vaccine: correlation between genetic stability and immunogenicity. J Virol 1997; 71:7841-7850.
193. Yim T-J, Tang S, Andino R. Poliovirus recombinants expressing Hepatitis B virus antigens elicited a humoral immune response in susceptible mice. Virology 1996; 218:61-70.
194. Andino R, Silvera D, Suggett SD et al. Engineering poliovirus as a vaccine vector for the expression of diverse antigens. Science 1995; 265:1448-1451.
195. Moldoveanu Z, Porter DC, Lu A et al. Immune responses induced by administration of encapsidated poliovirus replicons which express HIV-1 gag and envelope proteins. Vaccine 1995; 13:1013-1022.
196. Altmeyer R, Escriou N, Girard M et al. Attenuated Mengo virus as a vector for immunogenic human immunodeficiency virus type 1 glycoprotein 120. Proc Natl Acad Sci USA 1994; 91:9775-9779.
197. Van der Ryst E, Nakasone T, Habel A et al. Study of the immunogenicity of different recombinant Mengo viruses expressing HIV1 and SIV epitopes. Res Virol 1998; 149:5-20.
198. Altmeyer R, Girard M, van der Werf S et al. Attenuated Mengo virus: A new vector for live recombinant vaccines. J Virol 1995; 69:3193-3196.
199. Mandl S, Sigal LJ, Rock KL et al. Poliovirus vaccine vectors elicit antigen-specifiic cytotoxic T cells and protect mice against lethal challenge with malignant melanoma cells expressing a model antigen. Proc Natl Acad Sci USA 1998; 95:8216-8221.
200. Porter DC, Wang J, Moldoveanu Z et al. Immunization of mice with poliovirus replicons expressing the C-fragment of tetanus toxin protects against lethal challenge with tetanus toxin. Vaccine 1997; 15:257-264.

201. Mueller S, Wimmer E. Expression of foreign proteins by poliovirus polyprotein fusion: analysis of genetic stability reveals rapid deletions and formation of cardiovirus like open reading frames. J Virol 1998; 72:20-31.
202. Chambers TJ, Hahn CS, Galler R et al. Flavivirus genome organization, expression, and replication. Annu Rev Microbiol 1990; 44:649-688.
203. Lai C-J, Zhao B, Hori H et al. Infectious RNA transcribed from stably cloned full-length cDNA of dengue type 4 virus. Proc Nat Acad Sci USA 1991; 88:5139-5143.
204. Sumiyoshi H, Hoke CH, Trent DW. Infectious Japanese encephalitis virus RNA can be synthesized from in vitro-ligated cDNA templates. J Virol 1992: 66(9):5425-5431.
205. Kapoor M, Zhang L, Mohan PM et al. Synthesis and characterization of an infectious dengue virus type-2 RNA genome (New Guinea C strain). Gene 1995; 162(2):175-180.
206. Khromykh AA, Westaway EG. Completion of Kunjin virus RNA sequence and recovery of an infectious RNA transcribed from stably cloned full-length cDNA. J Virol 1994; 68(7):4580-4588.
207. Meyers G, Thiel H-J, Rumenapf T. Classical swine fever virus: Recovery of infectious viruses from cDNA constructs and generation of recombinant cytopthogenic defective interfering particles. J Virol 1996; 70(3):1588-1595.
208. Moormann RJM, van Gennip HGP, Miedema GKW et al. Infectious RNA transcribed from an engineered full-length cDNA template of the genome of a pestivirus. J Virol 1996; 70(2):763-770.
209. Mandl CW, Ecker M, Holzmann H et al. Infectious cDNA clones of tick-borne encephalitis virus European subtype prototypic strain Neudoerfl and high virulence strain Hypr. J Gen Virol 1997; 78:1049-1057.
210. Rice CM, Grakoui A, Galler R et al. Transcription of infectious yellow fever RNA from full-length cDNA templates produced by in vitro ligation. New Biol 1989; 1(3):285-296.
211. Yanagi M, Purcell RH, Emerson, SU et al. Transcripts from a single full-length cDNA clone of hepatitis C virus are infectious when directly transfected into the liver of a chimpanzee. Proc Nat Acad Sci USA 1997; 94:8738-8743.
212. Meyers G, Tautz N, Becher P et al. Recovery of cytopathogenic and noncytopathogenic bovine viral diarrhea viruses from cDNA constructs. J Virol 1996; 70:8606-8613.
213. Meyers G, Tautz N, Becher P et al. Recovery of cytopathogenic and noncytopathogenic bovine viral diarrhea viruses from cDNA constructs. J Virol 1997; 71:1735.
214. Vassilev VB, Collett MS, Donis RO. Authentic and chimeric full-length genomic cDNA clones of bovine viral diarrhea virus that yield infectious transcripts. J Virol 1997; 71:471-478.
215. Pletnev AG, Men R. Attenuation of the Langat tick-borne flavivirus by chimerization with mosquito-borne flavivirus dengue type 4. Proc Nat Acad Sci USA 1998; 95:1746-1751.
216. Chambers TJ, Nestorowicz A, Mason PW et al. Yellow fever/Japanese encephalitis chimeric viruses: Construction and biological properties. J Virol 73(4):3095-3101.
217. Monath TP, Soike K, Levenbook I et al. Recombinant, chimeric live, attenuated vaccine (ChimeriVax^{TM}) incorporating the envelope genes of Japanese encephalitis (SA14-14-2) virus and the capsid and nonstructural genes of yellow fever (17D) virus is safe, immunogenic and protective in non-human primates. Vaccine 1999; 17:1869-1882.
218. Conzelmann KK. Genetic manipulation of non-segmented negative-strand RNA viruses. J Gen Virol 1996; 77:381-389.
219. Conzelmann KK, Meyers G. Genetic engineering of animal RNA viruses. Trends in Micro 1996; 4 (10):386-393.
220. Palese P, Zheng H, Engelhardt OG et al. Negative-strand RNA viruses: Genetic engineering and applications. Proc Nat Acad Sci USA 1996; 93:11354-11358.
221. Radecke F, Billeter MA. Reverse genetics meets the nonsegmented negative-strand RNA viruses. Rev in Med Virol 1997; 7:49-63.
222. Roberts A, Rose JK. Recovery of negative-strand RNA viruses from plasmid DNAs: A positive approach revitalizes a negative field. Virology 1998; 247:1-6.
223. Garcia-Sastre A. Negative-strand RNA viruses: applications to biotechnology. TIBTECH 1998; 16:230-235.
224. Luytjes W, Krystal M, Enami M et al. Amplification, expression, and packaging of a foreign gene by influenza virus. Cell 1989; 59(6):1107-1113.

225. Enami M, Luytjes W, Krystal M et al. Introduction of site-specific mutations into the genome of influenza virus. Proc Nat Acad Sci USA 1990; 87(10):3802-3805.
226. Schnell MJ, Mebatsion T, Conzelmann K-K. Infectious rabies virus from cloned cDNA. EMBO Jnl 1994; 13(18):4195-4203.
227. Lawson ND, Stillman EA, Whitts MA et al. Recombinant vesicular stomatitis viruses from DNA. Proc Nat Acad Sci USA 1995; 92:4477-4481.
228. Whelan SP, Ball LA, Barr JN et al. Efficient recovery of infectious vesicular stomatitis virus entirely from cDNA clones. Proc Nat Acad Sci USA 1995; 92:8388-8392.
229. Radecke F, Spielhofer P, Schneider H et al. Rescue of measles virus from cloned DNA. EMBO Jnl 1995; 14(23):5773-5784.
230. Collins PL, Hill MG, Camargo E et al. Production of infectious human respiratory syncytial virus from cloned cDNA confirms an essential role for the transcription elongation factor from the 5' proximal open reading frame of the M2 mRNA in gene expression and provides a capability for vaccine development. Proc Nat Acad Sci USA 1995; 92:11563-11567.
231. Garcin D, Pelet T, Calain P et al. A highly recombinogenic system for the recovery of infectious Sendai paramyxovirus from cDNA: generation of a novel copy-back nondefective interfering virus. EMBO Jnl 1995; 14(24):6087-6094.
232. Kato A, Sakai Y, Shioda, T et al. Initiation of Sendai virus multiplication from transfected cDNA with negative or positive sense. Genes Cells 1996; 1(6):569-579.
233. Bridgen A, Elliott RM. Rescue of segmented negative-strand RNA virus entirely from cloned complementary DNAs. Proc Nat Acad Sci USA 1996; 93:15400-15404.
234. Baron MD, Barrett T. Rescue of rinderpest virus from cloned cDNA. J Virol 1997; 71(2):1265-1271.
235. Hoffman MA, Banerjee AK. An infectious clone of human parainfluenza virus type 3. J Virol 1997; 71(6):4272-4277.
236. Durbin AP, Hall SL, Siew JW et al. Recovery of infectious parainfluenza virus type 3 from cDNA. Virology 1997; 235(2):323-332.
237. He B, Paterson RG, Ward CD et al. Recovery of infectious SV5 from cloned DNA and expression of a foreign gene. Virology 1997; 237(2):249-260.
238. Peeters BP, de Leeuw OS, Koch G et al. Rescue of Newcastle disease virus from cloned cDNA: Evidence that cleavability of the fusion protein is a major determinant for virulence. J Virol 1999; 73(6):5001-5009.
239. Neumann G, Watanabe T, Ito H et al. Generation of influenza A viruses entirely from cloned cDNAs. Proc Nat Acad Sci USA 1999; 96:9345-9350.
240. Pekosz A, He B, Lamb RA. Reverse genetics of negative-strand RNA viruses: Closing the circle. Proc Nat Acad Sci USA 1999; 96:8804-8806.
241. Li S, Rodrigues M, Rodriguez D et al. Priming with recombinant influenza virus followed by administration of recombinant vaccinia virus induces $CD8^{+}$T-cell-mediated protective immunity against malaria. Proc Nat Acad Sci USA 1993; 90:5214-5218.
242. Rodrigues M, Li S, Murata K et al. Influenza and vaccinia viruses expressing malaria $CD8^{+}$T and B cell epitopes. Comparison of their immunogenicity and capacity to induce protective immunity. J Immunol 1994; 153:4636-4648.
243. Castrucci MR, Hou S, Doherty PC et al. Protection against lethal lymphocytic choriomeningitis virus (LCMV) infection by immunization of mice with an influenza virus containing an LCMV epitope recognized by cytotoxic T lymphocytes. J Virol 1994; 68(6):3486-3490.
244. Stazek J, Gilleland HE Jr, Gilleland LB et al. A chimeric influenza virus expressing an epitope of outer membrane protein F of *Pseudomonas aeruginosa* affords protection against challenge with *P. aeruginosa* in a murine model of chronic pulmonary infection. Infect Immun 1998; 66(8):3990-3994.
245. Roberts A, Kretzschmar E, Perkins AS et al. Vaccination with a recombinant vesicular stomatitis virus expressing an influenza virus hemagglutinin provides complete protection from influenza virus challenge. J Virol 1998; 72 (6):4704-4711.
246. Zhou Y, Konig M, Hobom G et al. Membrane-anchored incorporation of a foreign protein in recombinant influenza virions. Virology 1998; 246:83-94.
247. Ferko B, Katinger D, Grassauer A et al. Chimeric influenza virus replicating predominantly in the murine upper respiratory tract induces local immune responses against human immunodeficiency virus type 1 in the genital tract. J Infect Dis 1998; 178:1359-1368.

248. Schnell MJ, Buonocore L, Kretzschmar E et al. Foreign glycoproteins expressed from recombinant vesicular stomatitis viruses are incorporated efficiently into virus particles. Proc Nat Acad Sci USA 1996; 93:11359-11365.
249. Schnell MJ, Johnson JE, Buonocore L et al. Construction of a novel virus that targets HIV-1-infected cells and controls HIV-1 infection. Cell 1997; 90:849-857.
250. Takada A, Robison C, Goto H et al. A system for functional analysis of Ebola virus glycoprotein. Proc Nat Acad Sci USA 1997; 94:14764-14769.
251. Kahn JS, Schnell MJ, Buonocore L et al. Recombinant vesicular stomatitis virus expressing respiratory syncytial virus (RSV) glycoproteins: RSV fusion protein can mediate infection and cell fusion. Virology 1999; 254:81-91.
252. Kretzschmar E, Buonocore L, Schnell MJ et al. High-efficiency incorporation of functional influenza glycoproteins into recombinant vesicular stomatitis viruses. J Virol 1997; 71(8):5982-5989.
253. Jin H, Clarke D, Zhou HZ et al. Recombinant human respiratory syncytial virus (RSV) from cDNA and construction of subgroup A and B chimeric RSV. Virology 1998; 251(1):206-214.
254. Tao T, Durbin AP, Whitehead SS et al. Recovery of a fully viable chimeric human parainfluenza virus (PIV) type 3 in which the hemagglutinin-neuraminidase and fusion glycoproteins have been replaced by those of PIV type 1. J Virol 1998; 72(4):2955-2961.
255. Tao T, Skiadopoulos MH, Durbin AP et al. A live attenuated chimeric recombinant parainfluenza virus (PIV) encoding the internal proteins of PIV type 3 and the surface glycoproteins of PIV 1 induces complete resistance to PIV1 challenge and partial resistance to PIV3 challenge. Vaccine 1999; 17(9-10):1100-1108.
256. Spielhofer P, Bachi T, Fehr T et al. Chimeric measles virus with a foreign envelope. J Virol 1998; 72(3):2150-2159.

CHAPTER 8

Inactivated Virus Vaccines

Andrew D. Murdin, Benjamin Rovinski, Suryaprakash Sambhara

Introduction

Inactivated virus vaccines have made a significant contribution to the control of infectious disease during the 20th century and will surely remain an important feature of vaccination strategies in the 21st century. Inactivated vaccines are currently widely available for poliomyelitis, influenza, rabies, hepatitis A, tick-borne encephalitis (TBE) and Japanese encephalitis (JE), and several other products are available for limited, primarily military, use. Inactivated vaccines have been proposed for several other viruses, most notably HIV (see below), hantaviruses[1-4] and dengue,[5-7] but are not yet licensed. For comprehensive and up-to-date reviews concerning specific products, the reader is referred to the various chapters of Plotkin and Orenstein's definitive work on vaccines.[8]

In the space available we clearly have to take a selective approach to the subject, so we have divided this chapter into two parts. The first part provides a brief overview of the major inactivated virus vaccines currently available, while the second part considers some of the issues presently facing inactivated virus vaccines, with a particular focus on poliovirus, influenza virus and HIV vaccines. We have chosen these vaccines because these can be considered to represent three different stages in the lifecycle of a vaccine. Poliovirus is likely to follow smallpox as the second human pathogen to be successfully eradicated, which influences and limits the way that polio vaccines will be used and developed in the future. Influenza vaccine is a mature and effective product, but one for which there is both scope and demand for improvement. HIV vaccines have been tested in clinical trials but are still very much a developing technology and as such provide an insight into the future of inactivated virus vaccines. All of these products illustrate the changes which can be expected in the use of inactivated viral vaccines over the next few years; changing patterns of use, increasing use in combination with other products and improving methods of production and formulation.

Current Inactivated Virus Vaccines

Tick-Borne Encephalitis

Tick-borne encephalitis represents a significant public health problem in parts of Europe and the former USSR. Vaccines are manufactured by Immuno AG, Austria and Chiron Behring, Germany. Vaccine virus is passaged in mouse brain then cultured in primary chick embryo cells, formaldehyde-inactivated, purified and formulated. Alternative production processes such as additional passaging in primary chick embryo cells or culture in Vero cells are being developed and offer the prospect of higher virus titers and a purer final product.[9]

New Vaccine Technologies, edited by Ronald W. Ellis. ©2001 Eurekah.com.

Japanese Encephalitis

Historically endemic to Japan, China and Korea, Japanese encephalitis (JE) is also a significant problem in Southeast Asia and India. Vaccination has essentially eliminated human disease in Japan, Korea and Taiwan, although enzootic transmission continues. Formalin-inactivated vaccines prepared from virus passaged in mouse brain are made by a number of manufacturers in Japan, Korea, Taiwan and Southeast Asia. A mouse-brain vaccine manufactured by Biken in Japan is distributed as JE-VAX by Aventis Pasteur in North America. An inactivated vaccine produced on primary hamster kidney cells is manufactured in China but is not available elsewhere, and experimental inactivated products produced on Vero cells have been described.[10]

Rabies

Rabies vaccines enjoy a special place in the history of inactivated viral vaccines. The vaccine tested by Pasteur in 1885 was the second human vaccine to be developed (after smallpox) and the first to be intentionally inactivated, at least partially.[11,12] The vaccine was prepared by allowing rabies-infected rabbit spinal cords to air dry for various periods of time, and immunization was achieved by successive inoculations with a suspension of dried cord material. Initial inoculations were made with cords dried sufficiently that they contained no detectable virulent virus, but later inoculations used only partially dried cords which still contained virulent material. Although there was clear scope for improvement the procedure worked, and it offers an interesting parallel to immunization schedules for polio vaccines in which use of a killed vaccine precedes use of a live attenuated vaccine (see below).

Vaccines containing phenol-inactivated nerve tissue-derived virus, in particular the Semple vaccine,[13] were introduced at the beginning of the 20th century, and versions of these products remain in use as inexpensive, locally manufactured products in many developing countries. However these products do give rise to a significant number of neurological complictions in vaccinees. The current products of choice are the safer and more potent tissue-culture derived rabies vaccines developed in the 1960s and 70s, in particular the human diploid cell vaccine (HDCV) developed at the Wistar.[14] To produce HDCV the Pitman-Moore strain of rabies virus WI-38 is grown on MRC-5 cells, and the virus is inactivated by treatment with β-propiolactone. This product is used worldwide, and is manufactured in by Aventis Pasteur (France and Canada), Chiron Behring (Germany) and Berna (Switzerland). Comparable products are also produced on Vero cells by Aventis Pasteur, on primary chick embryo fibroblasts by Chiron Behring, in embryonated duck eggs by Berna and on fetal rhesus monkey lung fibroblasts by the Michigan State health department and Smithkline Beecham (reviewed in refs. 12 and 15). The cost of a full intramuscular immunization schedule using these products is a significant factor, and has led to the use of intradermal immunization with a reduced dose in certain circumstances.[16]

Hepatitis A

Inactivated HAV vaccines were developed in the 1980s following the successful growth of HAV in tissue culture.[17-20] Production technology is similar to that used to produce IPV, in that virus is grown on a substrate of MRC-5 cells, purified, inactivated with formaldehyde and formulated.[21-24] The first commercially available product (Havrix, produced by SmithKline Beecham) was licenced in Europe in 1991. Similar products are now manufactured by Merck (USA), Aventis Pasteur (France and Canada) and Berna (Switzerland).[21-24] The Berna product is notable in that it is adjuvanted with influenza virosomes.[22,25-27] All are safe and effective; see ref. 28 for a full review.

Poliovirus

It is sobering to reflect that it is now more than 40 years since the introduction of inactivated poliovirus vaccine (IPV) by Salk in the USA[29] and by Lepine in France.[30] Various forms of IPV, in particular enhanced potency versions of the Salk product (eIPV; unless indicated otherwise references to IPV should be taken to include eIPV), have remained in use worldwide ever since and have been found to be safe and effective.[31,32] Use of IPV on a country by country basis since 1955 has been discussed elsewhere[31] and will not be addressed in detail here. It should be recognised that IPV is more usually administered as a multivalent combination product containing, for example diphtheria, tetanus, pertussis and polio antigens (DTP-IPV).[33] Since there is no evidence that formulation into a multivalent product affects the performance of the IPV component, this chapter does not distinguish between IPV alone and combination products containing IPV unless otherwise indicated. IPV is currently manufactured by Aventis Pasteur in France and Canada, by SmithKline Beecham in Belgium, by Rijksinstitut voor Volksgezondheid en Mileuhygiene in The Netherlands, by Statens Seruminstitut in Denmark, and by the Swedish National Biological Laboratory in Sweden.

In conjunction with Sabin's live, oral, attenuated poliovirus vaccine (OPV), these products have effectively controlled poliovirus infections in many parts of the world since the late 1950s. The concurrent use of OPV and IPV led to a vigorous and continuing debate on the relative merits of the two products,[34] some aspects of which will be addressed below. Both IPV and OPV have made significant contributions to the imminent worldwide eradication of poliovirus.[32,35] Consequently, the present generation of IPV probably represents the final form of the vaccine, since the foreseeable eradication of poliovirus means that it is highly unlikely that a vaccine manufacturer would find it economical to develop, test and market a "next-generation" IPV given the safety and efficacy of current products. Nevertheless these vaccines will continue in use for some time to come even in a post-polio world, there will always be scope for improvement in manufacturing processes, and the use of combination vaccines containing IPV will evolve even though the basic product may not.

Influenza Virus

Influenza, an acute respiratory illness caused by infection with influenza viruses, is one of the oldest and most common diseases known to mankind and potentially one of the deadliest.[36,37] The primary method to control influenza is preventive vaccination. Inactivated vaccines currently licensed for use contain either killed whole virus or split (detergent-treated) virus. Despite the availability of effective influenza vaccines only about 40-50% of the target populations are immunized.[38] Furthermore, the efficacy of current vaccines in the target groups at greatest risk is generally lower than that observed for healthy adults and the young. Vaccinating people in target groups every year before the onset of influenza season is the most effective way of limiting the impact of influenza epidemics. Detailed descriptions of the influenza virus, viral proteins and pathogenesis are beyond the scope of this chapter and can be found elsewhere.[39,40]

Current vaccines consist of three virus strains selected for each season during February of each year. The strains are usually two type A (H1N1 and H3N2) and one type B that are currently in circulation, according to the recommendations of the WHO. These recommendations are based on the analysis of data coming out of several laboratories worldwide that are part of the influenza global surveillance program. For example, trivalent influenza vaccine prepared for the 1999-2000 season will include A/Beijing/262/95-like (H1N1), A/Sydney/5/97-like (H3N2), and B/Beijing/184/93-like hemagglutinin antigens (HA). For the B/Beijing/184/93-like antigen, U.S. manufacturers will use the antigenically equivalent B/Yamanashi/166/98 virus because of its growth properties and because it is representative of currently circulating B viruses.[41] Note that immunity to influenza virus is predominantly mediated by antibody responses to HA. Although anti-HA antibodies neutralize the virus, they are usually strain-specific and are only capable of preventing infection by virus variants within the same subtype. Anti-HA antibodies are less effective at protecting against infection by a different subtype.

Influenza viruses are grown in the allantoic cavity of embryonated hen eggs (see below). Egg grown viruses are highly purified by zonal cenrifugation or chromatography, then inactivated with either formaldehyde or beta-propiolactone. At this stage the virus preparation can be formulated as whole virus inactivated vaccine. The inactivated virus can be further treated, or split, with a detergent such as Triton X-100 or Triton N-101 to disrupt the lipid membrane thus solubilizing membrane glycoprotens. The resulting preparation contains aggregates of monomeric HA, neuraminidase, and internal proteins such as the nucleoprotein and has the advantage of being less reactogenic than the whole virus vaccine. In North America the major supplier of whole virus vaccine is Aventis Pasteur, while split products are available from Aventis Pasteur, Evans and American Home Products (Wyeth Ayherst).

Challenges to the future control of influenza include improving the immunization rates in the target populations, obtaining further insights into the short term nature of immunity to influenza vaccine, understanding the basic aspects of immunosenescence and developing strategies to improve the efficacy from the current 30-70% level in the target populations. Potentially even more important would be an expansion of the effective and low cost global surveillance system for influenza viruses combined with a strategy for the use of vaccines or anti-viral agents to contain new strains as they appear in circulation.

Human Immunodeficiency Virus Type-1 (HIV-1)

In the early stages of the AIDS epidemic, most HIV-1 vaccine approaches were focused on the development of whole-inactivated viruses since this strategy can safely provide long-lasting protection in humans against several viruses, including polio, influenza, hepatitis A, and rabies. Several virus challenge studies were conducted in non-human primates immunized with various forms of formalin-inactivated, simian immunodeficiency virus (SIV). In most cases, vaccinated animals were protected against pathogenic SIV challenges.[42-44] However, these promising results were subsequently challenged by the demonstration that in most cases the protection seen in the vaccinated animals was mediated, at least in part, by the immune response to proteins of cellular origin incorporated into the vaccine and challenge virus preparations.[45] Despite this initial failure to demonstrate that immune responses against viral antigens elicited by formalin-inactivated SIV vaccines could induce protection against viral challenge, novel approaches to develop inactivated HIV-1 vaccine candidates have recently been explored. The advantages of these new approaches, as well as the scientific obstacles and technical challenges that must be overcome, provide an insight into the future of inactivated virus vaccines which will be discussed further in the next section.

Issues Affecting the Use of Inactivated Virus Vaccines

Given that there is a need for a vaccine and that a vaccine is available, what factors affect the use of that vaccine? The most important factors are those which influence the safety of the product (an all-important consideration for any vaccine), its cost and availability, and its attractiveness relative to any alternative product. In the case of inactivated virus vaccines these factors would include purity of the product, efficiency of the inactivation process, yield on acceptable cell substrates, and advantages or disadvantages in comparison to other viral vaccines. This section considers these factors with particular reference to the three chosen examples, poliovirus, influenza virus and HIV.

Purity

Purity is in many ways a product-specific concept, and is usually best determined by methods appropriate for each vaccine.[46] In the case of viruses shed or released into the cell culture supernatant, such as rabies virus or poliovirus, an acceptable degree of purity can be obtained using well established technologies such as filtration or chromatography to separate the virus particles on the basis of size. The major issue is usually to retain an adequate yield while achieving the required degree of purity. Principal contaminants are likely to be host cell proteins or nucleic acids, materials added during the production process (e.g., antifoaming compounds), or adventitious viruses. The size differential between the virus particles and host cell or production process-derived contaminants is such that separation is relatively easy to achieve. Removal of potential adventitious viruses may be less easy to achieve, but even if such viruses were present their titer would be relatively low and it is likely that the inactivation process would reduce the titer to the point that risk to vaccinees was minimal. The use of defined, virus-free, cell lines for the production of vaccine virus is the simplest way to remove the risk of contamination with adventitious agents.[47,48]

Residual levels of host cell proteins are usually too low to be reactogenic or immunogenic. For example, influenza virus, which is grown in embryonated chicken eggs, can be readily purified from contaminating egg proteins by zonal centrifugation even though there is some overlap between the ovalbumin and virus peaks. Purity can be increased, though at the expense of virus yield, by excluding virus-containing fractions which overlap the ovalbumin peak or by repeating the zonal centrifugation step. It should be recognised that residual host cell proteins might retain biological activity even when present at very low levels. The production of cytokines and chemokines, for example, by cell substrates is not always well characterised, and the production of these and other soluble factors in response to virus growth may deserve more attention.[49] In the case of inactivated vaccines it is possible that the inactivation step may reduce or eliminate unwanted biological activity of any contaminating proteins.

Residual levels of host cell nucleic acids are also usually very low. In the case of Hepatitis A virus, which is not released into the culture medium but accumulates within infected cells,[23,24] the purification of virus from a lysate of infected cells requires the removal of substantial quantities of cell-derived contaminants including DNA. The addition of a nuclease treatment step during the purification of hepatitis A virus for the manufacture of the Merck vaccine, VAQTA, was found to both enhance product purity by faciliting the removal of nucleic acid and increase virus recovery by releasing virus from virus-nucleic acid complexes.[50,51] Similar processes could facilitate the removal of residual host-derived DNA from other virus vaccines if necessary.

Although DNA content can be reduced to picogram amounts per dose,[52] determining a safe level for DNA content is not always straightforward. Total DNA content can be determined relatively easily but may not be informative. For example, fragmented DNA from apoptotic cells may pose a much smaller risk than substantially intact DNA from viable cells. Current technology is already adequate to measure the number of copies of specific DNA sequences per dose, and the issue is not the measurement of contaminating DNA but the establishment of a useful threshold for accepting a product as safe.[48,53] This will likely need to be determined on a case-by-case basis as new products are licenced.

Inactivation

Efficient inactivation is obviously critical to producing an inactivated virus vaccine. The Cutter incident provides the best (or worst) example of the consequences of inadequate inactivation. In early 1955, children immunized with IPV produced by Cutter Laboratories, as well as family contacts, began to develop paralytic poliomyelitis. Prompt investigation determined that several incompletely inactivated virus lots had been used to prepare the vaccine, but by the end of the incident 69 children and 89 family contacts with paralytic disease attributable to the vaccine had been identified.[54] The cause was determined to be non-linear inactivation of the virus by the formaldehyde-treatment process used at the time, coupled with the presence of virus aggregates which shielded some particles from the inactivating agent. Extended inactivation plus filtration to remove virus aggregates was sufficient to solve the problem, but at the cost of reduced antigenicity of the product.[55]

Current inactivation technologies offer a greater range of choices for achieving effective inactivation while retaining antigenicity of the product, and these are well illustrated in the context of HIV-1 vaccines. In principle, inactivated HIV-1 virions would present the immune system with most of the viral proteins in their native configuration. Therefore, such a vaccine should elicit a broader spectrum of immune responses than other vaccines based on single recombinant viral antigens. Furthermore, the use of a mixture of inactivated HIV-1 particles could help achieve broad protection against most circulating strains of HIV-1. The demonstration of the efficacy of inactivated vaccines against retroviruses comes from studies in the field of veterinary vaccines, where inactivated virions account for the most successful vaccines against retroviral diseases.[56-58] In addition, the observation that at least one of these vaccines elicits protective cytotoxic T-cells (CTLs) contradicts the widely-held belief that inactivated vaccines generate only humoral immune responses.[57]

Several hurdles must be overcome to achieve the development of a safe and efficacious inactivated HIV-1 vaccine candidate. First, it is important to develop inactivating procedures that maintain the native configuration of the antigens and do not destroy their immunogenicity. Second, these procedures must result in a very high margin of safety. Third, since the envelope glycoprotein of some HIV-1 strains is easily shed, development of inactivation and purification procedures which prevent the loss of the viral envelope represents a major technical challenge. Finally, in order to produce large scale quantities of inactivated viruses, it is necessary to examine different cell substrates, which must be properly characterized to satisfy stringent safety and regulatory issues.[59]

Several viral inactivating methodologies have been used in the preparation of SIV and HIV-1 vaccine candidates, including the use of formalin, β-propiolactone (BPL) alone or with either cobalt 60-irradiation or binary ethylenimine (BEI), and psoralen exposure followed by ultraviolet (UV) radiation.[43,44,60-71] A potential problem with using formalin for viral inactivation is that this involves a series of chemical reactions that produce reactive products that can induce cross-linking of viral proteins and aggregation of virus particles. This could hamper the inactivating efficiency of the formalin and could also result in the partial destruction of the immunogenicity of the vaccine.

Inactivation of HIV-1 with a combination of BPL and cobalt 60-irradiation led to the development of a gp120 envelope-depleted HIV-1 immunogen that is presently being tested in efficacy trials as a therapeutic vaccine.[69,72] Although some encouraging results have been obtained with this vaccine candidate, the loss of the surface envelope glycoprotein gp120 following this inactivating procedure might compromise the efficacy of the vaccine. Inactivation with UV radiation and psoralen exposure, although capable of providing adequate safety margins (6.5 $\log_{10}$ drop in viral titers), is not really suitable for large-scale manufacturing.[59]

A multistep procedure for the chemical inactivation of HIV-1 developed by Race et al[73,74] was used to prepare whole HIV-1 particles that retain most of the external envelope glycoproteins and are highly immunogenic in small animals. This procedure included treatment with 0.2% BPL followed by purification. The viral particles were then deaggregated with 0.05% sodium cholate and further inactivated with 10 mM BEI and 0.02% formaldehyde. These conditions resulted in a reduction of virus infectivity by at least 20 $\log_{10}$ (10^{20} $TCID_{50}$/ml), which provides a very high margin of safety.

Another inactivation procedure that preserves the structural and functional integrity of the viral and virion-associated cellular proteins on the viral membrane, while also providing a wide margin of safety, was recently developed.[75] The chemical compound 2, 2'-dithiodipyridine, (Aldrithiol-2 or AT-2), which inactivates retroviral infectivity by covalently modifying the essential nucleocapsid zinc finger motifs in the nucleocapsid proteins, was shown to completely inactivate HIV-1 at low concentrations (>5 log units of inactivation at 1 mM) without affecting the conformational and functional integrity of viral and host cell-derived proteins retained on the virion surface. Furthermore, it was also shown that this inactivation procedure can be used for other viruses (SIV) and can be scaled-up to very large volumes.[75] Presently, non-human primate studies are in progress to evaluate the immunogenicity of AT-2-inactivated SIV and HIV-1 retroviral vaccine candidates. In order to increase the margin of safety of these inactivated immunogens, the same investigators have proposed to include a second inactivation step, such as the treatment with physical agents like some of the ones described above. Alternatively, it might also be possible to combine the chemical inactivation step with novel methodologies for inactivating viruses, such as a procedure in which virions are first infiltrated with highly compressed liquids called "superfluids" and subsequently exposed to decompressed liquids, effectively leading to the formation of virions ruptured at their weakest point but remaining structurally intact.[76]

Nevertheless, there are other issues regarding viral inactivation of HIV-1 that remain to be addressed, such as the difficulties in establishing the minimal acceptable margin of safety due to uncertainties about the smallest quantity of HIV-1 that represents a human infectious dose. In addition, the potential influence of viral strain on inactivation, and the extrapolation of results on infectious titers obtained by different assays to infectivity in humans, are complex issues that add to the uncertainties of clearly defining an acceptable margin of safety for an inactivated HIV-1 vaccine candidate.

An alternative to the use of retroviruses inactivated by chemical and/or physical procedures as vaccine immunogens is to produce recombinant, noninfectious, retrovirus-like particles. We and others have previously demonstrated the feasibility of producing nonreplicating, noninfectious, immunogenic HIV-1-like particles as potential inactivated candidate immunogens for the development of an HIV-1 vaccine.[77-84] This strategy has several advantages: 1) since the virus-like particles are genetically engineered, genetic elements contributing to viral infectivity and pathogenicity can be mutated or deleted;[78] 2) particles incorporating most of the HIV-1 antigens in their native conformation can be produced in cell substrates approved for vaccine production;[79,80] and 3) it is feasible to produce HIV-1-like particles with antigenic markers that may allow to serologically distinguish vaccinees from HIV-1 infectees.[77]

Several issues, however, remain to be resolved before this approach can be commercially exploited. For example, the first generation of expression vectors that we used to establish permanent cell lines which secrete HIV-1-like particles contain inducible promoters. A major drawback to the use of these vectors is that production yields are suboptimal, and the inducing agents are cytotoxic.[77,79,80] Large-scale production of HIV-1-like particles in mammalian cells will require establishment of cell lines stably transfected with expression vectors that contain constitutive promoters to enhance production yields. Other issues to be addressed are the potential contamination of HIV-1-like particles with nucleic acids from the producer cell lines, as well as endogenous viruses or retroviral-like elements. Therefore, it might be necessary after purification to further inactivate the recombinant virus-like particles with some of the procedures previously described.

Yield

Yield of virus from the culture after inactivation and purification is a critically important factor in determining whether it is feasible to produce a particular vaccine. Historically, one of the factors determining the preference for IPV over OPV in the United States was the difficulty of producing sufficient IPV using the technology available in the 1950s.[31] It remains true that fewer doses of IPV than OPV can be produced per unit volume of culture, but in the 50s and 60s IPV was produced on a substrate of primary monkey kidney cells, which were to some extent a limiting resource. Current enhanced potency products (eIPV) are produced in high yield on a substrate of MRC-5 or Vero cells[32] and production capacity is adequate to meet all requirements for a high-potency product.

Influenza virus vaccines provide a particular example of the critical importance of obtaining an adequate yield from the production process. As noted above, manufacturers have about six months from the selection of the year's vaccine virus strains to produce sufficient product to meet the demands of the year's flu season. Although influenza viruses can be grown in abundance in the allantoic cavity of embryonated hen eggs, newly isolated influenza viruses do not usually grow as well in eggs as egg-adapted strains. This could potentially cause a problem when changing vaccine strains each year and in producing sufficient virus to meet the demand for the vaccine. However, when chicken embryos are inoculated with a high yielding, egg-adapted, donor virus and a low yielding circulating strain that is antigenically different than selected in the presence of antibodies to the surface antigens of the donor virus, the resultant re-assortant viruses have the surface antigens of the low yielding virus but retain the growth characteristics of the donor viruses. This technique facilitates the generation and characterization of re-assortants fairly quickly and permits the scale up their production for vaccine use. Laboratory adapted strains such as A/PR/8/34 (H1N1) or its derivative X-31 (H3N2) are typically used as donor strains.

Since the production of vaccine requires the growth of large quantities of virus in embryonated chicken eggs, lack of a reliable supply of high quality eggs may limit the production of vaccine. Furthermore, concerns about the presence of adventitious viruses in eggs used for influenza vaccine production and about the presence of egg proteins in the vaccine which could cause serious problems in individuals allergic to eggs have focussed attention on growing influenza viruses in other host systems. Influenza viruses can be grown in mammalian cell lines and cell lines such as Madin-Darby canine kidney (MDCK) cells and Vero cells have been used to grow influenza viruses for vaccine production. In a recent clinical study a bivalent influenza vaccine produced from MDCK cell-grown viruses was shown to be as immunogenic as egg-derived virus vaccine.[85] The tissue culture derived virus vaccine was as safe as egg derived virus vaccine, as the overall local and systemic reactions post-immunization with tissue culture derived vaccine were similar to or lower than those induced by conventional egg derived virus vaccine. There were no differences between the groups immunized with either tissue culture-derived or egg-derived vaccines in terms of anti-HA titers. In pre-clinical studies, Vero cell derived virus vaccine has been shown to be as immunogenic as egg derived vaccines.[86] Tissue culture grown virus offers the additional advantages of scaling up the manufacturing process independently of egg supply and may obviate the need to generate re-assortants of the wild-type virus, provided the wild type viruses can infect the chosen cell substrate efficiently.

The switch from eggs to a cell substrate may be useful for influenza vaccine production, and illustrates some of the issues facing contemporary products which are not yet produced in well-defined substrates. The production of HIV-1 vaccine virus-like particles illustrates issues which may face the next generation of inactivated virus vaccines. As always, yield will be critical in providing an affordable product, particularly as immunization in developing countries becomes critical for disease eradication or containment. Production of higher levels of virus-like particles for use as inactivated vaccines could be achieved by taking advantage of new biotechnological applications of recombinant DNA technologies using mammalian cells. Most such applications make use of cell lines with stably integrated genes under the control of strong viral or cellular promoters.[87] The genes encoding the desired viral antigens are expressed from a DNA plasmid stably transfected into the cells, and expression levels are regulated by interactions of cis- and trans-acting elements. Cis-acting elements are special DNA sequence segments present on the plasmid construct and consist of promoters, enhancers, introns, terminators (termination and polyadenylation signal sequences), translational control sequences, and other DNA elements. Trans-acting elements are transcription factors and other cellular protein factors that interact with the cis-acting elements or other protein factors to regulate expression. While the trans-acting factors are made by the cell and are therefore difficult to regulate, much effort has been made toward the identification and manipulation of cis-regulatory elements to achieve high level and long-term gene expression.

The cytomegalovirus (CMV) promoter, when juxtaposed to the CMV enhancer, is probably the strongest eukaryotic promoter identified so far and has been used widely for gene expression.[87-90] The cellular elongation factor (EF1a) promoter has also been widely used for high level expression of recombinant proteins.[87] These promoters, although able to direct high level gene expression in many cell lines, exhibit transcriptional activities that are highly dependent on the cellular levels of the relevant transcription factors and on the chromatin structure at the integration site. In most cases, the inclusion of an intron in a plasmid vector is also beneficial for expression, although there are conflicting results.[91-93] This could be attributed to a potential functional role for introns in improving the efficiency of RNA processing and transport to the cytoplasm, stabilizing RNA, and activating transcription. In addition, transcriptional enhancers have been found in some introns, such as the immunoglobulin heavy chain gene intron. The presence of the CMV intron A region in a vector has been shown to increase expression significantly,[87,88] as well as the inclusion of the rabbit β-globin IVS2 sequence.[94] Expression levels can also be up-regulated by modifying the terminator elements. Transcription terminators control RNA transcript processing and polyadenylation at the 3'-end of a gene, and in turn regulate gene expression. The efficiency of this process varies between terminators, but in general the simian virus 40 (SV40) and the bovine growth hormone (BGH) terminators are the most widely used.[95,96] Finally, control of translational efficiency is equally important in regulating gene expression. For the most part, translational efficiency is regulated through the control of translational initiation, and cis-acting elements required for translational initiation have been studied in detail.[97] Inclusion in plasmid constructs of the consensus sequence for initiation of translation (Kozak sequence) can substantially improve gene expression.[97]

Advantages and Disadvantages in Comparison to Alternative Products

Inactivated virus vaccines have a number of features which distinguish them from other vaccines, particularly live attenuated vaccines. The purification and inactivation processes can provide a greater assurance of safety, not only from the presence of virulent vaccine viruses but from substrate-derived contaminants and from virulent adventitious viruses. However, the same processes can decrease the yield of the product resulting in limited supply or increased production costs. Inactivated products can be highly immunogenic, eliciting strong and long lasting antibody responses. This is advantageous to the extent that antibody-mediated immunity is protective; in those situations where mucosal or cell-mediated immunity is important live virus vaccines may be preferred. A final advantage of inactivated vaccines over live products may prove to be the most significant of all. It is much easier to formulate an inactivated product with an adjuvant or in combination with other antigens. The real need for multivalent vaccines for pediatric use and the development of effective adjuvants for use in humans will favor the use of inactivated virus vaccines over live attenuated equivalents.

The interplay of these factors is exemplified by the history of poliovirus vaccination in the United States. The USA was the first country to make use of IPV, in 1955, but use of the vaccine was essentially abandoned in favour of OPV by 1968, and OPV remained the recommended vaccine in the USA until very recently.[29,34,98] The reasons for the change to and preference for OPV illustrate some of the real or perceived differences between the two products which continue to fuel the debate on the relative merits of OPV and IPV.[34] At the time of its introduction IPV was in limited supply. To extend the available stock the required potency of the product was reduced below that of the initial product and of similar products in use elsewhere at the time (reviewed in 31), leading to a perception that the product was poorly immunogenic. As noted above, production capacity is no longer an issue.

A second perceived advantage of OPV was its ability to provide superior community protection; that is to protect unvaccinated members of a community from disease. OPV induces strong mucosal immunity in vaccinees,[99,100] greatly reducing their ability to excrete wild virus from the intestine. Furthermore, OPV will spread from vaccinees to unvaccinated contacts,[99,100] immunizing them also. However, OPV predictably reverts to a virulent form as it replicates in the gut of vaccinees,[101] and so vaccinees and infected contacts are at risk of developing vaccine-associated paralysis (VAP) at a rate of one case for every few million doses of OPV distributed.[102,103]

In contrast, IPV elicits a strong humoral antibody response which is sufficient to limit but not prevent infection of the nasopharynx and gut, although dissemination of the virus to the CNS and consequent paralytic disease is completely prevented (reviewed in ref. 31). Therefore IPV was held to be relatively ineffective at preventing the circulation of wild-type poliovirus and protecting unvaccinated members of the general population.[99,100] However, IPV and eIPV can induce a substantial degree of mucosal immunity with a demonstrable protective effect.[33,104-108] Fecal shedding, one of the major routes of transmission, was reduced in an IPV-immunized group following challenge with a live virus in comparison to an unimmunized group.[101,104] IPV was inferior to OPV in diminishing post-challenge shedding in faeces by vaccinees, but comparable to OPV in diminishing shedding from the pharynx.[99,104,105] Overall, IPV-immunized individuals who experience a subsequent infection shed less virus than infected unimmunized individuals. This must diminish the spread of a virus in an IPV-immunized community, and reduce the risk to unimmunized community members. IPV is proposed to have decreased community transmission of wild polioviruses in the US in the 1950s because of its ability to reduce pharyngeal shedding[105,107,108] and IPV has provided effective protection to the general population in Scandinavia, The Netherlands and Canada following outbreaks in religious groups whose members choose to remain unimmunized (reviewed in ref. 31).

While OPV has clearly been effective in eradicating poliovirus from the Americas,[109] the residuum of VAP cases in the USA has prompted a re-examination of the relative merits of IPV and OPV for use in the USA. In the absence of circulating wild-type virus and in the face of relatively rare introductions from abroad as the world-wide eradication of poliovirus progresses, the occurrence of several cases of VAP per year in the USA has become unacceptable. The initial response was to recommend a combined IPV/OPV schedule, in which two doses of IPV at ages 2 and 4 months preceded the use of two doses of OPV at 12-18 months and 4-6 years, as an alternative to the usual all-OPV schedule.[110] Experience with similar schedules in Denmark, Israel and the Canadian province of Prince Edward Island indicate that mixed schedules induce strong and protective immune responses and essentially eliminate the risk of VAP.[31] The new schedule was well received, and was implemented without an adverse impact on vaccine coverage.[111] IPV accounted for 29% of poliovirus vaccine doses distributed in the US in 1997 indicating that a majority of children born that year were receiving IPV for their first two doses.[112] Four cases of VAP were reported in 1997, down from about ten cases per year seen previously, and all were associated with first or second doses of OPV in an all OPV schedule.[111,112] To further reduce the occurrence of VAP, in 1999 OPV was no longer recommended for the first two doses of the immunization schedule,[113] and the Advisory Committee on

Immunisation Practices has voted to recommend that an all-IPV schedule should be adopted in 2000.[114,115]

One issue not illustrated by the poliovirus example is the potential to use adjuvants to modulate the immune response to an inactivated virus vaccine. While some inactivated vaccines elicit strong and durable protective immune responses (e.g., poliovirus[116] and hepatitis A[117]), others may be less effective. Alternatively, they may be targeted to a population such as the elderly with a reduced capacity to mount an effective immune response. In these cases the use of an adjuvant may offer a solution.

For example, the development of improved influenza vaccines that confer better protection against circulating strains and new strains of influenza viruses is the goal of immunologists, vaccinologists and public health professionals. The desired properties of an improved influenza vaccine are production of increased titers of virus neutralizing antibodies against influenza hemagglutinin in the serum (IgG) and at mucosal surfaces (sIgA) to prevent infection and induction of virus cross-reactive T lymphocytes ($CD8^+$ and/or $CD4^+$) and/or their cytokines to facilitate recovery from infection. Such characteristics should increase the proportion of vaccinated individuals who are protected against infection with homologous influenza virus. Furthermore, the ability of a vaccine to induce cytotoxic T lymphocytes (CTLs) that recognize highly conserved epitopes within proteins across the various influenza virus subtypes may reduce the severity of illness and promote faster recovery when the vaccine and circulating strains are serologically distinct.

In addition to HA and NA influenza vaccine contains significant amounts of NP and other internal proteins that are conserved among subtypes and which could potentially induce or re-call CTL responses. Soluble antigens are processed and the T-cell epitopes thus generated are presented in the context of class II MHC to activate specific $CD4^+$ T-cells, which are predominantly T-helper cells, in order to induce or re-call CTL responses against soluble antigens. To induce or re-call CTL responses against intracellular viral antigens the vaccine needs to be formulated so that the antigens are delivered directly into the cytosol of antigen presenting cells, where they can be processed and presented in the context of class I MHC molecules. So far very few delivery systems have been reported to deliver antigens successfully for presentation with class I MHC, besides enhancing antibody responses. Among such systems immunostimulatory complexes (ISCOMs), liposomes, and immunopotentiating reconstituted influenza virosomes (IRIVs) have been extensively studied with various antigens. ISCOMs are prepared by combining saponin, cholesterol, phospholipids and antigen and have been shown to elicit antibody responses of both the IgG2a and IgG1 subtypes and to enhance IL-2 and IFN-γ synthesis compared with soluble antigen. ISCOMs induce enhanced antibody titers and CTL responses in experimental animals against a variety of immunogenic proteins.[118-122] We and others have demonstrated that influenza antigens presented as ISCOMs induce up to 10-fold higher titers of virus neutralizing antibodies and offer improved and more durable protection compared with current influenza vaccines.[123-128] In a phase 1 clinical trial there were no statistically significant differences between groups that received vaccine or Flu-ISCOMs or vaccine mixed with ISCOM matrix (i.e., ISCOMs made without antigen) in terms of percent of vaccinees with >4-fold increase in HAI titers or individuals with protective HAI titer of ≥40 (unpublished data). Although influenza vaccine alone failed to induce CTL activity against H1N1 and H3N2 strains, the same vaccine formulated as ISCOMs or mixed with matrix induced CTL responses in over 50% the vaccinees.[129]

Influenza vaccine formulated as IRIVs has been shown to be well tolerated.[130-131] Furthermore, several trivalent IRIV-formulated vaccines tested in both healthy adults and the elderly have been shown to be more immunogenic than the conventional vaccines. The IRIV vaccine was found to be highly immunogenic especially when the pre-vaccination titers were low, consistent with the findings of other formulated influenza vaccines.

Formulation of influenza vaccine with poly[di-carboxylatophenoxy phosphazene] (PCPP) has been shown to enhance influenza virus-specific hemagglutination inhibition-titers over 10-fold when compared to vaccine in pre-clinical studies, and the results of Phase 1 trials are awaited.[132] Pre-clinical studies with influenza vaccines formulated with MF59 adjuvant emulsion have indicated that the adjuvanted vaccine was more immunogenic than the unadjuvanted vaccine.[133] Furthermore, in a Phase II trial in the elderly, vaccinees who received the MF59-adjuvanted vaccine showed stronger and longer lasting responses than those who received conventional vaccine.[134] Although the vaccine was generally well tolerated subjects that received the MF59-adjuvanted vaccine exhibited more frequent local reactions after the first dose but not after subsequent immunizations. This may have been due to hypotonicity of the adjuvanted vaccine used for the first injection.

A final problem affecting the use of all vaccines is intrinsic to the continuing success of vaccine development activities. As more vaccines become available it becomes increasingly difficult to fit them into childhood immunization schedules. An effective solution is to combine several vaccines into a multivalent combination product, and this is much more easily achieved with an inactivated virus vaccine than with a live virus vaccine, for example. The future use of a given vaccine will be heavily influenced by its availability in combination products.

As noted above IPV is more usually administered as a multivalent combination product such as DTP-IPV. France and Canada have for several years used a pentavalent product which contains antigens of diphtheria, tetanus, pertussis and *Haemophilus influenzae* B in addition to IPV.[33] More recently similar pentavalent products formulated using an acellular pertussis vaccine have been developed.[135,136] Hexavalent combinations with hepatitis B vaccine have been tested in clinical trials, and at least one manufacturer has plans to develop a heptavalent combination with hepatitis A and B.[137]

This availability of a combination product will be an important factor in determining the incorporation of a hepatitis A vaccine into a pediatric immunization schedule. A combination hepatitis A plus hepatitis B vaccine is now available in Europe (Twinrix, produced by SmithKline Beecham),[138] and a combination of a hepatitis A vaccine with a typhoid vaccine has been described.[139] In a small study the combination was well tolerated and serological responses were the same as those elicited by the monovalent products. A pentavalent vaccine combining antigens of hepatitis A, hepatitis B, diphtheria, tetanus and influenza A/B has been studied but optimization of the relative antigen contents was required to avoid suppression of the anti-HAV response.[140]

While clearly attractive from a marketing perspective, and to the fortunate child who can receive one less injection when it's time to be immunized, combination products are technically unattractive to develop. The formulation of multiple antigens into a single product poses a number of technical problems ranging from incompatibilities of stabilizers used for the individual components to immunological interference between the various antigens.[136,137,141] For example, even in the case of DTP-IPV, a product which has been used safely and effectively for many years, some studies have reported diminished antibody responses to either poliovirus or pertussis antigens (reviewed in ref. 137), and thimerosal, used as a stabilizer for some DTP vaccines, has been reported to diminish the potency of IPV combined with thimerosal-stabilized DTP.[142] There is probably little, if any, clinical significance of these particular effects, but the general issues will usually require that clinical trials be conducted to verify that the formulation of novel combination products does not compromise the potency or safety of the individual components, and the size and cost of these trials can increase dramatically as additional antigens are added to a product. Nevertheless there is significant competitive pressure between manufacturers to bring combination products to the market, and the availability of combination vaccines will be a significant factor driving the sale of IPV for the forseeable future.

Conclusion

Inactivated virus vaccines are in widespread use for several diseases and have proven to be safe and effective products. Changes to the use of these products can be foreseen in response to changing patterns of disease and to satisfy demands for safer, better defined, more effective but cheaper products. Improvements in vaccine technology make it likely that many of these demands will in fact be met and that inactivated virus vaccines will continue to be an important public health tool during the 21st century.

Acknowledgment

We wish to thank Dr. Larry Arthur for critically reading and providing suggestions to the sections dealing with inactivated HIV-1 vaccines, Dr. Shixian Cao for his contribution to the section dealing with the improved production of virus-like particles, Dr. Greg Slusaw for helpful comments on influenza vaccine production, the staff of the Aventis Pasteur Vaccine Information Service for their prompt and courteous responses to last-minute requests for information, and the staff of the Aventis Pasteur Library for their assistance with literature searches.

References

1. Yamanishi K, Tanishita O, Tamura M et al. Development of inactivated vaccine against virus causing haemorrhagic fever with renal syndrome. Vaccine 1988; 6:278-82.
2. Song G, Huang YC, Hang CS et al. Preliminary human trial of inactivated golden hamster kidney cell (GHKC) vaccine against haemorrhagic fever with renal syndrome (HFRS). Vaccine 1992; 10:214-6.
3. Lu Q, Zhu Z, Weng J. Immune responses to inactivated vaccine in people naturally infected with hantaviruses. J Med Virol 1996; 49:333-5.
4. Yoo YC, Yoshimatsu K, Koike Y et al. Adjuvant activity of muramyl dipeptide derivatives to enhance immunogenicity of a hantavirus-inactivated vaccine. Vaccine 1998; 16:216-24.
5. Putnak R, Cassidy K, Conforti N et al. Immunogenic and protective response in mice immunized with a purified, inactivated, Dengue-2 virus vaccine prototype made in fetal rhesus lung cells. Am J Trop Med Hyg 1996; 55:504-10.
6. Putnak R, Barvir DA, Burrous JM et al. Development of a purified, inactivated, dengue-2 virus vaccine prototype in Vero cells: immunogenicity and protection in mice and rhesus monkeys. J Infect Dis 1996; 174:1176-84.
7. Barrett AD. Japanese encephalitis and dengue vaccines. Biologicals 1997; 25:27-34.
8. Plotkin SA, Orenstein WA, eds. Vaccines, 3rd ed. Philadelphia: W. B. Saunders., 1999.
9. Barret PN, Dorner F, Plotkin SA. Tick-Borne encephalitis vaccine. In: Plotkin SA, Orenstein WA, eds. Vaccines, 3rd ed. Philadelphia: W. B. Saunders 1999: 767-780.
10. Tsai TF, Chang GJ, Yu YX. Japanese encephalitis vaccines. In: Plotkin SA, Orenstein WA, eds. Vaccines, 3rd ed. Philadelphia: W. B. Saunders 1999:672-710.
11. Pasteur L. Methode pour prevenir la rage apres morsure. C R Acad Sci 1885; 101:765-772
12. Plotkin SA, Rupprecht CE, Koprowski H. Rabies Vaccine. In: Plotkin SA, Orenstein WA, eds. Vaccines, 3rd ed. Philadelphia: W. B. Saunders 1999:743-766.
13. Semple D. The preparation of a safe and efficient antirabic vaccine. Sci Mem Med Sanit Dep India, No. 44, 1911.
14. Wiktor TJ, Plotkin SA, Grella DW. Human cell culture rabies vaccine. JAMA 1973; 224:1170-1171.
15. Dreesen DW. A global review of rabies vaccines for human use. Vaccine 1997; 15 Suppl:S2-S6.
16. Dreesen DW, Hanlon CA. Current recommendations for the prophylaxis and treatment of rabies. Drugs 1998; 5:801-809.
17. Siegl G, deChastonay J, Kronauer G. Propagation and assay of hepatitis A virus in vitro. J Virol Methods 1984; 9:53-67.
18. Provost PJ, Hilleman MR. Propagation of human hepatitis A virus in cell culture in vitro. Proc Soc Exp Biol Med 1979; 160:213-221.
19. Daemer RJ, Feinstone SM, Gust ID et al. Propagation of human hepatitis A virus in African green monkey kidney cell culture: primary isolation and serial passage. Infect Immun 1981; 32:388-393.

20. Frosner GG, Deinhardt F, Scheid R et al. Propagation of human hepatitis A virus in a hepatoma cell line. Infection 1979; 7:303-305.
21. Vidor E, Fritzell B, Plotkin S. Clinical development of a new inactivated hepatitis A vaccine. Infection 1996; 24:447-458.
22. Gluck R, Mischler R, Brantschen S et al. Immunopotentiating reconstituted influenza virus virosome vaccine delivery system for immunization against hepatitis A. J Clin Invest 1992; 90:2491-2495.
23. Peetermans J. Production, quality control and characterization of an inactivated hepatitis A vaccine. Vaccine 1992; 10 Suppl 2:S99-101.
24. Armstrong ME, Giesa PA, Davide JP et al. Development of the formalin-inactivated hepatitis A vaccine, VAQTA from the live attenuated virus strain CR326F. J Hepatol 1993; 18 Suppl 2:S20-26.
25. Gluck R. Adjuvant activity of immunopotentiating reconstituted influenza virosomes (IRIVs). Vaccine 1999; 17:1782-1787.
26. Ambrosch F, Wiedermann G, Jonas S et al. Immunogenicity and protectivity of a new liposomal hepatitis A vaccine. Vaccine 1997; 15:1209-1213.
27. Poovorawan Y, Theamboonlers A, Chumdermpadetsuk S et al. Safety, immunogenicity, and kinetics of the immune response to a single dose of virosome-formulated hepatitis A vaccine in Thais. Vaccine 1995; 13:891-893.
28. Feinstone SM, Gust ID. Hepatitis A vaccine. In: Plotkin SA, Orenstein WA, eds. Vaccines, 3rd ed. Philadelphia: W. B. Saunders 1999:650-671.
29. Salk JE. Considerations in the preparation and use of poliomyelitis virus vaccine. JAMA 1955; 158:123-1248.
30. Martin R, Roger F, Damas JP et al. Deux ans et demi de vaccination antipoliomyélitique (vaccin Lépine) a l'hopital de l'Institut Pasteur. Résultats immunologiques. Importance de l'injection de rappel. Annales de l'Institut Pasteur 1959; 97:757-779.
31. Murdin AD, Barreto L, Plotkin SA. Inactivated poliovirus vaccine: past and present experience. Vaccine 1996; 14:735-746.
32. Plotkin SA, Murdin A, Vidor E. Inactivated polio vaccine. In: Plotkin SA, Orenstein WA, eds. Vaccines, 3rd ed. Philadelphia: W. B. Saunders 1999: 345-363
33. Vidor E, Caudrelier P, Plotkin S. The place of DTP/eIPV vaccine in routine pediatric vaccination. Rev Med Virol 1994; 4:261-277.
34. Robbins FC. The history of polio vaccine development. In: Plotkin SA, Orenstein WA, eds. Vaccines, 3rd ed. Philadelphia: W. B. Saunders 1999: 13-27.
35. Cochi SL, Hull SL, Sutter RW et al. Global poliomyelitis eradication initiative: Status report. J Infect Dis 1997; 175 suppl. 1.
36. Murphy BR, Webster RG. Orthomyxoviruses. In Fields BN, Knipe DM, eds. Virology, 2nd ed, New York, Raven Press 1990:1091-1152.
37. De Jong JC, Claas ECJ, Osterhaus ADME et al. A pandemic warning? Nature1997; 389:554.
38. GAO. Immunization. HHS could do more to increase vaccination among older adults. GAO/ PEMD 1995; 95-14:2-66.
40. Bender BS, Small, PA. Influenza pathogenesis and defence mechanisms. Seminars in Respiratory Infections 1992; 7:38-45.
41. Kilbourne ED, Arden NH. Inactivated influenza vaccines. In: Plotkin SA, Orenstein WA, eds. Vaccines, 3rd ed. Philadelphia: W. B. Saunders 1999:531-551.
42. CDC. Prevention and control of influenza. MMWR 1999; 48(RR-04):1-28.
43. Hulskotte EGJ, Geretti A-M, Osterhaus ADME. Towards an HIV-1 vaccine: Lessons from studies in macaque models. Vaccine 1998; 16:904-915.
44. Hunsmann G, Stahl-Hennig C, Dormont D et al. Protection of macaques against simian immunodeficiency virus infection with inactivated vaccines: comparison of adjuvants, doses and challenge viruses. The European Concerted Action on Macaque Models for AIDS Research. Vaccine 1995; 13:295-300.
45. Warren JT, Levinson MA. Fourth annual survey of worldwide HIV, SIV, and SHIV challenge studies in vaccinated nonhuman primates. J Med Primatol 1995; 24:150-168.

46. Stott EJ. Anti-cell antibody in macaques. Nature 1991; 353:393.
47. Paoletti, LC and McInnes, PM, eds. Vaccines: From Concept to Clinic; A guide to the development and clinical testing of vaccines for human use. Boca Raton, Florida, 1998: CRC Press.
48. Stacey GN, Phillips P. Quality assurance for cell substrates. Dev Biol Stand 1999; 98:141-151.
49. Hilleman MR History, precedent, and progress in the development of mammalian cell culture systems for preparing vaccines: safety considerations revisited. J Med Virol 1990; 31:5-12.
50. Haykawa T, Robertson J. Safety of biological products prepared from mammalian cell culture. Host-cell proteins. Dev Biol Stand 1998; 93:139-40.
51. Hagen AJ, Oliver CN, Sitrin RD Optimization of poly(ethylene glycol) precipitation of hepatitis A virus used to prepare VAQTA, a highly purified inactivated vacc ine. Biotechnol Prog 1996; 12:406-12.
52. Hagen AJ, Aboud RA, DePhillips PA et al. Use of nuclease enzyme in the purification of VAQTA, a hepatitis A vaccine. Biotechnol Appl Biochem 1996; 23:209-15.
53. Hennessey JP Jr, Oswald CB, Dong Z et al. Evaluation of the purity of a purified, inactivated hepatitis A vaccine. Vaccine 1999; 17:2830-5.
54. Schild D, Furminger IG. Safety of biological products prepared from mammalian cell culture. DNA. Dev Biol Stand 1998; 93:136-8.
55. Nathanson N and Langmuir AD. Poliomyelitis following formaldehyde inactivated poliovirus vaccination in the United States during the spring of 1955. Am J Hyg 1963; 78:16-28.
56. Melnick J. Virus inactivation:Lessons from the past. Dev Biol Stand 1991; 75:29-36.
57. Pu R, Tellier MC, Yamamoto JK. Mechanism(s) of FIV vaccine protection. Leukemia 1997; 3:98-101.
58. Flynn JN, Keating P, Hosie MJ et al. Env-specific CTL predominate in cats protected from feline immunodeficiency virus infection by vaccination. J Immunol 1996; 157:3658-3665.
59. Hoover EA, Mullins JI, Chu HJ et al. Efficacy of an inactivated feline leukemia virus vaccine. AIDS Res Hum Retroviruses. 1996; 12:379-383.
60. Sheets RL, Goldenthal KL. Traditional approach preventive HIV vaccines: what are the cell substrate and inactivation issues? AIDS Res Hum Retroviruses 1998; 14:627-633.
61. de Vries P, Heeney JL, Boes J et al. Protection of rhesus macaques from SIV infection by immunization with different experimental SIV vaccines. Vaccine 1994; 12:1443-1452.
62. Putkonen P, Nilsson C, Hild K et al. Whole inactivated SIV vaccine grown on human cells fails to protect against homologous SIV grown on simian cells. J Med Primatol 1993; 22:100-103.
63. Cranage MP, Polyanskaya N, McBride B et al. Studies on the specificity of the vaccine effect elicited by inactivated simian immunodeficiency virus. AIDS Res Hum Retroviruses 1993; 9:13-22.
64. Marx PA, Compans RW, Gettie A et al. Protection against vaginal SIV transmission with microencapsulated vaccine. Science 1993; 260:1323-1327.
65. Cranage MP, Baskerville A, Ashworth LAE et al. Intrarectal challenge of macaques vaccinated with formalin-inactivated simian immunodeficiency virus. The Lancet 1992; 339:273-274.
66. Hartung S, Norley SG, Ennen J et al. Vaccine protection against SIV_{mac} infection by high-but not low-dose whole inactivated virus immunogen. J AIDS 1992; 5:461-468.
67. Murphey-Corb M, Martin LN, Davison-Fairburn BD et al. A formalin-inactivated whole SIV vaccine confers protection in macaques. Science 1989; 246:1293-1297.
68. Desrosiers RC, Wyand MS, Kodama T et al. Vaccine protection against simian immunodeficiency virus infection. Proc Natl Acad Sci USA 1989; 86:6353-6357.
69. Norley S, Beer B, Konig H et al. SIV_{MAC} vaccine studies using whole inactivated virus antigen sequentially depleted of viral proteins. J Med Primatol 1998; 27:184-192.
70. Turner JL, Trauger RJ, Daigle AE et al. HIV-1 immunogen induction of HIV-1-specific delayed-type hypersensitivity: Results of a double-blind, adjuvant-controlled, dose-ranging trial. AIDS 1994; 8:1429-1435.
71. Carlson JR, McGraw TP, Keddie E et al. Vaccine protection of rhesus macaques against simian immunodeficiency virus infection. AIDS Res Hum Retroviruses 1990; 6:1239-1246.
72. Sutjipto S, Pedersen NC, Miller CJ et al. Inactivated simian immunodeficiency virus vaccine failed to protect rhesus macaques from intravenous or genital mucosal infection but delayed disease in intravenously exposed animals. J Virol 1990; 64:2290-2297.

73. Moss RB, Trauger RJ, Giermakowska WK et al. Effect of immunization with an inactivated gp120-depleted HIV-1 immunogen on b-chemokine and cytokine production in subjects with HIV-1 infection. J AIDS 1997; 14:343-350.
74. Race E, Stein CA, Wigg MD et al. A multistep procedure for the chemical inactivation of human immunodeficiency virus for use as an experimental vaccine. Vaccine 1995; 13:1567-1575.
75. Race E, Frezza P, Stephens DM et al. An experimental chemically inactivated HIV-1 vaccine induces antibodies that neutralize homologous and heterologous viruses. Vaccine 1995; 13:54-60.
76. Rossio JL, Esser MT, Suryanarayana K et al. Inactivation of human immunodeficiency virus type 1 infectivity with preservation of conformational and functional integrity of virion surface proteins. J Virol 1998; 14:7992-8001.
77. Castor TP, Lander AD, Cosman MD et al. Viral inactivation method using near critical, supercritical or critical fluids. [US Patent 5,877,005] 1999.
78. Rovinski B, Dekaban GA, Cao SX et al. Engineering of noninfectious HIV-1-like particles containing mutant gp41 glycoproteins as vaccine candidates that allow vaccinees to be distinguished from HIV-1 infectees. Virology 1999; 257:438-448.
79. Persson R, Cao SX, Cates G et al. Modifications of HIV-1 retrovirus-like particles to enhance safety and immunogenicity. Biologicals 1998; 26:255-265.
80. Rovinski B, Rodrigues L, Cao SX et al. Induction of HIV type 1 neutralizing and -CD4 blocking antibodies by immunization with genetically engineered HIV type 1-like particles containing unprocessed gp160 glycoproteins. AIDS Res Hum Retroviruses 1995; 11:1187-1195.
81. Rovinski B, Haynes JR, Cao SX et al. Expression and characterization of genetically engineered human immunodeficiency virus-like particles containing modified envelope glycoproteins: Implications for development of a cross-protective AIDS vaccine. J Virol 1992; 66:4003-4012.
82. Wagner R, Deml L, Notka F et al. Safety and immunogenicity of recombinant human immunodeficiency virus-like particles in rodents and rhesus macaques. Intervirology 1996; 39:93-103.
83. Tobin GJ, Nagashima K, Gonda MA. Synthesis and assembly of chimeric human immunodeficiency virus gag pseudovirions. Intervirology 1996; 39:40-48.
84. Haffar OK, Smithgall MD, Moran PA et al. HIV-specific humoral and cellular immunity in rabbits vaccinated with recombinant human immunodeficiency virus-like particles. Virology 1991; 183:487-495.
85. Shioda T, Shibuta H. Production of human immunodeficiency virus (HIV)-like particles from cells infected with recombinant vaccinia viruses carrying the *gag* gene of HIV. Virology 1990; 175:139-148.
86. Halperin SA, Nestruck AC, Eastwood BJ. Safety and immunogenicity of a new influenza vaccine grown in mammalian cell culture. Vaccine 1998; 16:1331-1335.
87. Kistner O, Barrett PN, Mundt W et al. Development of a mammalian cell (Vero) derived candidate influenza virus vaccine. Vaccine 1998; 16:960-968.
88. Kaufman R. Methods Enzymol 1990; 185:487-512.
89. Simari RD, Yang ZY, Ling X et al. Requirements for enhanced transgene expression by untranslated sequences from the human cytomegalovirus immediate-early gene. Mol Med 1998; 4:700-706.
90. Chapman BS, Thayer R, Vincent KA et al. Effect of intron A from human cytomegalovirus (Towne) immediate-early gene on heterologous expression in mammalian cells. Nucleic Acids Res 1991; 19:3979-3986.
91. Fussenegger M, Bailey JE, Hauser H et al. Genetic optimization of recombinant glycoprotein production by mammalian cells. Trends Biotechnol 1999; 17:35-42.
92. Evans MJ and Scarpulla RC. Introns in the 3'-untranslated region can inhibit chimeric CAT and b-galactosidase gene expression. Gene 1989; 84:135-142.
93. Gross MK, Kainz MS, and Merrill GF. Introns are inconsequential to efficient formation of cellular thymidine kinase mRNA in mouse L cells. Mol Cell Biol 1990; 7:4576-4581.
94. Huang M and Gorman CM. The simian virus 40 small-t intron, present in many common expression vectors, leads to aberrant splicing. Mol Cell Biol 1990; 10:1805-1810.
95. Buchman A. and Berg P. Comparison of intron-dependent and intron-independent gene expression. Mol Cell Biol 1988; 8:4395-4405.
96. Hartikka J, Sawdey M, Jensen FC et al. An improved plasmid DNA expression vector for direct injection into skeletal muscle. Hum. Gene Ther 1996; 7:1205-1217.

97. Goodwin EC and Rottman FM. The 3'-flanking sequence of the bovine growth hormone gene contains novel elements required for efficient and accurate polyadenylation. J Biol Chem 1992; 297:16330-16334.
98. Kozak M. Initiation of translation in prokaryotes and eukaryotes. Gene 1999; 234:187-208.
99. Salk J, Salk D. Control of influenza and poliomyelitis with killed virus vaccines. Science 1977; 195:834-847.
100. Fox JP, Modes of action of poliovirus vaccines and relation to resulting immunity. Rev Infect Dis 1984; 6(Supplement 2):S352-S355.
101. Melnick JL. Advantages and disadvantages of killed and live poliomyelitis vaccines. Bull. WHO 1978; 56(1):21-38.
102. Dunn G, Begg NT, Cammack N et al. Virus excretion and mutation by infants following primary vaccination with live oral poliovaccine from two sources. J Med Virol 1990; 32:92-95.
103. Nkowane BM, Wassilak SGF, Orenstein WA et al. Vaccine-associated paralytic poliomyelitis. United States: 1973 through 1984. JAMA 1987; 257(10):1335-1340.
104. Prevots DR, Sutter RW, Strebel PM et al. Completeness of Reporting for Paralytic Poliomyelitis, United States, 1980 through 1991. Arch Pediatr Adoles Med 1994;20:479-485.
105. Onorato IM, Modlin JF, McBean AM et al. Mucosal immunity induced by enhanced-potency inactivated and oral polio vaccines. J Infect Dis 1991; 163:1-6.
106. Marine WM, Chin TDY, Gravelle CR. Limitation of fecal and pharyngeal poliovirus excretion in salk-vaccinated children. Am J Hyg 1962; 76:173-195.
107. Salk D. Herd effect and virus eradication with us of killed poliovirus vaccine. Dev Biol Stand 1981; 47:247-255.
108. Dick GWA, Dane DS, McAlister J et al. Vaccination against poliomyelitis with live virus vaccines. Brit Med J 1961; 2:266-269.
109. Glezen WP, McCollough RH, Lamb GA et al. Quantitative relationship of preexisting homotypic antibodies to excretion of poliovirus types 1,2 and 3 following the feeding of trivalent attenuated poliovirus vaccine. Am J Epidemiol 1969; 90(2):146-156.
110. CDC. Certification of Poliomyelitis Eradication – the Americas, 1994. MMWR 1994;43:720-722.
111. CDC. Poliomyelitis prevention in the United States: introduction of a sequential vaccination schedule of inactivated poliovirus vaccine followed by oral poliovirus vaccine; Recommendations of the Advisory Committee on Immunization Practices (ACIP). MMWR 1997; 46(RR-3);1-25.
112. CDC. Impact of the Sequential IPV/OPV Schedule on Vaccination Coverage Levels -- United States. MMWR 1998; 47:1017-1019.
113. American Academy of Pediatrics Committee on Infectious Diseases. Poliomyelitis prevention: revised recommendations for use of inactivated and live oral poliovirus vaccines. Pediatrics 1999; 103:171-172.
114. CDC. Notice to Readers Recommended Childhood Immunization Schedule -- United States. MMWR 1999; 48:8-16.
115. American Academy of Pediatrics, Committee on Infectious Diseases: Poliomyelitis prevention: Revised recommendations for use of inactivated and live oral poliovirus vaccine. 1999; 103:171-2.
116. Centers for Disease Control and Prevention: Recommendations of the Advisory Committee on Immunization Practices: Revised recommendations for routine poliomyelitis vaccination. 1999; 48:590.
117. Bottiger M. Polio immunity to killed vaccine: an 18-year follow-up. Vaccine 1990; 8:443-445.
118. Langer BC, Lovestad A, Frosner GG. High immunogenicity and good tolerability of a new hepatitis A vaccine candidate. Vaccine 1996; 14:1089-91.
119. Villacres-Eriksson M, Bergström-Mollaoglu M, Kåberg H. et al. Involvement of interleukin-2 and interferon-gamma in the immune response induced by influenza virus ISCOMs. Scand J Immunol 1992; 36:421-426.
120. Hoglund S, Dalsgaard K, Lovgren K et al. ISCOMs and immunostimulation with viral antigens. In: J.R. Harris, ed. Subcellular Biochemistry. New York and London: Plenum Press 1989:39-68.
121. Takahishi H, Takeshita T, Morein B et al. Induction of CD8+ cytotoxic T cells by immunization with purified HIV-1 envelope proteins in ISCOMS. Nature 1990; 344:873-875.
122. Morein B, Lovgren K, Ronnberg B et al. Immunostimulating complexes. Clinical potential in vaccine development. Clin Immunother 1995; 3:461-475.

123. Newman MJ, Wu JY, Gardner BH et al. Saponin adjuvant induction of ovalbumin specific CD8+ cytotoxic T lymphocyte responses. J Immunol 1992; 148:2357-2362.
124. Sambhara S, Woods S, Arpino R et al. Influenza (H1N1)-ISCOMs enhance immune responses and protection in aged mice. Mech Age Dev 1997; 96:157-169.
125. Sambhara S, Woods S, Arpino R et al. Induction of CTL and protective immunity in mice using Influenza Immunostimulatory complexes (ISCOMS). A comparison with influenza vaccine. J Infect Dis 1998; 177:1266-1274.
126. Sambhara S, Kurichh A, Miranda R et al. Up-regulation of CD86 by influenza-ISCOMs correlates with enhanced immune responses and resistance against infection in aged mice. Vaccine 1998; 16:1698-1704.
127. Rimmelzwaan GF, Osterhaus ADME. Cytotoxic T cell memory: role in cross-protective immunity against influenza? Vaccine 1995; 13:703-706.
128. Jones PD, Hla R, Morein B et al. Cellular immune responses to local immunization in the murine lung with influenza A virus glycoproteins in micelles and immunostimulatory complexes (Iscoms). Scand J Immunol 1988; 27:645-652.
129. Coulter A, Wong T, Drane D et al. Studies on experimental adjuvanted influenza vaccines: comparison of immune stimulating complexes (Iscoms^{TM}) and oil-in-water vaccines. Vaccine 1998; 16:1243-1253.
130. Ennis FA, Cruz J, Jameson J et al. Augmentation of human influenza A virus-specific cytotoxic T lymphocyte memory by influenza vaccine and adjuvanted carriers (ISCOMS). Virology 1999 In press.
131. Gluck R. Liposomal presentation of antigens for human vaccines. In: Powell MR, Newman MJ, eds. Vaccine Design: The Subunit and Adjuvant Approach. New York and London: Plenum Press, 1995:325-345.
132. Gluck R. Adjuvant activity of immunopotentiating reconstituted influenza virosomes (IRIVs). Vaccine 1999; 17:1782-1787.
133. Lendon GP, Jenkins SA, Woods AL et al. Poly[di(carboxylatophenoxy)phosphazene] (PCPP) is a potent immunoadjuvant for an influenza vaccine. Vaccine 1998; 16:92-98.
134. Ott G, Barchfeld GL, Cherno D et al. MF59: design and evaluation of a safe and potent adjuvant for human vaccines. In: Powell MR, Newman MJ, eds. Vaccine Design: The Subunit and Adjuvant Approach. New York and London: Plenum Press, 1995:277-296.
135. Minutello M, Senatore F, Cecchinelli G et al. Safety and immunogenicity of an inactivated subunit influenza virus vaccine combined with MF59 adjuvant emulsion in elderly subjects, immunized for three consecutive influenza seasons. Vaccine 1999; 17:99-104.
136. Halperin SA, Davies HD, Barreto L et al. Safety and immunogenicity of two inactivated poliovirus vaccines in combination with an acellular pertussis vaccine and diphtheria and tetanus toxoids in seventeen- to nineteen-month-old infants. J Pediatr 1997; 130:525-531.
137. Pines E, Barrand M, Fabre P et al. New acellular pertussis-containing paediatric combined vaccines. Vaccine 1999; 17:1650-1656.
138. Decker MD, Edwards KM. Combination vaccines. In: Plotkin SA, Orenstein WA, eds. Vaccines, 3rd ed. Philadelphia: W. B. Saunders 1999:508-530.
139. Thoelen S, Van Damme P, Leentvaar-Kuypers et al. The first combined vaccine against hepatitis A and B: an overview. Vaccine 1999; 26;17:1657-1662.
140. Van Hoecke C, Lebacq E, Beran J et al. Concomitant vaccination against hepatitis A and typhoid fever. J Travel Med 1998; 5:116-120.
141. Mengiardi B, Berger R, Just M et al. Virosomes as carriers for combined vaccines. Vaccine 1995; 13:1306-1315.
142. Ellis RW. Development of combination vaccines. Vaccine 1999; 17:1635-1642.
143. Sawyer LA, McInnis J, Patel A et al. Deleterious effect of thimerosal on the potency of inactivated poliovirus vaccine. Vaccine 1994; 12:851-856.

CHAPTER 9

Live Attenuated Bacterial Vaccines

Kevin P. Killeen and Victor J. DiRita

Introduction

Immunization is the most effective public health tool used to control infectious disease. Moreover, it is extremely cost effective given that treatment of disease is far more expensive than disease prevention. The cost of vaccines and their administration from birth to age 16 is estimated by the Centers for Disease Control (CDC) to be US$500. Each US$1 spent on vaccination saves US$16 in avoiding costly drug therapies and hospitalizations[1] ultimately saving ca. US$8,000 per vaccinated individual. Further, phenomena such as herd immunity can provide protection to a community when only a minority of the total population has been vaccinated. Ideally, vaccination leads to the total eradication of an infectious agent that has no alternative hosts or environmental reservoirs, e.g., smallpox and soon for polio.[2]

Although commercialized vaccines are very safe, examples of adverse reactions have been reported for each vaccine that has ever been widely used. In some cases, such as the DTP (diphtheria, tetanus, and pertussis) vaccine, widespread fears concerning potential harmful side-effects have resulted in declining use and a resultant resurgence of disease.[3]

During the past decade, recombinant DNA technology has made it possible to design an improved generation of vaccines. Although the first recombinant non-living vaccine, hepatitis B, has been an enormous success, we are only beginning to see the fruits of biotechnology in the arena of live attenuated vaccines. A recent series of live attenuated recombinant vaccines against cholera, typhoid, and shigellosis are at various stages of human clinical trials. The application of recombinant DNA technology to these vaccines will be emphasized throughout this chapter.

Types of Vaccines

Three types of commercialized vaccines exist: antigen subunits, inactivated organisms, and live attenuated organisms. Vaccines composed of subunits (i.e., purified inactivated proteins/peptides, carbohydrate capsules, or cross-linked conjugates of these) are generally considered safest because they are used in small quantities, are chemically well defined, and do not replicate and thus cannot spread to the environment and other non-vaccinated or immunocompromised individuals. However, subunit vaccines are often expensive to manufacture and have a limited ability to induce immune responses, requiring adjuvants and multiple doses. Inactivated organisms (i.e., whole bacteria/viruses killed by heat or chemical treatment), although considered very safe, suffer from other drawbacks, such as ill-defined molecular characteristics and poor immunogenicity (i.e., large quantities of vaccine are required to elicit protective immune responses). Inactivated vaccines are relatively easy to manufacture and can possess inherent immunomodulatory activity (e.g., the whole cell pertussis component of DTP), although the

New Vaccine Technologies, edited by Ronald W. Ellis. ©2001 Eurekah.com.

spectrum of immune response induced is usually limited to the humoral arm of the immune system.[2]

In contrast, live attenuated vaccines (i.e., living viruses/bacteria that carry mutations rendering them avirulent or significantly reduced in virulence) offer significant advantages in terms of manufacture and immunogenicity. For example, a single inoculation of live vaccine at a modest dose replicates in vivo to a very large immunogenic dose and expresses the majority of immunogens seen during natural disease. Moreover, the processing and presentation of these antigens more closely resembles natural infection, thereby often inducing 'convalescent-level' immune responses that endure and protect the vaccinee for long periods of time. Live vaccines can induce mucosal immune responses, sometimes required for long-lived protection, which are not elicited by systemically administered vaccines, and the mode of delivery of live vaccines can be quite simple (e.g., ingestion). Finally, because attenuated vaccines are living organisms they may be genetically engineered to express heterologous antigens, thus providing protection from more than one disease.

Thirty-two commercialized vaccines were widely available and distributed in 1994 (Fig. 9.1). Seventeen were directed against viral pathogens and 15 were directed against bacterial pathogens. An interesting contrast emerges when one examines the nature of these vaccines. Of the 17 viral vaccines, 8 are subunit or inactivated (e.g., hepatitis A and B) and 9 are live attenuated (e.g., mumps, measles, rubella (MMR) and oral polio) compared with 15 bacterial vaccines where 13 are subunit or inactivated (e.g., DPT) and only two (tuberculosis (BCG) and typhoid fever (Ty21a)) are live attenuated (Fig. 9.2). Thus, there are similar numbers of subunit/inactivated and live attenuated viral vaccines; however, there are a disproportionate number of subunit/inactivated bacterial vaccines compared to live attenuated. This disparity continues given that a recently commercialized Lyme disease vaccine (directed against the bacterium *Borrelia burgdorferi*) is subunit-based and a new rotavirus vaccine is live attenuated.

The principle reason for this contrast lies in the timing of critical basic scientific discoveries and their application to the arena of vaccinology. In the 1950s, virologists and vaccinologists began to isolate viruses from infected individuals and serially passage them in vitro in eucaryotic cell lines. This approach yielded several live attenuated vaccines including, most notably, polio and MMR. In the bacterial arena, the 1980s marked the dawning of cloning DNA and the understanding of the various functions that specific genes and gene-products serve in the complex process of pathogenesis. Only within the past decade has this technology manifested and have recombinant bacterial strains emerged which are genetically defined and thus specifically attenuated. Given that these genetic tools of molecular biology have only recently been applied toward the generation of live attenuated bacteria, it is predicted that the number of live attenuated bacterial vaccines will increase sharply in the next decade paralleling what occurred in the 1950s and 1960s with viral vaccines.

In 1996, there were 17 improved vaccine candidates and 45 novel vaccine candidates at various stages of Research and Development at vaccine companies throughout the world.[4] Of the 62 total vaccine candidates, 28 were directed against bacterial pathogens, 24 against viral pathogens, and 10 against fungal or parasitic pathogens. This large number of vaccines under commercial development underscores the need to minimize the number of doses administered to vaccinees. The technology to produce safe, effective live attenuated vaccines is available and these vaccines often confer protection following a single dose. Therefore, live attenuated bacterial vaccines that are safe and efficacious will play a central role as public health tools in the 21st century.

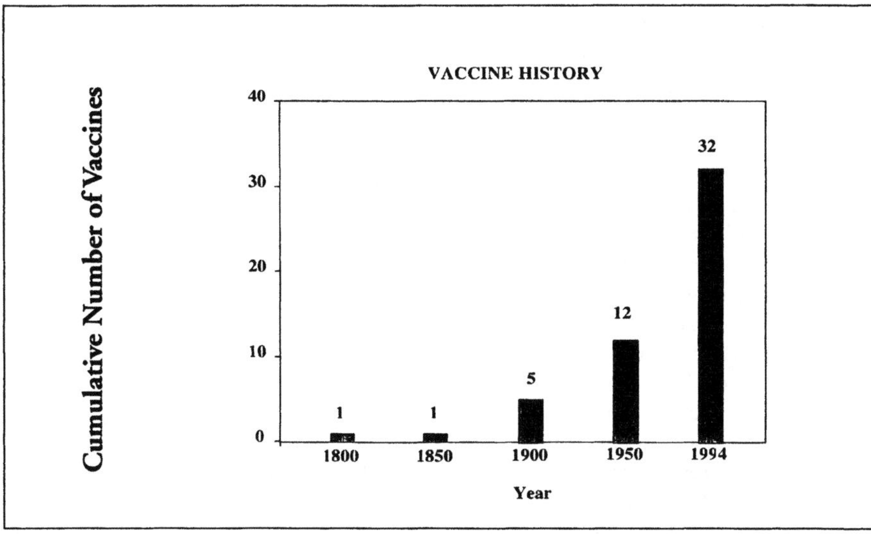

Fig. 9.1. Cumulative number of bacterial and viral vaccines as a function of time.

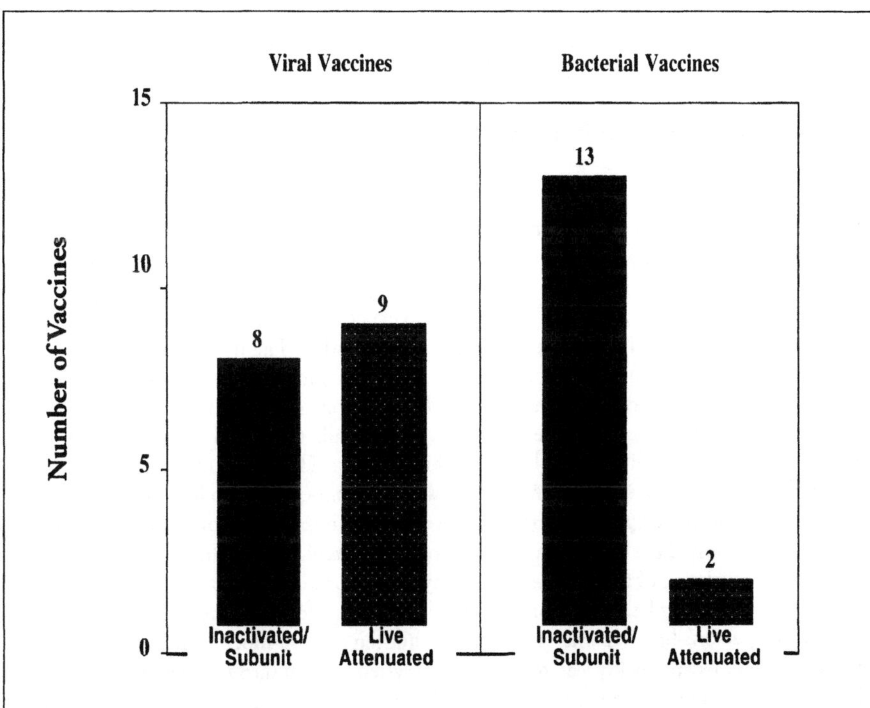

Fig. 9.2. Comparison of the number of viral and bacterial vaccines that are inactivated or subunit versus live-attenuated.

Live Attenuated Bacterial Vaccines

The most intensely investigated live attenuated bacteria include Bacille Calmette-Guérin (BCG), *Salmonella typhi*, *Shigella* spp.., and *Vibrio cholerae*. Each of these attenuated bacteria has been evaluated for safety, immunogenicity, and efficacy in humans as a vaccine against tuberculosis, typhoid fever, shigellosis, and cholera, respectively. In this chapter, we will focus on these live vaccines chiefly because they have been extensively studied in human clinical trials or are licensed, commercialized vaccines. Further, each of these live attenuated vaccines have been examined for their capacity to deliver foreign antigens as multivalent vaccine vectors in extensive pre-clinical and Phase I human studies. This subject is discussed in detail in a separate chapter (see Chapter 10).

Throughout this chapter we will emphasize several themes and their degree of importance in the research and development of each live attenuated bacterial vaccine. Chronologically, the first consideration for attenuating a bacterial pathogen is the selection of the progenitor strain. Was the strain initially isolated from a human or another host? What was its virulence in the original host? How does it compare to the virulence of other known strains? Second, is attenuation achieved through introduction of defined or undefined mutations? A defined mutation is specifically introduced into a strain overtly deleting a gene encoding a crucial enzyme involved in metabolism, a toxin, or a virulence factor. If the mutation is defined there exists a rational, molecular basis for the attenuation. Undefined mutations are induced by either chemical mutagenesis or serial passage in vitro which often renders avirulent strains that have accumulated spontaneous mutations and the nature of the DNA aberrations remain unknown. Thirdly, what are the pre-clinical models and what is their validity for predicting safety, immunogenicity, and protection in humans? Finally, are human challenge models available for determining the protective capacity of a live vaccine candidate?

Live Attenuated *Mycobacterium bovis* (BCG Tuberculosis Vaccine)

Tuberculosis

Tuberculosis has been a frightful disease throughout history, as in its advanced stages it is severely damaging causing a range of symptoms that include fever, a cough productive of bloody sputum, and chest pain. However, an overwhelming majority (greater than 90%) of healthy persons who become infected with *M. tuberculosis* experience only mild flu-like symptoms but otherwise remain well for a long period of time afterward.[5,6] The disease may take on one of two distinct manifestations: primary tuberculosis and secondary tuberculosis.

In primary tuberculosis, infection of alveolar macrophages results in a large number of the microbes being killed, but this infection nevertheless leads to limited intracellular growth of the microbe that results in localized lesions within the lung. The host controls infection through cell-mediated immunity, in which macrophages and lymphocytes act to form a walled off structure that contains the organisms; antibodies are thought to play little or no role in controlling TB infection. Individuals with the primary form of tuberculosis have few symptoms although granulomas consisting of a variety of host cell-types and bacteria may occur at this stage. These lesions may eventually heal to become calcified and in this form will be present for the rest of the infected person's life. In some cases, such as children or immunocompromised patients, the primary infection can quickly lead to a dangerous systemic form through hematogenous spread. This can be manifested as meningitis, if the infection takes a route into the meninges or brain, or as miliary tuberculosis, characterized by infected lesions in organs throughout the body.[6]

Secondary TB arises through reactivation of organisms that persist during the initially controlled infection. This may occur in up to 10% of individuals who had primary TB and is often the result of a compromise to cell-mediated immunity in the infected person. Infection in secondary TB often results in extensive lung involvement, as reactivated organisms multiply and necrosis of both resident cells and infiltrating macrophages occurs. This process causes necrotic ("caseous") lesions similar to those formed in primary TB, but which can combine with one another to form even larger areas of damage. In secondary TB, the microbe often disseminates to other sites in the body, where similar caseating lesions may occur to debilitating effect. In addition, persons with secondary TB who have organisms growing within damaged lungs may cough up infectious particles and therefore pose significant threat of infection to others with whom they are in close contact.[6]

Virulence Factors

The primary mode of pathogenesis of *M. tuberculosis* is its ability to survive and persist within the host in the face of a cell-mediated and humoral immune response. Accordingly, many of the factors produced by *M. tuberculosis* associated with pathogenesis play roles in altering the regulation of the host immune response. These factors are often components of the mycobacterial cell wall, which differs significantly from typical bacterial cell walls. A well-studied factor implicated in virulence is called "cord factor", called that for its ability to cause cord-like growth of organisms in culture.[7,6] It is a compound called trehalose 6,6'-dimycolate (TDM) and is found in virulent *M. tuberculosis* strains and can cause granuloma formation when injected by itself into tissue and also confers the ability of strains to persist in the lungs. TDM is also a potent inducer of TNF-α and this effect may play a role in some of the clinical features of tuberculosis.[7]

Other surface molecules that may be virulence factors include compounds called sulfatides, which consist of a trehalose sugar linked through an ester bond to a sulfate and several fatty acids. Sulfatides appear to inhibit the function of activated monocytes and to stimulate granuloma formation. Finally, a molecule that limits the host cell mediated immune response is liparabinomannan (LAM), a component of the unusual cell wall of *M. tuberculosis*. LAM blocks the effects of interferon-γ (IFN-γ) in activating macrophages.[7] In addition, mannose-capped LAM (ManLAM) from some virulent strains of *M. tuberculosis* decreases the capacity of the strain producing it for stimulating synthesis of TNF-α, a cytokine with multiple important affects on granuloma formation.[7]

Animal Models of Pathogenesis

A number of animal models are used in the study of pathogenesis by *M. tuberculosis* including monkeys, rabbits, guinea pigs and mice. Granuloma formation of the type seen in humans is particularly well-observed in rabbits and guinea pigs and can lead to necrotic lesions that ultimately lead to the death of the animal.[8] As the disease is characterized by a complex interplay of cytokine and cellular responses, the use of gene knockout technology in mice has recently contributed greatly to the understanding of infection, immunity and vaccinology.[9,10]

Unfortunately, use of animal models has had mixed utility in predicting the effectiveness of vaccines against tuberculosis.[11] There is variation in the responses to vaccines based on a number of parameters in the models, such as age, sex, route of infection, and others[12] leading to concern about the reproducibility of data from animal studies and their ability to confirm the efficacy of vaccine preparations.[12,11]

Vaccine Efforts

BCG is widely used throughout the world as a live vaccine for tuberculosis and in the treatment of certain cancers. BCG, actually a set of strains rather than a single strain, was originally isolated from a mastitis producing strain of *M. bovis*, a bovine pathogen closely related to *M. tuberculosis*. Calmette and Guérin began research into use of this strain as a potential TB vaccine in 1908, subjecting it to passage in vitro on a medium consisting of potato slices mixed with a mixture of glycerol and bile, the latter substance being included partly for its ability to prevent the organisms from clumping, which inhibited infection in guinea pigs.[13,14] Disaggregation with this method also aided in the ability to separate the culture for passaging.[15] Passage in this fashion continued for 13 years, culminating in 230 total passages, with periodic testing for reduced virulence in a variety of animals including guinea pigs, rabbits, cattle and horses.[16,13]

In 1921 the BCG vaccine strain was finally tested in humans. It was administered to infants as an oral vaccine because of the thought at the time that this was the normal route of infection.[16] Several hundred inoculations of children were accomplished through 1924, and statistical research by Calmette and Guérin provided evidence that the vaccine was providing protection against TB among vaccinated populations.[16] This led to great interest elsewhere for vaccinating with BCG and distribution of the strain began, probably first with distribution to Russia in 1924.[14] In the ensuing years the strain was passaged over 1000 additional times and distributed all over the world at different points during the passage history.

The molecular basis for attenuation in the BCG strains is unknown. Since the early period of strain passage, the presence of colony morphotypes has been well-documented,[14,13,11] and whether these morphotypes confer different in vivo phenotypes is controversial.[11] Less controversial is the conclusion that what is termed "BCG" represents a heterogeneous collection of genetically different strains that cluster phylogenetically in three different clades related to the timing of their original isolation.[14] Furthermore, using genome microarray technology, Small and colleagues have shown convincingly that BCG is significantly different genetically than current *M. bovis* strains. The extent of the difference is impressive, in that 38 open reading frames that are present in *M. bovis* are lacking to varying degrees in all BCG strains.[17] In addition, the comparison between virulent *M. tuberculosis* and virulent *M. bovis* shows large differences in genetic makeup, with 11 regions of the *M. tuberculosis* genome, accounting for 91 open reading frames, found lacking in that of *M. bovis*.

BCG has been used as a vaccine in more people than any other vaccine with the estimate of over 1 billion people having been immunized.[18,14] Nevetheless, it is considered a failure in the control of tuberculosis infection throughout the world.[18] Indeed, the results of numerous trials of BCG over the years since it was first given in 1921 as a vaccine have been mixed, demonstrating a range of efficacy from 0% to 80%.[19,11] The extensive variation in results has been attributed to many factors, perhaps the most compelling of which being heterologous immunity from prior exposure to environmental species of *Mycobacterium*. There is significant evidence, both theoretical and empirical, supporting this hypothesis.

One of the more intriguing arguments for natural variation comes from the "latitude effect", a term that describes the observation that the geographical latitude where vaccination takes place appears to correlate with vaccine efficacy with less promising results found closer to equatorial latitudes. Thus, a trial in two different places with the same strain, the BCG-Copenhagen isolate, revealed nearly 80% protective efficacy in the UK 15 years after vaccination and no protection in South India after the same time period.[11] While there is great variation, both genotypic and phenotypic, among BCG vaccines that have arisen during the more than 1000 passages of the original 1921 strain,[17,14,13] variation such as that observed with a vaccine preparation thought to be a single strain points to something in the population or in the way that the strains are being handled locally. To explain this outcome and similar trends in the literature of BCG vaccine trials, Fine hypothesized a higher level of exposure to environmental *Mycobacterium* species—and a resulting higher level of immunization by cross-reactive antigens—at lower latitudes.[19] Experimental data show that such exposure can predispose to protection against subsequent challenge with virulent strains.[20,21] This suggests that cross-reactive epitopes from environmental isolates may confer protection against bona fide infection with virulent *M. tuberculosis*. But such exposure may also contribute to poor induction of immunity by the BCG vaccine. Relevant to this point is that vaccine trials that have been very strict in excluding those individuals whose tuberculin test indicates prior exposure have resulted in demonstrations of very high efficacy for the vaccine.[19]

The great variation seen with BCG in trials around the world and the heterogeneity of the strain itself has maintained interest in developing better versions of the vaccine. Genetic manipulation of *Mycobacterium* spp has been made significantly more tenable in recent years and development of strains with defined lesions associated with survival in the host is now feasible. As a result of this, mutagenesis of metabolic genes is currently being investigated as a method for virulence attenuation for developing a new generation of vaccines against *Mycobacterium*.[22] Additionally, vaccine strains engineered to boost the immune response by expressing cytokines are also currently under investigation.[23]

Immune Correlates of Protection

Use of mice carrying defined lesions in cell-mediated immunity should provide important new knowledge regarding the immune response to *M. tuberculosis* infection. This, in turn, may lead to a better understanding of how to define protection immunologically. Classically, protection was thought to correlate with conversion to positivity in the tuberculin skin test using purified protein derivative (PPD). This test measures the delayed type hypersensitivity which is thought to be a contributing factor in immunity against *M. tuberculosis*.[24] However, data from animal experiments points to a lack of causality between this response and immunity, calling into question how important a positive PPD reaction is for predicting immunity. A reliable, predictive test for the efficacy of TB vaccines in humans, of the type that has been useful in development of vaccines for other infectious agents such as those discussed below, is still needed.

Live Attenuated *Salmonella typhi* (Typhoid Fever Vaccines and Vaccine Candidates)

Typhoid (Enteric) Fever

Following oral infection with *S. typhi*, the bacteria pass through the lymphoid tissues of the pharynx (tonsils) or the intestine (Peyer's patches) into the bloodstream. Subsequently, bacteria are removed by the cells of the reticuloendothelial system: liver, spleen, mesenteric lymph nodes, and bone marrow. The organisms multiply, particularly in the spleen and liver and by the tenth to fourteenth day, spill over into the bloodstream. This stage marks the end of the incubation period and the manifestation of clinical disease, or typhoid fever. Liberation of endotoxin from lysed bacteria causes fever and other symptoms such as headache, loss of appetite, weakness, diarrhea, and rose spots on the abdomen. The disease progresses for several weeks, and fever subsides when antibodies appear in circulation. Recovery occurs in about nine out of ten cases.[25]

Animal Model of Pathogenesis

S. typhimurium, the causative agent of mouse typhoid fever, is the generally accepted and validated model of infection for *S. typhi* in humans. *S. typhimurium* strains which possess given deletions are injected intraperitoneally (ip) into mice and measured for LD_{50} compared to progenitor strains in order to measure the degree of attenuation. Typically, the LD_{50} is ca. 10^1 colony forming units (cfu) of a wild-type strain introduced by the ip route. Appropriately attenuated strains of *S. typhimurium* possess LD_{50}s of ca. 10^5 cfu.[26] If attenuated *S. typhimurium* strains possess LD_{50}s in the 10^5 range then the identical deletions in *S. typhi* are likely to be viable typhoid fever vaccine candidates.

Vaccine Efforts

In the early 1970s, a streptomycin-dependent *S. typhi* was proven safe, immunogenic, and effective as a typhoid fever vaccine in preliminary human clinical studies.[27,28] However, when formulated for lyophilization, this vaccine candidate failed to reproducibly protect volunteers, arresting its further clinical development.[28] Importantly, however, proof of principle was established that live attenuated *S. typhi* vaccines could be effective in combating typhoid fever.

In the mid-1970s, Germanier and Furer developed *S. typhi* Ty21a.[29] Nitrosoguanidine was used as a chemical mutagen to induce random genetic lesions in a *S. typhi* Ty2 parent strain with selection of a *galE* mutation. The *galE* gene encodes UDP-glucose-4-epimerase that interconverts UDP-galactose and UDP-glucose. Conceptually, Ty21a would be attenuated because an abundance of galactose resulting in accumulation of gal-1-phosphate and UDP-galactose would lead to autolysis.[29] The process of random chemical-induced mutation caused other dysfunctions of Ty21a including nutritional auxotrophies, reduced growth rate, an inability to produce H_2S, and no production of Vi, a capsular surface antigen. Nonetheless, the basis of Ty21a's attenuation was its presumed galactose sensitivity in vivo. Retrospectively, a *galE* mutant of *S. typhi* was constructed and found to cause typhoid fever in 2 of 4 volunteers[30] indicating that the basis of Ty21a's attenuation is associated with other aberrations caused by chemical mutagenesis.[30] Despite the lack of understanding of Ty21a's specific attenuation, it is the sole example of a chemically-induced mutagenic process successfully employed in a commercialized bacterial vaccine product.

Ty21a's first field trial evaluation took place in Alexandria, Egypt from 1978-1981.[31,32] Approximately 32,000 6-7 year olds received either placebo or Ty21a administered in three oral doses on alternating days. The vaccine was very well tolerated and over a three-year surveillance period was associated with a 96% efficacy level. A subsequent field study in Santiago, Chile using an enteric-coated capsule formulation of Ty21a conferred significant protection to volunteers (67% protective efficacy). The encouraging efficacy studies of Ty21a promoted its commerciability as a typhoid fever vaccine in the developing world as well as a traveler's vaccine.

Ty21a vaccination activates both arms of the immune system eliciting humoral (IgG and sIgA) and cell-mediated immunity (CMI). It is unknown whether Ty21a-induced antibodies or CMI, or a combination of both, confer protection. However, the level of IgG against *S. typhi* O-antigen increases following each dose administered and the seroconversion rate correlates with protection. Therefore, anti-O-antigen IgG titers are considered an excellent surrogate marker of protection.[33]

Defined Genetic Attenuation

In direct contrast to Ty21a, there are several live attenuated *S. typhi* strains in human clinical trials that possess genetically defined attenuating mutations. Defined genetic attenuation can be dissected into two functional subclasses: gene(s) disruption affecting metabolism/regulation and gene(s) disruption affecting virulence. Each vaccine strain and its mode of genetic attenuation is described below.

Metabolically Attenuated *S. typhi* Strains

The elucidation of bacterial metabolic pathways coupled with the identification of the genes which encode enzymes in metabolic pathways enabled the concept of generating metabolically attenuated live bacteria. Disruption of genes encoding essential metabolic functions was the initial rational mode of attenuation applied to live bacteria to generate defined genetic aberrations.

A *S. typhi* strain harboring *aroA* and *purA* mutations (genes which encode essential biosynthetic enzymes in the aromatic amino acid pathway and DNA biosynthesis pathway, respectively), termed 541Ty, was constructed.[34] 541Ty was evaluated in human clinical trials and although well tolerated, was less immunogenic than Ty21a[35] and not considered for further clinical development. The fact that 541Ty, defective in two distinct central biochemical pathways, was overly attenuated suggested that a *S. typhi* strain defective in one biochemical pathway may generate a viable typhoid fever vaccine candidate. Two different wild-type *S. typhi* parent strains, Ty2 and ISP1820, were attenuated by deleting two different genes in the biochemical pathway for aromatic acid biosynthesis. CVD 906 (derivative of ISP1820) and CVD 908 (derivative of Ty2), each possessing *aroCD* mutations, were constructed. The reason for deleting two different genes which encode enzymes in the same biochemical pathway is if reversion occurs in one gene then the remaining defective gene prevents the restoration of the functional biochemical pathway. Each vaccine candidate was evaluated for safety and immunogenicity in human volunteers. Although the data were not statistically significant, the volunteer study trends indicated that CVD 906 was more reactogenic than CVD 908.[36] These results emphasize the degree of importance in the selection of the progenitor strain. This theme is highlighted later in this chapter in the attenuated *V. cholerae* vaccine section. In subsequent CVD 908 vaccine studies it was determined that single, oral dose levels required for immunogenicity caused transient bacteremia impeding the vaccine candidate's clinical development.[37] Theoretically, CVD 908's inability to produce aromatic amino acids de novo should preclude its ability to cause bacteremia. The fact that the strain replicates in vivo to cause bacteremia indicates that the bioavailability of essential metabolic substrates in vivo is very difficult to predict during an attenuated live vaccines complex course of colonization in humans.

Following a different approach, Curtiss and colleagues exploited metabolic/regulatory genes as targets for attenuation. They constructed *S. typhi* Chi^{3927}, a cya^-,crp^- (genes encoding adenylate cyclase and cAMP receptor protein, respectively) typhoid fever vaccine candidate. However, Chi^{3927} caused unacceptable levels of high fever and bacteremia thwarting its clinical development.[38] It is interesting to note that Chi^{3927} is a derivative of *S. typhi* Ty2, as are Ty21a and CVD 908, described above. Each vaccine candidate displayed a different clinical safety profile in human studies. This observation emphasizes the importance of the appropriate combination of the attenuating mutations in the specific bacterial progenitor strain towards the generation of a safe, immunogenic live attenuated vaccine.

Combination Regulatory/Virulence and Metabolic/Virulence Attenuated *S. typhi*

The attenuated *S. typhi* vaccine field evolved from the largely unsuccessful metabolic- or regulatory-attenuated candidates described above towards combinations of regulatory/virulence-, metabolic/virulence-, or solely virulence-attenuated vaccine strains. Attenuated *S. typhi* Chi^{4073}(Chi^{3927}+ cdt^- (gene which participates in dissemination of *S. typhi* from gut-associated lymphoid tissue to deeper organs of the reticuloendothelial system)), a combination regulatory- and virulence-attenuated strain, was constructed and evaluated for safety and immunogenicity in human volunteers. Chi^{4073} was generally well tolerated and immunogenic at 10^7-10^8 cfu doses.[39] The conceptual importance of this study was the demonstration that the addition of a virulence attenuation in a regulatory-attenuated background substantially improves the safety profile of a live attenuated vaccine candidate.

S. typhi TyLH445, a combined $aroA^-$, $phoP^-/Q^-$ (genes which encode virulence regulation) strain was evaluated in Phase I/II human volunteer studies. Although the vaccine was safe, it was ineffective at provoking significant immune responses after two oral doses.[26] Thus, in this particular case, the addition of another attenuating mutation which affected virulence into a metabolically-attenuated background yielded a strain that was overly attenuated. Separately, CVD 908$htrA^-$ (gene encoding a bifunctional stress protein/serine protease) was constructed and evaluated in human volunteers. The strain was well tolerated and immunogenic in the dose ranges tested (10^7-10^9cfu), however compromised in immunogenicity compared to its progenitor, CVD 908.[39]

Virulence-Attenuated *S. typhi* Strain Ty800

Given that the combination metabolic- and virulence-attenuated *S. typhi* TyLH445 was overly attenuated, a *phoP⁻/Q⁻ S. typhi* strain, Ty800, was constructed. Ty800 is the first example of an attenuated *Salmonella* strain that is metabolically intact and attenuated solely for virulence. Ty800 was evaluated in human studies and discovered to be safe and immunogenic after single, oral doses ranging from 10^7 to 10^{10} cfu.[40] Ty800 set precedence in the live attenuated *Salmonella* vaccine field that metabolically intact strains that are attenuated solely for virulence could be successfully generated. Perhaps the reason for the success of Ty800 is that the majority of virulence genes encode dispensable functions unique to the organism which do not affect replication in vivo thereby minimizing any negative impact on immunogenicity. In contrast, metabolic attenuation generally compromises live bacteria's ability to produce essential metabolites thereby reducing replication in vivo ultimately resulting in decreased immunogenicity.

A recently discovered novel metabolic attenuation is a *dam* deletion in *Salmonella*. Methylation at adenine bases by Dam (DNA adenine methylase) controls the timing and targeting of important biological processes such as DNA replication, methyl-directed mismatch repair, and transposition.[41] In addition, Dam regulates the expression of operons such as pyelonephritis-associated pili (pap), which are an important virulence determinant in upper respiratory tract infections. *dam* deletion mutants were demonstrated to be highly attenuated (>10,000 times the wild-type LD_{50}) when introduced orally to mice. When mice were orally immunized with *dam- S. typhimurium*, 8/8 were resistant to virulent *S. typhimurium* challenge compared to 0/12 unvaccinated mice. *dam* mutation may serve as a stand alone attenuation of *S. typhi* for evaluation in humans or can be combined with other virulence or metabolic attenuations.[41]

Experimental challenges of human volunteers with virulent *Shigella flexneri* and *V. cholerae* facilitated the development of shigella and cholera vaccines. Given the invasive and systemic nature of *S. typhi*, a safe, validated human challenge is lacking. However studies are being planned to develop a human challenge model for *S. typhi*. Thus, vaccinations with the several live attenuated *S. typhi* vaccine candidates described above followed by challenge can provide an early prediction of success in Phase III studies, expediting the commercialization of oral, single-dose, typhoid fever vaccines.

Live Attenuated *Shigella* spp. (Shigellosis Vaccine Candidates)

Shigellosis (Bacillary Dysentery)

Dysentery is an acute infectious disease of the lower ileum and colon. The incubation period following ingestion of *Shigella* spp. is typically about 48 hours, and the disease begins suddenly with fever, abdominal pain, vomiting, and diarrhea. There is often ulceration and marked inflammation of the intestinal mucosa, and the watery excreta frequently contain blood and mucus. Prostration is due largely to loss of fluid and electrolytes resulting from the purging diarrhea.[25]

Shigellae are extraordinarily adept intestinal pathogens, as evidenced by the small infectious dose.[42] The bacteria do not pass through the intestine wall but remain localized within the intestine or in cells lining the intestinal mucosa. Organisms invade enterocytes through the basolateral membrane and internalized bacteria subsequently spread within infected cells by organizing host cell actin into a cytoskeleton-based motor spreading into contiguous epithelial cells via membrane protrusions.[43]

S. dysenteriae is the most virulent of the *Shigella* species. It produces a highly potent virulence factor termed shiga-toxin, which attacks the endothelial cells of blood vessels.[44,45] *S. dysenteriae* is associated with the most severe cases of dysentery and the highest rates of complications, e.g., hemolytic-uremic syndrome, hemorrhagic colitis, sepsis, and purpura.[46] *S. sonnei and flexneri* are not as virulent as *S. dysenteriae*; however, these species produce and shed endotoxin during infection, which is responsible for intestinal hemorrhage, loss of weight and other general symptoms associated with bacterial endotoxins. Mortality due to bacillary dysentery varies greatly according to the species of infecting organism and the population infected. *S. dysenteriae* infection can by as high as 20 percent, however most species are associated with significantly lower mortality. In fact, many cases are mild and are not recognized as bacillary dysentery.[25]

Animal Models

Study of the pathogenic mechanism of shigellosis has been complicated by the innate resistance of most animals to oral infection by shigellae.[47] However, animal models can reproduce some aspects of human shigellosis. Oral challenge of starved, opiated guinea pigs with *S. flexneri* 2a first established the invasive nature of *Shigella* infections and oral challenge of rhesus monkeys, which are susceptible to naturally acquired *Shigella* infection, confirmed the invasive pathogenic mechanims. Injection of shigellae into ligated rabbit ileal loops causing active fluid secretion has been used as a model of bacterial invasion. A non-surgical rabbit model of enteric *Shigella* infection was developed for studying the pathogenesis and immunology of shigellosis and for evaluating *Shigella* vaccine candidates.[48] Complementing these animal models of intestinal shigellosis is the Sereny test, which assesses the ability of shigellae to invade the corneal epithelium and to elicit keratoconjunctivitis in the eyes of rabbits, guinea pigs, or mice.[47]

Vaccine Efforts

A significant obstacle in the generation of vaccines against shigellae is that there exists greater than 30 serotypes. However, a limited number of serotypes or a specific species are often endemic in certain regions.[49] For example, *S. sonnei* predominates in industrialized countries, *S. flexneri* 2a and 3 predominate less-developed countries, and *S. dysenteriae* serotype 1 causes epidemic dysentery in Central America, Africa, and the Indian subcontinent.[49]

A live, oral *Shigella* vaccine candidate was constructed by transfer of the 140-MDa invasiveness plasmid from *S. flexneri* 5 and the chromosomal genes encoding the group- and type-specific O antigen of *S. flexneri* 2a to *E. coli* K-12, yielding EcSf2a-1. In human clinical studies, this vaccine candidate produced a high level of adverse reactions including fever, diarrhea, or dysentery in volunteers that received a single oral dose of $1x10^9$ cfu. Further, vaccinees who received more tolerable doses, $5x10^6$ cfu or $5x10^7$ cfu, demonstrated no statistically significant level of protection against challenge with virulent *S. flexneri* 2a.[50] In an attempt to reduce reactogenicity levels an *aroD* mutation was created in EcSf2a-1, yielding EcSf2a-2. This strain was also evaluated in Phase I safety and immunogenicity human studies. Three oral doses of ca. $2x10^9$ cfu were required to elicit significant levels of seroconversion (61% developed serum antibodies against LPS and 44% against invasiveness plasmid antigens (Ipa)). Despite this degree of immunogenicity, the vaccine candidate regimen conferred only 36% protection against illness induced by experimental challenge.[50] In a subsequent study, EcSf2a-2 was orally administered in four doses of $7x10^8$ cfu to 21 adults on days 1, 3, 14, and 17. There were no serious adverse reactions although one volunteer developed fever and another developed mild diarrhea. All vaccinees developed a modest antibody secreting cell (ASC) against LPS and 19% developed a serum IgG response against LPS. Despite the four dose regimen, upon challenge with virulent *S. flexneri* 2a only 27% volunteers were protected against disease.[51]

Historically, the attenuation strategy for *Shigella* spp. strongly parallels that of *S. typhi.* That is to say, mutation of genes which encode metabolic function were inadequate in attenuating *Shigella* spp. and combinations of mutations of genes which encode metabolic/virulence functions were ultimately required to generate safer vaccine candidates. *Shigella flexneri* SFL1 is a virulent strain isolated from a tourist returning to Sweden with dysentery after vacation in Asia.[52] The strain was orally administered to human volunteers at doses of $7x10^3$ cfu. These volunteers developed overt signs of shigellosis confirming the virulence of the strain. This study also indicated that although SFL1 is virulent, it was less so than previously tested *S. flexneri* strains.[52]

An *aroD* deletion was constructed in the *S. flexneri* SFL1 parent, generating strain SFL124. In a Sereny test, it was observed that SFL1 caused keratoconjunctivitis in guinea-pigs and SFL124 did not, indicating that the aromatic amino acid auxotrophy impairs intracellular multiplication and attenuates the strain.[52] Following pre-clinical analyses, SFL 124 was evaluated in Vietnamese children. When orally administered as a single dose of 10^7, 10^8 or 10^9 cfu SFL124 caused mild symptoms. Local mucosal immune responses against *S. flexneri* LPS and Ipa were elicited in a dose-dependent fashion. High titers of of serum antibodies to LPS and Ipa were found in all children before ingestion of SFL124 that elicited increases in serum antibody titers in only a few volunteers. Thus, the immune response patterns elicited indicate a booster rather than a primary response.[53]

Subsequently, an *aroD* deletion was constructed in a *S. flexneri* 2a isolate, yielding SFL1070.[54] This vaccine candidate was orally administered to adult volunteers who received three doses of $1x10^5$, $1x10^7$, $1x10^8$, or $1x10^9$ cfu within 5 days. Volunteers who received the 10^7,10^8, or 10^9 cfu doses reported symptoms ranging from abdominal pain to watery diarrhea following the initial dose. Although the 10^5 cfu dose was well tolerated, only 1 of 9 volunteers mounted an immune response.[54] The poor safety profile of SFL124 and SFL1070 in human volunteers directed live attenuated *Shigella* vaccine development towards the concept of combining metabolic- and virulence–attenuations. In vitro studies had defined many important virulence factors of shigellae.[55] Thus, *S. flexneri* vaccine candidates that were combination metabolic- and virulence-attenuated were constructed.

There are two examples of combination, metabolic- and virulence-attenuated vaccine candidates, CVD 1203 and SC602. CVD1203 is a *S. flexneri* 2a strain attenuated by deletion of *aroA* (encodes another gene-product in the aromatic acid biosynthesis pathway) and *virG* (encodes gene-product which enhances intracellular replication and intra- and intercellular spread to adjacent cells). In pre-clinical studies, CVD 1203 was orogastrically administered to guinea pigs and elicited anti-IgA *S. flexneri* 2a antibodies and protected against conjunctival sac challenge.[56] In clinical studies, CVD 1203 was administered in two oral doses (primed with $2x10^8$ cfu or $2x10^9$ cfu, followed by $1x10^8$ cfu boost) and was evaluated for safety and immunogenicity. A separate group of individuals received a single, oral dose of 10^6 cfu CVD 1203. Briefly, objective symptoms including fever, diarrhea, dysentery, were observed following the first dose in some volunteers in all groups. In addition, seroconversion rates were low, ranging from 30-45%. The authors concluded that less reactogenic vaccines were needed.[57]

S. flexneri 2a strain SC602, a *virG⁻/icsA⁻* (causing intra-and intercellular migration defects) and iuc⁻/iut⁻ (causing iron acquisition defect) was constructed and tested in volunteers. Dose escalation studies revealed that SC602 caused shigellosis in a majority of volunteers when $3x10^8$ or $2x10^6$ cfu were ingested.[58] Even a low dose of $1x10^4$ cfu was associated with transient fever or mild diarrhea in 2 of 15 volunteers. Eight weeks after SC602 vaccination a subset of 7 of 15 volunteers and 7 unvaccinated volunteers were challenged with $2x10^3$ cfu virulent *S. flexneri* 2a. Six of 7 control volunteers developed shigellosis with fever and sever diarrhea or dysentery, while none of the vaccinees had fever, dysentery, or severe symptoms.[58] These efficacy results are a significant improvement over the EcSf2a-2 efficacy results and suggest that live attenuated *S. flexneri* 2a vaccines that confer protection to humans can be generated.

Live attenuated shigellae vaccine development has been stymied by the lack of reliable animal models and the generation of safe, immunogenic strains. Thus, immunological correlates of immunity had been lacking. However, field studies have demonstrated that an injectable vaccine consisting of *S. sonnei* O-antigen conjugated to *Pseudomonas aeruginosa* recombinant exoprotein A elicits high-titer anti-LPS antibodies that protect against disease.[59,60] This surrogate marker should serve as a standard in the future for the evaluation of live attenuated *Shigella* vaccines.

Live Virulence-Attenuated *Vibrio cholerae* (Cholera Vaccines)

Cholera

In contrast to the pathogenesis of tuberculosis, *S. typhi,* and *Shigella* spp., disease caused by *Vibrio cholerae* is not associated with invasion and dissemination of bacteria within host cells. Rather, disease is due principally to the action of cholera toxin (CT), an ADP-ribosylating enzyme that modifies the regulatory subunit (Gsα) of a heterotrimeric G protein complex controlling adenylate cyclase activity. Enzymatic modification of Gsα results in constitutive adenylate cyclase activity, leading to elevated levels of cAMP in enterocytes followed by a protein kinase A-mediated modification of chloride channel function. Chloride and other ions leave the cell, followed by water, and the intestinal lumen thus fills with the watery, isotonic stool characteristic of cholera. Secretion of water and electrolytes is accompanied by disruption of absorption in villus cells, which contributes to fluid loss. Diarrhea can be very profuse in this disease leading to severe loss of electrolytes and ultimately to death in untreated individuals.[61]

Oral rehydration with a solution of glucose, sodium chloride, sodium bicarbonate and potassium chloride is an effective treatment designed to maintain electrolyte balance during the self-limiting infection with *V. cholerae.* Although easily administered and inexpensive, oral rehydration therapy nevertheless requires a fairly long course of treatment with large volumes of solution. A vaccine to prevent or control cholera in endemic areas is therefore viewed as highly desirable. Because individuals infected with *V. cholerae* exhibit long term protection against subsequent infection, there is a strong opinion in the public health community that a live vaccine against cholera has the potential to ameliorate significant morbidity from this disease throughout the world. Before describing the approaches taken to achieve this aim, it is necessary to describe further the epidemic and pathogenic nature of *V. cholerae.*

Biotype and Serotype

Epidemic strains of *V. cholerae* are predominantly of the O1 serogroup. Within this serogroup there exists two further classifications of epidemic *V. cholerae*, biotype and serotype. Biotype is determined by laboratory tests including phage typing, the production of a specific metabolic intermediate, hemolysin and hemagglutinin production and sensitivity to polymixin B. The two major biotypes of epidemic *V. cholerae* are the El Tor and classical biotypes. The current cholera pandemic, which began in 1961, is the first of the seven recorded pandemics dating back to the early 19th century known to be caused by El Tor bacteria.

Either biotype may be one of two serotypes, Ogawa and Inaba, which are determined by a specific modification of the lipopolysaccharide (LPS) structure. A major alteration in the LPS of *V. cholerae* was identified in 1992 among isolates from outbreaks of cholera in India and Bangladesh. Strains otherwise characterized as El Tor isolates were shown to have a non-O1 LPS structure and these were subsequently classified as O139 Bengal. Prior infection with O1 *V. cholerae* was not protective against subsequent O139 Bengal infection, highlighting the efficacy of anti-LPS antibodies in immunizing against cholera. It is widely accepted that enduring immunity against cholera is due to induction of antibodies of the sIgA class directed against lipopolysaccharide and outer membrane proteins. Antibodies directed against the host-cell binding subunit of CT, CT-B, are also associated with protection.[62,63]

Other Virulence Factors

Although CT production is a pre-requisite for *V. cholerae* to cause disease, other virulence factors are very important for establishing successful interaction with the host. Principal among these is the toxin-coregulated pilus (TCP), a surface organelle required for colonizing in both animal models and human volunteer studies.[64] Strains lacking TCP do not elicit an immune response when fed to human volunteers, suggesting that colonization mediated by TCP is required for infecting organisms to be recognized by the immune system.[64] As its name suggests, TCP is expressed in coordination with CT by a complex regulatory system that also controls production of potentially immunogenic outer membrane proteins.[65]

Recent observations regarding the elements encoding both CT and TCP have significant impact on understanding how virulent isolates of *V. cholerae* may have evolved. These are the discoveries that both factors are encoded from the genomes of filamentous bacteriophage, CTXϕ and VPIϕ, respectively, and that the former appears to use the latter as its receptor.[66,67] Two other genes, *ace* and *zot* whose products have an effect on the physiology of intestinal tissue in vitro are also encoded by CTXϕ and, based on sequence homologies with other such proteins in the data base, these two proteins appear to be the phage coat protein.[68,69,66] For VPIϕ, the major subunit of TCP, called TcpA, may be the coat protein.[67]

Animal Model

The infant, or suckling, mouse model of *V. cholerae* infection is a widely accepted measure of reactogenicity and immunogenicity in humans. Several O1 and O139 cholera vaccine candidates were evaluated using this model. For example, Bang-3 and Bah-3 vaccine candidates colonized suckling mice ca. 10-fold more than Peru-3 or Peru-5 and nearly 1000-fold more than Peru-14. The in vivo murine model demonstrated there is a strong correlation between colonization level in mice and reactogenicity and immunogenicity in humans, elaborated in the next section.[70]

Vaccine Efforts

The knowledge of the molecular basis of cholera pathogenesis made it possible to establish parameters for construction of live, oral vaccines against disease. The major symptoms of cholera are due to the enzymatically active moiety of CT (CT-A), which is encoded by a separate gene (*ctxA*) from the immunogenic binding domain (CT-B; *ctxB*), thus enabling construction of virulence-attenuated strains. The major premise of virulence attenuation is that virulence determinants are often non-essential, or dispensable, functions of bacterial pathogens. Thus, in direct contrast to metabolic attenuation, the attenuation of virulence has no direct manifestation on replication/colonization in vivo, yielding a more fit live vaccine candidate. Attenuating *V. cholerae* in this fashion marked a new era for the construction of live bacterial vaccine strains that are metabolically intact and attenuated solely for virulence. Initially, a rational process was used, in which a deletion of *ctxA* was introduced as the sole attenuating mutation.[71,72]

In volunteer studies, these early vaccines were shown to be highly reactogenic although a single dose could induce long-lasting immunity to experimental wild-type cholera challenge.[73,64] The basis of reactogenicity, in which vaccinees may exhibit any number of symptoms ranging from loose stools to headache, nausea or vomiting, is not well understood. Until recently, only one recombinant strain, *V. cholerae* CVD 103-HgR, had been sufficiently tolerated by vaccinees to be considered for further development.[73] In experimental challenge models, volunteers receiving CVD 103-HgR show 87% protection against challenge with homologous classical biotype strains, but only 62% against El Tor strains.[74] Thus, a need existed for the development of vaccines that provide better protection against the El Tor biotype of *V. cholerae* as well as recently emerged 0139 strains. Another potential drawback of CVD 103-HgR—from the standpoint of rational vaccine design—is that its progenitor is the classical Inaba strain, 569B, which carries at least one mutation in the regulatory system governing CT and TCP and outer membrane protein production[75] and therefore behaves differently in vitro, and probably in vivo, than other *V. cholerae* strains.

An attenuated El Tor derivative called CVD 111 has recently been produced and tested in human volunteers. This strain, a derivative of the El Tor, Ogawa strain N16961, is deleted for the *ctx* genes as well as for two genes encoded near *ctx*, called *ace* and *zot*, which have toxic effects on intestinal tissue in vitro. CVD 111 showed significant protective efficacy in U.S. volunteers (over 80%) but also caused mild diarrhea in an unacceptably high percentage of them (12%).[76] A bivalent vaccine preparation consisting of both CVD 103-HgR and CVD 111 fared better in volunteer studies, causing rises in anti-LPS levels against both strains with no reactogenicity.[77]

A second generation of vaccine candidates has been constructed by genetically engineering geographically distinct field isolates of *V. cholerae* with a variety of lesions predicted to attenuate the organism as well as to reduce the level of reactogenicity. Several examples of such genetically defined, live El Tor strains, of both O1 and O139 serogroups have been made and tested in humans for safety and efficacy.[70,78,79]

The first of these vaccine candidates extensively analyzed in vivo were strains named Peru-3, Bang-3 and Bah-3, derived from strains isolated in Peru, Bangladesh and Bahrain, respectively.[70] They have deletions of CTXϕ, and *ctxA*, as well as a deletion/insertion mutation at the *recA* locus that introduces the *ctxB* gene under the control of a heat shock promoter. A filamentous, motility-deficient derivative of Peru-3, called Peru-14, was isolated to test the hypothesis that reactogenicity is associated with a motility-related colonization phenotype. In addition, Peru-5 was constructed in which the *recA* gene was left intact and the *ctxB* gene was instead introduced into the *V. cholerae lacZ* gene under a natural *ctx* promoter.

All six volunteers who ingested ca. 10^6 cfu Bang-3 or Bah-3 experienced symptoms such as abdominal cramping and nausea, and three of the six had diarrhea. Peru-3, in contrast, caused no abdominal cramping in any of three volunteers and diarrhea in only one, even at doses 100 times higher than those used for Bah-3 and Bang-3. The *recA*$^+$ strain Peru-5 caused abdominal cramps in two out of three volunteers but diarrhea in only one, indicating a potential role for RecA in reactogenicity.[70]

The best results in terms of reactogenicity were with Peru-14, the filamentous non-motile strain that colonized suckling mice very poorly. This strain caused diarrhea in only one of 12 volunteers given either ~10^6 or ~10^8 cfu and no cramping at either dose. In fact, even when given at 10^9 cfu, it caused no diarrhea in nine volunteers and cramping in only two of the nine. Most promising for the potential of these strains as vaccines, particularly the low-reactogenic strains Peru-3 and Peru-14, is that they induced strong serologic responses and protection against challenge with wild type *V. cholerae*. Peru-3 induced a significant rise in vibriocidal serum antibody, a standard measure of the effectiveness of the immune response, in two out of three volunteers given 4×10^6 cfu and in all three volunteers given 100 times that amount. Likewise, Peru-14 caused a vibriocidal serum response in all 21 volunteers who ingested doses ranging from 2×10^6 to 10^9 cfu. Volunteers vaccinated with both the Peru-3,5 or Peru-15 strains were protected from infection with 3×10^6 cfu of an El Tor, Inaba strain (N16961) with an efficacy of 87% and 80%, respectively. Subsequent study of an independently derived non-motile vaccine candidate called Peru-15 displayed a similar safety profile and high levels of immunogenicity and protection.[78] Additional studies of Peru-15 formulated with different buffers demonstrated that 30/30 volunteers that ingested a single oral dose developed high titer vibriocidal antibodies.[80]

Similar mutations were engineered into *V. cholerae* O139 isolates shortly after the outbreak with this serogroup occurred in 1992-1993. The most promising of these, Bengal-15, is deleted for the CTXϕ and *recA* and carries a motility defect as well. Bengal-15, administered as a 10^8 cfu single oral dose was well tolerated and highly immunogenic; 9 of 10 volunteers developed vibriocidal antibodies, in human volunteers. In challenge studies with virulent O139, Bengal-15 showed an efficacy of 83%.[79]

Each attenuated *Vibrio* strain described above is metabolically fit and therefore fully competent for replication in the human host. The levels of protection they confer prove the utility of rapidly constructing genetically defined, virulence attenuated vaccine candidates from geographical isolates of *V. cholerae*. The success of these live attenuated *V. cholerae* vaccines set a precedent for the generation of additional bacterial vaccines attenuated specifically for virulence.

Technologies Enabling "Next-Generation" Live Attenuated Bacterial Vaccines

Three technologies have recently emerged that will enable the creation of improved live attenuated bacterial vaccines. Those technologies include in vivo expression technology (IVET), signature tagged mutagenesis (STM), and whole-genome random DNA sequencing (WGRS).

IVET (in vivo expression technology) is a technology that identifies genes that are induced during a pathogen's course of infection in vivo.[81] Briefly, a gene encoding an essential DNA biosynthetic enzyme, *purA*, (the same gene used to attenuate *S. typhi* 541Ty above) is deleted from the genome of *S. typhimurium*. These *purA*⁻ strains are highly attenuated and fail to colonize the spleen of infected mice when introduced by the ip route. Separately, chromosomal DNA from *S. typhimurium* is digested and randomly cloned upstream of a promoterless *purA* gene on a plasmid which recombines onto the chromosome of the *S. typhimurium purA* mutant. Recombinants are introduced ip into mice and those that have in vivo promoters that turn on expression of *purA* survive and colonize the spleen. Bacteria from spleens are harvested and the DNA upstream of *purA* is sequenced to determine the gene which is "naturally" expressed by the in vivo promoter. Initially, 15 *ivi* fusions (in vivo induced genes) yielded were studied for their in vivo induction level and a subset of three were evaluated for decreased LD_{50}. All three strains were greatly reduced in virulence and one (an insertion within the *pheST/himA* operon) conferred protection against an oral challenge of virulent *S. typhimurium*.[81] Since IVET's conception in 1993, >100 *S. typhimurium ivis* have been identified which are expressed either in mice or in mouse cultured macrophages.[82] Most of these *ivis* await extensive analyses to determine their protective capacity as live attenuated bacterial vaccines.

STM is a genetic selection system that identifies virulence genes which are potential knockout targets for attenuation of bacteria.[83] STM is a negative selection method for virulence gene identification based on random transposon induced mutation. In vitro, microbial genes are tagged by the insertion of a transposon which contains a unique sequence of DNA. A pool of mutant bacteria are introduced into an animal and are collected during their course of pathogenesis. DNA hybridization techniques are applied to discern the input pool from the recovered pool of bacteria. Each gene mutation that causes attenuation of the bacteria can be detected by the nature of the hybridization technique and the unique DNA tag permits the identification of the specific gene mutated. The technology was applied to *S. typhimurium* in the murine model of pathogenesis and revealed three distinct subsets of genes.[83] A subset of 13 *S. typhimurium* genes which were formerly identified as virulence genes, a second subset of 6 genes which were new sequences similar to known genes, and a subset of 9 genes which did not correspond with any database entries. Since STM's inception, the technology has also been successfully applied to *Staphylocococcus aureus*[84] and *V. cholerae*[85] identifying important virulence genes in each pathogen.

The first published report of whole-genome random sequencing (WGRS) of a bacterial pathogen was first described in 1995 using *Haemophilus influenza*.[86] Remarkably, within three years 20 bacterial genomes were completed and greater than 45 bacterial genomes were in progress (http://www.fp.mcs.gov/gaasterland/genomes.html). Examples of completed genomes include *Mycoplasma pneumoniae*, *Helicobacter pylori*, *Borrelia burgdorferi*, and *Chlamydia pneumoniae*. Pertinent examples of genomes in progress include *Salmonella typhi* and *typhimurium*, *Shigella flexneri* 2a, and *Vibrio cholerae* (http://www.tigr.org/tdb/mdb/mdb.html). WGRS will undoubtedly identify a significant number of genes which encode virulence factors of these pathogens. As additional genes which encode virulence factors are elucidated by IVET, STM and WGST, safer and more effective live bacterial vaccines will continue to emerge.

Acknowledgment

Thanks to Dr. Michael Mahan (University of California, Santa Barbara) and Dr. Larry Hale (WRAIR) for sharing helpful manuscripts with authors in advance of their publication.

References

1. Fettner A. A Crack in the Shield—Our Unvaccinated Children, A Report on the Colloquium, Cold Spring Harbor Laboratory December 11-14, 1994.
2. Killeen KP, Spriggs DR, Mekalanos JJ. Bacterial vaccines. In Defense of Mucosal Surfaces: Pathogenesis, Immunity and Vaccines. Curr Topics Microbiol Immunol. Berlin, Heidelberg, New York, Tokyo: Springer-Verlag Press 1999.
3. Brennan MJ, Burns DL, Meade BD et al. Recent development in the development of pertussis vaccines. In: Ellis RW, ed. Vaccines: New Approaches to Immunological Problems. Boston: Butterworth-Heinemann 1992:23-52.
4. Jordan Report, Accelerated Development of Vaccines 1996.
5. Bloom BR, Murray CJL. Tuberculosis: Commentary on a Re-emergent Killer. Science 1992; 257:1055-1064.
6. Spitznagel JK. Mycobacteria: Tuberculosis and Leprosy. In: Shaecter, M., Engleberg, NC, Eisenstein, BI, and Medoff, G (eds.) Mechanisms of Microbial Disease, 3rd Edition. Baltimore, MD: Williams and Wilkins 1998:230-242.
7. Kraehenbuhl JL. Role of mycobacterial constituents in regulation of macrophage effector function. In: Roth JA, Bolin CA, Brogden KA et al, eds. Virulence Mechanisms of Bacterial Pathogens. 2nd ed. Washington, DC: ASM Press, 1995:97-114.
8. Orme IW. The immunopathogenesis of tuberculosis: A new working hypothesis. Trends Microbiol 1998; 6:94-97.
9. Flynn JL, Goldstein MM, Chan J et al. Tumor necrosis factor-alpha is required in the protective immune response against *Mycobacterium tuberculosis* in mice. Immunity 1995a; 2:561-72.
10. Flynn JL, Goldstein MM, Triebold KJ et al. IL-12 increases resistance of BALB/c mice to *Mycobacterium tuberculosis* infection. J Immunol 1995b; 155:2515-24.
11. Brewer TF, Colditz GA. Relationship between Bacille Calmette-Guérin (BCG) strains and the efficacy of BCG vaccine in the prevention of tuberculosis. CID 1995; 20:126-135.
12. Wiegeshaus E, Balasubramanian V, Smith DW. Immunity to tuberculosis from the perspective of pathogenesis. Infect Immun 1989; 57:3671-3676.
13. Grange JM, Gibson J, Osborn TW et al. What is BCG? Tubercle 1983; 64:129-139.
14. Behr MA, Small PM. A historical and molecular phylogeny of BCG strains. Vaccine 1999; 17:915-922.
15. Osborn TW. Changes in BCG strains. Tubercle 1983; 64:1-13.
16. Sakula A. BCG: Who were Calmette and Guérin? Thorax 1983; 38:806-812.
17. Behr MA, Wilson MA, Gill WP et al. Comparative genomics of BCG vaccines by whole-genome DNA microarray. Science 1999; 284:1520-1523.
18. Fine PEM. Bacille Calmette-Guérin Vaccines: A Rough Guide. CID 1995; 20:11-14.
19. Fine PEM. Variation in protection by BCG: Implications of and for heterologous immunity. Lancet 1995; 346:1339-1345.
20. Palmer CE, Long MW. Effects of infection with atypcial mycobacteria on BCG vaccination and tuberculosis. Am Rev Respir Dis 1966; 94:553-568.
21. Orme IM, Collins FM. Efficacy of *Mycobacterium bovis* BCG vaccination in mice undergoing prior pulmonary infection with atypical mycobacteria. Infect Immun 1984; 44:28-32.

22. Jackson M, Phalen SW, Lagranderie M et al. Persistence and protective efficacy of a *Mycobacterium tuberculosis* auxotrophic vaccine. Infect Immun 1999; 67:2867-2873.
23. Murray PJ, Aldovini A, Young RA. Manipulation and potentiation of antimycobacterial immunity using recombinant Bacille Calmette-Guérin strains that secrete cytokines. Proc Natl Acad Sci USA 1996; 93:934-9.
24. Dannenberg AM, Jr. Delayed-type hypersensitivity and cell-mediated immunity in the pathogenesis of tuberculosis. Immunol Today 1991; 12:228-233.
25. Carpenter PL. Microbiology. Fourth Edition. Philadelphia, London, Toronto: W.B. Saunders 1977.
26. Hohmann EL, Oletta CA, Miller SI. Evaluation of a *phoP/phoQ* deleted, *aroA*-deleted live oral *Salmonella typhi* vaccine strain in human volunteers. Vaccine 1996; 14:19-24.
27. DuPont HL, Hornick RB, Snyder MJ et al. Immunity in typhoid fever: Evaluation of live streptomycin-dependent vaccine. Antimicrob Agents Chemother 1970; 1971:236.
28. Levine MM, DuPont HL, Hornick RB et al. Attenuated streptomycin-dependent *Salmonella typhi* oral vaccine: potential deleterious effects of lyophilization. J Infect Dis 1976; 133:424.
29. Germanier R, Furer E. Isolation and characterization of *galE* mutant Ty21a of *Salmonella typhi*: A candiate strain for a live oral typhoid vaccine. J infect Dis 1975, 141:553.
30. Hone DM, Attridge SR, Forrest B et al. A *galE* via (Vi Antigen-Negative) mutant of *Salmonella typhi* Ty2 retains virulence in humans. Infect Immun 1988; 56:1326-1333.
31. Wahdan MH, Serie C, Cerisier Y et al. A controlled field trial of live oral typhoid vaccine Ty21a. Bull WHO 1980; 58:469.
32. Wahdan MH, Serie C, Cerisier Y et al. A controlled field trial of live *Salmonella typhi* strain Ty21a oral vaccine against typhoid: Three year results. J Infect Dis 1982; 145:292.
33. Levine MM, Ferreccio C, Black RE, Chilean Typhoid Commission et al. Progress in vaccines against typhoid fever. Rev Infect Dis 1989; supplement:S552.
34. Edwards MF, Stocker BAD. Construction of *ΔaroA his ΔpurA* strains of *Salmonella typhi*. J Baceriol 1984; 170:3991-3995.
35. Levine MM, Ferreccio C, Black RE et al. Large-scale field trial of Ty21a live oral typhoid vaccine in enteric-coated capsule formulation. Lancet 1987; 1:1049-1052.
36. Tacket CO, Hone DM, Losonsky GA et al. Clinical acceptability and immunogenicity of CVD 908 *Salmonella typhi* vaccine strain. Vaccine 1992; 10:443-446.
37. Sztein MB, Wasserman SS, Tacket CO et al. Cytokine production patterns and lymphoproliferative responses in volunteers orally immunized with attenuated vaccine strains of *Salmonella typhi*. J Infect Dis 1994; 170:1508-1517.
38. Tacket CO, Hone DM, Curtiss R, III et al. Comparison of the safety and immunogenicity of *ΔaroC ΔaroD* and *ΔcyaΔcrp Salmonella typhi* strains in adult volunteers. Infect Immun 1992; 60:536-541.
39. Tacket CO, Sztein MB, Losonsky GA et al. Safety of live oral *Salmonella typhi* vaccine strains with deletions in *htrA* and *aroC, aroD* and immune response in humans. Infect Immun 1997; 65(2):452-456.
40. Hohmann EL, Oletta CA, Killeen KP et al. *phoP/phoQ* deleted *Salmonella typhi* (Ty800) is a safe and immunogenic single-dose typhoid fever vaccine in volunteers. J Infect Dis 1996; 173:1408-1414.
41. Heithoff DM, Sinsheimer, RL, Low DA et al. An essential role for DNA adenine methylation in bacterial virulence. Science 1999; 284:967-70.
42. DuPont HL, Levine MM, Hornick RB et al. Inoculum size in shigellosis and implications for expected mode of transmission J Infect Dis 1989; 159:1126-1128.
43. Goldberg M, Sansonetti PJ. *Shigella* subversion of the cytoskeletion: A strategy for epithelial colonization. Infection and Immunity 1994; 61:4941-4946.
44. Fontaine A, Arondel J, Sansonetti PJ. Role of shiga toxin in the pathogenesis of bacillary dysentery, studied by using a Tox mutant of *Shigella dysenteriae* 1. Infect Immun 1988; 56:3099-3109.
45. Lopez EL, Diaz M, Grinstein S et al. Hemolytic uremic syndrome and diarrhea in Argentine children; the role of Shiga-like toxins. J Infect Dis 1989; 160:469-475.
46. Levine MM. Bacillary dysentery, mechanisms and treatment. Med Clin North Am 1982; 66:623-638.
47. Hale TL. Genetic basis of virulence in *Shigella* species. Microbiol Rev June 1991:206-224.
48. Etheridge ME, Hoque ATMS, Sack DA. Pathologic study of a rabbit model for shigellosis. Laboratory Animal Science 1996; 46:61.

49. Hale TL, Formal SB. Live oral vaccines consisting of *Escherichia coli* or *Salmonella typhi* expressing Shigella antigens. In: Woodrow GC, Levine MM, eds. New Generation Vaccines. New York: Marcel Dekker 1990:667-677.
50. Kotloff KI, Herrington DA, Hale TL et al. Safety, immunogenicity, and efficacy in monkeys and humans of invasive *Escherichia coli* K-12 hybrid vaccine candidates expressing *Shigella flexneri* 2a somatic antigen. Infect Immun June 1992:2218-2224.
51. Kotloff KI, Losonsky GA, Nataro JP et al. Evaluation of the safety, immunogenicity, and efficacy in healthy adults of four doses of live oral hybrid *Escherichia coli—Shigella flexneri* 2a vaccine strain EcSf2a-2. Vaccine 1995; 13(5):495-502.
52. Lindberg AA, Karnell A, Stocker BAD: Aromatic-dependent *Shigella* strains as live oral vaccines. In: Woodrow GC, Levine MM, eds. New Generation Vaccines New York: Marcel Dekker 1990: 677-687.
53. Li A, Cam PD, Islam D et al. Immune responses in Vietnamese children after a single dose of the auxotrophic, live *Shigella flexneri* Y vaccine strain SFL124. J Infect 1994 Jan; 28(1):11-23.
54. Karnell A, Li A, Zhao CR et al. Safety and immunogenicity study of the auxotrophic *Shigella flexneri* 2a vaccine SFL1070 with a deleted *aroD* gene in adult Swedish volunteers. Vaccine 1995; 13(1):88-99.
55. Fontaine A, Arondel J, Sansonetti PJ. Construction and evaluation of live attenuated vaccine strains of *Shigella flexneri* and *Shigella dysenteriae* 1. Res Microbiol 1990; 141:907-912.
56. Noriega FR, Wang JY, Losonsky G et al. Construction and characterization of attenuated Δ*aroA* Δ*virG Shigella flexneri* 2a strain CVD 1203, a prototype live oral vaccine. Infect Immun 1994:5168-5172.
57. Kotloff KI, Noriega F, Losonsky GA et al. Safety, immunogenicity, and transmissibility in humans of CVD 1203, a live oral *Shigella flexneri* 2a vacine candidate attenuated by deletions in *aroA* and *virG*. Infect Immun 1996; 64(11):4542-8.
58. Coster TS, Hoge CW, VanDeVerg LL et al. Vaccination against shigellosis with attenuated *Shigella flexneri* 2a strain SC602. Infect Immun 1999 July; 67(7):3437-43.
59. Cohen D, Ashkenazi S, Green MS et al. Double-blind vaccine-controlled radomised efficacy trial of an investigational *Shigella sonnei* conjugate vaccine in young adults. Lancet 1997; 18;349(9046):155-99.
60. Robin G, Keisari Y, Slepon R et al. Quantitative analysis of IgG class and subclass and IgA serum response to *Shigella sonnei* and *Shigella flexneri* 2a polysaccharides following vaccination with *Shigella* conjugate vaccines. Vaccine 1999; 17(23-24):3109-15.
61. Bennish ML. Cholera: Pathophysiology, clinical features, and treatment. In: Wachsmuth IK, Blake PA, Olsvik Ø, eds. *Vibrio cholerae* and cholera. Washington, D.C.: ASM Press 1994:229-255.
62. Svennerholm AM, Jertborn M, Gothefors L et al. Mucosal antitoxic and antibacterial immunity after cholera disease and after immunization with a combined B subunit-whole cell vaccine. J Infect Dis 1984; 149:884-893.
63. Kraehenbuhl J, Neutra MR. Molecular and cellular basis of immune protection of mucosal surfaces. Phys Rev 1992; 72:853-879.
64. Herrington DA, Hall RH, Losonsky G et al. Toxin, toxin-coregulated Pili, and the *tox*R regulon are essential for *Vibrio cholerae* pathogenesis in humans. J Exp Med 1988; 168:1487-1492.
65. DiRita, VJ, Parsot C, Jander G et al. Regulatory cascade controls virulence in *Vibrio cholerae*. Proc Natl Acad Sci U.S.A. 1991; 88:5403-5407.
66. Waldor MK, Mekalanos JJ. Lysogenic conversion by a filamentous phage encoding cholera toxin. Science 1996b; 272:1910-14
67. Karaolis DKR, Somara S, Manaeval Jr DR et al. A bacteriophage encoding a pathogenicity island, a type-IV pilus and a phage receptor in cholera bacteria. Nature 1999; 399:375-379.
68. Fasano A, Baudry B, Pumplin DW et al. *Vibrio cholerae* produces a second enterotoxin, which affects intestinal tight junctions. Proc Natl Acad Sci 1991; 284:5242-5246.
69. Trucksis M, Galen JE, Michalski J et al. Accessory cholera enterotoxin (Ace), the third toxin of a *Vibrio cholerae* virulence cassette. Proc Natl Acad Sci 1993; 90:5267-5271.
70. Taylor DN, Killeen KP, Hack DC et al. Development of a live, oral, attenuated vaccine against El Tor cholera. J Infect Dis 1994; 170:1518-1523.

71. Mekalanos JJ, Swartz DJ, Pearson GD et al. Cholera Toxin Genes: Nucleotide Sequence, Deletion Analysis and Vaccine Development. Nature 1983; 306:551-557.
72. Kaper JB, Lockman H, Balini M et al. Recombinant nontoxinogenic *Vibrio cholerae* strains as attenuated cholera vaccine candidates. Nature 1984; 308:655-658.
73. Levine MM, Kaper JB, Herrington D et al. Volunteer studies of deletion mutants of *Vibrio cholerae* 01 prepared by recombinant techniques. Infect Immun 1988; 56:161-167.
74. Levine MM, Kaper JB, Herrington D et al. Safety, immunogenicity, and efficacy of recombinant live oral cholera vaccines, CVD 103 and CVD 103-HgR. Lancet 1988; ii:467-470.
75. Pearson GD, DiRita VJ, Boyko SA et al. New attenuated derivatives of *Vibrio cholerae*. Res Microbiol 1990; 141:893-899.
76. Tacket CO, Kotloff KL, Losonsky G et al. Volunteer studies investigating the safety and efficacy of live oral El Tor *Vibrio cholerae* 01 vaccine strain CVD 111. Am J Trop Med Hyg 1997; 56(5):533-7.
77. Taylor DN, Tacket CO, Losonsky G et al. Evaluation of a bivalent (CVD 103-HgR/CVD 111) live oral cholera vaccine in adult volunteers from the United States and Peru. Infect Immun 1997; 65:3852-3856.
78. Kenner JR, Coster TS, Taylor DN et al. Peru-15, an improved live attenuated oral vaccine candidate for *Vibrio cholerae* 01. J Infect Dis 1995; 172:1126-9.
79. Coster TS, Killeen KP, Waldor MK et al. Safety, immunogenicity, and efficacy of live attenuated *Vibrio cholerae* 0139 vaccine prototype. Lancet 1995; 345:959-952.
80. Sack DA, Shimko J, Sack RB et al. Comparison of alternative bufers for use with a new live oral cholera vaccine, Peru-15, in outpatient volunteers. Infect Immun 1997b; 65:2107-11.
81. Mahan M, Slauch JM, Mekalanos JJ. Selection of bacterial virulence genes that are specifically induced in host tissues. Science 1993; 259:686-688.
82. Heithoff DM, Conner CP, Hanna PC et al. Bacterial infection as assessed by in vivo gene expression. Proc Natl Acad Sci USA 1997; 94:934-939.
83. Hensel M, Shea JJE, Gleeson C et al. Simultaneous identification of bacterial virulence genes by negative selection. Science 1995; 269:400-403.
84. Mei JM, Nourbakhsh F, Ford CW et al. Identification of *Staphylococcus aureus* virulence genes in a murine model of bacteraemia using signature-tagged mutagenesis. Mol Microbiol 1997; 26(2):399-407.
85. Chiang SL, Mekalanos JJ. Use of signature-tagged transposon mutagenesis to identify *Vibrio cholerae* genes critical for colonization in Mol Micro 1998; 27(4):797-805.
86. Fleischmann RD, Adams MD, White O et al. Whole-genome sequencing and assembly of Haemophilus influenzae Rd. Science 1995; 269(5223):496-512.

Chapter 10

Live Attenuated Bacterial Vectors

Sims K. Kochi and Kevin P. Killeen

Introduction

It is doubtful that Louis Pasteur could have anticipated the significance in 1881 that his discovery of bacterial attenuation would have on the use of microorganisms to protect against infectious diseases. Now, more than a century later, live attenuated bacterial vaccines have been developed that elicit sustained and protective immune responses (for a review see ref. 1). A novel application of these vaccines has been their use as vectors to deliver foreign antigens. This type of delivery is attractive for a number of reasons. For example, a live replicating vector can express in situ both heterologous antigens and native immunomodulatory factors, potentially obviating the need to add an adjuvant in cases where one might ordinarily be required. Another benefit is the exploitation of a vector's natural route of entry to deliver a foreign antigen. Presentation of an antigen in this manner often mimics natural infection and would presumably result in a broader spectrum of both mucosal and systemic immune responses. Finally, advances in molecular biology have enabled the genetic manipulation of most bacterial vectors and provide an opportunity to deliver a wide variety of protective immunogens.

Advances made in producing safe and protective live vaccines have facilitated the development of safe bacterial vectors. Live attenuated strains like *Salmonella typhi* Ty21a[2] and *Mycobacterium bovis* BCG[3] were the first vaccines to be used as vectors and are attenuated as a result of undefined mutations. The elucidation of microbial biosynthetic pathways facilitated the development of second-generation vectors that were rationally attenuated by the disruption of genes encoding metabolic functions. Vectors like *S. typhi* CVD 908[4] and Chi[4073, 5] were created based on this strategy and have 1 or more defined deletions in essential biosynthetic genes. However, some vaccines of this type are affected in their capacity to replicate and colonize an immunized host and therefore diminished in their utility as vectors.

To address this issue, live bacterial vectors were created by deleting virulence-associated genes. For example, mutant strains of *Salmonella* and *Shigella flexneri* deleted in the *phoP phoQ* virulence regulon and the virulence-associated *virG* gene, respectively, are highly attenuated with respect to their capacity to cause disease.[6-8] These recombinant strains presumably replicate in vivo as well as their wild-type homologues and are more fit than a similar strain attenuated by the deletion of a gene or genes involved in biosynthesis. As we will discuss below, the nature of these types of mutations and their impact on vector fitness in vivo may contribute, at least in part, to the capacity of a recombinant strain to elicit an immune response to vectored foreign antigens.

In this chapter we will present live attenuated bacterial vectors that we believe show the greatest promise for use in humans and emphasize traits that contribute to their effectiveness. This will include the basis of a vector's attenuation and the effect that attenuation has on the metabolism or the virulence of the strain. In addition, we will discuss the results of studies evaluating the immunogenicity and efficacy of live attenuated vectors delivering heterologous antigens in preclinical animal models and the utility of such models in predicting safety, immunogenicity and protection in humans.

Salmonella Vectors

Enteric diseases like typhoid fever are a major cause of morbidity and mortality worldwide, particularly in developing countries where sanitation conditions are poor. The live attenuated *S. typhi* strain Ty21a was created to vaccinate against typhoid fever and has since become the most intensely studied typhoid fever vaccine.[2] While it is approved for use in humans, the three-dose immunization regimen makes it impractical as a bacterial vector. More recent live *S. typhi*–based typhoid fever vaccines have been developed based on the results of studies of *S. typhimurium* conducted in animal models.[9] *S. typhimurium* infection in mice, for example, produces symptoms similar to typhoid fever in humans and is widely accepted as a model for typhoid fever. The availability of this model has facilitated the identification of genes in *S. typhimurium* involved in virulence and cell metabolism and has resulted in the creation of strains that are highly attenuated and capable of protecting mice against lethal *Salmonella* challenge. The development of *S. typhi* strains with similar mutations has resulted in safe, attenuated vectors for delivering foreign antigens to humans.

Live attenuated *S. typhimurium* strains have been used to deliver more than 50 different bacterial, viral and parasite antigens in animals. Many of these studies showed that mice immunized with *Salmonella* expressing a heterologous antigen developed immune responses that protected against challenge (for a review see ref. 10). These promising preclinical results catalyzed the development of several live attenuated strains of *S. typhi* expressing foreign antigens: the malarial circumsporozoite protein (CSP;11), a modified hepatitis B virus (HBV) surface antigen,[5] the *Helicobacter pylori* urease A and B subunits.[12,13] An attenuated *S. typhimurium* strain expressing CSP was shown to colonize the liver, induce antigen-specific cell-mediated immunity (CMI) and protect mice against sporozoite challenge in the absence of anti-sporozoite antibodies.[14] These results prompted a similar study in humans with the vaccine candidate, CVD 908, an attenuated strain of *S. typhi* deleted of the *aroC* and *aroD* genes that encode enzymes involved in the biosynthesis of aromatic amino acids. For this study, CVD 908 was further modified to express a fragment of the *Plasmodium falciparum* CSP from a single gene copy on its chromosome. Volunteers fed a single, oral dose tolerated the immunization with no serious adverse events and 7 of 10 vaccines shed the vector over a period of several days, indicating prolonged colonization.[11] As anticipated, all of the volunteers seroconverted against the vector; however, only 20% of the vaccinees developed a significant anti-sporozoite antibody response and only 10% developed CSP-specific $CD8^+$ cytotoxic T lymphocyte (CTL) activity. In spite of its modest clinical success, the results of this study are significant because they were the first to show that an attenuated strain of *S. typhi* delivering a heterologous antigen in humans is capable of inducing an antibody or CMI response to that antigen (Table 10.1).

The inability of CVD 908 to deliver a single gene copy of CSP effectively in humans renewed efforts to improve *S. typhi* as a vector. One approach taken was the use of multi-copy plasmids to increase the level of foreign antigen expression. An increased level of antigen expression should induce an immune response that is more prominent than that elicited by a bacterial vector expressing the same antigen from a single-copy gene on its chromosome. For vaccines intended for humans, the balanced-lethal plasmid system shows promise, as it allows the maintenance and high-level expression of plasmid-borne foreign antigens in the absence of antibiotic selection: the vector has a lethal mutation that is rescued by a bifunctional plasmid that both complements the mutation and encodes the heterologous antigen. Balanced-lethal

Table 10.1. Characteristics of live vectors used to deliver heterologous antigens in human clinical studies

Vector strain	Attenuating mutations	Foreign antigen(s)	Immune response to vectored antigen
S. typhi CVD 908	Δ*aroCD*	CSP	antibody, CTL
S. typhi Chi^{4632}	Δ*cya* Δ*crp* Δ*cdt* Δ*asd*	HBc-preS1	none detected
S. typhi Ty1033	Δ*phoPQ* Δ*purB*	UreA/B	none detected
S. typhimurium LH1160	Δ*phoPQ* Δ*purB*	UreA/B	antibody
M. bovis BCG	undefined	OspA	none detected
V. cholerae CH22CP	Δ*ctxA*	O-antigen	antibody
V. cholerae JRB-10	Δ*attRS1*Δ*irgA*	$StxB_1$	none detected

systems have been used with some success in *Salmonella* and have been constructed by the introduction of deletions in the chromosomal genes involved in cell wall biosynthesis (e.g., *asd*; 15) and purine biosynthesis (*purB*; 16). One of the first balanced-lethal vectors that yielded positive results in animals was Chi^{4064} (pNS27-53PS2), a recombinant strain of *S. typhimurium* deleted of the cAMP receptor protein (*crp*), adenylate cyclase (*cya*), and *asd* genes and transformed with a plasmid expressing Asd and an HBV nucleocapsid (HBc)-envelope protein (preS1) fusion protein.[17] Mice immunized with Chi^{4064} (pNS27-53PS2) developed anti-preS1 and anti-LPS serum IgG antibodies after a single, oral dose. These promising preclinical results provided an impetus to test an analogous strain of *S. typhi* for its capacity to elicit an immune response in humans. In this study, volunteers were fed multiple doses of *S. typhi* Chi^{4632} (pYA3167), a Δ*cya*Δ(*crp*-cdt)Δ*asd* strain transformed with a balanced-lethal plasmid encoding the genes for Asd and the HBc-preS1 fusion protein described above.[18] Six of the 7 volunteers immunized orally and 1 of 6 volunteers vaccinated rectally seroconverted against *S. typhi* LPS; several volunteers also developed anti-LPS sIgA responses. However, only 1 vaccinated LPS-seropositive volunteer developed a serum anti-preS1 antibody response. In a subsequent study, *S. typhi* Chi^{4632}(pYA3167) was given orally to 10 human volunteers. While most of the volunteers developed *S. typhi*-specific serum antibody responses and produced anti-*S. typhi*-specific antibody-secreting cells, none of the vaccinees developed serum antibody to the HBV preS1 envelope protein.[5] The investigators in these studies suggested that a more fit vector might better express the foreign antigen and induce an immune response.

One approach to address the issue of vector fitness was the development of live *Salmonella* vectors that were attenuated by the deletion of virulence-associated genes rather than genes involved in vector biosynthesis. Ty800, for example, was produced by the deletion of the *S. typhi phoP/Q* virulence regulon.[6] The immunogenicity of Ty800 delivering a foreign antigen was evaluated in four volunteers who were fed a Δ*asd* strain of *S. typhi* Ty800 carrying plasmid pYA3167, the same balanced-lethal plasmid described above that encodes *asd* and the gene for the HBc-preS1 fusion protein. The results of this study showed that three of four volunteers seroconverted against the carrier, but none developed an anti-HBV immune response (K. Killeen, unpublished observations). The outcome of this study and those evaluating *S. typhi* Chi^{4632} suggest that the failure of vaccinees to develop specific immune responses to the foreign antigen is probably not related to the nature of the attenuating mutation in the vector. Rather, these results may be more related to issues of poor expression of the fusion gene in situ, degradation of the cloned gene products in vivo before immune priming can occur,[5] or an inaccessibility of the antigen for immune processing due to improper cellular localization.

More recently, Δ*phoP/Q*-attenuated *Salmonella* strains were used to deliver an antigen from *H. pylori*. These studies were designed in part to exploit the vector's natural route of entry to mimic infection. In a preclinical study, the immunogenicity of *S. typhimurium* LH961 (Δ*phoP/Q* Δ*purB*) transformed with a plasmid encoding PurB and the *H. pylori* UreA and UreB protein subunits was tested in mice. Animals immunized with three oral doses of LH961 shed UreA/B-positive *S. typhimurium* recombinants for 2 weeks, demonstrating the stability of the balanced-lethal plasmid-complemented strain.[16] Immunized mice also developed strong serum and mucosal anti-UreA/B antibody responses. These results prompted the development of a Δ*phoP/Q* strain of *S. typhi* for use in humans. Ty1033 is the *S. typhi* equivalent of LH961 that was modified for use in a balanced-lethal expression system. Ty1033 (Δ*phoP/Q*Δ*purB*) carries a plasmid that complements the Δ*purB* mutation and directs the expression of UreA and UreB.[16] The immunogenicity of *S. typhi* Ty1033 was evaluated in eight adult volunteers, each of whom received a single, oral dose. The vaccine was safe, and immunized volunteers developed strong immune responses against *S. typhi* antigens, as expected. However, none of the volunteers developed a detectable mucosal or humoral antibody response to either UreA or UreB. To determine if a booster immunization could generate an immune response to the vectored antigens, three of the volunteers received an oral booster vaccination consisting of purified UreA/B protein with an adjuvant (*E. coli* heat-labile toxin). The results of the study showed no evidence of an immunological boost as none of the volunteers developed a detectable anti-UreA/B humoral or mucosal antibody response.

The inability of existing attenuated strains of *S. typhi* to elicit antibody responses to foreign antigens in humans suggested that novel strategies may be required to express and deliver antigens using *Salmonella*. One approach is the use of *S. typhimurium* to vector foreign antigens in humans. Recently, *S. typhimurium* LH1160 was used in humans to induce an antibody response to *H. pylori* UreA/B.[19] *S. typhimurium* LH1160 is a Δ*phoP/Q* attenuated strain that harbors a balanced-lethal plasmid complementing the strain's PurB auxotrophy and coexpresses recombinant *H. pylori* UreA/B. The amount of UreA/B protein produced in vitro by *S. typhimurium* LH1160 is equivalent to that produced by *S. typhi* Ty1033. *S. typhimurium* LH1160 was relatively well tolerated by the six volunteers who received a single, oral dose and most developed *S. typhimurium*–specific anti-LPS and anti-flagella mucosal and humoral immune responses. More importantly, half of the volunteers developed a detectable antibody response to UreA/B. The results of this study are significant because they were the first to show that an attenuated strain of *S. typhimurium* can be used in humans to elicit an immune response to a vectored foreign antigen. While the number of volunteers in this study was limited, it is interesting to note the differences between *Salmonella* vectors with regard to their capacity to deliver UreA/B. The balanced-lethal plasmid expressing the UreA/B protein, for example, was more stable in a *S. typhimurium* genetic background and resulted in volunteers who were colonized longer than those immunized with a similarly modified strain of *S. typhi*. This presumably resulted in *S. typhimurium*–vaccinated volunteers receiving a larger in vivo "dose" of the UreA/B antigen. Remarkably, LH1160 elicited an immune response after a single immunization and at a dose 1/100th of that used in the *S. typhi* Ty1033 study. These results demonstrate that the delivery of some foreign antigens is dependent not only upon the antigen in question but also the manner in which the antigen is presented by the vector to the immune system. It remains to be determined whether the increased capacity of *S. typhimurium* to deliver foreign antigens relative to that of *S. typhi* is due to the characteristics discussed above or some other factor or factors intrinsic to the biology of *S. typhimurium* in the human host. Additional testing of disease relevant antigens delivered by *S. typhimurium* needs to be performed to determine its suitability as a generalized vector.

Bacille Calmette-Guérin

Live, attenuated *M. bovis* bacille Calmette-Guérin (BCG) is used to immunize humans against tuberculosis. The BCG vaccine has a number of features that make it an attractive vector for delivering foreign antigens in humans. First, the vaccine has an excellent safety record, having been used to immunize more than 3 billion people worldwide. Second, BCG persists in the macrophages of an infected host in spite of a vigorous host immune response, a trait that may be beneficial in vectoring foreign antigens for therapeutic immunization purposes. A third consideration is that recombinant strains of BCG (rBCG) expressing various antigens have been shown in preclinical studies to elicit humoral as well as CTL immune responses by several routes of immunization, including oral, parenteral, rectal, and vaginal administration.[20] Finally, the BCG vaccine is inexpensive to produce, making any recombinant strain suitable for use in developing world countries.

Recombinant BCG strains vectoring various viral antigens have been shown to be effective in eliciting both antibody and CTL responses in animal models.[21] For example, a rBCG strain expressing a simian immunodeficiency virus (SIV) nucleoprotein (NP) epitope was shown to induce a CTL response in macaques, although the response was not protective.[22] However, parenteral immunization with a rBCG strain expressing an epitope from the V3 loop of HIV-1 gp120 was shown to elicit both CMI as well as antibody responses that were protective against viral infection in a severe combined immunodeficiency (SCID) mouse model.[23] Immunization of mice with a cocktail of three different rBCG strains each expressing either the SIV *nef*, *env*, or *gag* genes was found to induce both neutralizing antibodies and a CTL response.[20] Specifically, mucosal vaccination with the rBCG strain cocktail resulted in a local IgA response and systemic IgG and CTL immune responses against the expressed SIV antigens. The responses from mucosal vaccination were biased toward the IgG_1 isotype suggestive of a T helper cell type 1 (Th1) immune response.

More recently, preclinical results were reported evaluating a Lyme disease vaccine based on BCG expression of the *Borrelia burgdorferi* outer surface protein OspA. These studies used a rBCG strain that expresses *ospA* genetically fused to a mycobacterial lipoprotein signal peptide to direct the export and localization of the chimeric lipoprotein to the cell membrane. Mice immunized with the OspA-lipoprotein expressing BCG strain developed a high-titered anti-vector and anti-OspA serum antibody response that was protective upon subsequent challenge with *B. burgdorferi*.[24] The encouraging preclinical profile of this study hastened the clinical testing of BCG as a vector to deliver OspA in humans. In a phase I study exploring the safety and immunogenicity of rBCG-OspA, 24 volunteers were given a single, intradermal injection of the vaccine. While many of the volunteers developed an immune response to BCG, the results of the study were generally disappointing since none of the volunteers developed detectable anti-OspA antibody responses.[25]

One of the limitations of using BCG as a vector is that it does not effectively stimulate a host $CD8^+$ T-cell response in humans.[26,27] To address this issue, rBCG strains have been created that facilitate the entry of a vectored antigen into the cytosol of an infected host cell. For example, a rBCG strain was constructed that secretes biologically active listeriolysin O (LLO), a cytolysin normally secreted by *Listeria monocytogenes* that is essential for the escape of *Listeria* from the phagosomal vacuole into the cytoplasm of the host cell.[26] It was reasoned that better access of both BCG and vectored antigens to the host cell cytoplasm would lead to better MHC class I presentation. A chimeric protein composed of LLO fused to a BCG-specific signal peptide was secreted by rBCG and localized in the cytoplasmic vacuoles of infected cultured macrophages. The rBCG/LLO strain was unable to escape from the macrophage phagosome but nevertheless facilitated the presentation of a model antigen, ovalbumin (OVA), to the MHC class I processing pathway. This was manifest in the capacity of rBCG/LLO-infected macrophages that had cophagocytosed OVA to stimulate an OVA-specific $CD8^+$ T-cell hybridoma. These results and others show the promise of using rBCG systems expressing foreign antigens to induce strong CTL as well as neutralizing antibody responses and reveal a potential strategy for overcoming the intrinsic inability of rBCG to stimulate an effective $CD8^+$ T-cell response.

Vibrio Vectors

The use of *Vibrio cholerae* as a live, attenuated vector is attractive because the bacterium is noninvasive but still elicits a prominent and long-lasting systemic and mucosal anti-vector immune response after a single, oral dose. If this potential can be transferred to foreign antigens it might be possible to construct vectors that can confer protective immunity by oral immunization against both mucosally- and parenterally-transmitted pathogens. Unlike the pathogenesis of other organisms discussed in this chapter, disease caused by *V. cholerae* is not associated with invasion of host cells but is due principally to the presence of cholera toxin (CT). For this reason, the most effective strategy to date for attenuating *V. cholerae* has been to engineer mutant strains in which the CT gene (*ctx*) or the CT genetic element[28] has been either partially or completely deleted. The live-vector strains CVD 103-HgR[29] and Peru-2[30] do not produce CT and have been shown to be effective in animals at eliciting significant levels of specific mucosal sIgA directed against both the vector and a heterologous antigen. For example, the Peru-2 derivative, JBR10, was created by inserting a single copy of the *Escherichia coli* Shiga-like toxin B-subunit (*stxB1*) into the *Vibrio* chromosome by in vivo marker exchange.[31] In a preclinical study, JRB10 colonized the gastrointestinal mucosa of orally immunized rabbits, inducing serum IgG and biliary IgA immune responses. These results prompted a small clinical trial in which volunteers were fed a single dose of live attenuated *V. cholerae*. The vector was found to be safe and immunogenic and vaccinees developed a serological immune response to vibrial antigens. However, none of the immunized volunteers seroconverted against $StxB_1$ (K. Killeen, unpublished observations).

Prompted in part by the results of the study above, Peru-2 was modified for use in balanced-lethal plasmid expression systems.[32] Peru-2Δ*glnA* carries a lethal deletion in *glnA* (glutamine synthesis) that can be rescued by transformation with a plasmid expressing a functional copy of the gene.[33] Mice immunized with Peru-2Δ*glnA* carrying a complementing plasmid expressing the CT B-subunit (CTB) fused to the dodecapeptide repeat of the serine-rich *Entamoeba histolytica* protein (SREHP) elicited a stronger serum and mucosal anti-CTB antibody response than mice immunized with Peru-2 expressing a chromosomal copy of the same gene. These results confirm the notion that higher-level antigen expression in vivo can lead to an increased immune response and support the continued development of balanced-lethal plasmid systems in *V. cholerae*.

Cholera toxin and the structurally related *E. coli* heat-labile toxin (LT) are potent mucosal adjuvants that elicit a vigorous systemic immune response, and their use to boost the immunological response to vectored antigens may be beneficial. However, the oral ingestion of even minute quantities of CT or LT is sufficient to induce severe diarrheal disease in humans,[34] precluding their use in native form. To address this problem, a recombinant strain of Peru-2 expressing a mutant form of LT was recently evaluated. Mice were fed Peru-2 expressing LT (R192G), a nonenterotoxic LT mutant that contains an arginine to glycine substitution at the toxin's active site yet retains adjuvant activity.[35] Animals immunized with Peru-2 expressing LT (R192G) developed an increased anti-vector antibody response compared to animals that received the vaccine strain alone. The benefit of coexpressing this adjuvant with a *V. cholerae*-vectored foreign protein was minima; however, in germ-free mice immunized with JRB-14/LT(R192G), a modified form of JRB-10 expressing 2 immunogens, $StxB_1$ and an outer membrane protein, EaeA, from enteropathogenic strains of *E. coli*. That is to say, none of the immunized mice developed an increased serum antibody response to either foreign antigen.[36]

Live, attenuated *V. cholerae* CVD 103-HgR also protects humans against cholera after a single oral dose,[29] and it has considerable potential as a bacterial vector. In preclinical studies, rabbits fed two oral doses of CVD 103-HgR expressing *E. coli* $StxB_1$ developed serum anti-$StxB_1$ antibody that neutralized holotoxin activity in an in vitro cytotoxicity assay.[37] In another study, the immunogenicity of CVD 103-HgR expressing an O-antigen moiety of *Shigella sonnei* LPS was evaluated in rabbits.[38] Animals were given multiple injections of heat-inactivated CVD 103-HgR transformed with a plasmid expressing *rfb* and *rfc*, genes encoding the heterologous O-antigen moiety. Animals immunized in this manner mounted significant antibody responses to *S. sonnei* and *V. cholerae* LPS. However, a number of the plasmid constructs expressing heterologous O-antigen in this study were unstable in the CVD 103-HgR genetic background. This prompted the construction of a recombinant form of CVD 103-HgR in which the *rfb/rfc* locus was cloned and genetically recombined into the chromosome. This recombinant strain, CH22CP, was subsequently evaluated in a small clinical study, in which 11 volunteers received a single, oral dose of CH22CP and 10 volunteers received a priming dose followed by two booster immunizations. In the single-dose group, 3 of 11 volunteers developed a moderate anti-*S. sonnei* LPS IgG immune response. In the multiple dose group, however, 7 of 10 volunteers mounted a LPS-specific IgG and sIgA response, demonstrating the potential of CH22CP as a vaccine in humans (D. Favre, personal communication).

In a follow-up study, the immunogenicity of CH22CP was compared to that of the Peru-2 variant, Peru-15,[39] that had been modified to express the *S. sonnei rfb/rfc* genes.[40] Rabbits orogastrically fed a single dose of CH22CP developed only a weak anti-O-antigen immune response. However, animals fed Peru-15 expressing *S. sonnei* O-antigen developed an immune response that was superior to that elicited by CVD103-HgR. The ability of Peru-15 to colonize the host better than CVD 103-HgR and its derivatives could account for the difference in immune responses, and suggests that recombinant Peru-15 expressing *S. sonnei* O-antigen may be a potential single dose vaccine against shigellosis in humans.

The results of preclinical and clinical studies with CVD 103-HgR and others like BCG demonstrate the difficulty of predicting a clinical outcome based on preclinical data. Furthermore, these studies show that caution is in order when expressing heterologous antigens in bacterial vectors. That is, different stains may express the same antigen but vary in their capacity to elicit an immune response.

Shigella Vectors

The development of live, attenuated *Shigella* vectors for the delivery of heterologous antigens has been hampered by the lack of reliable animal models and the existence of safe immunogenic vaccine strains. The first generation of live *S. flexneri* vectors was attenuated by the deletion of the *aroA* and *aroD* biosynthetic genes. These attenuated strains lyse in situ due to the lack of complementing metabolites and are particularly well suited for use as vectors for DNA vaccines (see below). More recently, studies have focused on designing live *Shigella* vaccine vectors that retain the bacterium's normal invasiveness but prevent intercellular spreading. Attenuated *Shigella* strains carrying deletions in the virulence-associated gene, *virG*, were constructed and show promise as vectors for delivering heterologous antigens in humans.

The *S. flexneri* 2a vaccine strain candidate, SC602, a *virG/iuc* (aerobactin) double mutant, was recently tested in humans but found to be reactogenic at a relatively low dose.[41] Nevertheless, immunization of volunteers with SC602 protected them completely against homologous wild-type *Shigella* challenge. *S. flexneri* SC602 has been evaluated only for its capacity to deliver foreign antigens in preclinical studies. For example, chimeric proteins composed of the *Shigella* antigen, IpaC, genetically fused to multiple copies of a neutralizing B-cell epitope (C3) of the VP1 protein from poliovirus have been vectored in SC602.[42] However, mice immunized with this strain developed only moderate serum IgG and mucosal IgA anti-C3 antibody responses.

The investigators in this study concluded that an improvement in antigen expression by SC602 is required to induce a more efficient immune response, suggesting that the use of a balanced-lethal plasmid expression system may be useful.

The *S. flexneri* 2a vaccine strain CVD 1203 has deletions in the *aroA* and *virG* genes. Oral administration of CVD 1203 to guinea pigs stimulates a mucosal IgA response and protects against *Shigella*–induced keratoconjuctivitis.[43] The utility of CVD 1203 as a vaccine vector has been shown in preclinical studies where it was used to deliver fimbrial surface antigens that are required for colonization by enterotoxigenic *E. coli* (ETEC; 44). CVD 1203 was transformed with the genes encoding *E. coli* LT-B and the most prevalent ETEC colonization factor antigens, CFA/1 and CS3. Immunization elicited serum IgG and IgA and mucosal IgA responses to CFA/1, CS3, and *Shigella*. While the protective capacity of the anti-CFA/1 and CS3 immune responses was not determined, this study hints at the possibility of creating a multivalent vaccine that could provide broad-spectrum protection against *Shigella* and ETEC.

One factor that has hindered the development of CVD 1203 as a vector is its residual diarrheal reactogenicity in humans.[45] To address this problem, *S. flexneri* CVD 1207 was constructed with deletions in the *virG, sen* (*Shigella* enterotoxin 2), *set* (*Shigella* enterotoxin 1), and *guaBA* (purine biosynthesis) loci.[46] CVD 1207 has been tested in human volunteers and was well tolerated at doses of up to 10^8 organisms. Associated with the increased attenuation, however, was a decrease in the volunteers' anti-LPS IgA response compared to CVD 1203. The significance of this decrease and its impact on protection await additional clinical evaluation.

Listeria Vectors

Listeria monocytogenes has served for many years as a model organism to study aspects of cellular immunity. After invasion of the host through the gut, *Listeria* may be found in large numbers in the cytosol of splenic antigen-presenting cells (APC) and particularly in macrophages where the bacteria actively replicate. It is this ability to propagate in the host cell cytosol that makes *L. monocytogenes* an attractive live vector candidate for inducing cell-mediated immunity to vectored foreign antigens.

Acquired immunity to listeriosis is multifactorial and includes the participation of phagocytes, several T-cell subpopulations, and a number of cytokines.[47-49] Antigens produced by *L. monocytogenes* during infection and released extracellularly, for example, may be taken up by APC, degraded in lysosomal vacuoles, and the resulting peptides loaded onto MHC class II molecules for subsequent presentation by APC. Processing in this manner leads to the production of antigen-specific MHC class-II-restricted $CD4^+$ T-helper cells that are primarily of the Th1 subclass.[50] Alternatively, antigens that are secreted by intracellular forms of *L. monocytogenes* into the host cell cytoplasm may enter the pathway for MHC class I processing and presentation[51] and induce a specific protective $CD8^+$ T-cell response. Finally, listeriosis in mice induces cytokine production that increases the expression of cell surface molecules and stimulates the expansion of MHC class-I and class-II-restricted effector cell populations.[52]

The ability of *L. monocytogenes* to provoke a protective $CD8^+$ T-cell immune response has been evaluated as a vaccine against viral infection. For example, recombinant listerial strains expressing the entire lymphocytic choriomeningitis virus (LCMV) NP or an immunoprotective H-2L^d–restricted CTL epitope ($NP_{118\text{-}126}$) were constructed and injected into naïve mice. Immunized animals were protected against subsequent challenge with virulent strains of LCMV that establish chronic infection in unvaccinated mice. Protection corresponded to the presence of $NP_{118\text{-}126}$-specific $CD8^+$ T cells, and in vivo depletion of $CD8^+$ T cells from vaccinated mice abrogated their ability to clear viral infection.[49] These results demonstrate that recombinant listerial vaccines can induce immunity to a viral antigen that effectively protects against infection in vivo.

The ability of *L. monocytogenes* to stimulate a vigorous $CD8^+$ T-cell response may also have potential as a therapy against some forms of cancer, since experimental evidence suggests that T cells, and especially MHC class-I-restricted $CD8^+$ lymphocytes, are a critical mediator of an effective anti-tumor response.[52] A recent study showed that a recombinant strain of *L. monocytogenes* inhibited tumor growth in a murine cancer model. Tumorigenic cells transduced with the influenza virus NP antigen were injected into mice to generate visible, subcutaneous tumors. Immunization of these mice with *L. monocytogenes* expressing NP inhibited tumor growth and ultimately stimulated the regression of macroscopic tumors.[53] Interestingly, this NP-expressing listerial vector was also capable of conferring $CD8^+$T-cell protection against a lethal challenge of renal carcinoma cells expressing the NP antigen.

There are relatively few published applications of live, attenuated, listerial vaccine vectors in animal models and none on their use to deliver foreign antigens in humans. In light of the recognized capacity of tumor cells to escape immune surveillance by down-regulating expression of MHC class I molecules, the therapeutic use of recombinant *L. monocytogenes* to treat cancer or other diseases or infections where immune suppression is a factor merits evaluation in human clinical trials.

Next Generation Bacterial Vectors

Many of the live attenuated vector systems discussed throughout this chapter employ high-level expression systems like balanced-lethal plasmids as a means to augment immune responses to vectored foreign antigens. However, increased antigen expression does not universally result in an enhanced host immune response. This may be the result of an inaccessibility of the antigen for processing, the elimination of critical epitopes due to improper protein folding, or toxicity of the expressed protein to the cell. Novel expression systems have therefore been developed that overcome these limitations by secreting antigens into the extracellular environment.[54] In its most simple form, this type of delivery system uses live vectors to express foreign peptide epitopes that are genetically fused to the nonenzymatic B-subunit of naturally secreted bacterial toxins like CT or LT.[55,56] Preclinical studies evaluating the immunogenicity of these vectors demonstrated that they could induce anti-fusion protein immune responses in animals. In one study, however, the immune response to the B-subunit carrier was several orders of magnitude higher than that to the foreign epitope, suggesting that presentation of the epitope was suboptimal.[55] The further use of this type of delivery system may also be limited in the size of the epitopes that can be fused to the toxin B-subunit without affecting functions such as receptor binding, translocation and antigen processing.

A vector system based on *E. coli* α-hemolysin (HlyA), a bacterial cytolysin, is being evaluated to deliver foreign antigens by secretion to the host immune system. HlyA is secreted by a multicomponent export system that includes HlyB, HlyD and TolC, proteins that are localized in the periplasm and cell membrane.[54] Secretion of HlyA is dependent on the carboxy-terminal 27 amino acids of the protein that are recognized by the HlyB-HlyD-TolC secretion machinery. Fusion of the 27-amino-acid "tag" ($HlyA_s$) to a somatic protein is sufficient for targeting to the extracellular milieu, when coexpressed on a plasmid encoding HlyB and HlyD.[57] $HlyA_s$ is capable of mediating protein export from bacterial species other than *E. coli,*[58] suggesting that in some cases other components of the system, like TolC, are also functionally conserved. For example, $HlyA_s$ fusion proteins have been expressed in attenuated strains of *Salmonella* and examined for antigen processing by APC. In one study, a *Salmonella* strain secreting a LLO-$HlyA_s$ fusion protein was superior to native LLO (which normally remains in the cytoplasm) at conferring protection to mice against challenge with a virulent strain of *L. monocytogenes.*[59] In a separate study, $HlyA_s$ was genetically fused to *L. monocytogenes* superoxide dismutase (SOD), a somatic antigen that is not normally protective. Remarkably, mice that were orally immunized with the SOD-$HlyA_s$-secreting *Salmonella* strain were subsequently protected against lethal challenge with *L. monocytogenes*.[60] These results illustrate that the manner in which antigens are presented to the host immune system by live vectors can have a dramatic impact on the nature of the immune response.

The HlyA secretion system has also been adapted for use in attenuated strains of *V. cholerae*. The attenuated *V. cholerae* classical biotype strain O395-NT was used to express a *Clostridium difficile* toxoid A-$HlyA_s$ fusion.[61] Orogastric immunization of rabbits produced a significant serum and mucosal anti-*Vibrio* and anti-toxoid A antibody response that protected against *Clostridium difficile* toxin A in an ileal loop challenge assay. However, the toxoid A-$HlyA_s$ fusion protein was secreted poorly or not at all by several *V. cholerae* El Tor or non-O1 biotype vaccine strains. A large difference in levels of antigen secretion was also observed when a UreB-$HlyA_s$ fusion was expressed in the same *Vibrio* biotypes (S. Kochi, unpublished observations). This contrast may be attributed to functional differences in proteins like, TolC, which are required for secretion and are encoded by the vector.[62] *Salmonella*, for example, has a homologue of the *E. coli tolC* gene,[63] and efficiently secretes a $HlyA_s$-UreB fusion protein (L. Hohmann, personal communication). In contrast, a putative *V. cholerae tolC* homologue can only partially complement a *tolC*$^-$ strain of *E. coli*,[64] suggesting functional dissimilarities between these genes. These results demonstrate the difficulties in developing a universal expression system and stress the empirical nature of delivering heterologous antigens in live, attenuated vectors.

The attenuation of some wild-type *S. typhi* strains has resulted in the creation of mutants like Ty800 that are deficient in their ability to escape from phagocytic vacuoles and enter the cytosol of APC. This defect could hinder attenuated *Salmonella* vectors from effectively delivering heterologous antigens for MHC class-I processing and inducing a cell-mediated immune response. One approach to addressing this potential problem has been the modification of a *Salmonella* secretion system that normally participates in *Salmonella* invasion of host cells. During infection, *Salmonella* binds to host cells and activates a specialized protein secretion system. This secretion system, classified as a type III secretion system, delivers a set of bacterial effector proteins to the host cell cytosol. Entry of these components into the cell activates cellular transcription factors and induces the reorganization of the cell's cytoskeleton. One of these effector proteins, SptP, was modified by fusing it to a murine-restricted CTL epitope from the influenza virus NP.[65] The chimeric SptP-NP epitope fusion was efficiently secreted in vitro by a *ΔaroAΔsptP* attenuated strain of *S. typhimurium* and was delivered into the cytosol of *Salmonella* infected epithelial cells. Mice that were fed the recombinant *S. typhimurium* strain expressing the fusion protein developed MHC class-I-restricted CTL that were epitope-specific and recognized endogenously processed antigen. In separate experiments, the same vector was used to deliver SptB genetically fused to a LCMV MHC class I epitope. Mice immunized with the recombinant vector expressing the SpB-LCMV fusion protein developed LCMV-specific CTL that correlated with protection against lethal LCMV challenge, supporting the notion that *S. typhimurium* can effectively elicit a MHC class-I-restricted immune response to heterologous antigens. In addition, both the α-hemolysin and *Salmonella* type III secretion system show promise as effective means to elicit protective immune responses. Additional studies of these and other bacterial secretion systems are ongoing.

DNA Delivery

Stimulating the immune system using genetic immunization is a promising new vaccine approach. The injection of purified plasmid DNA into muscle or skin cells leads to antigen expression if transcription is regulated by a eukaryotic promoter. Proper presentation of an expressed foreign antigen can lead to B- and T-cell stimulation and protective immune responses.[66] It has recently been demonstrated that live, attenuated bacteria are also capable of delivering DNA.[67] This type of delivery takes advantage of the natural ability of many organisms to enter and grow intracellularly or to be phagocytosed by APC such as macrophages. Inside the host cell, the bacterial vector may reside in an endocytic or phagocytic vacuole or may escape from a vacuole and reside in the host cell cytosol. In either case, the escape or subsequent breakdown of the vector in the vacuole has been shown to be a viable method to deliver plasmid DNA to the cell cytoplasm.[67]

The use of a live attenuated bacterium as a delivery vehicle has several advantages over the direct injection of naked DNA. For example, all of the bacterial vectors that have been discussed in this chapter can be administered orally and, in some cases, induce long-lived protection after a single dose. The use of live vectors for DNA delivery may also exploit a vector's natural tropism for M cells, APC, or other cell types or compartments targeting them to key inductive sites of the immune system.[66,68] In contrast, it remains unclear which cells function as APC when plasmid DNA is injected, and the number of APC at the site of injection that can be transformed with naked DNA may be limited. Finally, an added benefit of a live bacterial carrier is the presence of vector-associated immunomodulatory components like LPS that may enhance the host's immune response to the plasmid-encoded foreign antigen.[66]

The natural ability of *S. flexneri* to enter intestinal epithelial cells or to be phagocytosed by APC makes it a logical choice to deliver plasmid DNA. Live attenuated strains of *S. flexneri* have been developed that are capable of delivering antigen-encoding plasmid DNA to cultured mammalian cells and in vivo.[69,70] *S. flexneri* 2a strain 15D has an *asd* deletion that impairs bacterial cell wall synthesis.[70] As a consequence, *S. flexneri* 15D lyses in situ due to the lack of diaminopamilic acid, releasing plasmid DNA into the host cell cytosol. Cultured mammalian cells that had been infected with 15D harboring a eukaryotic expression plasmid encoding β-galactosidase under the control of a cytomegalovirus (CMV) promoter (15D/pCMVb) efficiently expressed the marker enzyme. Immunization of mice with twp intranasal doses of 15D/pCMVb at a 4-week internal resulted in strong CMI and antibody responses against β-galactosidase.[71] However, the utility of this system, awaits the development of DNA vectors expressing virulence-associated genes that stimulate immune responses in animal models where protection from virulent challenge can be assessed.

Like *Shigella*, the ability of *L. monocytogenes* to enter the cytosol after phagocytosis and deliver plasmid DNA directly to the cytoplasm makes it an equally attractive DNA delivery candidate to induce CMI to foreign antigens. An attenuated strain of *L. monocytogenes*, Δ2, was constructed by deleting the lecithinase operon that is required for cell-to-cell spreading after escape from the phagosome.[68] Delta2 is able to escape from the phagosome into the cytosol and then autolyses due to the expression of an intracellularly activated listerial bacteriophage cytolysin. Cell lysis results in the release of the foreign antigen-encoding plasmid DNA into the host cell cytosol. Cultured mouse macrophages infected with Δ2 expressing a fluorescent marker were evident in cultures at 3 days postinfection. Bone marrow-derived macrophages infected with Δ2 harboring an expression vector encoding the OVA epitope were able to stimulate an $OVA_{257-264}$-specific CTL clone to secrete Interleukin-2. This demonstrated the ability of Δ2 to present an epitope to target cells in the context of MHC class I.

S. typhimurium remains in the phagosomal compartment of the host cell (unlike *Shigella* and *Listeria* which readily escape) and has been used to deliver eukaryotic expression vectors to cultured cells and in vivo. For example, a mutant *S. typhimurium* strain was constructed that harbored a plasmid encoding the marker enzyme β-galactosidase under the control of a CMV promoter (pCMVβ;66). Experiments showed that in vitro cultures of peritoneal macrophages infected with *S. typhimurium* (pCMVβ) expressed functional β-galactosidase. An attenuated strain of *S. typhimurium* vectoring listerial virulence factors was tested for its ability to elicit an immune response against the foreign antigens. Mice were fed a *S. typhimurium* recombinant harboring a plasmid encoding the gene for either ActA or LLO, 2 proteins required for the virulence of *L. monocytogenes*. Immunized animals developed strong IgG2a antibody-biased responses against both ActA and LLO, following either single or multiple immunizations. The antibody response corresponded with the appearance of specific CTL to murine-restricted ActA and LLO epitopes. Immunized mice were subsequently challenged with a virulent strain of *L. monocytogenes* at 10 times the murine LD_{50}. Animals immunized with multiple doses of *S. typhimurium* delivering LLO were completely protected, while 60% of those immunized with one dose survived the challenge. Interestingly, mice immunized with *S. typhimurium* vectoring ActA were not protected against lethal challenge at any dose, in spite of vigorous antibody and CMI responses to ActA. These promising preclinical data support the merit of using live, attenuated strains of *Salmonella* to deliver plasmid DNA in humans.

Summary

A great deal of data in animal models supports the utility of live attenuated bacterial vectors for generating immune responses against assorted viral, bacterial, and parasite antigens. A number of studies have demonstrated that oral administration of a vector like *Salmonella* delivering a foreign antigen can immunize at the level of the gut mucosa and induce the production of serological as well as mucosal antibody responses. Likewise, the expression of a heterologous gene on a eukaryotic expression plasmid in the same vector can induce a potent $CD8^+$ T-cell response. However, despite these promising preclinical results, the use of animal models to predict the clinical outcomes of these vaccines in humans remains elusive. Yet, preliminary results from human studies evaluating recombinant strains like CH22CP and CVD 908 or employing *S. typhimurium* expressing UreA/B show considerable promise and suggest that the next generation of live attenuated vectors for clinical use will be a result of novel approaches to vaccine design.

References

1. Killeen K, DiRita V. Live attenuated bacterial vaccines. In: Ellis R, ed. New Vaccine Technologies. Georgetown, TX: Landes Bioscience 2000:171.
2. Germanier R, Furer E. Isolation and characterization of GalE mutant Ty21a of *Salmonella typhi*: A candidate strain for a live, oral typhoid vaccine. J Infect Dis 1975; 131:553-558.
3. Guleria I, Teitelbaum R, McAdam RA et al. Auxotrophic vaccines for tuberculosis. Nat Med 1996; 2:334-337.
4. Hone DM, Harris AM, Chatfield S et al. Construction of genetically defined double *aro* mutants of *Salmonella typhi*. Vaccine 1991; 9:810-816.
5. Tacket, CO, Kelly SM, Schodel F et al. Safety and immunogenicity in humans of an attenuated *Salmonella typhi* vaccine vector strain expressing plasmid-encoded hepatitis B antigens stabilized by the Asd-balanced lethal vector system. Infect Immun 1997; 65:3381-3385.
6. Hohmann EL, Oletta CA, Killeen KP et al. *PhoP/phoQ*-deleted *Salmonella typhi* (TY800) is a safe and immunogenic single-dose typhoid fever vaccine in volunteers. J Infect Dis 1996; 173:1408-1414.
7. Sansonetti PJ, Arondel J, Fontaine A et al. *OmpB* (osmo-regulation) and *icsA* (cell-to-cell spread) mutants of *Shigella flexneri*: Vaccine candidates and probes to study the pathogenesis of shigellosis. Vaccine 1991; 9:416-422.
8. Miller SI, Kukral AM, Mekalanos JJ. A two-component regulatory system (*phoP phoQ*) controls *Salmonella typhimurium* virulence. Proc Natl Acad Sci USA 1989; 86:5054-5058.
9. Sirard JC, Niedergang F, Kraehenbuhl JP. Live attenuated *Salmonella*: A paradigm of mucosal vaccines. Immunol Rev 1999; 171:5-26.
10. Kaufmann SHE. Concepts in vaccine development. In: Kaufmann SHE ed. Berlin: Walter de Gruyter 1996.

11. Gonzalez C, Hone D, Noriega FR et al. *Salmonella typhi* vaccine strain CVD 908 expressing circumsporozoite protein of *Plasmodium falciparum*: Strain construction and safety and immunogenicity in humans. J Infect Dis 1994; 169:927-931.
12. Corthesy-Theulaz IE, Hopkins S, Bachmann D et al. Mice are protected from *Helicobacter pylori* infection by nasal immunization with attenuated *Salmonella typhimurium phoPc* expressing urease A and B subunits. Infect Immun 1998; 66:581-586.
13. Gomez-Duarte OG, Lucas B, Yan ZX et al. Protection of mice against gastric colonization by *Helicobacter pylori* by single oral dose immunization with attenuated *Salmonella typhimurium* producing urease subunits A and B. Vaccine 1998; 16:460-471.
14. Sadoff JC, Ballou WR, Baron LS et al. Oral *Salmonella typhimurium* vaccine expressing circumsporozoite protein protects against malaria. Science 1988; 240:336-338.
15. Nakayama K, Kelly SM, Curtiss III R. Construction of an *asd*+ expression-cloning vector: stable maintenance and high level expression of cloned genes in a *Salmonella* vaccine strain. Nat Biotechnol 1988; 6:693-697.
16. DiPetrillo MD, Tibbetts T, Kleanthous H et al. Safety and immunogenicity of *phoP/phoQ*-deleted *Salmonella typhi* expressing *Helicobacter pylori* urease in adult volunteers. Vaccine 2000; 18:449-459.
17. Schodel F, Kelly SM, Peterson DL et al. Hybrid hepatitis B virus core-preS proteins synthesized in avirulent *Salmonella typhimurium* and *Salmonella typhi* for oral vaccination. Infect Immun 1994; 62:1669-1676.
18. Nardelli-Haefliger D, Kraehenbuhl JP, Curtiss III R et al. Oral and rectal immunization of adult female volunteers with a recombinant attenuated *Salmonella typhi* vaccine strain. Infect Immun 1996; 64:5219-5224.
19. Angelakopoulos H, Hohmann EL. Pilot study of *phoP/phoQ*-deleted *Salmonella enterica* serovar typhimurium expressing *Helicobacter pylori* urease in adult volunteers. Infect Immun 2000; 68:2135-2141.
20. Lagranderie M, Winter N, Balazuc AM et al. A cocktail of *Mycobacterium bovis* BCG recombinants expressing the SIV Nef, Env, Gag antigens induces antibody and cytotoxic responses in mice vaccinated by different mucosal routes. AIDS Res Hum Retroviruses 1998; 14:1625-1633.
21. Stover CK, de la Cruz VF, Fuerst TR et al. New use of BCG for recombinant vaccines. Nature 1991; 351:456-460.
22. Yasutomi Y, Koenig S, Woods RM et al. A vaccine-elicited, single viral epitope-specific cytotoxic T lymphocyte response does not protect against intravenous cell-free simian immunodeficiency virus challenge. J Virol 1995; 69:2279-2284.
23. Honda M, Matsuo K, Nakasone T et al. Protective immune responses induced by secretion of a chimeric soluble protein from a recombinant *Mycobacterium bovis* bacillus Calmetete-Guerin vector candidate vaccine for human immunodeficiency virus type I in small animals. Proc Natl Acad Sci USA 1995; 92:10693-10697.
24. Stover CK, Bansal GP, Hanson MS et al. Protective immunity elicited by recombinant Bacille Calmette-Guerin (BCG) expressing outer surface protein A (OspA) lipoprotein: A candidate Lyme disease vaccine. J Exp Med 1993; 178:197-209.
25. Edelman R, Palmer K, Russ KG et al. Safety and immunogenicity of recombinant Bacille Calmette-Guerin (rBCG) expressing *Borrelia burgdorferi* outer surface protein A (OspA) lipoprotein in adult volunteers: a candidate Lyme disease vaccine. Vaccine 1998; 17:904-914.
26. Hess J, Miko D, Catic A et al. *Mycobacterium bovis* bacille Calmette-Guerin strains secreting listeriolysin of *Listeria monocytogenes*. Proc Natl Acad Sci USA 1998; 95:5299-5304.
27. Kaufmann SHE. In:Paul WE ed. Fundamental Immunology. New York: Raven 1997.
28. Pearson GD, Woods A, Chiang SL et al. CTX genetic element encodes a site-specific recombination system and an intestinal colonization factor. Proc Natl Acad Sci USA 1993,90:3750-3754.
29. Levine MM, Kaper JB, Herrington D et al. Safety, immunogenicity, and efficacy of recombinant live oral cholera vaccines, CVD 103 and CVD 103-HgR. Lancet 1988; 2:467-470.
30. Taylor DN, Killeen KP, Hack DC et al. Development of a live, oral, attenuated vaccine against El Tor cholera. J Infect Dis 1994; 170:1518-1523.
31. Butterton JR, Beattie DT, Gardel CL et al. Heterologous antigen expression in *Vibrio cholerae* vector strains. Infect Immun 1995; 63:2689-2696.

32. Klose KE, Novik V, Mekalanos JJ. Identification of multiple sigma54-dependent transcriptional activators in *Vibrio cholerae*. J Bacteriol 1998; 180:5256-5259.
33. Ryan ET, Crean TI, Kochi SK et al. Development of a Δ*glnA* balanced lethal plasmid system for expression of heterologous antigens by attenuated vaccine vector strains of *Vibrio cholerae*. Infect Immun 2000; 68:221-226.
34. Levine MM, Kaper JB, Black RE et al. New knowledge on pathogenesis of bacterial enteric infections as applied to vaccine development. Microbiol Rev 1983; 47:510-550.
35. Ryan ET, Crean TI, John M et al. In vivo expression and immunoadjuvancy of a mutant of heat-labile enterotoxin of *Escherichia coli* in vaccine and vector strains of *Vibrio cholerae*. Infect Immun 1999; 67:1694-1701.
36. Butterton JR, Ryan ET, Acheson DWK et al. Coexpression of the B subunit of Shiga toxin 1 and EaeA from enterohemorrhagic *Escherichia coli* in *Vibrio cholerae* vaccine strains. Infect Immun 1997; 65:2127-2135.
37. Acheson DWK, Levine MM, Kaper JB et al. Protective immunity to Shiga-like toxin I following oral immunization with Shiga-like toxin I B-subunit-producing *Vibrio cholerae* CVD 103-HgR. Infect Immun 1996; 64:355-357.
38. Viret J-F, Cryz Jr SJ, Lang AB et al. Molecular cloning and characterization of the genetic determinants that express the complete *Shigella* serotype D (*Shigella sonnei*) lipopolysaccharide in heterologous live attenuated vaccine strains. Mol Microbiol 1993; 7:239-252.
39. Kenner JR, Coster TS, Taylor DN et al. Peru-15, an improved live attenuated oral vaccine for *Vibrio cholerae* O1. J Infect Dis 1995; 172:1126-1129.
40. Favre D, Que JU, Haider K et al. Expression of *Shigella sonnei* O-antigen in Classical and El Tor *Vibrio cholerae* vaccine strains and comparison of their immunogenicity in rabbits following mucosal immunization. Submitted for publication.
41. Coster TS, Hoge CW, VanDeVerg LL et al. Vaccination against shigellosis with attenuated *Shigella flexneri* 2a strain SC602. Infect Immun 1999; 67:3437-3443.
42. Barzu S, Arondel J, Guillot S et al. Immunogenicity of IpaC-hybrid proteins expressed in the *Shigella flexneri* 2a vaccine candidate SC602. Infect Immun 1998; 66:77-82.
43. Noriega FR, Wang JY, Losonsky G et al. Construction and characterization of attenuated Δ*aroA* Δ*virG Shigella flexneri* 2a strain CVD 1203, a prototype live oral vaccine. Infect Immun 1994; 62:5168-5172.
44. Noriega FR, Losonsky G, Wang JY et al. Further characterization of Δ*aroA* and Δ*virG Shigella flexneri* 2a strain CVD 1203 as a mucosal *Shigella* vaccine and as a live-vector vaccine for delivering antigens of enterotoxigenic *Escherichia coli*. Infect Immun 1996; 64:23-27.
45. Kotloff KL, Noriega F, Losonsky GA et al. Safety, Immunogneicity, and transmissibility in humans of CVD 1203, a live oral *Shigella flexneri* 2a vaccine candidate attenuated by deletions in *aroA* and *virG*. Infect Immun 1996; 64:4542-4548.
46. Kotloff KL, Noriega FR, Samandari T et al. *Shigella flexneri* 2a strain CVD 1207, with specific deletions in *virG, sen, set,* and *guaBA*, is highly attenuated in humans. Infect Immun 2000; 68:1034-1039.
47. Kaufmann SH. Immunity against intracellular bacteria. Annu Rev Immunol 1993; 11:129-163.
48. Czuprynski CJ, Brown JF. Effects of purified anti-Lyt-2 mAb treatment on murine listeriosis: comparative roles of Lyt-2+ and L3T4+ cells in resistance to primary and secondary infection, delayed-type hypersensitivity and adoptive transfer of resistance. Immunology 1990; 71:107-112.
49. Shen H, Slifka MK, Matloubian M et al. Recombinant *Listeria monocytogenes* as a live vaccine vehicle for the induction of protective anti-viral cell-mediated immunity. Proc Natl Acad Sci USA 1995; 92:3987-3991.
50. Maggee, DM, Wing EJ. Cloned L3T4+ T lymphocytes protect mice against *Listeria monocytogenes* by secreting IFN-gamma. J Immunol 1988; 141:3203-3207.
51. Ikonomidis G, Paterson Y, Kos F et al. Delivery of a viral antigen to the class I processing and presentation pathway by *Listeria monocytogenes*. J Exp Med 1994; 180:2209-2218.
52. Paterson Y, Ikonomidis G. Recombinant *Listeria monocytogenes* cancer vaccines. Curr Opin Immunol 1996; 8:664-669.

53. Pan ZK, Ikonomidis G, Lazenby A et al. A recombinant *Listeria monocytogenes* vaccine expressing a model tumor antigen protects mice against lethal tumor challenge and causes regression of established tumors. Nat Med 1995; 1:471-477.
54. Blight MA, Holland IB. Heterologous protein secretion and the versatile *Escherichia coli* haemolysin translocator. Trends Biotechnol 1994; 12:450-453.
55. Schodel F, Hans W, Johansson S et al. Synthesis in *Vibrio cholerae* and secretion of hepatitis B virus antigens fused to *Escherichia coli* heat-labile enterotoxin subunit B. Gene1991; 99:255-259.
56. Backstrom M, Lebens M, Schodel F et al. Insertion of a HIV-1-neutralizing epitope in a surface-exposed internal region of the cholera toxin B-subunit. Gene 1994; 149:211-217.
57. Gentschev I, Hess J, Goebel W. Change in the cellular localization of alkaline phosphatase by alteration of its carboxy-terminal sequence. Mol Gen Genet 1990; 222:211-216.
58. Tzschaschel BD, Guzman CA, Timmis KN et al. An *Escherichia coli* hemolysin transport system-based vector for the export of polypeptides: export of Shiga-like toxin IIeB subunit by *Salmonella typhimurium aroA*. Nat Biotechnol 1996; 14:765-769.
59. Hess K, Gentschev I, Miko D et al. Superior efficacy of secreted over somatic antigen display in recombinant salmonella vaccine induced protection against listeriosis. Proc Natl Acad Sci USA 1996; 93:1458-1463.
60. Hess J, Dietrich G, Gentschev I et al. Protection against murine listeriosis by an attenuated recombinant *Salmonella typhimurium* vaccine strain that secretes the naturally somatic antigen superoxide dismutase. Infect Immun 1997; 65:1286-1292.
61. Ryan ET, Butterton JR, Smith RN et al. Protective immunity against *Clostridium difficile* toxin A induced by oral immunization with a live, attenuated *Vibrio cholerae* vector strain. Infect Immun 1997; 65:2941-2949.
62. Wandersman C, Delepelaire P. TolC, an *Escherichia coli* outer membrane protein required for hemolysin secretion. Proc Natl Acad Sci USA 1990; 87:4776-4780.
63. Stone BJ, Miller VL. *Salmonella enteritidis* has a homologue of *tolC* that is required for virulence in BALB/c mice. Mol Microbiol 1995; 17:701-712.
64. Colmer JA, Fralick JA, Hamood AN. Isolation and characterization of a putative multidrug resistance pump from *Vibrio cholerae*. Mol Microbiol 1998; 27:63-72.
65. Russmann H,. Shams H, Poblete F, Fu Y, Galan JE, Donis RO. Delivery of epitopes by the *Salmonella* type III secretion system for vaccine development. Science 1998; 281:565-568.
66. Darji A, Guzman CA, Gerstel B et al. Oral somatic transgene vaccination using attenuated *S. typhimurium*. Cell 1997; 91:765-775.
67. Dietrich G, Gentschev I, Hess J et al. Delivery of DNA vaccines by attenuated intracellular bacteria. Immunol Today 1999; 20:251-253.
68. Dietrich G, Bubert A, Gentschev I et al. Delivery of antigen-encoding plasmid DNA into the cytosol of macrophages by attentuated suicide *Listeria monocytogenes*. Nat Biotechnol 1998; 16:181-185.
69. Courvalin P, Goussard S, Grillot-Courvalin C. Gene transfer from bacteria to mammalian cells. C R Acad Sci III 1995; 318:1207-1212.
70. Sizemore DR, Branstrom A, Sadoff JC. Attenuated *Shigella* as a DNA delivery vehicle for DNA-mediated immunization. Science 1995; 270: 299-302.
71. Sizemore DR, Branstrom A, Sadoff JC. Attenuated bacteria as a DNA delivery vehicle for DNA-mediated immunization. Vaccine 1997; 15:804-807.

CHAPTER 11

Protein-Based Vaccines

Sheena M. Loosmore, Gavin R. Zealey, and Raafat E.F. Fahim

Introduction

The introduction of inactivated vaccines for viral or bacterial diseases such as smallpox, polio, pertussis, tuberculosis, measles, mumps and rubella has led to the eradication or control of these diseases. The early vaccines, though effective, were often crude and poorly characterized. The advent of purified subunit vaccines using native or recombinant proteins has led to the development of modern well-characterized vaccines for diseases such as hepatitis B and pertussis. There are still many infectious diseases for which new or improved vaccines are needed, e.g., AIDS, diarrheal diseases, hemorrhagic fevers, malaria and other parasitic diseases, tuberculosis and other respiratory diseases, and prion diseases. Immunotherapy for, or prevention of, cancer and autoimmune diseases is a new and exciting application of vaccinology. This chapter has been organized into the following topics: pediatric vaccines, adult vaccines, vaccines against nosocomial infections, cancer vaccines, vaccines against autoimmune diseases, current technologies, and emerging technologies.

Pediatric Vaccines

Neonates are susceptible to a variety of infectious diseases due to the immaturity of their immune systems. Maternal immunization with tetanus toxoid or influenza vaccines has demonstrated that it is feasible to protect a newborn through passive transfer of maternal antibodies in utero. Several other inactivated, attenuated, polysaccharide, conjugate or subunit vaccines are being tested or considered as vaccines for pregnant women.[1]

The main cause of life-threatening bacterial infections in neonates is Group B Streptococci (GBS). There are nine serotypes of GBS capsular polysaccharide, and it has been shown that the capsule confers protective immunity. Conjugate vaccines, including a maternally administered multivalent vaccine, have proven efficacious in animal models.[2] A bivalent alum-adjuvanted protein vaccine, based upon purified cell surface proteins Rib and a, has also been shown to be protective in animal models.[3]

Group A Streptococci (GAS) are responsible for a wide spectrum of diseases including pharyngitis, strep throat, rheumatic fever, bacteremia, toxic shock syndrome, necrotizing fasciitis, and even death. The M protein is the major protective antigen, and each of the approximately 80 serotypes of GAS expresses a unique subtype of M. The development of a vaccine based upon the M protein is further complicated by the presence of epitopes that cross-react with human tissues, especially the myocardium. A multivalent vaccine in which immunogenic fragments of the M protein, derived from multiple serotypes, are arranged in tandem in a recombinant protein, was protective in animal models.[4] Recently, a new streptococcal protective antigen (Spa) was identified that elicits protective antibodies that are cross-reactive in an opsonization assay.[5] The major cause of GAS toxic shock syndrome is the toxic shock syndrome toxin

New Vaccine Technologies, edited by Ronald W. Ellis. ©2001 Eurekah.com.

(TSST-1), which is related to the superantigenic staphylococcal enterotoxins (SEA, SEB, SEC, SED, and SEE). All of these toxins can induce high fever and lethal shock. Using molecular modeling and site-directed mutagenesis, nontoxic SEA and SEB mutants were generated that were protective in animals.[6] The SEB vaccine was also partially protective against TSST.

Bordetella pertussis is the main cause of whooping cough, a serious, potentially life-threatening disease of young children. Although effective whole-cell vaccines were developed in the 1940s, fears of reactogenicity led to a decline in their use, with a concomitant increase in epidemic pertussis. Thus, there are still ca. 100,000 deaths per year worldwide from whooping cough, mainly in the developing world. Acellular or component pertussis vaccines have been developed that comprise one to five purified antigens: pertussis toxin (PT), filamentous hemagglutinin (FHA), pertactin (69 kDa or P69) and agglutinogens 2 and 3, with chemical detoxification of some components to ensure safety. A detailed description of the technologies used to produce subunit pertussis vaccine is found below. Subunit vaccines have substantially reduced local and systemic adverse events relative to the whole-cell pertussis vaccines and are highly efficacious.[7] In recent years it has been shown that adolescents and adults may be carriers of *Bordetella*, in which case a safe and effective vaccine for this age group would be beneficial. An adult pertussis vaccine based upon the five component pertussis vaccine (PT, FHA, 69 kDa, agglutinogens 2 and 3), has recently been licensed. Recombinant pertussis vaccines have also been developed, based upon the genetic detoxification of the pertussis toxin (PT) gene.[8] An acellular pertussis vaccine that utilizes purified subunits from a recombinant strain expressing the S1 K^9G^{129} mutant of PT is in use in Italy.[9]

Bacterial meningitis is a serious disease in children and adolescents. The advent of the *Haemophilus influenzae* type b (Hib) conjugate vaccines has resulted in a dramatic decrease in Hib-related meningitis, mainly due to a decrease in carriage.[10] A second major cause of meningitis is *Neisseria meningitidis*. Effective polysaccharide and/or conjugate vaccines have been developed against *N. meningitidis* serotypes A, C, Y and W135, but the polysaccharide of serotype B is not immunogenic, presumably due to its similarity to glycoproteins found in normal human tissues. Clinical trial results suggest that intranasal delivery of a vaccine prepared from serotype B outer membrane vesicles (OMV) may be effective.[11] A new OMV vaccine that includes six PorA proteins derived from two recombinant *N. meningitidis* strains is also under development.[12] Other promising protein antigens include the lactoferrin binding protein B (LbpB)[13] and the transferrin binding protein B (TbpB),[14] both of which are utilized for iron acquisition. The variability of the *N. meningitidis* TbpB and LbpB proteins makes it difficult to produce a broadly protective vaccine, but the use of truncated and chimeric TbpB proteins may improve the cross-protection afforded by this antigen.[15,16] The highly conserved surface protein NspA has been shown to elicit protective antibody responses against serogroups A, B and C in mice.[17] Another potential vaccine antigen is the recently discovered T-cell stimulating protein A (TspA).[18]

Seventy to eighty percent of all children will have at least one occurrence of middle ear infection or acute otitis media (AOM) by the time they are six years of age. AOM is caused by bacterial and/or viral pathogens. The three major bacterial organisms[19] are *Streptococcus pneumoniae*, nontypable *H. influenzae*, and *Moraxella catarrhali*. The major viral causes[20] are respiratory syncytial virus (RSV), parainfluenza virus (PIV), influenza virus, rhinovirus, adenovirus and enterovirus. Of particular concern is the high and increasing incidence of antibiotic-resistant causative bacteria. There are ca. 90 serotypes of *S. pneumoniae*, and currently licensed vaccines contain 23 capsular polysaccharides. However, since polysaccharide vaccines are known to be poorly immunogenic in young children, a new generation of conjugate vaccines has been developed. These contain 4-11 capsular polysaccharides conjugated to diphtheria toxoid, CRM_{197}, meningococcal protein complex (OMPC), or tetanus toxoid and are currently in efficacy trials. Preliminary data suggest that vaccination may result in substitution of colonizing serotypes,[21] thus *S. pneumoniae* protein-based subunit vaccines are being reevaluated. Native or recombinant versions of *S. pneumoniae* neuraminidase,[22] pneumolysin,[23] pneumococcal surface adhesin A (PsaA),[24] pneumococcal surface protein A (PspA),[25] and combinations[26] have all shown potential as vaccine antigens in animal models. A novel *S. pneumoniae* secretory IgA binding protein (SpsA) has also been identified as a possible vaccine candidate.[27] One of the emerging concerns for conjugate vaccines is the issue of carrier overload, but the use of genetically detoxified pneumolysin as a carrier protein for *S. pneumoniae* capsular polysaccharide may enable the development of autologous conjugate vaccines with reduced antigen content.[28] Pneumococcal pneumonia is the leading cause of childhood mortality in most parts of the developing world[29] and invasive pneumococcal diseases include meningitis and endocarditis. Thus, any pneumococcal vaccines should protect against multiple disease states.

A number of protein subunits have been studied for protection against *H. influenzae* disease, including otitis media. OMP D15,[30] fimbriae,[31] Hap/IgA protease,[32] Hia,[33] Hsf,[34] high molecular weight (HMW) proteins,[35] HtrA,[36] OapA,[37] OMP26,[38] P2,[39] P4,[40] P6,[41] PCP,[42] Protein D,[43] TbpB,[44] and combinations of antigens[45] have all been studied as possible vaccine candidates. The variability of many of these proteins has made the development of a cross-protective vaccine difficult. Several promising *M. catarrhalis* adhesins or OMPs have been identified, including CopB,[46] CD,[47] E and G,[48,49] LbpB,[50] TbpB,[51] and UspA.[52] A 200-kilodalton (kDa) adhesin that is involved in hemagglutination has also been identified recently.[53] A final vaccine to protect against bacterial otitis media caused by the three principal pathogens will probably be quite complex; however, if effective, it should reduce the use of antibiotics for this common complaint and hopefully reduce the increase of antibiotic resistant organisms.

Respiratory syncytial virus (RSV) is a major cause of AOM and of serious lower respiratory tract disease, particularly in infants under the age of 3 months and in the elderly. Worldwide, RSV causes 5 million deaths annually, and young children may develop severe disease even in the presence of maternal antibodies. A formalin-inactivated vaccine was developed in the 1960s, but immunization resulted in potentiation of disease following natural exposure.[54] This approach has been abandoned, and new strategies include live-attenuated strains of RSV,[55] vaccinia MVA-based vaccines,[56] protein subunits and synthetic peptides.[57] Vaccines containing purified fusion (F) protein have been shown to be immunogenic and nonreactogenic in seropositive children.[58] In animal studies, the immune response to a recombinant chimeric F-G protein, derived from the F protein and the attachment glycoprotein G, could be altered depending upon the adjuvant used.[59] Another chimeric protein between the streptococcal G protein and a 100-amino acid polypeptide derived from the RSV G protein was also protective in animal models.[60] Immunization with chimeric proteins based upon the RSV F protein and the PIV3 hemagglutinin-neuraminidase (HN) protein was found to afford protection against both diseases in animal models.[61]

Rotavirus infection is the leading cause of viral gastroenteritis in young children worldwide, leading to more than 500,000 deaths each year in developing countries. The worldwide morbidity and mortality associated with rotaviral infections has led to a global initiative to develop a vaccine. Although an oral live-attenuated reassortant rotavirus vaccine was developed[62] and recommended for routine use, it has now been withdrawn from use, pending investigation of unanticipated adverse events. Cross-protective recombinant virus-like particle (VLP) subunit vaccines have been developed for parenteral delivery.[63] Intranasal immunization with recombinant subunit vaccines fused to maltose binding protein also afforded some protection in animal models.[64]

Cytomegalovirus (CMV) is the leading infectious cause of congenital central nervous system disease and is also a serious problem for the immunocompromised, such as AIDS patients or organ transplant recipients. Live-attenuated vaccines have been developed, but more defined subunit vaccines are deemed safer for the target populations. Glycoprotein B (gB) is involved in cell-surface attachment, virion penetration, and intercellular spread of infection and is the major protein that induces virus neutralizing antibodies during infection. A subunit vaccine based upon recombinant gB, produced in CHO cells, has been shown to be safe and immunogenic in Phase I human clinical trials.[65]

Adult Vaccines

Respiratory Tract Diseases

It is estimated that one-third of the world's population is infected with *Mycobacterium tuberculosis*, the causative agent of tuberculosis. Thirty million people suffer from active TB, with 8 million new cases per year, and 3 million deaths. There is also an alarming increase in multi-drug-resistant bacteria and patients suffering from both HIV and tuberculosis.[66] The only TB vaccine currently available is the attenuated strain of *M. bovis*, Bacille Calmette-Guérin (BCG), which although efficacious in animal models, has shown 0-80% efficacy in human field trials. Approaches under investigation to develop an improved TB vaccine include DNA immunization,[67] recombinant BCG expressing cytokines,[68] recombinant *Salmonella* expressing *M. tuberculosis antigen* Ag85B,[69] and subunit vaccines.[70]

Enteric Diseases

Vaccines against enteric diseases such as cholera, enterotoxigenic *E. coli*, shigellosis, and rotavirus are needed in endemic regions and also as travelers' vaccines. Cholera is an important cause of disease in developing countries, responsible for more than 120,000 deaths per year. A parenteral whole-cell vaccine was found to have limited efficacy and was supplanted by a highly effective whole-cell vaccine containing added cholera toxin B (CTB) subunit.[71] To improve the yield of vaccine, the CTB component can be produced from a genetically-engineered strain of *Vibrio cholerae* that does not contain the cholera toxin A gene.[72] Recently, it was found that transcutaneous immunization of mice with native cholera toxin (CT) could protect against a lethal challenge with toxin and was not toxic.[73] Enterotoxigenic *Escherichia coli* (ETEC) is responsible for 800,000 deaths annually. The toxins produced by *Vibrio cholerae* and ETEC are structurally and genetically related and are their main virulence factors. Hybrid toxins have been engineered by mutating regions of the CT gene and replacing sequence with that of the *E. coli* heat-labile enterotoxin (LT) gene.[74] The resultant hybrid toxins were cross-neutralizing and show potential for development of a cross-reactive multi-potent vaccine.

Shigella species cause epidemic bacillary dysentery or shigellosis, that affects an estimated 200 million people worldwide, with 650,000 deaths annually. There are several species of *Shigella* and an effective vaccine would need to protect against *S. dysenteriae*, *S. sonnei*, and *S. flexneri*.[75] Most efforts have focused on a live attenuated approach and recently a *S. flexneri* candidate vaccine was developed that showed some efficacy in a human challenge trial.[76] Another approach has been O-specific polysaccharide conjugate vaccines, which have also shown some efficacy in human trials.[77] Shiga toxins (STX) and Shiga-like toxins (SLT) are related cytotoxins elaborated by Shigella and/or enterohemorrhagic strains of *Escherichia coli* (EHEC). Immunization with the nontoxic B subunit of Shiga toxin elicited toxin-neutralizing antibodies,[78] and a conjugate of *E. coli* O-specific polysaccharide and Shiga toxin B subunit was shown to also elicit bactericidal antibodies to EHEC strain O157.[79]

Parasitic and Vector-Borne Diseases

Vaccines against parasitic and vector-borne diseases such as malaria, schistosomiasis, leishmania, and Dengue fever are desperately needed in many parts of the world and as travelers' vaccines. There are an estimated 500 million cases of malaria each year, with 2.5 million deaths, mostly in children under the age of five. Most malaria is caused by *Plasmodium falciparum*, and drug-resistant parasites are widespread. Because of the complex multistage life cycle of *P. falciparum*, the development of an effective vaccine has been very difficult and many approaches have been tried. Vaccines based on live attenuated parasites delivered by mosquito bite are effective but not practical, while vaccines derived from whole killed organisms were not completely effective. Single-component subunit vaccines incorporating the circumsporozoite (CS) gp190 precursor (or fragments thereof),[80] the Pfs25 antigen,[81] the N-terminal fragment of the serine repeat antigen (SERA),[82] or the C-terminal 19-kDa fragment of the merozoite surface protein-1 ($MSP\text{-}1_{19}$)[83] have all shown some protection in animal models. A synthetic peptide polymer vaccine (Spf66) containing four different peptides based upon the circumsporozoite protein, the merozoite surface protein, and two blood-stage antigens, has shown mixed efficacy in clinical trials.[84] A recombinant polyepitope vaccine encoding multiple B- and T-cell epitopes from nine stage-specific antigens has shown promise in small animals and in vitro.[85] A poxvirus vaccine expressing seven stage-specific antigens showed a low rate of efficacy in clinical trials.[86] A DNA vaccine encoding the CS protein afforded partial protection in an animal model.[87] A vaccine based upon self-replicating hepatitis B virus core particles carrying CS epitopes was effective in animals.[88] A series of vaccines based upon self-replicating hepatitis B virus surface antigen (HbsAg) particles in different adjuvants showed some efficacy in humans.[89] Despite these efforts, the final composition of an effective malaria vaccine remains uncertain.

Shistosomiasis is caused by infection with the helminth *Schistosoma mansoni* and affects several hundred million people in tropical areas. The morbidity of the disease is associated with the inflammatory response at the site of deposition of the parasite eggs. Immunization with attenuated parasites or protein antigens such as Sm28GST,[90] Sm97,[91] and a 74 kDa antigen[92] have proved partially effective. Intranasal administration of a conjugate Sm28GST-cholera toxoid vaccine has been shown to protect infected mice from disease-associated immunopathology, suggesting the possibility of a therapeutic vaccine.[93]

There are an estimated 12 million cases of leishmaniasis worldwide caused by members of the protozoan *Leishmania* family. The disease ranges from a self-curing cutaneous infection to a severe and potentially fatal visceral leishmaniasis, and organisms are becoming increasingly drug-resistant. There is evidence in animal models that some antigens or immunization protocols may exacerbate disease.[94] A mixture of the *L. mexicana mexicana* promastigote surface antigen gp63 plus the promastigote lipophosphoglycan was protective in mouse models when encapsulated in liposomes.[95] The M-2 glycoprotein (gp46) of *L. amazonensis* promastigotes, mixed with various adjuvants, was partially protective in mouse models.[96] A recombinant *L. major* promastigote-derived protein with homology to receptors for activated protein kinase C (LACK), was quite protective in mouse models when administered with IL-12.[97] Similarly, a recombinant *L. major* promastigote-derived protein with homology to eukaryotic thiol-specific-antioxidant (TSA) proteins was highly protective in animal models when administered with IL-12.[98] The *L. major* promastigote-derived surface antigen (PSA-2) was found to be partially protective against challenge when the native protein or a recombinant protein produced from a Leishmania host was administered.[99] The recombinant *L. braziliensis* promastigote-derived protein (LeIF) was found to induce an IL-12 cytokine profile and to partially protect mice against challenge.[100] The p4 and p8 antigens purified from the amastigote stage of *L. pifanoi* were partially protective in mouse models when adjuvanted with *Corynebacterium parvum.*[101]

There are four serotypes of Dengue viruses, DEN 1-4. They are mosquito-borne and cause *ca.* 50 million infections annually, with disease symptoms ranging from mild dengue fever (DF) to severe dengue hemorrhagic fever (DHF) and fatal dengue shock syndrome (DSS). Although effective vaccines against other flaviviruses such as yellow fever virus, Japaneses encephalitis virus and tick-borne encephalitis virus have been produced, it has been much more difficult to develop a vaccine against DEN. During disease, patients develop protective immunity against the infecting serotype but can be infected later by a different DEN serotype. An antibody-dependent immunopotentiation of disease has been observed in patients upon secondary infection, leading to increased incidence of DHF and DSS, the more severe forms of disease. It is thought that disease during a secondary infection is enhanced by the presence of nonneutralizing cross-reactive antibodies acquired during the primary infection. An effective vaccine thus should be able to protect against multiple serotypes without inducing cross-reactive antibodies. The viral envelope E protein induces virus-neutralizing antibodies and has been the focus of studies for subunit vaccines. A recombinant subunit vaccine based upon the E protein from DEN-2 was partially protective in animals.[102] A vaccine derived from the B-binding domain of the E protein was also partially protective in animals and did not induce cross-reactive antibodies.[103] Immunization of animals with a hybrid DEN-2/DEN-3 E protein-based vaccine elicited antibodies to both serotypes, and these antibodies were also able to inhibit virus binding to human cells.[104]

AIDS and other Sexually-Transmitted Diseases

One of the major challenges of the last 15 years has been to develop a vaccine against AIDS. The human immunodeficiency virus (HIV) is expected to infect 1% of the world population by the year 2000. Although significant progress has been made in the treatment of AIDS,[105] especially the use of combination therapy with reverse transcriptase and protease inhibitors, these treatments are lifelong, costly, and not without side-effects. Inactivated HIV vaccines have been studied, but concerns about safety have made this approach unlikely. A therapeutic gp120-deleted, inactivated HIV-1 vaccine (REMUNE) has been tested in clinical trials and shows some promise.[106] Live attenuated SIV vaccines showed promise in macaques[107] and the discovery of a group of long-term disease-nonprogressers that had all received a nef-deleted HIV strain from contaminated blood,[108] led to research into live attenuated HIV vaccines, but this approach is also unlikely due to safety concerns. Envelope subunits of gp120 or gp160 have been extensively studied with generally disappointing results,[109,110] although they have protected in some cases.[111] The failures are probably due to a requirement for a native oligomeric structure of the envelop glycoproteins for protection. The recently solved crystal structure of HIV-1 gp120 interacting with CD4 and a neutralizing antibody should advance the rational development of env-based subunit vaccines.[112] Modification of gp120 by epitope insertion has recently been shown to improve its immunogenicity.[113] Another way to overcome conformational limitations of surface glycoproteins is by the use of VLPs.[114] A p24 VLP has recently been shown to be safe and immunogenic in asymptomatic HIV patients.[115] Recombinant Tat toxoid has been proposed as a component of a preventive vaccine based upon the hypothesis that anti-Tat antibodies could prevent Tat-induced immune suppression, leading to protection.[116] A novel approach to HIV vaccination is the use of a heat shock protein (hsp70)-p24 fusion protein which gave rise to both humoral and cellular responses.[117] Live vectors such as poxvirus,[118] adenovirus,[119] poliovirus,[120] alphavirus,[121] BCG,[122] or *Listeria*[123] have great promise because of their ability to direct the immune response. DNA immunization was found to be safe and immunogenic, but probably insufficient for protection.[124] Peptides[125] and polyepitopes[126] have also shown some promise. However, the best approaches seem to be prime/boost protocols combining different strategies such as DNA prime/DNA + protein boost,[127] DNA prime/canarypoxvirus boost,[128] DNA prime/vaccinia virus boost/protein boost,[129] canarypoxvirus prime/protein boost,[130] canarypoxvirus prime/peptide boost,[131] or adenovirus prime/protein boost.[132] Although HIV-2 has not been as extensively studied as HIV-1, the canarypoxvirus prime/protein boost approach also looks promising for this disease.[133]

Thus, while protein-based HIV vaccines may not be effective on their own, they do comprise an important part of the AIDS therapeutic and preventive vaccine repertoire as components of prime/boost protocols.

For 1995, the WHO estimated the global incidence of sexually-transmitted diseases (STDs) such as chlamydia, gonorrhea, syphilis, and trichomoniasis at >300 million new cases.[134] Hepatitis B virus (HBV), HIV, human papilloma virus (HPV), and herpes simplex virus (HSV) also cause STDs, yet an effective vaccine exists for only one of these viruses, HBV. A hepatitis B vaccine produced in yeast was the first licensed recombinant protein vaccine.[135] This HBsAg vaccine has been widely used and is very effective.[136]

Infection with *Chlamydia trachomatis* is a major cause of sexually transmitted genital tract disease throughout the world and of ocular disease in the developing world. Although genital tract disease can be asymptomatic or mild, in up to 40% of infected women the bacteria spreads to the upper genital tract and is a major cause of pelvic inflammatory disease which is associated with infertility and ectopic pregnancy. Natural infection does not provide protection against subsequent infection, and there is evidence that the generation of an inappropriate immune response by vaccination may result in more severe disease upon subsequent infection. An acellular vaccine, based upon the outer membrane complex, was protective in animal models.[137] The major outer membrane protein (MOMP) is the immunodominant *Chlamydia* antigen and has been the focus of defined vaccine development. Animals immunized with adjuvanted MOMP or its peptides were partially protected, suggesting that additional or alternative immunogens may be required to provide an efficacious preparation.[138] The elucidation of the sequence of the genome of *C. trachomatis* may assist in the identification of additional protective antigens against *Chlamydia* infection.[139] A novel approach for vaccine development has been the use of hybrid polioviruses expressing MOMP epitopes. These constructs were able to induce neutralizing antibodies against multiple serovars.[140]

Infection with *Neisseria gonorrhoeae* is a global problem, with ca. 800,000 cases per year in the US alone and an increase in antibiotic resistant strains. Several *N. gonorrhoeae* surface proteins have been identified as potential vaccine components but an effective vaccine has not been developed, in large part due to the phase and antigenic variability of *N. gonorrhoeae* species. Pilus-based vaccines have been tested in males and provided some protection against homologous, but not heterologous, challenge strains. The major OMP, Por or PI, is a stable protein that has two main structural subclasses, PIA and PIB. Anti-PI antibodies elicited during infection may confer some protection, making it an attractive vaccine candidate.[141] The opacity protein (Opa) or PII, is a variable protein that is important for colonization; however, it has multiple gene loci and a strain may express none or several versions of this protein. Human challenge studies have shown that there is no preferred form of Opa expressed during early colonization, which would have simplified the use of this antigen as a vaccine component.[142] Like *N. meningitidis*, *N. gonorrhoeae* utilizes transferrin and lactoferrin binding proteins to acquire nutrient iron. It has been demonstrated that in order to infect humans, the transferrin receptor must be expressed, suggesting that it is a virulence factor.[143] Recently, the genes encoding the lactoferrin binding proteins have been cloned and sequenced.[144,145] The Neisserial surface protein (NspA) is a protective antigen first identified in *N. meningitidis* and recently identified in *N. gonorrhoeae*.[146] Production of immunoglobulin A is a major host defense mechanism for mucosal pathogens. *N. gonorrhoeae* produces an IgA1 protease which enhances intracellular survival of the organism.[147] Some patients produce antibodies to IgA protease, indicating that it may be useful as a vaccine antigen. Thus, there are a number of promising antigens that may form part of a subunit vaccine against *N. gonorrhoeae*.

By the 19th century, syphilis was one of the most common diseases in Europe and the US, and as recently as 1990, there were 134,000 cases of syphilis reported during an outbreak in the US. Syphilis is caused by infection with the spirochete *Treponema pallidum*. In adults, primary infection is localized and resolves without antibiotic treatment. Similarly, secondary infection can resolve without treatment; however there may be widespread dissemination of the organisms, which become latent. Tertiary syphilis, although rare, is a multi-organ, potentially fatal disease. Congenital syphilis results in a high incidence of still births and neonatal deaths. *T. pallidum* cannot be cultured continuously in vitro, which has hampered the development of a vaccine. The membrane proteins TpN17 and TpN47, as well as the endoflagellar sheath protein TpN37 all induced strong proliferative T-cell responses during infection, making them potential vaccine candidates.[148] The complete genome sequence of *T. pallidum* has recently been determined, and these data may aid in the development of a vaccine.[149]

Herpes simplex virus types-1 and -2 (HSV-1, HSV-2) are responsible for a variety of illnesses, including oral, facial, ocular, and genital infections as well as herpes encephalitis and neonatal herpes. Most genital herpes infections are caused by HSV-2. There are several major envelope glycoproteins expressed on herpes viruses, including glycoproteins B and D (gB and gD), both of which induce virus-neutralizing antibodies. Both gB and gD have protein sequences that are highly conserved between HSV-1 and HSV-2 and induce antibodies that recognize both viral types. Immunization with recombinant HSV-1 gB or gD protected guinea pigs from intravaginal challenge with HSV-2.[150] Parenteral immunization with recombinant HSV-1 gD protected mice from lethal intraperitoneal challenge with HSV-1 or HSV-2.[151] Intranasal immunization of guinea pigs with recombinant HSV-2 gD and a genetically-detoxified heat-labile enterotoxin (LTK63) as adjuvant, resulted in a reduction of disease severity, but only intramuscular immunization in the presence of the MF59 adjuvant afforded protection against disease incidence.[152] Despite the promising results in animals, Phase III clinical trials of a combination of recombinant gB2 and gD2 showed the vaccine to be safe and immunogenic, but not protective.[153]

Contraceptive Vaccines

The WHO recently announced the estimated birth of the 6 billionth human on the planet. At current rates, the world population is predicted to reach 10 billion by 2050, a number that may not be sustainable. Although effective barrier and hormonal birth control methods are available, issues of compliance and side effects have spurred the search for immunocontraceptive methods. Unlike vaccines against infectious disease, a human contraceptive vaccine should be reversible. The first contraceptive vaccine tested in humans was based upon human chorionic gonadotrophin (HCG), an early hormone necessary to establish and maintain pregnancy. HCG is a heterodimeric glycoprotein composed of alpha and beta subunits (CGα and CGβ). CGα is nearly identical to the α-subunit of other hormones of pituitary origin such as follicle-stimulating hormone (FSH), luteinizing hormone (LH), and thyroid stimulating hormone (TSH). CGβ is very similar to the beta chain of LH, except that it contains an additional carboxy-terminal peptide of 30 residues. A vaccine was developed that contained a heterospecies dimer (HSD) of hCGb and the a subunit of ovine LH coupled to tetanus toxoid or diphtheria toxoid.[154] The vaccine was found to be safe in women and to prevent pregnancy if anti-HCG antibody levels were > 50 ng/mL. The anti-fertility effect was also reversible, resulting in healthy babies.[155] This vaccine was manufactured by combining native hCGβ, purified from the urine of pregnant women, with native LH, purified from the pituitary glands of sheep. A recombinant CGβ protein has been shown to be safe and immunogenic in animals and offers the possibility of consistent large-scale manufacture of a HCGβ-based contraceptive vaccine.[156] The first contraceptive vaccine tested in males was based upon FSH, a hormone that is important for the development of sperm. The vaccine was derived from lyophilized sheep pituitary glands and was shown to be safe in men and to cause sterility in monkeys.[157] The effect does not appear to be due to a lack of sperm but to production of poor quality sperm, which may not be adequate or desirable for a human vaccine. Antibody raised to the ovine FSH beta subunit (FSHβ) binds to and neutralizes primate FSH, making it a viable alternative to the full-length protein. A recombinant ovine FSHβ vaccine is being developed as a consistent and safe source of large amounts of immunogen. Other approaches based upon sperm or zona pellucida antigens or peptides are being evaluated in animals.

Vaccines Against Nosocomial Infections

The immunocompromised patients with chronic disease or hospitalized patients are especially susceptible to infections. *Pseudomonas aeruginosa* is an opportunistic pathogen that can cause fatal lung infections in cystic fibrosis patients and systemic infections in burn patients. A Phase III trial of a flagella-based vaccine is currently underway.[158] A *P. aeruginosa* purified OMP vaccine has recently been tested in humans and found to be safe and immunogenic, with immune sera affording protection in a passive model.[159] A recombinant vaccine using the outer membrane protein OprI was tested in humans and found to be safe and immunogenic.[160] The OprF protein was also found to be effective in mice.[161] Recent studies have shown that immunization with recombinant PcrV resulted in protection in a mouse model.[162] A novel *P. aeruginosa* vaccine containing a hybrid protein consisting of the N-terminus of the mature OprI protein fused to the C-terminus of OprF was shown to be protective in mice [163] and safe and immunogenic in humans.[164]

Staphylococcus aureus causes a variety of diseases ranging from mild skin infections, to pneumonia, meningitis, bacterial arthritis, endocarditis, toxic shock syndrome and death. It is a significant cause of nosocomial infections, particularly in patients having heart valve surgery, hip or knee replacement, or kidney dialysis, and the incidence of multidrug resistant organisms is increasing. Whole-cell vaccines have been tested in human clinical trials, but were not very effective.[165] Vaccines based upon capsular polysaccharide or conjugates are under development.[166,167] *S. aureus* elaborates a number of adhesion factors that bind to extracellular matrix components such as collagen, fibronectin, and fibrinogen.[168] A subunit vaccine based upon a recombinant fragment of the collagen adhesin protects animals against septic death.[169] *S. aureus* also produces a family of enterotoxins termed SEA, SEB, SEC, SED and toxic shock syndrome toxin-1 (TSST-1), of which SEA and TSST-1 are known to function as superantigens. Vaccination of mice with recombinant SEA that had been mutated to delete its superantigen activity protected them from septicemia.[170] A novel approach to vaccine development has been to study the protective effect of a protein called RAP (RNA III activating protein) that regulates the production of virulence factors such as the toxins. Mice immunized with purified RAP were protected from *S. aureus* pathology.[171]

Cancer Vaccines

Until recently, chemotherapy, radiation therapy, surgery, and combinations of these have been the only treatments for malignancies. The potential use of vaccines in cancer therapy and in special instances as prophylaxis is an exciting development. A few cancers are viral in origin, such as hepatic cancer caused by HBV, cervical cancer caused by HPV, and adult T-cell leukemia caused by HTLV-1. At least one cancer has a bacterial origin: gastric cancer caused by *Helicobacter pylori*. These cancers caused by microorganisms have the greatest potential for development of prophylactic vaccines.

Hepatocellular carcinoma (HCC) is one of the most common cancers, especially in some parts of Asia, where it ranks first or second in incidence. It can be caused by viral infection with hepatitis B or C viruses. Worldwide, there are 350 million carriers of HBV, with an annual death rate of about 1 million. The advent of HBV vaccines and increasingly universal vaccination has led to a dramatic decrease in carriage rates and a concomitant decrease in HCC.[172]

Cervical cancer is the second most common cancer among women worldwide, with an estimated 450,000 new cases diagnosed per year and 200,000 deaths. There are over 80 different HPV genotypes, of which types 16, 18, 31 and 45 can be detected in nearly 80% of cervical carcinomas. In animals, HPV VLPs comprised of the late proteins L1 and L2 induce virus neutralizing antibodies and a protective immune response.[173] This approach is favored for prophylaxis, as the vaccine does not incorporate any oncogenic material. Expression of the early proteins E6 and E7 appears to be necessary for the neoplastic state,[174] and they are thus good vaccine targets for a therapeutic vaccine. Chimeric VLPs comprised of L1 plus E7 fused to L2 were protective in a challenge model.[175] Immunization with recombinant E7 protein in incomplete Freund's adjuvant also induced protection in an animal model.[176] Thus, the prospects for safe and effective prophylactic and therapeutic vaccines against cervical cancer are very high.

Adult T-cell leukemia (ATL) is one of the many diseases caused by infection with the human retrovirus HTLV-1. The virus tends to be geographically localized and is most prevalent in Japan, where 1-2 million people are infected. Only a small proportion of HTLV-1-positive patients develop ATL, and that occurs after a long latency.[177] There is no effective therapy against HTLV-1-induced leukemia. Recombinant vaccinia viruses expressing HTLV-1 env[178] can induce a neutralizing antibody response, but a better response is observed using a prime-boost protocol whereby the vaccinia virus constructs are boosted with recombinant protein.[179] Peptide vaccines are being developed based upon the gp46 protein.[180] The HTLV-1 Tax gene product is thought to transactivate cellular genes, resulting in proliferation and development of leukemia. It is a major target of CTLs and a number of epitopes have been identified that have potential for development of peptide vaccines.[181]

H. pylori is a spiral-shaped Gram-negative bacterium that colonizes human gastric epithelial cell surfaces and the overlying mucous layer. Approximately 50% of the world population is infected by *H. pylori*, with rates of >90% occurring in some developing countries. Disease ranges from asymptomatic gastric inflammation, to gastric or duodenal ulcers, and antral adenocarcinoma or mucosa-associated lymphoid tissue (MALT) lymphoma. Although therapies involving combinations of antibiotics, proton-pump inhibitors and bismuth compounds are somewhat efficacious, they do not prevent recurrence and are expensive. Animal models have demonstrated that mucosal immunization with protein antigens such as urease,[182] catalase,[183] and the heat shock proteins HspA and HspB[184] can be effective in prophylactic vaccines. In addition, therapeutic vaccination with recombinant VacA or CagA with a mucosal adjuvant eradicated chronic infection in mice.[185] Recently, immunodominant antigens were identified using patient sera and proteomic technology, although pathology-specific antigens were not found.[186]

For cancers that are not caused by microorganisms, complex vaccination protocols are being tried in which patients may receive autologous tumor-infiltrating lymphocytes (TILs) or dendritic cells, poxvirus, plasmid DNA, or recombinant BCG, *Listeria* or *Salmonella* vaccines expressing tumor associated antigens (TAAs), with or without immunostimulatory molecules. In addition to these specialized delivery platforms, patients are also being immunized with recombinant antigens, peptides, modified peptides, carbohydrate antigens, anti-idiotypic antibodies, and monoclonal antibodies (Mabs).[187] Immunotoxins, composed of a tumor targeting moiety and an effector molecule, are also important components of the anti-cancer arsenal. Novel anti-angiogenesis or apoptosis-inducing strategies may lead to generic cancer therapies.

In the US, there are an estimated 134,000 new cases of colorectal cancer each year. The most advanced immunotherapies for colorectal cancer involve the use of carcinoembryonic antigen (CEA) or the KSA antigen (also called EpCAM, GA733-2 or CO17-1A). CEA belongs to a family of heavily glycosylated 175-200 kDa proteins found in normal fetal tissues and in colorectal, breast and lung carcinomas. Patients immunized with recombinant CEA mount a weak immune response, which is enhanced by coimmunization with GM-CSF.[188] When CEA is expressed in vaccinia vectors, its immunogenicity is improved and it elicits CEA-specific CTLs.[189] Another approach has been to immunize patients with an anti-idiotype MAb mimicking CEA.[190] This vaccine elicited potent anti-CEA humoral and cellular immune responses, resulting in clinical responses. KSA is an epithelium-specific cell-cell adhesion molecule, expressed on the majority of colorectal carcinoma cells. In animal studies, recombinant KSA was shown to elicit humoral and cellular immune responses.[191] Monoclonal antibody and anti-idiotype antibody therapies are also being explored.[192] A novel approach using an anti-KSA Fab-SEA superantigen fusion was shown to be effective against established, disseminated and heterogenic tumors in a transgenic animal model, and the effectiveness of this vaccine was further improved when coadministered with a Fab-IL-2 fusion.[193]

There are an estimated 42,000 new cases of melanoma diagnosed in the US each year, and the incidence is increasing. If treated at an early stage, this cancer has a 95% survival rate, but at later stages there is a 50-90% mortality rate. Melanoma patients have been immunized with semi-purified polyvalent vaccines prepared from antigens shed into the culture medium by a pool of melanoma cell lines. These vaccines contain MAGE-1, MAGE-3, Melan-A/MART-1, tyrosinase, gp100, and other melanoma-associated antigens, and have proven efficacious.[194] Current vaccination strategies against melanoma are focused on peptides, viral vectors, or cell-based approaches, with little or no protein-based vaccine development.[195] An immunoconjugate has been created by linking an scFv, specific for melanoma chondroitin sulfate proteoglycan, and the Fc effector domain of human IgG1.[196] An in vitro cytotoxicity assay demonstrated that it was able to target melanoma cells for lysis by NK cells and complement.

Without a family history of disease, a woman's lifetime risk of developing breast cancer is about 1 in 8, and for ovarian cancer it is 1 in 70. One of the most promising vaccine candidates appears to be HER-2/neu, which is a proto-oncogene encoding a 185 kDa transmembrane glycoprotein with tyrosine kinase activity and structural similarity to the epidermal growth factor receptor. It is amplified and overexpressed in approximately 30-40% of breast and ovarian cancers and is present in primary and metastatic lesions. Rats immunized with the homologous, but not identical, human HER-2/neu protein, were able to overcome tolerance and generate humoral and cell mediated responses.[197] This strategy of immunization with highly homologous foreign proteins may be an effective vaccine strategy for "self" tumor antigens. In an animal model, immunization with a recombinant protein representing the extracellular domain of HER-2/neu, was partially effective at preventing tumor development.[198] HER-2/neu peptide-based vaccines have also showed promise in clinical trials.[199] Passive immunization with anti-Her-2/neu specific antibodies was effective in animal models[200] and showed some promise in human clinical trials.[201] A vaccine based upon the sialyl-Tn carbohydrate antigen (STn-KLH, Theratope) is also showing promise in clinical trials.[202] An immunotoxin comprised of the *Pseudomonas aeruginosa* exotoxin A enzymatic fragment fused to a Le^y antigen-specific MAb was shown to cure human breast carcinoma xenografts in animal models.[203] Diphtheria toxin (DT) is a potent killer of cells, but it is nonspecific and most people have circulating antibodies. DT was protected from circulating antibodies and targeted to tumor cells by encapsulation in immunoliposomes. In vitro assays indicated that the DT-immunoliposomes were able to kill ovarian tumor cells, even in the presence of anti-DT antibodies.[204]

Prostate cancer is the most common type of cancer in men and the second leading cause of cancer deaths in men after lung cancer. In the US, there are an estimated 317,000 new cases of prostate cancer diagnosed per year and >40,000 deaths. Chemotherapy, surgery and/or radiation therapy are effective for early-stage disease, but the average survival is only 2-3 years after diagnosis of metastatic disease. Several TAAs have been identified that may be useful for subunit vaccine development.[205] Elevated expression of prostate-specific antigen (PSA) is widely used as a clinical marker of prostatic disease, including prostate cancer. At least two different vaccines based upon this antigen, have entered clinical trials. OncoVax-P is a vaccine consisting of liposome-encapsulated rPSA and lipid A, which has been shown to be safe and immunogenic in clinical trials.[206] A vaccinia-PSA vaccine has also been found to be safe in clinical trials.[207] Unlike PSA, which is a secreted protein, prostate-specific membrane antigen (PSMA) is an integral membrane protein, making it useful for targeted therapy. A vaccine based upon dendritic cells pulsed with two PSMA peptides showed promise in Phase II trials,[208] and recombinant PSMA has recently been produced in baculovirus.[209] Prostate stem cell antigen (PSCA) has been identified as another prostate-specific cell surface antigen that is strongly expressed in a tumor cell line and may have vaccine potential.[210] Prostatic acid phosphatase (PAP) is a secreted prostate antigen. Immunization of rats with recombinant human or rat prostatic acid phosphatase (PAP), resulted in a specific antibody response, but no CTL response. Immunization with recombinant vaccinia expressing human PAP, but not rat PAP, elicited CTLs, but no detectable antibody response.[211] Prostatic steroid binding protein (PSBP) is the major protein produced and secreted into seminal fluid by the ventral prostate. Immunization with purified PSBP has been shown to induce experimental autoimmune prostatitis in rats, which may indicate its potential as a vaccine antigen for prostate cancer patients.[212]

Angiogenesis is a critical requirement for tumor survival and anti-angiogenic therapies offer a new approach to cancer treatment.[213] Angiostatin is an internal fragment of plasminogen, initially isolated from mice bearing Lewis lung carcinoma. Immunization with recombinant angiostatin inhibited establishment of experimental primary and metastatic tumors.[214] Endostatin is a C-terminal fragment of collagen 18a and in animal models, immunization with recombinant endostatin led to regression of primary tumors to dormant microscopic lesions.[215] The Apo2 ligand (Apo2L) is one of several molecules involved in cellular apoptosis. Apo2L was cytostatic or cytotoxic in vitro for 32/39 cell lines from colon, lung, breast, kidney, brain and skin cancer. Treatment of tumor-bearing animals with recombinant Apo2L resulted in tumor cell apoptosis, suppressed tumor progression and improved survival.[216]

Vaccines Against Autoimmune Diseases

Treatment or prevention of autoimmune diseases such as arthritis, diabetes, and multiple sclerosis is a promising new application of vaccine technology. Oral administration of antigens has long been recognized as a method to induce systemic immunologic tolerance and this approach is being actively pursued for vaccines against these three diseases.[217]

Type I diabetes or insulin-dependent diabetes mellitus (IDDM) is a multifactorial autoimmune disease for which susceptibility is determined by both genetic and environmental factors. The disease is caused by T-cell-mediated destruction of the insulin-producing beta cells of the pancreatic islets of Langerhans. The nonobese diabetic (NOD) mouse spontaneously develops IDDM with many of the same immunological and pathological characteristics as the human disease, making this model widely used as a research tool. When milligram quantities of insulin were delivered orally to NOD mice, it was found that there was reduced incidence of IDDM and a delayed onset of disease.[218] When insulin was conjugated to cholera toxin B (CTB), a mucosal adjuvant, oral administration of microgram quantities of the insulin-CTB conjugate protected mice against spontaneous autoimmune diabetes.[219] As the research to develop edible vaccines progresses, recombinant potatoes that express insulin-cholera toxin B fusion protein have been developed and the results of animal studies are promising.[220]

Multiple sclerosis (MS) is a chronic neurological disease that affects young adults, with onset usually occurring between 20 and 40 years of age. There are two types of disease, relapsing-remitting MS and chronic progressive MS. Although there appears to be a genetic predisposition toward development of disease, there are also environmental factors involved.[221]

MS is believed to be an autoimmune disease caused by reaction against myelin proteins. Experimentally induced allergic encephalomyelitis (EAE) is a model of MS in which mice are immunized with myelin basic protein (MBP) in complete Freund's adjuvant to induce disease. Oral administration of milligram quantities of MBP has been shown to suppress chronic relapsing EAE.[222] In a small clinical trial involving patients with relapsing-remitting MS, given daily oral doses of 300 mg bovine MBP for a year, the treated group was found to have less severe disease than a control group, offering hope that this approach may be feasible.[223] Oral administration of microgram quantities of MBP conjugated to CTB has been shown to protect animals in an EAE model.[224] Most importantly, it not only prevented disease but it ameliorated existing disease, making it a candidate therapeutic approach.

Rheumatoid arthritis (RA) is an autoimmune disease characterized by chronic inflammation of the synovial joints. It occurs in about 1% of the population, with a higher frequency in women. The autoantigen is thought to be collagen and an animal model of collagen-induced arthritis (CIA) is used to investigate disease and develop treatments. Intragastric administration of 500 mg of type II collagen has been shown to suppress the onset and severity of CIA in mice.[225] Nasal administration of microgram quantities of a type II collagen-CTB conjugate significantly reduced the prevalence and severity of CIA in mice, even when given after onset of disease.[226] In human clinical trials involving patients with severe active RA given daily oral doses of solubilized chicken type II collagen for three months, the treated group was found to have a decrease in the number of swollen and tender joints, relative to the control group.[227] Four of 30 treated patients experienced a complete remission of disease, demonstrating efficacy for this approach.

Current Technologies

In the 1970s and 1980s there was a global initiative to develop acellular and component pertussis vaccines. An early extract pertussis vaccine called Tri-Solgen was produced by Eli Lilly and was used extensively in the U.S.A. in the 1970s. It was in Japan however, that acellular pertussis vaccines were first adopted for routine immunization. These vaccines vary in specific composition, but the predominant pertussis antigens are pertussis toxin (PT) and filamentous hemagglutinin (FHA). The production of one of these vaccines is described by Sato et al.[228] Briefly, phase I *B. pertussis* is cultivated in defined Stainer Scholte medium and the PT and FHA are released into the growth medium. These secreted proteins are first concentrated by fractional ammonium sulphate precipitation. This crude extract, although high in hemagglutinating activity, still contains large quantities of bacterial endotoxin. The endotoxin has a molecular weight of about one million Daltons and can be readily separated from the PT and FHA components by sucrose density gradient centrifugation. Following separation, the fractions that contain HA activity but no endotoxin activity are pooled. Following detoxification, the pertussis vaccine is typically combined with other antigens, such as diphtheria and tetanus toxoids, and adjuvanted with alum. These vaccines have high potency in the mouse intracerebral challenge assay, show reduced local and general side-effects compared to the whole-cell vaccine and have demonstrated efficacy in humans. There are essentially two types of these acellular vaccines. The Takeda-type vaccines have a ratio of FHA:PT of about 90:10 and contain small amounts of agglutinogen 2 and the 69-kDa protein. The Biken-type vaccines contain only FHA and PT in an equal weight ratio. Each of these types of acellular pertussis vaccines has been licensed in the U.S.A. Thus, ACEL-IMUNE contains the acellular pertussis manufactured by Takeda Chemical Laboratories (Osaka, Japan) and diphtheria and tetanus toxoids manufactured by Lederle Laboratories. Tripedia is produced by Aventis Pasteur (U.S.) and combines the acellular pertussis vaccine of Biken and Tanabe Corporation (Osaka, Japan) with the diphtheria and tetanus toxoids manufactured by Aventis Pasteur.

Component pertussis vaccines are prepared by the separate isolation and purification of protective antigens and their formulation in defined ratios. Certiva is the simplest component pertussis DT vaccine containing only detoxified PT. The PT is isolated by fetuin affinity chromatography and is detoxified using hydrogen peroxide. The D and T components are produced by Statens Seruminstitut, Copenhagen, Denmark and the PT production and vaccine formulation are performed by North American Vaccine Inc. A 2-component pertussis vaccine is produced by Aventis Pasteur in Lyon, France and contains the pertussis antigens PT and FHA. When combined with DT, it is sold as Triavax. Tetravac further contains a Hib conjugate vaccine and Pentavac adds inactivated polio vaccine. Infanrix, produced by SmithKline Beecham Biologicals in Belgium, contains three pertussis antigens, PT, FHA and 69-kDa. PT and FHA are isolated from the *B. pertussis* culture by adsorption onto hydroxylapatite, followed by hydrophobic, affinity and size exclusion chromatographies. The PT is detoxified using formaldehyde and glutaraldehyde. The 69-kDa protein is isolated from the cells by heat treatment, flocculation with barium chloride, followed by ion-exchange, hydrophobic and size-exclusion chromatographies. The purified FHA and 69-kDa proteins are treated with formaldehyde. The acellular pertussis components are produced by SmithKline Beecham Biologicals and are combined with diphtheria and tetanus toxoids produced by Chiron Behring (Marburg, Germany). Acelluvax is produced by Chiron Vaccine (Siena, Italy). It contains FHA, 69-kDa, and genetically detoxified PT. Aventis Pasteur Canada produces a 5-component (PT, FHA, 69-kDa and fimbrial agglutinogens 2 and 3) vaccine. The PT and FHA are produced by sequential chromatography on perlite and hydroxylapatite.[229] The PT is detoxified with glutaraldehyde and the FHA is detoxified with formaldehyde. The 69-kDa protein is precipitated with ammonium sulphate from culture supernatants, after the PT and FHA have been removed, then purified by chromatography on hydroxylapatite and Q-Sepharose. The agglutinogens are isolated from cells by extraction with 4M urea, heat treatment of the cell lysate and precipitation with PEG-8000. Further purification is achieved by passage over PEI silica. The acellular pertussis components are combined with diphtheria and tetanus toxoids to produce Tripacel. The further addition of a Hib conjugate vaccine produces Quadracel, and when inactivated polio vaccine is also added, the product is Pentacel. Details of many of the licensed acellular pertussis and component vaccines[230] are shown in Table 11.1.

Emerging Technologies

Emerging technologies for protein-based vaccines include epitope display on self-assembling viral particles or bacterial toxins, making use of toxin or transferrin transport for delivery of epitopes, and using heat shock proteins to deliver epitopes.

The expression of epitopes on self-assembling particles has been explored as a novel way to present antigens or generate cross-protective vaccines. One of the earliest examples is a HbsAg-HSV-1 gD chimera expressed in yeast.[231] The ability of the hepatitis B core (HBc) protein to accept foreign epitopes at diverse locations and still form functional particles, has made it a very good scaffold for epitope display.[232] Poliovirus has been used to express epitopes from *C. trachomatis* MOMP,[140] *H. pylori* urease,[233] or HIV-1 gag and env.[120] Plant viruses have also been explored as carriers for epitopes. The tobacco mosaic virus (TMV) has been used to present epitopes from poliovirus,[234] the murine zona pellucida ZP3 protein,[235] malaria,[236] influenza virus and HIV env.[237] Cowpea mosaic virus (CPMV) was used to express epitopes from HIV-1 gp41,[238] mink enteritis virus,[239] *S. aureus* fibronectin-binding protein,[240] and *P. aeruginosa* OprF.[241] A chimeric alfalfa mosaic virus conferred protection against rabies.[242]

The crystal structures of many bacterial toxins have been determined, leading to the development of chimeric toxins that can express foreign epitopes. Chimeric pertussis toxins that incorporate CTL epitopes from lymphocytic choriomeningitis virus (LCMV) or *Plasmo-*

Table 11.1. Composition and efficacy of licensed component pertussis vaccines. (adapted from ref. 230)

Vaccine	Manufacturer	Composition (μg/dose) PT	FHA	69 kDa	Agg	Method of detoxification	Preservative	Absolute Efficacy (95% CI)[4]
Acelluvax	Chiron Vaccines	5[1]	2.5[1]	2.5[1]	0	Genetic	Thimerosal	99
Infanrix	SB Biologicals	25	25[2]	8[3]	0	Formaldehyde and Glutaraldehyde	Phenoxyethanol	84 (76-90) 89 (77-79)
Tripedia	Aventis Pasteur (US)	25	25	0	0	Formaldehyde	Thimerosal	93 (63-99)
Triavax Tetravac Pentavac	Aventis Pasteur (France)	25	25	0	0	Glutaraldehyde	Thimerosal	85 (66-93)
Tripacel Quadracel Pentacel	Aventis Pasteur (Canada)	20	20[2]	3	5 (Agg2 and Agg3)	Glutaraldehyde	Phenoxyethanol	85 (81-89)[5]
ACEL-IMUNE	AHP Wyeth Lederle	3.5	35	2	0.8 (Agg2)	Formaldehyde	Thimerosal	78 (60-88)
Certiva	North American Vaccine					Hydrogen Peroxide	Thimerosal	71 (63-78)
Japanese National Institute of Health	JNIH-6	23	23	0	0	Formaldehyde	Thimerosal	81 (61-90)

1 Proteins are treated with formaldehyde for stabilization.
2 FHA is treated with formaldehyde to remove any residual PT activity.
3 69 kDa protein is treated with formaldehyde.
4 21 days of spasmodic cough with laboratory confirmed infection; although differences in trial design prevent direct comparisons.
5 Duration of protection against mild disease is maximal for all vaccines shown.

dium yoelii[243] have been generated from recombinant *B. pertussis* strains. The purified chimers were found to lyse target cells in an MHC class I restricted manner. The *B. pertussis* adenylate cyclase toxin (CyaA) has a unique mechanism of entry into cells, by which the catalytic domain is directly translocated across the plasma membrane of the target cell, independent of an endocytic pathway.[244] When animals were immunized with a nontoxic CyaA carrying a CTL epitope from LCMV, they were protected from challenge.[245] Similarly, a chimeric CyaA toxin carrying a CTL epitope from OVA protected mice from challenge with tumor cells expressing OVA.[246] These studies demonstrate the utility of CyaA chimeric toxins to induce protective antiviral and antitumor immunity. The anthrax toxin protective antigen has been combined with recombinant lethal factor containing epitopes from the *Listeria monocytogenes* lysteriolysin O antigen or the nucleoprotein of LCMV. When mice were immunized with the chimeric toxin proteins, they raised CTLs to both organisms.[247] Shiga-like toxin 1 (SLT1) is a member of a family of cytotoxic proteins that are endocytosed from the cell surface to the endoplasmic reticulum, before being transported to the cytoplasm.[248] This phenomenon of retrograde transport has been utilized to deliver an epitope from the influenza virus matrix protein to the MHC class I presentation pathway.[249] A fusion between the nontoxic B subunit of Shiga toxin and an epitope from the MAGE-1 tumor antigen, also presented the epitope to the MHC class I pathway.[250] Fusion proteins based upon bacterial toxin subunits have led to cross-protective vaccines. A fusion between diphtheria toxin A and the C180 N-terminal fragment of pertussis toxin S1 subunit resulted in a nontoxic protein that was able to elicit neutralizing antibodies against both DT and PT.[251] Similarly, a fusion between a fragment of pertussis toxin S1 and the protective fragment C of tetanus toxin was protective against both pertussis and tetanus in animal models.[252] The CTB subunit has been used extensively as a fusion partner with epitopes linked at the N- or C-termini or at permissive internal sites.[253] As part of a fusion protein, the *E. coli* heat labile enterotoxin B subunit (LTB) has been used to target an epitope of HSV 1 DNA polymerase to the nucleus of an infected cell.[254]

Human serum transferrin (Tf) regulates the transport and uptake of iron into cells. The Tf receptor binds Tf, which is internalized to deliver the iron to the cell. The Tf receptor is expressed on most cells, but is overexpressed in tumor cells. Recently, recombinant Tf proteins were produced that contained an epitope cleavable by the HIV-1 protease.[255] The chimeric proteins were found to retain their Tf function and to express the epitope on the surface where it could be cleaved by the protease. It is hoped that this technology can be developed into a means to deliver therapeutic agents to tumor cells.

The discovery that heat shock proteins (HSPs) can carry noncovalently linked peptides derived from tumor antigens, has led to their investigation as cancer vaccines. The protective effect of HSP-peptide complexes derived from experimental tumors suggests the possible development of autologous tumor vaccines.[256] In animal models, immunization with a HSP70-OVA fusion protein was shown to protect against challenge with tumor cells expressing OVA.[257] This HSP fusion protein technology can be also applied to infectious diseases as demonstrated when immunization with a fusion of hsp71-HIV p24[117] or HSP65-influenza virus nucleoprotein[258] elicited specific CTLs. These results indicate the general utility of heat shock proteins as peptide carriers or fusion partners in vaccine development.

Summary

Today, protein antigens are key components of prophylactic vaccines against infectious diseases and in the future they will continue to be major constituents of therapeutic vaccines against infectious, autoimmune, and malignant diseases. While most protein antigens are currently produced from native or recombinant organisms, future vaccines may be produced from transgenic plants or animals, then combined with immunostimulatory molecules and delivered mucosally. Current combination vaccines may be supplanted in part by multipotent vaccines prepared from chimeric proteins, multi-epitope expression or epitope display. Such advances will require an improved understanding of host-pathogen interactions and the protective immune response associated with specific diseases.

References

1. Glezen WP and M Alpers. Maternal immunization. Clin Infect Dis 1999; 28:219-224.
2. Paoletti LC, Wessels MR, Rodewald AK et al. Neonatal mouse protection against infection with multiple group B streptococcal (GBS) serotypes by maternal immunization with a tetravalent GBS polysaccharide-tetanus toxoid conjugate vaccine. Infect Immun 1994; 62:3236-3243.
3. Larsson C, Stalhammar-Carlemalm M, and Lindahl G. Protection against experimental infection with group B streptococcus by immunization with a bivalent protein vaccine. Vaccine 1999;17:454-458.
4. Dale JB. Multivalent group A streptococcal vaccine designed to optimize the immunogenicity of six tandem M protein fragments. Vaccine 1999; 17:193-200.
5. Dale JB, Chiang EY, Liu S et al. New protective antigen of group A streptococci. J Clin Invest 1999; 103:1261-1268.
6. Ulrich RG, Olson MA, and Bavari S. Development of engineered vaccines effective against structurally related bacterial superantigens. Vaccine 1998; 16:1857-1864.
7. Gustafsson L, Hallander HO, Olin P et al. A controlled trial of a two-component acellular, a five-component acellular, and a whole-cell pertussis vaccine. New Engl J Med 1996; 334:349-355.
8. Loosmore SM, Zealey GR, Boux HA et al. Engineering of genetically detoxified pertussis toxin analogs for development of a recombinant whooping cough vaccine. Infect Immun 1990; 58:3653-3662.
9. Di Tommaso A, Bartalini M, Peppoloni S et al. Acellular pertussis vaccines containing genetically detoxified pertussis toxin induce long-lasting humoral and cellular responses in adults. Vaccine 1997; 15:1218-1224.
10. Barbour ML, Mayon-White RT, Coles C et al. The impact of conjugate vaccine on carriage of *Haemophilus influenzae* type b. J Infect Dis 1995; 171:93-98.
11. Drabick JJ, Brandt BL, Moran EE et al. Safety and immunogenicity testing of an intranasal group B meningococcal native outer membrane vesicle vaccine in healthy volunteers. Vaccine 2000; 18:160-172.
12. Cartwright K, Morris R, Rumke H et al. Immunogenicity and reactogenicity in UK infants of a novel meningococcal vesicle vaccine containing multiple class 1 (PorA) outer membrane proteins. Vaccine 1999; 17:2612-2619.
13. Pettersson A, Prinz T, Umar A et al. Molecular characterization of LbpB, the second lactoferrin-binding protein of *Neisseria meningitidis*. Molec Microbiol 1998; 27:599-610.
14. Ala'Aldeen DAA and Borriello SP. The meningococcal transferrin-binding proteins 1 and 2 are both surface exposed and generate bactericidal antibodies capable of killing homologous and heterologous strains. Vaccine 1996; 14:49-53.
15. Legrain M, Findeli A, Villeval D et al. Molecular characterization of hybrid Tbp2 proteins from *Neisseria meningitidis*. Molec Microbiol 1996; 19:159-169.
16. Rokbi B, Mignon M, MaitreWilmotte G et al. Evaluation of recombinant transferrin-binding protein B variants from *Neisseria meningitidis* for their ability to induce cross-reactive and bactericidal antibodies against a genetically diverse collection of serogroup B strains. Infect Immun 1997; 65:55-63.
17. Martin D, Cadieux N, Hamel J et al. Highly conserved *Neisseria meningitidis* surface protein confers protection against experimental infection. J Exp Med 1997; 185:1173-1183.
18. Kizil G, Todd I, Atta M et al. Identification and characterization of TspA, a major CD4+ T-cell- and B-cell-stimulating *Neisseria*-specific antigen. Infect Immun 1999; 67:3533-3541.
19. Clary RA, Bahadori RS, Muntz HR et al. Bacteria in the middle ear and nasopharynx during tympanostomy tube insertion. Am J Otolaryngol 1998; 19:301-304.
20. Uhari M, Hietals J, and Tuokko H. Risk of acute otitis media in relation to the viral etiology of infections in children. Clin Infect Dis 1995; 20:521-524.

21. Mbelle N, Huebner RE, Wasas AD et al. Immunogenicity and impact on nasopharyngeal carriage of a nonavalent pneumococcal conjugate vaccine. J Infect Dis 1999; 180:1171-1176.
22. Lock RA, Paton JC, and Hansman D. Comparative efficacy of pneumococcal neuraminidase and pneumolysin as immunogens protective against *Streptococcus pneumoniae.* Microb Pathog 1988; 5:461-467.
23. Alexander JE, Lock RA, Peeters CCAM, et al. Immunization of mice with pneumolysin toxoid confers a significant degree of protection against at least nine serotypes of *Streptococcus pneumoniae.* Infect Immun 1994; 62:5683-5688.
24. Talkington DF, Brown BG, Tharpe JA et al. Protection of mice against fatal pneumococcal challenge by immunization with pneumococcal surface adhesin A (PsaA). Microbial Pathog 1996; 21:17-22.
25. Briles DE, King JD, Gray MA et al. PspA, a protection-eliciting pneumococcal protein: Immunogenicity of isolated native PspA in mice. Vaccine 1996; 14:858-867.
26. Briles DE, Ades E, Paton JC et al. Intranasal immunization of mice with a mixture of the pneumococcal proteins PsaA and PspA is highly protective against nasopharyngeal carriage of *Streptococcus pneumoniae.* Infect Immun 2000; 68:796-800.
27. Hammerschmidt S, Talay SR, Brandtzaeg P et al. SpsA, a novel pneumococcal surface protein with specific binding to secretory Immunoglobulin A and secretory component. Molec Microbiol 1997; 25:1113-1124.
28. Michon F, Fusco PC, Minetti CASA et al. Multivalent pneumococcal capsular polysaccharide conjugate vaccines employing genetically detoxified pneumolysin as a carrier protein. Vaccine 1998; 16:1732-1741.
29. Gove S. Integrated management of childhood illness by outpatient health workers: technical basis and overview. Bull WHO 1997; 75 (Suppl 1): 7-24.
30. Loosmore SM, Yang Y-P, Coleman DC et al. Outer membrane protein D15 is conserved among *Haemophilus influenzae* species and may represent a universal protective antigen against invasive disease. Infect Immun 1997; 65:3701-3707.
31. Sirakova T, Kolattukudy PE, Murwin D et al. Role of fimbriae expressed by nontypeable *Haemophilus influenzae* in pathogenesis of and protection against otitis media and relatedness of the fimbrin subunit to outer membrane protein A. Infect Immun 1994; 62:2002-2020.
32. St. Geme JW III, de la Morena ML, and Falkow S. A *Haemophilus influenzae* IgA protease-like protein promotes intimate interaction with human epithelial cells. Mol Microbiol 1994; 14:217-233.
33. Barenkamp SJ and St. Geme JW III Identification of a second family of high-molecular-weight adhesion proteins expressed by nontypable *Haemophilus influenzae.* Molec Microbiol 1996; 19:1215-1223
34. St. Geme JW III and Cutter D. Evidence that surface fibrils expressed by *Haemophilus influenzae* type b promote attachment to human epithelial cells. Mol Microbiol 1995; 15:77-85.
35. Barenkamp SJ. Immunization with High-Molecular-Weight adhesion proteins of nontypeable *Haemophilus influenzae* modifies experimental otitis media in chinchillas. Infect Immun 1996; 64:1246-1251
36. Loosmore SM, Yang Y-P, Oomen R et al. The *Haemophilus influenzae* HtrA protein is a protective antigen. Infect Immun 1998; 66:899-906.
37. Prasadarao NV, Lysenko E, Wass CA et al. Opacity-associated protein A contributes to the binding of *Haemophilus influenzae* to Chang epithelial cells. Infect Immun 1999; 67:4153-4160.
38. El-Adhami W, Kyd JM, Bastin DA et al. Characterization of the gene encoding a 26-kilodalton protein (OMP26) from nontypeable *Haemophilus influenzae* and immune responses to the recombinant protein. Infect Immun 1999; 67:1935-1942.
39. Murphy TF, Campagnari AA, Nelson B et al. Somatic antigens of *Haemophilus influenzae* as vaccine components. Pediatr Infect Dis J 1989; 8:S66-S68.
40. Green BA, Farley JE, Quinn-Dey T et al. The *e* (P4) outer membrane protein of *Haemophilus influenzae*: biologic activity of anti-*e* serum and cloning and sequencing of the structural gene. Infect Immun 1991; 59:3191-3198.
41. DeMaria TF, Murwin DM, and Leake ER. Immunization with outer membrane protein P6 from nontypeable *Haemophilus influenzae* induces bactericidal antibody and affords protection in the chinchilla model of otitis media. Infect Immun 1996; 64:5187-5192.

42. Deich RA, Anilionis A, Fulginiti J et al. Antigenic conservation of the 15,000-Dalton outer membrane lipoprotein PCP of *Haemophilus influenzae* and biologic activity of anti-PCP antisera. Infect Immun 1990; 58:3388-3393.
43. Akkoyunlu M and Forsgren A. Local and systemic antibody levels against protein D of *Haemophilus influenzae* following immunization and infection in rats. APMIS 1996; 104:709-717.
44. Loosmore SM, Yang Y-P, Coleman DC et al. Cloning and expression of the *Haemophilus influenzae* transferrin receptor genes. Molec Microbiol 1996; 19:575-586.
45. Green BA, Vazquez ME, Zlotnick GW et al. Evaluation of mixtures of purified *Haemophilus influenzae* outer membrane proteins in protection against challenge with nontypeable *H. influenzae* in the chinchilla otitis media model. Infect Immun 1993; 61:1950-1957.
46. Helminen ME, Maciver I, Latimer JL et al. A major outer membrane protein of *Moraxella catarrhalis* is a target for antibodies that enhance pulmonary clearance of the pathogen in an animal model. Infect Immun 1993; 61:2003-2010.
47. Murphy TF, Kyd JM, John A et al. Enhancement of pulmonary clearance of *Moraxella (Branhamella) catarrhalis* following immunization with outer membrane protein CD in a mouse model. J Infect Dis 1998; 178:1667-1675.
48. Murphy TF and Bartos LC. Surface-exposed and antigenically conserved determinants of outer membrane proteins of *Branhamella catarrhalis*. Infect Immun 1989; 57:2938-2941.
49. Bhushan R, Kirkham C, Sethi S et al. Antigenic characterization and analysis of the human immune response to outer membrane protein E of *Branhamella catarrhalis*. Infect Immun 1997; 65:2668-2675.
50. Du R-P, Wang Q, YangY-P et al. Cloning and expression of the *Moraxella catarrhalis* lactoferrin receptor genes. Infect Immun 1998; 66:3656-3665.
51. Myers LE, Yang Y-P, Du R-P et al. The transferrin binding protein B of *Moraxella catarrhalis* elicits bactericidal antibodies and is a potential vaccine antigen. Infect Immun 1998; 66:4183-4192.
52. Chen D, Barniak V, VanDerMeid KR et al. The levels and bactericidal capacity of antibodies directed against the UspA1 and UspA2 outer membrane proteins of *Moraxella (Branhamella) catarrhalis* in adults and children. Infect Immun 1999; 67:1310-1316.
53. Fitzgerald M, Mulcahy R, Murphy S et al. A 200 kDa protein is associated with haemagglutinating isolates of *Moraxella (Branhamella) catarrhalis*. FEMS Immunol Med Microbiol 1997; 18:209-216.
54. Kapikian AZ, Mitchell RH, Chanock RM et al. An epidemiologic study of altered clinical reactivity to respiratory syncytial (RS) virus infection in children previously vaccinated with an inactivated RS virus vaccine. Am J Epidemiol 1969; 89:405-421.
55. Juhasz K, Whitehead SS, Boulanger CA et al. The two amino acid substitutions in the L protein of *cpts*530/1009, a live-attenuated respiratory syncytial virus candidate vaccine, are independent temperaturesensitive and attenuation mutations. Vaccine 1999; 17:1416-1424.
56. Wyatt LS, Whitehead SS, Venanzi KA et al. Priming and boosting immunity to respiratory syncytial virus by recombinant replication-defective vaccinia virus MVA. Vaccine 2000; 18:392-397.
57. Dudas RA and Karron RA. Respiratory Syncytial Virus vaccines. Clin Microbiol Rev 1998; 11:430-439.
58. Groothuis JR, King SJ, Hogerman DA et al. Safety and immunogenicity of a purified F protein Respiratory Syncytial Virus (PFP-2) vaccine in seropositive children with bronchopulmonary dysplasia. J Infect Dis 1998; 177:467-9.
59. Neuzil KM, Johnson JE, Tang Y-W et al. Adjuvants influence the quantitative and qualitative immune response in BALB/c mice immunized with respiratory syncytial virus FG subunit vaccine. Vaccine 1997; 15:525-532.
60. Corvaia N, Tournier P, Nguyen TN et al. Challenge of BALB/c mice with Respiratory Syncytial Virus does not enhance Th 2 pathway induced after immunization with a recombinant G fusion protein, BBG2NA, in aluminum hydroxide. J Infect Dis 1997; 175:560-569.
61. Homa FL, Brideau RJ, Lehman DJ et al. Development of a novel subunit vaccine that protects cotton rats against both human respiratory syncytial virus and human parainfluenza virus type 3. J Gen Virol 1993; 74:1995-1999.
62. Kapikian AZ, Hoshino Y, Chanock RM et al. Efficacy of a quadrivalent rhesus rotavirus-based human rotavirus vaccine aimed at preventing severe rotavirus diarrhea in infants and young children. J Infect Dis 1996; 174 (suppl1): S65-S72.

63. Madore HP, Estes MK, Zarley CD et al. Biochemical and immunologic comparison of virus-like particles for a rotavirus subunit vaccine. Vaccine 1999; 17:2461-2471.
64. Choi AH-C, Basu M, McNeal MM et al. Antibody-independent protection against rotavirus infection of mice stimulated by intranasal immunization with chimeric VP4 or VP6 protein. J Virol 1999; 73:7574-7581.
65. Pass RF, Duliege A-M, Boppana S et al. A subunit cytomegalovirus vaccine based upon recombinant envelope glycoprotein B and a new adjuvant. J Infect Dis 1999; 180:970-975.
66. Nolan CM. Nosocomial multidrug-resistant tuberculosis—global spread of the third epidemic. J Infect Dis 1997; 176:748-751.
67. Kamath AT, Feng CG, MacDonald M et al. Differential protective efficacy of DNA vaccines expressing secreted proteins of *Mycobacterium tuberculosis*. Infect Immun 1999; 67:1702-1707.
68. Slobbe L, Lockhart E, O'Donnell MA et al. An in vivo comparison of Bacillus Calmette-Guerin (BCG) and cytokine-secreting BCG vaccines. Immunology 1999; 96:517-523.
69. Hess J and Kaufmann SHE. Live antigen carriers as tools for improved anti-tuberculosis vaccines. FEMS Immunol Med Microbiol 1999; 23:165-173.
70. Elhay MJ and Andersen P. Immunological requirements for a subunit vaccine against tuberculosis. Immunol Cell Biol 1997; 75:595-603.
71. Holmgren J, Svennerholm A-M, Lonnroth I et al. Development of improved cholera vaccine based on subunit toxoid. Nature 1977; 269:602-604.
72. Sanchez J and Holmgren J. Recombinant system for overexpression of cholera toxin B subunit in *Vibrio cholerae* as a basis for vaccine development. Proc Natl Acad Sci USA 1989; 86:481-485.
73. Glenn GM, Scharton-Kersten T, Vassell R et al. Transcutaneous immunization with cholera toxin protects mice against lethal mucosal toxin challenge. J Immunol 1998; 161:3211-3214.
74. Lebens M, Shahabi V, Backstrom M et al. Synthesis of hybrid molecules between heat-labile enterotoxin and cholera toxin B subunits: Potential for use in a broad-spectrum vaccine. Infect Immun 1996; 64:2144-2150.
75. Lindberg AA. Vaccination against enteric pathogens: From science to vaccine trials. Curr Opin Microbiol 1998; 1:116-124.
76. Coster TS, Hoge CW, VanDeVerg LL et al. Vaccination against shigellosis with attenuated *Shigella flexneri* 2a strain SC602. Infect Immun 1999; 67:3437-3443.
77. Cohen D, Ashkenazi S, Green MS et al. Double-blind vaccine-controlled randomised efficacy trial of an investigational *Shigella sonnei* conjugate vaccine in young adults. Lancet 1997; 349:155-159.
78. Boyd B, Richardson S, and Gariepy J. Serological responses to the B subunit of Shiga-like toxin 1 and its peptide fragments indicate that the B subunit is a vaccine candidate to counter the action of the toxin. Infect Immun 1991; 59:750-757.
79. Konadu E, Donohue-Rolfe A, Calderwood SB et al. Syntheses and immunologic properties of *Escherichia coli* O157 O-specific polysaccharide and shiga toxin 1 B subunit conjugates in mice. Infect Immun 1999; 67:6191-6193.
80. Etlinger HM, Caspers P, Matile H et al. Ability of recombinant or native proteins to protect mokeys against heterologous challenge with *Plasmodium falciparum*. Infect Immun 1991; 59:3498-3503.
81. Barr PJ, Green KM, Gibson HL et al. Recombinant Pfs25 protein of *Plasmodium falciparum* elicits malaria transmission-blocking immunity in experimental animals. J Exp Med 1991; 174:1203-1208.
82. Inselburg J, Bathurst IC, Kansopon J et al. Protective immunity induced in Aotus monkeys by a recombinant SERA protein of *Plasmodium falciparum*: further studies using SERA 1 and MF75.2 adjuvant. Infect Immun 1993; 61:2048-2052.
83. Kumar S, Yadava A, Keister DB et al. Immunogenicity and in vivo efficacy of recombinant *Plasmodium falciparum* merozoite surface protein-1 in Aotus monkeys. Mol Med 1995; 3:325-332.
84. Nosten F, Luxemburger C, Kyle DE et al. Randomised double-blind placebo-controlled trial of Spf66 malaria vaccine in children in northwestern Thailand. Lancet 1996; 348:701-707.
85. Shi YP, Hasnain SE, Sacci JB et al. Immunogenicity and in vitro protective efficacy of a recombinant multistage *Plasmodium falciparum* candidate vaccine. Proc Natl Acad Sci USA 1999; 96:1615-1620.

86. Ockenhouse CF, Sun P-F, Lanar DE et al. Phase I/IIa safety, immunogenicity, and efficacy trial of NYVAC-Pf7, a pox-vectored, multiantigen, multistage vaccine candidate for *Plasmodium falciparum* malaria. J Infect Dis 1998; 177:1664-1673.
87. Sedegah M, Hedstrom R, Hobart P et al. Protection against malaria by immunization with plasmid DNA encoding circumsporozoite protein. Proc Natl Acad Sci USA 1994; 91:9866-9870.
88. Schodel F, Wirtz R, Peterson D et al. Immunity to malaria elicited by hybrid hepatitis B virus core particles carrying circumsporozoite protein epitopes. J Exp Med 1994; 180:1037-1046.
89. Stoute JA, Kester KE, Krzych U et al. Long-term efficacy and immune responses following immunization with the RTS,S malaria vaccine. J Infect Dis 1998; 178:1139-1144.
90. Boulanger D, Reid GDF, Sturrock RF et al. Immunization of mice and baboons with the recombinant Sm28GST affects both worm viability and fecundity after experimental infection with *Schistosoma mansoni*. Parasite Immunol 1991; 13:473-490.
91. Pearce EJ James, SL, Hieny S et al. Induction of protective immunity against *Schistosoma mansoni* by vaccination with schistosome paramyosin (Sm97), a nonsurface parasite antigen. Proc Natl Acad Sci USA 1988; 85:5678- 5682.
92. Attallah AM, Attia H, Ismail H et al. Vaccination against *Schistosoma mansoni* infection using 74 kDa *Schistosoma* protein antigen. Vaccine 1999; 17:2786-2791.
93. Sun JB, Mielcarek N, Lakew M et al. Intranasal administration of a *Schistosoma mansoni* glutathione S-transferase-cholera toxoid conjugate vaccine evokes antiparasitic and antipathological immunity in mice. J Immunol 1999; 163:1045-1052.
94. Liew FY, Singleton A, Cillari E et al. Prophylactic immunization against experimental leishmaniasis. V. Mechanism of the anti-protective blocking effect induced by subcutaneous immunization against *Leishmania major* infection. J Immunol 1985; 135:2102-2107.
95. Russell DG and Alexander J. Effective immunization against cutaneous leishmaniasis with defined membrane antigens reconstituted into liposomes. J Immunol 1988; 140:1274-1279.
96. Champsi J and McMahon-Pratt D. Membrane glycoprotein M-2 protects against *Leishmania amazonensis* infection. Infect Immun 1988; 52:3272-3279.
97. Mougneau E, Altare F, Wakil AE et al. Expression cloning of a protective *Leishmania* antigen. Science 1995; 268:563-566.
98. Webb JR, Campos-Neto A, Ovendale PJ et al. Human and murine immune responses to a novel *Leishmania major* recombinant protein encoded by members of a multicopy gene family. Infect Immun 1998; 66:3279-3289.
99. Handman E, Symons FM, Baldwin TM et al. Protective vaccination with promastigote surface antigen 2 from *Leishmania major* is mediated by a Th1 type of immune response. Infect Immun 1995; 63:4261-4267.
100. Skeiky YAW, Kennedy M, Kaufman D et al. LeIF: A recombinant *Leishmania* protein that induces an IL-12-mediated Th1 cytokine profile. J Immunol 1998; 161:6171-6179.
101. Soong L, Duboise SM, Kima P et al. *Leishmania pifanoi* amastigote antigens protect mice against cutaneous leishmaniasis. Infect Immun 1995; 63:3559-3566.
102. Velzing J, Groen J, Drouet MT et al. Induction of protective immunity against Dengue virus type 2: Comparison of candidate live attenuated and recombinant vaccines. Vaccine 1999; 17:1312-1320.
103. Simmons M, Nelson WM, Wu SJL et al. Evaluation of the protective efficacy of a recombinant Dengue envelope B domain fusion protein against Dengue 2 virus infection in mice. Am J Trop Med Hyg 1998; 58:655-662.
104. Bielefeldt-Ohmann H, Beasley DWC, Fitzpatrick DR et al. Analysis of a recombinant dengue-2 virus-dengue-3 virus hybrid envelope protein expressed in a secretory baculovirus system. J Gen Virol 1997; 78:2723-2733.
105. Palella FJ, Delaney KM, Moorman AC et al. Declining morbidity and mortality among patients with advanced human immunodeficiency virus infection. N Engl J Med 1998; 338:853-860.
106. Churdboonchart V, Moss RB, Sirawaraporn W et al. Effect of HIV-specific immune-based therapy in subjects infected with HIV-1 subtype E in Thailand. AIDS 1998; 12:1521-1527.
107. Johnson RP, Lifson JD, Czajak SC et al. Highly attenuated vaccine strains of Simian Immunodeficiency Virus protect against vaginal challenge: Inverse relationship of degree of protection with level of attenuation. J Virol 1999; 73:4952-4961.

108. Deacon NJ, Tsykin A, Solomon A et al. Genomic structure of an attenuated quasi species of HIV-1 from a blood transfusion donor and recipients. Science 1995; 270:988-991.
109. Goebel F-D, Mannhalter JW, Belshe RB et al. Recombinant gp160 as a therapeutic vaccine for HIV-infection: results of a large randomized, controlled trial. AIDS 1999; 13:1461-1468.
110. Mascola JR, Snyder SW, Weislow OS et al. Immunization with envelope subunit vaccine products elicits neutralizing antibodies against laboratory-adapted but not primary isolates of human immunodeficiency virus type 1. J Infect Dis 1996; 173:340-348.
111. Mooij P, van der Kolk M, Bogers WMJM et al. A clinically relevant HIV-1 subunit vaccine protects rhesus macaques from in vivo passaged simian-human immunodeficiency virus infection. AIDS 1998; 12:F15-F22.
112. Kwong PD, Wyatt R, Robinson J et al. Structure of an HIV gp120 envelope glycoprotein in complex with the CD4 receptor and a neutralizing human antibody. Nature 1998; 393:648-659.
113. Liang X, Munshi S, Shendure J et al. Epitope insertion into variable loops of HIV-1 gp120 as a potential means to improve immunogenicity of viral envelope protein. Vaccine 1999; 17: 2862-2872.
114. Rovinski B, Rodrigues L, Cao SX et al. Induction of HIV type 1 neutralizing and env-CD4 blocking antibodies by immunization with genetically engineered HIV type 1-like particles containing unprocessed gp160 glycoproteins. AIDS Res Hum Retroviruses 1995; 11:1187-1195.
115. Kelleher AD, Roggensack M, Jaramillo AB et al. Safety and immunogenicity of a candidate therapeutic vaccine, p24 virus-like particle, combined with zidovudine, in asymptomatic subjects. AIDS 1998; 12:175-182.
116. Gringeri A, Santagostino E, Muca-Perja M et al. Tat toxoid as a component of a preventive vaccine in seronegative subjects. J Acquir Immune Defic Syndr Hum Retrovirol 1999; 20:371-375.
117. Suzue K and Young RA. Adjuvant-free hsp70 fusion protein system elicits humoral and cellular immune responses to HIV-1 p24. J Immunol 1996; 156:873-879.
118. Evans TG, Keefer MC, Weinhold KJ et al. A canarypox vaccine expressing multiple human immunodeficiency virus type 1 genes given alone or with rgp120 elicits broad and durable CD8+ cytotoxic T lymphocyte responses in seronegative volunteers. J Infect Dis 1999; 180:290-298.
119. Natuk RJ, Lubeck MD, Chanda PK et al. Immunogenicity of recombinant human adenovirus-human immunodeficiency virus vaccines in chimpanzees. AIDS Res Hum Retrovirus 1993; 9:395-404.
120. Moldoveanu Z, Porter DC, Lu A et al. Immune responses induced by administration of encapsidated poliovirus replicons which express HIV-1 gag and envelope proteins. Vaccine 1995; 13:1013-1022.
121. Osterhaus ADME, van Baalen CA, Gruters RA et al. Vaccination with Rev and Tat against AIDS. Vaccine 1999; 17:2713-2714.
122. Yasutomi Y, Koenig S, Haun SS et al. Immunization with recombinant BCG-SIV elicits SIV-specific cytotoxic T lymphocytes in rhesus monkeys. J Immunol 1993; 150:3101-3107.
123. Mata M and Paterson Y. Th1 T cell responses to HIV-1 gag protein delivered by a *Listeria monocytogenes* vaccine are similar to those induced by endogenous Listerial antigens. J Immunol 1999; 163:1449-1456.
124. MacGregor RR, Boyer JD, Ugen KE et al. First human trial of a DNA-based vaccine for treatment of human immunodeficiency virus type 1 infection: safety and host response. J Infect Dis 1998; 178:92-100.
125. Bartlett JA, Wasserman SS, Hicks CB et al. Safety and immunogenicity of an HLA-based HIV envelope polyvalent synthetic peptide immunogen. AIDS 1998; 12:1291-1300.
126. Gomez CE, Navea L, Lobaina L et al. The V3 loop based multi-epitope polypeptide TAB9 adjuvated with montanide ISA720 is highly immunogenic in nonhuman primates and induces neutralizing antibodies against five HIV-1 isolates. Vaccine 1999; 17:2311-2319.
127. Letvin NL, Montefiori DC, Yasutomi Y et al. Potent, protective anti-HIV immune responses generated by bimodal HIV envelope DNA plus protein vaccination. Proc Natl Acad Sci USA 1997; 94:9378-9383.
128. Robinson HL, Montefiori DC, Johnson RP et al. Neutralizing antibody-independent containment of immunodeficiency virus challenges by DNA priming and recombinant pox virus booster immunizations. Nature Med 1999; 5:526-534.
129. Caver TE, Lockey TD, Srinivas RV et al. A novel vaccine regimen utilizing DNA, vaccinia virus and protein immunizations for HIV-1 envelope presentation. Vaccine 1999; 17:1567-1572.

130. Clements-Mann ML, Weinhold K, Matthews TJ et al. Immune responses to human immunodeficiency virus (HIV) type 1 induced by canarypox expressing HIV-1_{MN} gp120, HIV-1 $_{SF2}$ recombinant gp120, or both vaccines in seronegative adults. J Infect Dis 1998; 177:1230-1246.
131. Salmon-Ceron D, Excler J-L, Finkielsztejn L et al. Safety and immunogenicity of a live recombinant canarypox virus expressing HIV type 1 gp120 MN tm/gag/protease LAI (ALVAC-HIV, vCP205) followed by a p24E-V3 MN synthetic peptide (CLTB-36) administered in healthy volunteers at low risk for HIV infection. AGIS Group and L'Agence Nationale de Recherches sur Le Sida. AIDS Res Hum Retroviruses 1999; 15:633-645.
132. Robert-Guroff M, Kaur H, Patterson LJ et al. Vaccine protection against a heterologous, nonsyncytium-inducing, primary human immunodeficiency virus. J Virol 1998; 72:10275-10280.
133. Franchini G, Robert-Guroff M, Tartaglia J et al. Highly attenuated HIV type 2 recombinant poxviruses, but not HIV-2 recombinant *Salmonella* vaccines, induce long-lasting protection in rhesus macaques. AIDS Res Hum Retroviruses 1995; 11:909-920.
134. Agacfidan A and Kohl P. Sexually transmitted diseases (STDs) in the world. FEMS Immunol Med Microbiol 1999; 24:431-435.
135. Valenzuela P, Medina A, Rutter WJ et al. Synthesis and assembly of hepatitis B virus surface antigen particles in yeast. Nature 1982; 298:347-350.
136. Assad S and Francis A. Over a decade of experience with a yeast recombinant hepatitis B vaccine. Vaccine 2000; 18:57-67.
137. Pal S, Theodor I, Peterson EM et al. Immunization with an acellular vaccine consisting of the outer membrane complex of *Chlamydia trachomatis* induces protection against a genital challenge. Infect Immun 1997; 65:3361-3369.
138. Stagg AJ. Vaccines against *Chlamydia*: approaches and progress. Molec Med Today 1988; April pp166-173.
139. Stephens RS, Kalman S, Lammel C et al. Genome sequence of an obligate intracellular pathogen of humans: *Chlamydia trachomatis*. Science 1998; 282:754-759.
140. Murdin AD, Su H, Klein MH et al. Poliovirus hybrids expressing neutralizing epitopes from variable domains I and IV of the major outer membrane protein of *Chlamydia trachomatis* elicit broadly cross-reactive *C. trachomatis*-neutralizing antibodies. Infect Immun 1995; 63:1116-1121.
141. Plummer FA, Simonsen JN, Chubb H et al. Epidemiologic evidence for the development of serovar-specific immunity after gonoccoal infection. J Clin Invest 1989; 83:1472-1476.
142. Jerse AE, Cohen MS, Drown PM et al. Multiple gonococcal opacity proteins are expressed during experimental urethral infection in the male. J Exp Med 1994; 179: 911-920.
143. Cornelissen CN, Kelley M, Hobbs MM et al. The transferrin receptor expressed by gonococcal strain FA1090 is required for the experimental infection of human male volunteers. Molec Microbiol 1998; 27:611-616.
144. Biswas GD and Sparling PF. Characterization of *lbpA*, the structural gene for a lactoferrin receptor in *Neisseria gonorrhoeae*. Infect Immun 1995; 63:2958-2967.
145. Biswas GD, Anderson JE, Chen C-J et al. Identification and functional characterization of the *Neisseria gonorrhoeae lbpB* gene product. Infect Immun 1999; 67:455-459.
146. Plante M, Cadieux N, Rioux CR et al. Antigenic and molecular conservation of the gonococcal NspA protein. Infect Immun 1999; 67:2855-2861.
147. Lin L, Ayala P, Larson J et al. The *Neisseria* type 2 IgA1 protease cleaves LAMP1 and promotes survival of bacteria within epithelial cells. Molec Microbiol 1997; 24:1083-1094.
148. Arroll TW, Centurion-Lara A, Lukehart SA et al. T-cell responses to *Treponema pallidum* subsp. *pallidum* antigens during the course of experimental syphilis infection. Infect Immun 1999; 67:4757-4763.
149. Fraser CM, Norris SJ, Weinstock Gmet al. Complete genome sequence of *Treponema pallidum*, the syphilis spirochete. Science 1998; 281:375-388.
150. Stanberry LR, Bernstein DI, Burke RL et al. Vaccination with recombinant herpes simplex virus glycoproteins: protection against initial and recurrent genital herpes. J Infect Dis 1987; 155:914-920.
151. Berman PW, Gregory T, Crase D et al. Protection from genital herpes simplex virus type 2 infection by vaccination with cloned type 1 glycoprotein D. Science 1985; 227:1490-1492.

152. O'Hagan D, Goldbeck C, Ugozzoli M et al. Intranasal immunization with recombinant gD2 reduces disease severity and mortality following genital challenge with herpes simplex virus type 2 in guinea pigs. Vaccine 1999; 17:2229-2236.
153. Corey L, Langenberg AGM, Ashley R et al. Recombinant glycoprotein vaccine for the prevention of genital HSV-2 infection. JAMA 1999; 282:331-340.
154. Talwar GP. Fertility regulating and immunotherapeutic vaccines reaching human trials stage. Hum Reprod Update 1997; 3:301-310.
155. Singh M, Das SK, Suri Set al. Regain of fertility and normality of progeny born during below protective threshold antibody titers in women immunized with the HSD-hCG vaccine. Am J Reprod Immunol 1998; 39:395-398.
156. Mukhopadhyay A, Bhatia PK, and Majumdar SS. Preliminary studies with recombinant chorionic gonadotrophic b-subunit produced in *Escherichia coli* for use as an antigen in a birth control vaccine. Am J Reprod Immunol 1998; 39:172-182.
157. Moudgal NR, Jeyakumar M, Krishnamurthy HN et al. Development of male contraceptive vaccine- a perspective. Hum Reprod Update 1997; 3:335-346.
158. Doring G and Dorner F. A multicenter vaccine trial using the *Pseudomonas aeruginosa* flagella vaccine IMMUNO in patients with cystic fibrosis. Behring Inst Mitt 1997; 98:338-344.
159. Jang I-J, Kim I-S Park WJ et al. Human immune response to a *Pseudomonas aeruginosa* outer membrane protein vaccine. Vaccine 1999; 17:158-168.
160. Von Specht B-U, Lucking HC, Blum B et al. Safety and immunogenicity of a *Pseudomonas aeruginosa* outer membrane protein I vaccine in human volunteers. Vaccine 1996; 14:1111-1117.
161. Matthews-Greer JM and Gilleland HE Jr. Outer membrane protein F (porin) preparation of *Pseudomonas aeruginosa* as a protective vaccine against heterologous immunotype strains in a burned mouse model. J Infect Dis 1987; 155:1282-1291.
162. Sawa T, Yahr TL, Ohara M et al. Active and passive immunization with the *Pseudomonas* V antigen protects against type III intoxication and lung injury. Nature Med 1999; 5:392-398.
163. Gabelsberger J, Knapp B, Bauersachs S et al. A hybrid outer membrane protein antigen for vaccination against *Pseudomonas aeruginosa*. Behr Inst Mitt 1997; 98:302-314.
164. Knapp B, Hundt E, Lenz U et al. A recombinant hybrid outer membrane protein for vaccination against *Pseudomonas aeruginosa*. Vaccine 1999; 17:1663-1666.
165. Lee JC. The prospects for developing a vaccine against *Staphylococcus aureus*. Trends Microbiol 1996; 4:162-166.
166. Fattom AI, Sarwar J, Ortiz A et al. A *Staphylococcus aureus* capsular polysaccharide (CP) vaccine and CP-specific antibodies protect mice against bacterial challenge. Infect Immun 1996; 64:1659-1665.
167. Herbelin C and Poutrel B. Immunogenicity in mice of multivalent conjugates composed of *Staphylococcus aureus* type 5 and 8 capsular polysaccharides and a- and b-toxins. Vaccine Res 1996; 5:15-28.
168. Foster TJ and Hook M. Surface protein adhesins of *Staphylococcus aureus*. Trends Microbiol 1998; 6:484-488.
169. Nilsson I-M, Patti JM, Bremell T et al. Vaccination with a recombinant fragment of collagen adhesin provides protection against *Staphylococcus aureus*-mediated septic death. J Clin Invest 1998; 101:2640-2649
170. Nilsson I-M, Verdrengh M, Ulrich RG et al. Protection against *Staphylococcus aureus* sepsis by vaccination with recombinant staphylococcal enterotoxin A devoid of superantigenicity. J Infect Dis 1999; 180:1370-1373.
171. Balaban N, Goldkorn T, Nhan RT et al. Autoinducer of virulence as a target for vaccine and therapy against *Staphylococcus aureus*. Science 1998; 280:438-440.
172. Blumberg BS. Hepatitis B virus, the vaccine, and the control of primary cancer of the liver. Proc Natl Acad Sci USA 1997; 94:7121-7125.
173. Rose RC, Lane C, Wilson S et al. Oral vaccination of mice with human papillomavirus virus-like particles induces systemic virus-neutralizing antibodies. Vaccine 1999; 17:2129-2135.
174. Munger K, Phelps WC, Bubb V et al. The E6 and E7 genes of the human papillomavirus type 16 together are necessary and sufficient for transformation of primary human keratinocytes. J Virol 1989; 63:4417-4421.

175. Greenstone HL, Nieland JD, de Visser KE et al. Chimeric papillomavirus virus-like particles elicit antitumor immunity against the E7 oncoprotein in an HPV16 tumor model. Proc Natl Acad Sci USA 1998; 95:1800-1805.
176. De Bruijn MLH, Schuurhuis DH, Vierboom MPM et al. Immunization with human papillomavirus type 16 (HPV16) oncoprotein-loaded dendritic cells as well as protein in adjuvant induces MHC class I-restricted protection to HPV16-induced tumor cells. Cancer Res 1998; 58:724-731.
177. Uchiyama T. Human T cell leukemia virus type 1 (HTLV-1) and human diseases. Annu Rev Immunol 1997; 15:15-37.
178. Ford CM, Arp J, Palker TJ et al. Characterization of the antibody response to three different versions of the HTLV-1 envelope protein expressed by recombinant vaccinia viruses: induction of neutralizing antibody. Virol 1992; 191:448-453.
179. Arp J, Ford CM, Palker TJ et al. Expression and immunogenicity of the entire human T cell leukaemia virus type I envelope protein produced in a baculovirus system. J Gen Virol 1993; 74:211-222.
180. Baba E, Nakamura M, Ohkuma K et al. A peptide-based human T cell leukemia virus type I vaccine containing T and B cell epitopes that induces high titers of neutralizing antibodies. J Immunol 1995; 154:399-412.
181. Pique C, Connan F, Levilain J-P et al. Among all human T-cell leukemia virus type 1 proteins, tax, polymerase, and envelope proteins are predicted as preferential targets for the HLA-A2-restricted cytotoxic T-cell response. J Virol 1996; 70:4919-4926.
182. Pappo J, Thomas WD Jr, Kabok Z et al. Effect of oral immunization with recombinant urease on murine *Helicobacter felis* gastritis. Infect Immun 1995; 63:1246-1252.
183. Radcliff FJ, Hazell SL, Kolesnikow T et al. Catalase, a novel antigen for *Helicobacter pylori* vaccination. Infect Immun 1997; 65:4668-4674.
184. Ferrero RL, Thiberge J-M, Kansau I et al. The GroES homolog of *Helicobacter pylori* confers protective immunity against mucosal infection in mice. Proc Natl Acad Sci USA 1995; 92:6499-6503.
185. Ghiara P, Rossi M, Marchetti M et al. Therapeutic intragastric vaccination against *Helicobacter pylori* in mice eradicates an otherwise chronic infection and confers protection against reinfection. Infect Immun 1997; 65:4996-5002.
186. Kimmel B, Bosserhoff A, Frank R et al. Identification of immunodominant antigens from *Helicobacter pylori* and evaluation of their reactivities with sera from patients with different gastroduodenal pathologies. Infect Immun 2000; 68:915-920.
187. Pardoll DM. Cancer vaccines. Nature Med 1998; 4:525-531.
188. Samanci A, Yi Q, Fagerberg J et al. Pharmacological administration of granulocyte/macrophage-colony-stimulating factor is of significant importance for the induction of a strong humoral and cellular response in patients immunized with recombinant carcinoembryonic antigen. Cancer Immunol Immunother 1998; 47:131-142.
189. Marshall JL, Hawkins MJ, Tsang KY et al. Phase I study in cancer patients of a replication-defective avipox recombinant vaccine that expresses human carcinoembryonic antigen. J Clin Oncol 1999; 17:332-337.
190. Foon KA, John WJ, Chakraborty M et al. Clinical and immune responses in resected colon cancer patients treated with anti-idiotype monoclonal antibody vaccine that mimics the carcinoembryonic antigen. J Clin Oncol 1999; 17:2889-2895.
200. Herlyn D, Somasundaram R, Zaloudik J et al. Anti-idiotype and recombinant antigen in immunotherapy of colorectal cancer. Cell Biophys 1994; 25:143-153.
201. Ragnhammar P, Fagerberg J, Frodin JE et al. Effect of monoclonal antibody 17-1A and GM-CSF in patients with advanced colorectal carcinoma – long-lasting, complete remissions can be induced. Int J Cancer 1993; 53:751-758
202. Rosendahl A, Kristensson K, Carlsson M et al. Long-term survival and complete cures of B16 melanoma-carrying animals after therapy with tumor-targeted IL-2 and SEA. Int J Cancer 1999; 81:156-163.
203. Hsueh EC, Nathanson L, Foshag LJ et al. Active specific immunotherapy with polyvalent melanoma cell vaccine for patients with in-transit melanoma metastases. Cancer 1999; 85:2160-2169.

204. Kawakami Y, Robbins PF, Wang RF et al. The use of melanosomal proteins in the immunotherapy of melanoma. J Immunotherapy 1999; 21:237-246.
205. Wang B, Chen Y-B, Ayalon O et al. Human single-chain Fv immunoconjugates targeted to a melanoma-associated chondroitin sulfate proteoglycan mediate specific lysis of human melanoma cells by natural killer cells and complement. Proc Natl Acad Sci USA 1999; 96:1627-1632.
206. Disis ML, Shiota FM, and Cheever MA. Human HER-2/neu protein immunization circumvents tolerance to rat neu: A vaccine strategy for "self" tumour antigens. Immunol 1998; 93:192-199.
207. Esserman LJ, Lopez T, Montes R et al. Vaccination with the extracellular domain of $p185^{neu}$ prevents mammary tumor development in neu transgenic mice. Cancer Immunol Immunother 1999; 47:337-342.
208. Disis ML, Grabstein KH, Sleath PR et al. Generation of immunity to the HER-2/neu oncogenic protein in patients with breast and ovarian cancer using a peptide-based vaccine. Clin Cancer Res 1999; 5:1289-1297.
209. Katsumata M, Okudaira T, Samanta A et al. Prevention of breast tumor development in vivo by downregulation of the $p185^{neu}$ receptor. Nature Med 1995; 1:644-648.
210. Baselga J, Tripathy D, Mendelsohn J et al. Phase II study of weekly intravenous recombinant humanized anti-p185HER2 monoclonal antibody in patients with HER2/neu-overexpressing metastatic breast cancer. J Clin Oncol 1996; 14:737-744.
211. Sandmaier BM, Oparin DV, Holmberg LA et al. Evidence of a cellular immune response against sialyl-Tn in breast and ovarian cancer patients after high-dose chemotherapy, stem cell rescue, and immunization with Theratope STn-KLH cancer vaccine. J Immunother 1999; 22:54-66.
212. Siegall CB, Chace D, Mixan B et al. A single-chain immunotoxin fusion protein that cures human breast carcinoma xenografts in athymic mice and rats. J Immunol 1994; 152:2377-2384
213. Vingerhoeds MH, Steerenberg PA, Hendriks JJGW et al. Targeted delivery of diphtheria toxin via immunoliposomes: efficient antitumor activity in the presence of inactivating anti-diphtheria toxin antibodies. FEBS Letts 1996; 395:245-250.
214. Zhang S, Zhang HS, Reuter VE et al. Expression of potential target antigens for immunotherapy on primary and metastatic prostate cancers. Clin Cancer Res 1998; 4:295-302.
215. Harris DT, Matyas GR, Gomella LG et al. Immunologic approaches to the treatment of prostate cancer. Semin Oncol 1999; 26:439-447
216. Sanda MG, Smith DC, Charles LG et al. Recombinant vaccinia-PSA (PROSTVAC) can induce a prostate-specific immune response in androgen-modulated human prostate cancer. Urology 1999; 53:260-266.
217. Tjoa BA, Simmons SJ, Elgamal A et al. Follow-up evaluation of a Phase II prostate cancer vaccine trial. Prostate 1999; 40:125-129.
218. Lodge PA, Childs RA, Monahan SJ et al. Expression and purification of prostate-specific membrane antigen in the baculovirus expression system and recognition by prostate-specific membrane antigen-specific T cells. J Immunother 1999; 22:346-355.
219. Reiter RE, Gu Z, Watabe T et al. Prostate stem cell antigen: A cell surface marker overexpressed in prostate cancer. Proc Natl Acad Sci USA 1998; 95:1735-1740.
220. Fong L, Ruegg CL, Brockstedt D et al. Induction of tissue-specific autoimmune prostatitis with prostatic acid phosphatase immunization—implications for immunotherapy of prostate cancer. J Immunol 1997; 159:3113-3117.
221. Liu K-J, Chatta GS, Twardzik DR et al. Identification of rat prostatic steroid-binding protein as a target antigen of experimental autoimmune prostatitis—implications for prostate cancer therapy. J Immunol 1997; 159:472-480.
222. Hanahan D and Folkman J. Patterns and emerging mechanisms of the angiogenic switch during tumorigenesis. Cell 1996; 86:353-364.
223. Sim BKL, O'Reilly MS, Liang H et al. A recombinant human angiostatin protein inhibits experimental primary and metastatic cancer. Cancer Res 1997; 57:1329-1334.
224. O'Reilly MS, Boehm T, Shing Y et al. Endostatin: An endogenous inhibitor of angiogenesis and tumor growth. Cell 1997; 88:277-285.
225. Ashkenazi A, Pai RC, Fong S et al. Safety and antitumor activity of recombinant soluble Apo2 ligand. J Clin Invest 1999; 104:155-162.

226. Weiner HL. Oral tolerance: immune mechanisms and treatment of autoimmune diseases. Immunology Today 1997; 18:335-343.
227. Zhang ZJ, Davidson L, Eisenbarth G et al. Suppression of diabetes in nonobese diabetic mice by oral administration of porcine insulin. Proc Natl Acad Sci USA 1991; 88:10252-10256
228. Bergerot I, Ploix C, Petersen J et al. A cholera toxoid-insulin conjugate as an oral vaccine against spontaneous autoimmune diabetes. Proc Natl Acad Sci USA 1997; 94:4610-4614.
229. Arakawa T, Yu J, Chong DKX et al. A plant-based cholera toxin B subunit-insulin fusion protein protects against the development of autoimmune diabetes. Nature Biotechnol 1998; 16:934-938.
230. Ewing C and Bernard CC. Insights into the aetiology and pathogenesis of multiple sclerosis. Immunol Cell Biol 1998; 76:47-54.
231. Meyer AL, Benson JM, Gienapp IE et al. Suppression of murine chronic relapsing autoimmune encephalomyelitis by the oral administration of myelin basic protein. J Immunol 1996; 157:4230-4238.
232. Weiner HL, Mackin GA., Matsui M et al. Double-blind pilot trial of oral tolerization with myelin antigens in multiple sclerosis. Science 1993; 259:1321-1324.
233. Sun J-B, Rask C, Olsson T et al. Treatment of experimental autoimmune encephalomyelitis by feeding myelin basic protein conjugated to cholera toxin B subunit. Proc Natl Acad Sci USA 1996; 93:7196-7201.
234. Nagler-Anderson C, Bober LA, Robinson ME et al. Suppression of type II collagen-induced arthritis by intragastric administration of soluble type II collagen. Proc Natl Acad Sci USA 1986; 83:7443-7446.
235. Tarkowski A, Sun J-B, Holmdahl R et al. Treatment of experimental autoimmune arthritis by nasal administration of a type II collagen-cholera toxoid conjugate vaccine. Arthritis & Rheumatism 1999; 42:1628-1634.
236. Trentham DE, Dynesius-Trentham RA, Orav EJ et al. Effects of oral administration of type II collagen on rheumatoid arthritis. Science 1993; 261:1727-1730.
237. Sato Y, Kimura M and Fukumi H. The development of a pertussis component vaccine in Japan. Lancet 1984; 1:122-126.
238. Tan LUL, Fahim REF, Jackson G et al. A novel process for preparing an acellular pertussis vaccine composed of nonpyrogenic toxoids of pertussis toxin and filamentous hemagglutinin. Mol Immunol 1991; 28:251-255.
239. Edwards KM, Decker MD, and Mortimer EA. Pertussis vaccines. In: Plotkin SA and Orenstein WA, eds. Vaccines. WB Saunders and Company. 1999; 293-344.
240. Valenzuela P, Coit D, Medina-Selby MA et al. Antigen engineering in yeast: Synthesis and assembly of hybrid hepatitis B surface antigen-herpes simplex 1 gD particles. Biotechnology 1985; 3:323-326.
241. Pumpens P and Grens E. Hepatitis B core particles as a universal display model: a structurefunction basis for development. FEBS Letts 1999; 442:1-6.
242. Novak MJ, Smythies LE, McPherson SA et al. Poliovirus replicons encoding the B subunit of *Helicobacter pylori* urease elicit a Th1 associated immune response. Vaccine 1999; 17:2384-2391.
243. Haynes JR, Cunningham J, von Seefried A et al. Development of a genetically-engineered, candidate polio vaccine employing the self-assembling properties of the tobacco mosaic virus coat protein. Biotechnology 1986; 4:637-641.
244. Fitchen J, Beachy RN, and Hein MB. Plant virus expressing hybrid coat protein with added murine epitope elicits autoantibody response. Vaccine 1995; 3:1051-1057.
245. Turpen TH, Reinl SJ, Charoenvit Y et al. Malarial epitopes expressed on the surface of recombinant tobacco mosaic virus. Biotechnology 1995; 13:53-57.
246. Sugiyama Y, Hamamoto H, Takemoto S et al. Systemic production of foreign peptides on the particle surface of tobacco mosaic virus. FEBS Letts 1995; 359:247-250.
247. Durrani Z, McInerney TL, McLain L et al. Intranasal immunization with a plant virus expressing a peptide from HIV-1 gp41 stimulates better mucosal and systemic HIV-1-specific IgA and IgG than oral immunization. J Immunol Methods 1998; 220:93-103.
248. Dalsgaard K, Uttenthal A, Jones TD et al. Plant-derived vaccine protects target animals against viral disease. Nature Biotechnol 1997; 15:248-252.

249. Brennan FR, Bellaby T, Helliwell SM et al. Chimeric plant virus particles administered nasally or orally induce systemic and mucosal immune responses in mice. J Virol 1999; 73:930-938.
250. Brennan FR, Gilleland LB, Staczek J et al. A chimaeric plant virus vaccine protects mice against bacterial infection. Microbiol1999; 145:2061-2067.
251. Modelska A, Dietzschold B, Sleysh N et al. Immunization against rabies with plant-derived antigen. Proc Natl Acad Sci USA 1998; 95:2481-2485.
252. Carbonetti NH, Irish, TJ, Chen CH et al. Intracellular delivery of a cytolytic T-lymphocyte epitope peptide by pertussis toxin to major histocompatibility complex class I without involvement of the cytosolic class I antigen processing pathway. Infect Immun 1999; 67:602-607.
253. Guermonprez P, Ladant D, Karimova G et al. Direct delivery of the *Bordetella pertussis* adenylate cyclase toxin to the MHC class I antigen presentation pathway. J Immunol 1999; 162:1910-1916.
254. Saron MF, Fayolle C, Sebo P et al. Antiviral protection conferred by recombinant adenylate cyclase toxins from *Bordetella pertussis* carrying a CD8+ T cell epitope from lymphocytic choriomeningitis virus. Proc Natl Acad Sci USA 1997; 94:3314-3319.
255. Fayolle C, Ladant D, Karimova G et al. Therapy of murine tumors with recombinant *Bordetella pertussis* adenylate cyclase carrying a cytotoxic T cell epitope. J Immunol 1999;162:4157-4162.
256. Doling AM, Ballard JD, Shen H et al. Cytotoxic T-lymphocyte epitopes fused to anthrax toxin induce protective antiviral immunity. Infect Immun 1999; 67:3290-3296.
257. Sandvig K, Garred O, Prydz K et al. Retrograde transport of endocytosed Shiga toxin to the endoplasmic reticulum. Nature 1992; 358:510-512.
258. Noakes KL, Teisserenc HT, Lord JM et al. Exploiting retrograde transport of Shiga-like toxin 1 for the delivery of exogenous antigens into the MHC class I presentation pathway. FEBS Letts 1999; 453:95-99.
259. Lee R-S, Tartour E, van der Bruggen P et al. Major histocompatibility complex class I presentation of exogenous soluble tumor antigen fused to the B-fragment of Shiga toxin. Eur J Immunol 1998; 28:2726-2737.
260. Barbieri JT, Armellini D, Molkentin J et al. Construction of a diphtheria toxin A fragment-C180 peptide fusion protein which elicits a neutralizing antibody response against diphtheria toxin and pertussis toxin. Infect Immun 1992; 60:5071-5077.
261. Boucher P, Sato H, Sato Y et al. Neutralizing antibodies and immunoprotection against pertussis and tetanus obtained by use of a recombinant pertussis toxin-tetanus toxin fusion protein. Infect Immun 1994; 62:449-456.
262. Backstrom M, Holmgren J, Schodel F et al. Characterization of an internal permissive site in the cholera toxin B-subunit and insertion of epitopes from human immunodeficiency virus-1, hepatitis B virus and enterotoxigenic *Escherichia coli*. Gene 1995; 165:163-171.
263. Loregian A, Papini E, Satin B et al. Intranuclear delivery of an antiviral peptide mediated by the B subunit of *Escherichia coli* heat-labile enterotoxin. Proc Natl Acad Sci USA 1999; 96: 5221-5226.
264. Ali SA, Joao HC, Hammerschmid F, et al. Transferrin trojan horses as a rational approach for the biological delivery of therapeutic peptide domains. J Biol Chem 1999; 274:24066-24073.
265. Yedavelli SP, Guo L, Daou ME et al. Preventive and therapeutic effect of tumor derived heat shock protein, gp96, in an experimental prostate cancer model. Int J Molec Med 1999; 4:243-248.
266. Suzue K, Zhou X, Eisen HN et al. Heat shock fusion proteins as vehicles for antigen delivery into the major histocompatibility complex class I presentation pathway. Proc Natl Acad Sci USA 1997; 94:13146-13151.
267. Anthony LSD, Wu U, Sweet H et al. Priming of CD8+ CTL effector cells in mice by immunization with a stress protein-influenza virus nucleoprotein fusion molecule. Vaccine 1999; 17:373-383.

CHAPTER 12

Peptide Vaccines

Damu Yang, Gregory E. Holt, Michael P. Rudolf, Markwin P. Velders, Remco M. P. Brandt, Eugene D. Kwon, and W. Martin Kast

Introduction

For most of the history of vaccine production, the development of a new vaccine involved producing inactivated organisms or crude components of the pathogen. There is now generally a need to develop more precisely defined and novel vaccines against cancers and some pathogens (e.g., HIV), for which traditional approaches are unavailable or do not work well. With developments in recombinant DNA technology and cell biology, it is possible to dissect out the epitopes from the tumor cells and pathogens recognized by B- and T-cell receptors. These regions may be essential and effective for eliciting protective responses through neutralizing antibodies and T cell mediated immunity. This chapter will deal with the different peptide-based immunization strategies and their characteristics.

Molecular Basis for the Development of Peptide Vaccines

Epitopes recognized by B cells or neutralizing antibodies are usually classified as either continuous, consisting of a short linear fragment of an antigen, or discontinuous, comprising an assembly of amino acid residues brought together by folding of the protein chain. Most B-cell epitopes are discontinuous or conformational, which means that such epitopes require the full folded polypeptide for their presentation. In contrast, other epitopes are linear, being fully antigenic as short linear sequences in the range of 6- to 20-amino acid oligopeptides. A variety of techniques have been used to identify B-cell epitopes. The linear B-cell epitopes can be mapped with synthetic peptides or direct sequencing of fragmented peptides recognized by the antibodies. Although discontinuous B-cell epitopes cannot be constructed within a short peptide, it is possible to identify small reactive peptides (mimotopes) that antigenically mimic the conformational immunogens by screening recombinant-based peptide libraries with antisera.[1-5] These linear epitopes and mimotopes are candidate vaccines for eliciting production of protective neutralizing antibodies by B cells.

While B cells recognize conformational or sequential epitopes on the surface of native proteins, T cells recognize only peptides derived from the processing of antigens in association with major histocompatibility complex (MHC) molecules.[6] However, there are major differences between the recognition of antigen by CD 4^+ and $CD8^+$ T cells, both in terms of the cellular site where the peptides are generated and the nature of MHC molecules required for their recognition.

$CD4^+$ T cells recognize peptides bound to MHC class II molecules on the surface of cells. Proteins derived from pathogens residing in intracellular vesicles are degraded by vesicular proteases into peptide fragments that bind to MHC II molecules for delivery to cell surface.[7,8]

$CD4^+$ T cells also recognize peptide fragments derived from exogenous pathogens and proteins that are internalized into similar intracellular vesicles.[9] It has been recently reported that exogenous proteins can be degraded into antigenic peptides extracellularly by proteases secreted by dentritic cells (DCs) and loaded onto empty or peptide-receptive class II MHC molecules on the surface of DCs.[10]

$CD8^+$ T cells recognize peptides associated with MHC class I molecules. These peptides are derived from cytosolic proteins,[11] which are cleaved by proteasomes and translocated to the endoplasmic reticulum (ER) by peptide transporters associated with antigen processing (TAP) before final association with MHC class I molecules. The peptide-MHC class I complexes are then transported to the cell membrane where they can be recognized by $CD8^+$ T cells.[12] Certain professional antigen presenting cells (APCs) can also take up and present exogenous antigens through the MHC class I pathway to $CD8^+$ cells.[13,14]

On the basis of this knowledge of peptide processing and presentation, vaccines containing peptide epitopes recognized by $CD4^+$ and $CD8^+$ T cells have been developed.

Advantages and Disadvantages of Peptide-Based Vaccines

The peptide-based vaccine has a number of advantages, which include: Peptides can be chemically defined products and are relatively stable. They are relatively easy to manufacture and store. No infectious agent is involved in its manufacture. Any potential oncogenic or deleterious biological activity associated with whole pathogens or recombinant vaccines is avoided. Different molecules can be linked with peptides to enhance their immunogenicity.

The limitations of the peptide vaccines are: Many B cell epitopes are discontinuous, and adjacent molecules contribute to the epitopes. The conformation of a B-cell epitope in a protein may differ markedly from its shape as a free peptide. For a T-cell vaccine, this agent will need to contain multiple epitopes to cover the HLA diversity of target population and to generate immunity for different epitope variants.

Adjuvants and Delivery Systems

Adjuvants have been used to increase the peptide-induced immune responses to the corresponding antigens. Although a number of adjuvants have been evaluated, only few, including incomplete Freund's adjuvant (IFA) and Montanide ISA, have been used in peptide-based clinical trials. Recently, molecularly defined agents have been shown to be promising adjuvants for peptide vaccines. Mouse models were used to evaluate large panels of molecularly defined adjuvants. These studies revealed that cytokines, IL-2 and IL-12,[15-17] are very potent in the ability to increase the efficacy of vaccines. GM-CSF also plays an important role in the induction and magnitude of cellular immune responses.[18] But in clinical trials, only peptide plus IL-2 was associated with an increase in clinical efficacy.[19,20] No enhancement in clinical efficacy was observed using GM-CSF or IL-12.[20]

Other non-cytokine immunomodulators involved in costimulation of T cells could be considered as candidate adjuvants. One of these, CD40 and its ligand (CD40L), has been shown to be important in DCs and B-cell activation, production of type 1 cytokines by T-helper cells, and generation of cytotoxic memory responses. The addition of CD40L to DNA vaccination was found to increase the antitumor efficacy.[21] The FLTs (tyrosine kinase receptor family) ligand can induce the apparent growth and differentiation of functional dendritic cells and has been reported to have antitumor effects.[22,23]

The B7-1 and B7-2 molecules expressed on the APCs play a critical role in controlling the activation or anergy of T cells. The engagement of B7 ligand, CD28, is associated with proliferation and differentiation, whereas an encounter with another B7 ligand, cytotoxic T-lymphocyte antigen-4 (CTLA-4), may trigger functional unresponsiveness.[24] In a recent report, soluble B7-IgG fusion proteins were shown to be effective in therapy of established tumors and as a vaccine adjuvant in four mouse tumor models.[25] The blockade of the engagement of CTLA-4 has also been reported to potentiate immune responses to tumor cells.[26]

The delivery system that might lead to prolonged or pulsatile release of the peptides will reduce the frequency of immunization and elicit comparable or greater immune responses. Keyhole limpet hemocyanin (KLH), which has been shown to recruit T helper cells and promote a memory cytotoxic T cell (CTL) response, has been used as a carrier in clinical trials. Liposomes that enable the introduction of lipid-soluble molecules or peptides to the immune system have been shown to induce both humoral and cell-mediated immune responses to a wide spectrum of antigens.[27] Poly-L-lactide co-glycolide (PLGA) is another suitable carrier to deliver peptides.[28] Immunostimulating complexes (ISCOMs) elicit humoral and cellular immunity, as well as CTL responses. Peptides incorporated into preformed ISCOMs-containing influenza virus-derived protein could stimulate specific immune responses.[29,30] Intact viruses, virus like particles (VLP), and recombinant plasmids also serve as carriers, which will be discussed in a later part of this chapter.

Design of Peptide Vaccines: Synthetic Peptides as B-Cell Vaccines

Initially, the development of peptides as possible vaccines was entirely directed to the production of neutralizing antibodies by the production of an epitope that would be recognized by B cells.[31,32] The concept was to identify and synthesize the epitope sequences of pathogen proteins that could form the candidate vaccines. In many cases, it has been possible to identify B-cell epitopes against which neutralizing antibodies are directed. The techniques of recombinant DNA combined with serological studies have enabled some epitopes to be mapped to precise amino acid residues. Linear B-cell epitopes of this type have been defined for the malarial circumsporozoite protein[33] and HIV-1 gp120.[34] Both of these polypeptides contain linear epitopes that are recognized by antibodies that neutralize the respective pathogens. However, some linear epitopes are only weakly immunogenic when presented in the context of full polypeptides. Such peptides would still be effective antigens if they were rendered more immunogenic.

The peptide can be conjugated to a carrier protein to increase its immunogenicity. The most commonly used carrier proteins in conjugates are bacterial proteins that humans commonly encounter, such as tetanus toxoid (TT), for which a conjugate with the malarial circumsporozoite epitope has been tested clinically.[35]

Increasing the number of the peptide groups in a conjugate can substantially increase immunogenicity particularly if the peptide epitopes are presented as a tandem array. A common approach is to form multiple antigenic peptides (MAP). A multimer of peptides from HIV gp120 having this structure was highly immunogenic.[36]

The immunogenicity of linear epitopes can also be increased by fusing the defined epitopes to a carrier protein that forms a large particle to improve the presentation of the peptide to cells of the immune system. The commonly used protein fusion partners of this type include HBsAg[37] and hepatitis B core antigen.[38]

The recombinant-based peptide libraries have been used to create the mimotopes for conformational B-cell epitopes that cannot be readily produced by recombinant or synthetic methods.[1-5] Peptide sequences were identified from the peptide library by the IgG antibodies (from patients with Graves' disease) that recognized the thyroid-stimulating-hormone receptor (TSHR).[39] These peptides were able to inhibit the cAMP synthesis that was induced by the IgGs from the same patient. Interestingly, these peptides do not resemble the linear sequence of TSHR and thus may mimic a spatial arrangement of the key antigenic residues. It has been recently reported that mimotopes can act as structural mimics of non-protein antigens and induce the production of anti-DNA antibodies.[40]

Peptide-Based T-Cell Vaccines. Identification of Peptide Epitopes Recognized by T Cells

The antigens recognized by T cells have been identified by the transfection of genomic DNA or cDNA library pools into cells expressing the appropriate MHC molecule and screened with antigen-specific T cells.[41,42] The candidate peptides from the antigens can be selected by computer programs[43,44] based on known MHC-binding motifs or synthesized as overlapping peptides if their MHC-binding motifs are unknown. Subsequently the peptides can be screened for their binding to MHC molecules.[45-47] The antigenic peptides are then determined by testing the ability of the binding peptides from the antigen to stimulate the cytokine secretion by the antigen-specific T cells. The drawback of this commonly used approach involves the requirement for determining the MHC restriction element for the antigen-specific T cells.

Antigenic peptides can be eluted with acid from either the antigen-expressing cell surface or purified peptide-MHC complexes, and subsequently separated by high-pressure liquid chromatography (HPLC). After pulsing onto APCs the eluted peptide fractions are tested for their ability to stimulate the antigen-specific T cells. The sequence of the peptide is then determined by Edman degradation or tandem mass spectrometry. A naturally processed epitope recognized by five melanoma-specific T-cell lines was identified by this method.[48] This approach has been limited by the need for highly specialized equipment and the requirement that peptides be present in sufficient quantity to enable their identification by these physical techniques.

Synthetic peptide libraries based on the MHC-binding motifs have been used for identification of antigenic peptides recognized by T cells.[49] However, the identified peptide may be a cross-reactive peptide, rather than a naturally processed peptide derived from an antigen.

Each of the three techniques mentioned above is dependent on the prior availability of T cells capable of recognizing the antigens, a requirement that often cannot be met. A new approach to the identification of antigenic peptides is involved in attempts to develop in vitro sensitization techniques.[50] Genes encoding candidate antigens can be transfected or transduced into APCs or synthesized peptides from candidate antigens based on known MHC-binding motifs can be pulsed onto APCs and used for in vitro sensitizations. If the generated T cells recognize the antigen-expressing cells, they can be used to identify the antigenic peptides. The peptides identified by this technique must have the ability to sensitize the T cells recognizing both the peptides themselves and the antigen-expressing cells.

A new possibility to identify T-cell epitopes is created by the use of HLA transgenic mice.[51] This approach involves screening the candidate peptides in the HLA transgenic mice, and subsequently testing these peptides in humans. Antigenic peptides have been identified by this techniques and results indicate that the immunogenicity of peptides in transgenic mice resembles the immune responses against these peptides in humans bearing the same HLA haplotypes.[52-54] However, the peptides which are immunogenic in transgenic mice are not always immunogenic in humans.

Synthetic Peptides as T-Cell Vaccines

Synthetic peptides corresponding to epitopes recognized by T cells represent an ideal safe vaccine. Protective CTL responses induced by vaccination with MHC class I binding peptides were first reported by Schulz et al[55] for lymphocytic choriomeningitis virus (LCMV) and independently by Kast et al[56] for Sendai virus. In these studies, vaccination with peptides in IFA elicited substantial antiviral immune responses. Adjuvants were used to create a depot after injection for a slow release of peptides in vivo in these approaches. Much of the literature emphasizes the requirement for T-cell help for generating antipeptide immune reactivity. Conversely, some peptides appeared to induce CTL reactivity in the absence of obvious $CD4^+$ T cell help.[57] If a helper T-cell response is required to obtain an efficient CTL response, the inclusion of general T-cell help agents (e.g., tetanus toxoid) or helper T cell epitopes (e.g., PADRE)[58,59] in peptide vaccines should be considered.

In hopes of improving the potency of free peptide vaccinations, many alterations of the original immunization protocols have been tested. It has been found that modification of certain residues of a peptide can enhance the immunogenicity of peptide through increasing its ability to bind MHC class I molecule without compromising the interaction of this complex with T-cell receptor.[60] Indeed, three groups have shown that such a modification enhanced the peptide's capacity for eliciting a CTL response.[61-63] In contrast, Clay et al[64] found that although the peptide analogues induced a greater CTL response, they were unable to cross react with the antigen-expressing cells. These results indicate that the use of modified peptide may only be beneficial if the resultant CTLs recognize not only the peptide analogue but also the cells expressing the authentic antigen.

Another advance in peptide vaccination involves the use of activated DCs to deliver the peptides. The DCs, expressing all costimulatory molecules necessary to efficiently initiate a cellular immune response, are by far the most potent antigen presenting cells.[65] DCs can be obtained by culturing either peripheral blood mononuclear cells or bone marrow cells with GM-CSF and IL-4. After the appropriate activation and loading of the DCs with peptides, they are infused back into the patient where the activated DCs home towards secondary lymphoid organs, interact with $CD8^+$ cells, and cause the subsequent induction of peptide reactive CTLs. The successful induction of protective immune responses by peptide loaded DCs have been shown in a number of studies.[66-70] In a recent clinical trial, 16 patients with metastatic melanoma were immunized with DCs pulsed either with immunodominant melanoma peptides or with melanoma lysates, and objective tumor regression was reported in five patients.[71]

Heat shock proteins (HSPs), which can bind the peptides and induce effective immune responses, have been used as chaperones to directly target the peptides to professional APCs.[72,73] CyaA, a detoxified cellular invasive *Bordetella pertussis* adenylate cyclase, has also been shown to have similar function.[74] These methods increase the efficacy of peptide vaccines probably through the direct targeting of peptides to the class I processing pathway. Other promising novel strategies may include the use of toxin-linked peptides,[75] T-helper sequence linked peptides,[76] and peptides attached to endoplasmic insertion signal sequences.[77]

It is important to realize that peptide vaccines do not always protect. Vaccination with certain peptides is associated with protective immunity in regimens[78-80] that can lead to tolerance in the case of other peptides[81-83] due to yet unknown mechanisms. When dealing with the peptides that are prone to induce tolerance, novel strategies (e.g., peptide-loaded DCs)[68] need to be considered.

Recombinant Vaccines Expressing T-Cell Epitopes

Minigenes encoding immunodominant peptide epitopes from the pathogens and tumor antigens can be engineered into recombinant virus and DNA. The injection of such recombinant constructs containing minigenes makes up a new way of epitope-based vaccination.

The recombinant viruses are engineered through the replacement of normal genes necessary for viral replication with minigenes encoding epitopes alone or in combination with endoplasmic reticulum insertion signals and immunomodulatory molecule sequences (such as cytokines and costimulatory molecules).[15,84-87] Thomson et al[88] created a recombinant vaccinia virus containing epitopes from different viruses, one parasite and the immunodominant peptide from ovalbumin. Vaccination with this recombinant virus showed potent CTL responses to each peptide and induced protective responses against challenges of viruses and tumor cells expressing corresponding epitopes encoded by inserted minigenes. As all viruses have the potential to cause problems, the safety of the recombinant viruses should be considered and might be ensured in a number of ways. For example, some of the recombinant vaccines contain the viruses that are incapable of replicating in mammalian cells because of their host range[89] or removal of their viral genes critical for viral replication,[90] and some are highly attenuated viruses.[91] Another problem is that recombinant vaccines based on vaccinia and adenoviruses are likely to suffer from the pre-existing immunity. In data from both mouse[92] and human[93] studies, a preexisting immunity to vaccinia reduced the magnitude of the induced immune responses after revaccination with the recombinant vaccinia vectors. This limiting effect of a pre-existing immunity was overcome through an alteration in the route of immunization performed with the recombinant viral vaccine.[94] Another way of circumventing the problem is the use of viruses whose natural hosts are non-mammalian, such as the avian poxviruses.[95] It has also been possible to use similar viruses from different species to deliver the included peptides without suffering from the neutralizing antibody response of a previously used viral vector (Kast et al,unpublished).

The T-cell epitopes can be delivered by recombinant plasmids containing the corresponding minigenes. Vaccination with recombinant plasmids coding for multiple T-cell epitopes have resulted in effective immune responses.[96,97] However, recombinant DNA is generally not as potent as recombinant viruses at eliciting effective immune responses. Important innovations concerning the design of these vectors include promoter optimization, enhancement of polyadenylation sequences, the removal of untranslated regions from the minigenes, and the use of intronic sequences to improve nuclear export. Other variations to augment the vaccine efficacy include the insertion of genes for IL-2, IL-12, IL-15, IFN-γ, and GM-CSF,[98-100] integration of DNA for B7.1 and B7.2,[101] inclusions of ubiquitin signals,[102] and incorporation of the "danger signal" sequence, unmethylated cytosine-guanine oligodinucleotides (CpG ODN).[103] Also DCs have been used to enhance the vaccination efficacy through directly transfecting DCs with epitope-containing plasmids.[104] Although no obvious adverse effects have been reported, these epitope-based DNA vaccines like all other DNA vaccines have the potential to induce anti-DNA autoimmune responses and integrate their DNA to the host genomes.

Adoptive Cellular Therapy

The immunogenic peptides can be used to stimulate the peripheral blood lymphocytes in vitro to induce disease-specific CTLs. Once induced, the specific population can be expanded and then reinfused into the patient. This therapeutic approach may be beneficial to the patient who is immunosuppressed. This therapeutic approach is relatively costly since it must be individualized and because lymphocyte expansion is labor-intensive. Clinical trials in humans using CTLs that are specific for CMV have been conducted.[105,106] The CTLs utilized in these trials were induced using virus-infected fibroblasts as APCs. Early results showed that these treatments are safe and efficacious in preventing CMV infection. Clinical trials of adoptive immunotherapy have been reported for prevention and treatment of Epstein-Barr virus, HIV and human cancers.[107-111] In these studies, the expanded cell population tends to be heterogeneous, and the specific CTL population varies from treatment to treatment. This inconsistency in CTL generation may account for the relatively low success rate of previous adoptive T-cell therapy approach.

New strategies in the peptide-based adoptive immunotherapy of cancers and infective diseases are now being explored. Lymphocytes from cancer or chronic virus-infected patients are stimulated in vitro with APCs that have been optimally loaded with the antigenic peptides of choice. Among APCs, the DCs have been shown to be effective in presenting the selected peptides to the CTL precursors.[66,67,70] The addition of cytokines such as IL-7 and IL-12 at the early stage of the culture might facilitate the expansion of the CTL precursors. Several cycles of antigen restimulation in the presence of IL-2 may be required to obtain the appropriate number of antigen-specific CTLs necessary for the adoptive transfer into the patients.

Another strategy is the use of modified peptides to sensitize CTLs for adoptive immunotherapy. The peptides, modified to increase their affinity for the MHC molecules, may have enhanced ability to stimulate the CTL precursor and preferably induce the CTLs with high avidity to antigens of pathogens and tumor cells. Indeed, the peripheral blood lymphocytes from melanoma patients stimulated with the modified gp100 peptide:g209-2M grow faster and have greater antitumor reactivity than T cells grown with the native g209 peptide.[60] A modified antigenic peptide derived from cyclophilin B was found to have increased ability in the induction of CTLs compared to the native peptides.[112]

The use of a clonal population of CTLs represents another new strategy. From the bulk population of induced CTLs, individual clonal populations of CTLs can be derived and assayed for their antigen activity. The T-cell clones with the highest apparent avidity can be expanded for adoptive transfer. This approach may have the advantage of generating a nearly uniform capacity to recognize the pathogens and tumor cells.

In addition to the use of a population of $CD8^+$ CTLs for adoptive transfer, $CD4^+$T cells can also be included in this therapy to enhance its efficacy. In studies using T cells specific for CMV, Waltel et al[113] have reported that the cytotoxic activity of adoptively transferred $CD8^+$ clones declined in patients deficient in helper $CD4^+$T cells specific for CMV. These results suggested the $CD4^+$T cell help is needed for the persistence of transferred $CD8^+$T cells. For this aim, epitopes from pathogens and tumor cells recognized by $CD4^+$T cells are required. Attempts to clone the molecular targets of $CD4^+$T cells have already met with considerable success.[114-118]

Summary and Perspectives

It has been shown in animal studies that peptide-based vaccines are capable of affording protection against infectious disease and cancer, as well as in control of these diseases once they have been established. The following challenge is to translate these results into prophylactic and therapeutic agents applicable to human diseases. We think that the use of selected antigenic peptides to elicit neutralizing antibodies and specific CTL responses will play an important role in vaccine development in certain fields.

For some infectious diseases and cancers, peptide-based preparations seem to offer the best hope for vaccination development. Some pathogens and tumor cells contain the epitopes recognized by neutralizing antibodies and T cells. But in many cases, it is very difficult to isolate the proteins containing the epitopes to use as immunogens for immunotherapy. In addition, peptides are relatively safe molecules. Administration of a short amino acid fragment derived from a pathogen or tumor cell offers fewer safety risks than the use of attenuated pathogens, full-length nucleic acids, or recombinant proteins, which are more likely to retain inherent biological activities.

The use of antigenic peptides also has a capacity to specifically manipulate the immune system. There are only a few potential epitopes within an antigen, and the capacity to deliver selected antigenic peptides at relatively high concentrations in immunogenic formulations is likely to be important in initiating and boosting an immune response where disease already exists. The peptide approach may also be useful for targeting immune responses to epitopes that are underrepresented or non-existent in the responses normally induced during infection or oncogenesis.

It is very important to test immunotherapy in combination with other therapeutic approaches in the management of human diseases. For instance, in the chronic viral disease settings, it may be important to use antigen-specific immunotherapy along with antiviral drugs that are capable of reducing the viral replication. Similarly, combining forms of immunotherapy may be important in certain disease settings. For example, in cancer therapy, the patients may first be primed with the antigenic peptides using suitable formations, their lymphocytes will be expanded to tumor-specific CTLs in vitro, delivered back, and finally be boosted periodically with peptide to maintain high level of anti-tumor immune responses. Distinct strategies may have unique potentials for satisfying the different requirements for induction of protective immunity. In a recent study, Schneider et al[119] were able to show the increased efficacy of protecting mice against malaria through the initial vaccination with a naked DNA construct followed by a boost with a recombinant vaccinia virus containing the malaria antigens. Reversal of the order of vaccination actually revoked the CTL induction potential. This suggested that optimization of the combination and the method of vaccination may enhance protective immune responses.

For peptide vaccines, more studies are required to define the ideal combination of peptides and the best antigenic formulation, as well as a more appropriate selection of the patients and optimization of the immune monitoring. In addition, active vaccination should be tried as early as possible in patients with cancers and some infectious diseases. Indeed, most of the preclinical studies in animal models and the results of preliminary clinical trials have demonstrated that active immunotherapy has more chance of success when patients bear minimal tumor or infection burden.

Acknowledgment

This review was partly based on studies supported by grants from the NIH (CA74397, CA/AI 78399, CA74182), the department of defense (PC970131) (all to W. M. Kast), the Illinois Department of Public Health (M. P. Rudolf), the Cancer Research Institute (M. P. Velders), and the American Foundation for Urologic Disease (D. Yang).

References

1. Lam KS, Salmon SE, Hersh EM et al. A new type of synthetic peptide library for identifying ligand-binding activity. Nature 1991; 354:82-84.
2. Houghten RA, Pinilla C, Blondelle SE et al. Generation and use of synthetic peptide combinatorial libraries for basic research and drug discovery. Nature 1991; 354:84-86.
3. Pinilla C, Appel JR, Campbell GD et al. All-D peptides recognized by an anticarbohydrate antibody identified from a positional scanning library. J Mol Biol 1998; 283:1013-1025.
4. Pinilla C, Martin R, Gran B et al. Exploring immunological specificity using synthetic peptide combinatorial libraries. Curr Opin Immunol 1999; 11:193-202.
5. Rudolf MP, Vogel M, Kricek F, Ruf C et al. Epitope-specific antibody response to IgE by mimotope immunization. J Immunol 1998; 160:3315-3321.
6. Babbitt BP, Allen PM, Matsueda G et al. Binding of immunogenic peptides to Ia histocompatibility molecules. Nature 1985; 317:359-362.
7. Nuchtern JG, Biddison WE, Klausner RD. Class II MHC molecules can use the endogenous pathway of antigen presentation. Nature 1990; 343:74-76.
8. Weiss S, Bogen B. MHC class II-restricted presentation of intracellular antigen. Cell 1991; 64:767-776.
9. Germain RN, Margulies DH. The biochemistry and cell biology of antigen processing and presentation. Annu Rev Immunol 1993; 11:403-450.
10. Santambrogio L, Sato AK, Carven GJ et al. Extracellular antigen processing and presentation by immature dendritic cells. Proc Natl Acad Sci U S A 1999; 96:15056-15061.
11. Rammensee HG, Falk K, Roetzsche O. Peptides naturally presented by MHC class I molecules. Annu Rev Immunol 1993; 11:213-244.
12. Monaco JJ, Cho S, Attaya M. Transport protein genes in the murine MHC: Possible implications for antigen processing. Science 1990; 250:1723-1726.
13. Rock KL, Gamble S, Rothstein L. Presentation of exogenous antigen with class I major histocompatibility complex molecules. Science 1990; 249:918-921.

14. Kovacsovics-Bankowski M, Rock KL. A phagosome-to-cytosol pathway for exogenous antigens presented on MHC class I molecules. Science 1995; 267:243-246.
15. Bronte V, Tsung K, Rao JB et al. IL-2 enhances the function of recombinant poxvirus-based vaccines in the treatment of established pulmonary metastases. J Immunol 1995; 154:5282-5292.
16. Irvine KR, Rao JB, Rosenberg SA et al. Cytokine enhancement of DNA immunization leads to effective treatment of established pulmonary metastases. J Immunol 1996; 156:224-231.
17. Rao JB, Chamberlain RS, Bronte V et al. Interlukin-12 is an effective adjuvant to recombinant vaccinia virus based tumor vaccine: Enhancement by simultaneous B7-1 expression. J Immunol 1996; 156:3357-3365.
18. Wada H, Noguchi Y, Marino MW et al. T cell functions in granulocyte/macrophage colony-stimulating factor in deficient mice. Proc Natl Acad Sci U S A 1997; 94:12557-12561.
19. Rosenberg SA, Yang JC, Schwartzentruber DJ et al. Immunologic and therapeutic evaluation of a synthetic tumor associated peptide vaccine for the treatment of patients with metastatic melanoma. Nat Med 1998; 4:321-328.
20. Rosenberg SA, Yang JC, Schwartzentruber DJ et al. Impact of cytokine administration on the generation of antitumor reactivity in patients with metastatic melanoma receiving a peptide vaccine. J Immunol 1999; 163:1690-16955.
21. Gurunathan S, Irvine KR, Wu CY et al. CD40 ligand/trimer DNA enhances both humoral and cellular immune responses and induces protective immunity to infectious and tumor challenge. J Immunol 1998; 161:4563-4571.
22. Esche C, Subbotin VM, Maliszewski C et al. FLT3 ligand administration inhibits tumor growth in murine melanoma and lymphoma. Cancer Res 1998; 58:380-383.
23. Peron JM, Esche C, Subbotin VM et al. FLT3-ligand administration inhibits liver metastases: Role of NK cells. J Immunol 1998; 161:6164-6170.
24. Van Parijs L, Abbas AK. Homeostasis and self-tolerance in the immune system: turning lymphocytes off. Science 1998; 280:243-248.
25. Sturmhoefel K, Lee K, Gray GS et al. Potent activity of soluble B7-IgG fusion proteins in therapy of established tumors and as vaccine adjuvant. Cancer Res 1999; 59:4964-4972.
26. Leach DR, Krummel MF, Allison JP. Enhancement of antitumor immunity by CTL-A4 blockade. Science 1996; 271:1734-1736.
27. Gregoriadis G, McCormack B, Obrenovic M et al. Vaccine entrapment in liposomes. Methods 1999; 19:156-162.
28. Jabbal-Gill I, Lin W, Jenkins P, Watts P et al. Potential of polymeric lamellar substrate particles (PLSP) as adjuvants for vaccines. Vaccine 1999; 18:238-250.
29. Ben Ahmeida E, Gregoriadis G, Potter CE et al. Immunopotentiation of local and systemic humoral immune responses by ISCOMs, liposomes and FCA: Role in protection against influenza A in mice. Vaccine 1993; 11:1302-309.
30. Barr IG, Mitchell. ISCOMs (immunostimulating complexes): The first decade. Immunol Cell Biol 1996; 74:8-25.
31. Ada GL Vaccines. In: Paul WE, ed. Fundamental immunology, 2nd ed. New York: Raven Press 1993:985-1032.
32. Brown F. The potential of peptides as vaccines. Semin Virol 1990; 1:67-74.
33. Zavala F, Cochrane AH, Nardin EH et al. Circumsporozoite proteins of malaria parasites contain a single immunodominant region with two or more identical epitopes. J Exp Med 1983; 157:1947-1957.
34. Javaherian K, Langlois AJ, McDanal C et al. Principal neutralizing domain of the human immunodeficiency virus type 1 envelope protein. Proc Natl Acad Sci U S A 1989; 86:6768-6772.
35. Herrington DA, Clyde DE, Losonsky G et al. Safety and immunogenicity in man of a synthetic peptide malaria vaccine against Plasmodium falciparum sporozoites. Nature 1987; 328:257-259.
36. Wang CY, Looney DJ, LI ML et al. Long-term high-titer neutralizing activity induced by octameric synthetic HIV-1 antigen. Science 1991; 254:285-288.
37. Vreden SG, Verhave JP, Oettinger T et al. Phase I clinical trial of a recombinant malaria vaccine consisting of the circumsporozoite repeat region of Plasmodium falciparum coupled to hepatitis B surface antigen. Am J Trop Med Hyg 1991; 45:533-538.

38. Schodel F, Peterson D, Hughes J et al. Hybrid hepatitis B virus core antigen as a vaccine carrier moiety: I. presentation of foreign epitopes. J Biotechnol 1996; 44:91-96.
39. Park JY, Kim IJ, Lee MH et al. Identification of the peptides that inhibit the stimulation of thyrotropin receptor by Graves' immunoglobulin G from peptide libraries. Endocrinology 1997; 138:617-626.
40. Putterman C, Diamond B. Immunization with a peptide surrogate for double-stranded DNA (dsDNA) induces autoantibody production and renal immunoglobulin deposition. J Exp Med 1998; 188:29-38.
41. Boon T. Tumor antigens recognized by cytolytic T lymphocytes: Present perspectives for specific immunotherapy. Int J Cancer 1993; 54:177-180.
42. Rosenberg SA. Development of cancer immunotherapies based on identification of the genes encoding cancer regression antigens. J Natl Cancer Inst 1996; 88:1635-1644.
43. Parker KC, Bednarek MA, Coligan JE. Scheme for ranking potential HLA-A2 binding peptides based on independent binding of individual peptide side-chains. J Immunol 1994; 152:163-175.
44. D'Amaro J, Houbiers JG, Drijfhout JW et al. A computer program for predicting possible cytotoxic T lymphocyte epitopes based on HLA class I peptide-binding motifs. Hum Immunol 1995; 43:13-18.
45. Kast WM, Offringa R, Peters PJ etal. Eradication of adenovirus E1-induced tumors by E1A-specific cytotoxic T lymphocytes. Cell 1989; 59:603-614.
46. Nijman HW, Houbiers JG, Vierboom MP et al. Identification of peptide sequences that potentially trigger HLA-A2.1-restricted cytotoxic T lymphocytes. Eur J Immunol 1993; 23:1215-1219.
47. Kast WM, Brandt RM, Sidney J et al. Role of HLA-A motifs in identification of potential CTL epitopes in human papillomavirus type 16 E6 and E7 proteins. J Immunol 1994; 152:3904-3912.
48. Cox AL, Skipper J, Chen Y et al. Identification of a peptide recognized by five melanoma-specific human cytotoxic T cell lines. Science 1994; 264:716-719.
49. Blake J, Johnston JV, Hellstrom KE et al. Use of combinatorial peptide libraries to construct functional mimics of tumor epitopes recognized by MHC class I-restricted cytolytic T lymphocytes. J Exp Med 1996; 184:121-130.
50. Parkhurst MR, Fitzgerald EB, Southwood S et al. Identification of a shared HLA-A*0201-restricted T-cell epitope from the melanoma antigen tyrosinase-related protein 2 (TRP2). Cancer Res 1998; 58:4895-4901.
51. Vitiello A, Marchesini D, Furze J et al. Analysis of the HLA-restricted influenza-specific cytotoxic T lymphocyte response in transgenic mice carrying a chimeric human-mouse class I major histocompatibility complex. J Exp Med 1991; 173:1007-10015.
52. Schonbach C, Nokihara K, Bangham CR et al. Identification of HTLV-1-specific CTL directed against synthetic and naturally processed peptides in HLA-B*3501 transgenic mice. Virology 1996; 226:102-112.
53. Ressing ME, Sette A, Brandt RM et al. Human CTL epitopes encoded by human papillomavirus type 16 E6 and E7 identified through in vivo and in vitro immunogenicity studies of HLA-A*0201-binding peptides. J Immunol 1995; 154:5934-5943.
54. Vissers JL, De Vries IJ, Schreurs MW et al. The renal cell carcinoma-associated antigen G250 encodes a human leukocyte antigen (HLA)-A2.1-restricted epitope recognized by cytotoxic T lymphocytes. Cancer Res 1999; 59:5554-5559.
55. Schulz M, Zinkernagel RM, Hengartner H et al. Peptide-induced antiviral protection by cytotoxic T cells. Proc Natl Acad Sci U S A 1991; 88:991-993.
56. Kast WM, Roux L, Curren J et al. Protection against lethal Sendai virus infection by in vivo priming of virus-specific cytotoxic T lymphocytes with a free synthetic peptide. Proc Natl Acad Sci USA 1991; 88:2283-2287.
57. Buller RM, Holmes KL, Hugin A et al. Induction of cytotoxic T-cell responses in vivo in the absence of CD4 helper cells. Nature 1987; 328:77-79.
58. Alexander J, Sidney J, Southwood S et al. Development of high potency universal DR-restricted helper epitopes by modification of high affinity DR-blocking peptides. Immunity 1994; 1:751-761.
59. Alexander J, Fikes J, Hoffman S et al. The optimization of helper T lymphocyte (HTL) function in vaccine development. Immunol Res 1998; 18:79-92.

60. Parkhurst M, Salgaller ML, Southwood S et al. Improved induction of melanoma-reactive CTL with peptides from the melanoma antigen gp100 modified at HLA-A*0201-binding residues. J Immunol 1996; 157:2539-2548.
61. Vierboom MP, Feltkamp MC, Neisig A et al. Peptide vaccination with an anchor-replaced CTL epitope protects against human papillomavirus type 16-induced tumors expressing the wild-type epitope. J Immunother 1998; 21:399-408.
62. Valmori D, Fonteneau JF, Lizana CM et al. Enhanced generation of specific tumor-reactive CTL in vitro by selected Melan-A/MART-1 immunodominant peptide analogues. J Immunol 1998; 160:1750-1758.
63. Men Y, Micronnet I, Valmori D et al. Assessment of immunogenicity of human Melan-A peptide analogues in HLA-A*0201/Kb transgenic mice. J Immunol 1999; 162:3566-3573.
64. Clay TM, Custer MC, McKee MD et al. Changes in the fine specificity of gp100(209-217)-reactive T cells in patients following vaccination with a peptide modified at an HLA-A2.1 anchor residue. J Immunol 1999; 162:1749-1755.
65. Banchereau J, Steinman RM. Dendritic cells and the control of immunity. Nature 1998; 392:245-52.
66. Porgador A, Gilboa E. Bone marrow-generated dendritic cells pulsed with a class I-restricted peptide are potent inducers of cytotoxic T lymphocytes. J Exp Med 1995; 182:255-260.
67. Celluzzi CM, Mayordomo JI, Storkus WJ et al. Peptide-pulsed dendritic cells induce antigen-specific CTL-mediated protective tumor immunity. J Exp Med 1996; 183:283-287.
68. Toes RE, van der Voort EI, Schoenberger SP et al. Enhancement of tumor outgrowth through CTL tolerization after peptide vaccination is avoided by peptide presentation on dendritic cells. J Immunol 1998; 160:4449-4456.
69. Cao X, Zhang W, He L et al. Lymphotactin gene-modified bone marrow dendritic cells act as more potent adjuvants for peptide delivery to induce specific antitumor immunity. J Immunol 1998; 161:6238-6244.
70. Subklewe M, Chahroudi A, Schmaljohn A et al. Induction of Epstein-Barr virus-specific cytotoxic T-lymphocyte responses using dendritic cells pulsed with EBNA-3A peptides or UV-inactivated, recombinant EBNA-3A vaccinia virus. Blood 1999; 94:1372-1381.
71. Nestle FO, Alijagic S, Gilliet M et al. Vaccination of melanoma patients with peptide- or tumor lysate-pulsed dendritic cells. Nat Med 1998; 4:328-332.
72. Suto R, Srivastava PK. A mechanism for the specific immunogenicity of heat shock protein-chaperoned peptides. Science 1995; 269:1585-1588.
73. Ishii T, Udono H, Yamano T et al. Isolation of MHC class I-restricted tumor antigen peptide and its precursors associated with heat shock proteins hsp70, hsp90, and gp96. J Immunol 1999; 162:1303-1309.
74. Fayolle C, Ladant D, Karimova G et al. Therapy of murine tumors with recombinant Bordetella pertussis adenylate cyclase carrying a cytotoxic T cell epitope. J Immunol 1999; 162:4157-4162.
75. Goletz TJ, Klimpel KR, Arora N et al. Targeting HIV proteins to the major histocompatibility complex class I processing pathway with a novel gp120-anthrax toxin fusion protein. Proc Natl Acad Sci U S A 1997; 94:12059-12064.
76. Shirai M, Pendleton CD, Ahlers J et al. Helper-cytotoxic T lymphocyte (CTL) determinant linkage required for priming of anti-HIV CD8+ CTL in vivo with peptide vaccine constructs. J Immunol 1994; 152:549-556.
77. Minev B, McFarland BJ, Spiess PJ et al. Insertion signal sequence fused to minimal peptides elicits specific CD8+ T-cell responses and prolongs survival of thymoma-bearing mice. Cancer Res 1994; 54:4155-4161.
78. Melief CJM, Kast WM. T-cell immunotherapy of tumors by adoptive transfer of cytotoxic T lymphocytes and by vaccination with minimal essential epitopes. Immunol Rev 1995; 145:167-177.
79. FeltkampMC, Smits HL, Vierboom MP et al. Vaccination with cytotoxic T lymphocyte epitope-containing peptide protects against a tumor induced by human papillomavirus type 16-transformed cells. Eur J Immunol 1993; 23:2242-2249.
80. Mandelboim O, Vadai E, Fridkin M et al. Regression of established murine carcinoma metastases following vaccination with tumour-associated antigen peptides. Nat Med 1995; 1:117911-83.

81. Toes RE, Blom RJ, Offringa R et al. Enhanced tumor outgrowth after peptide vaccination. Functional deletion of tumor-specific CTL induced by peptide vaccination can lead to the inability to reject tumors. J Immunol 1996; 156:3911-3918.
82. Toes RE, Offringa R, Blom RJ et al. Peptide vaccination can lead to enhanced tumor growth through specific T-cell tolerance induction. Proc Natl Acad Sci U S A 1996; 93:7855-7860.
83. Nieland JD, Da Silva DM, Velders MP et al. Chimeric papillomavirus virus-like particles induce a murine self-antigen-specific protective and therapeutic antitumor immune response. J Cell Biochem 1999; 73:145-152.
84. Restifo NP, Bacik I, Irvine KR et al. Antigen processing in vivo and the elicitation of primary CTL responses. J Immunol 1995; 154:4414-4422.
85. Irvine KR, McCabe BJ, Resenberg SA et al. Synthetic oligonucleotide expressed by a recombinant vaccinia virus elicits therapeutic CTL. J Immunol 1995; 154:4651-4657.
86. McCabe BJ, Irvine KR, Nishimura MI et al. Minimal determinant expressed by a recombinant vaccinia virus elicits therapeutic antitumor cytolytic T lymphocyte responses. Cancer Res 1995; 55:1741-1747.
87. Chamberlain RS, Carroll MW, Bronte V et al. Costimulation enhances the active immunotherapy effect of recombinant anticancer vaccines. Cancer Res 1996; 56:2832-2836.
88. Thomson SA, Sherritt MA, Medveczky J et al. Delivery of multiple CD8 cytotoxic T cell epitopes by DNA vaccination. J Immunol 1998; 160:1717-1723.
89. Wang M, Bronte V, Chen PW et al. Active immunotherapy of cancer with a nonreplicating recombinant fowlpox virus encoding a model tumor-associated antigen. J Immunol 1995; 154:4685-4692.
90. Chen PW, Wang M, Bronte V et al. Therapeutic antitumor response after immunization with a recombinant adenovirus encoding a model tumor-associated antigen. J Immunol 1996; 156:224-231.
91. Restifo NP, Surman DR, Zhen H et al. Transfectant influenza A viruses are effective recombinant immunogens in the treatment of experimental cancer. Virology 1998; 249:89-97.
92. Rooney JF, Wohlenberg C, Cremer K et al. Immunization with a vaccinia virus recombinant expressing herpes simplex virus type 1 glycoprotein D: Long-term protection and effect of revaccination. J Virol 1988; 62:1530-1534.
93. Cooney EL, Collier AC, Greenberg PD et al. Safety of and immunological response to a recombinant vaccinia virus vaccine expressing HIV envelope glycoprotein. Lancet 1991; 337:567-572.
94. Belyakov IM, Moss B, Strober W et al. Mucosal vaccination overcomes the barrier to recombinant vaccinia immunization caused by preexisting poxvirus immunity. Proc Natl Acad Sci U S A 1999; 96:4512-457.
95. Breitburd F, Kirnbauer R, Hubbert NL et al. Immunization with viruslike particles from cottontail rabbit papillomavirus (CRPV) can protect against experimental CRPV infection. J Virol 1995; 69:3959-3963.
96. Thomson SA, Sherritt MA, Medveczky J et al. Delivery of multiple CD8 cytotoxic T cell epitopes by DNA vaccination. J Immunol 1998; 160:1717-1723.
97. Ishioka GY, Fikes J, Hermanson G et al. Utilization of MHC class I transgenic mice for development of minigene DNA vaccines encoding multiple HLA-restricted CTL epitopes. J Immunol 1999; 162:3915-3925.
98. Barough DH, Santra S, Steenbeke TD et al. Augmentation and suppression of immune responses to an HIV-1 DNA vaccine by plasmid cytokine/Ig administration. J Immunol 1998; 161:1875-1882.
99. Kim JJ, Trivedi NN, Nottingham LK et al. Modulation of amplitude and direction of in vivo immune responses by co-administration of cytokine gene expression cassettes with DNA immunogens. Eur J Immunol 1998; 28:1089-1103.
100. Chow YH, Chiang BL, Lee YL et al. Development of Th1 and Th2 populations and the nature of immune responses to hepatitis B virus DNA vaccines can be modulated by codelivery of various cytokine genes. J Immunol 1998; 160:1320-1329.
101. Corr M, Tighe H, Lee D et al. Costimulation provided by DNA immunization enhances antitumor immunity. J Immunol 1997; 159:4999-5004.
102. Wu Y, Kipps TJ. Deoxyribonucleic acid vaccines encoding antigens with rapid proteasome-dependent degradation are highly efficient inducers of cytolytic T lymphocytes. J Immunol 1997; 159:6037-6043.

103. Roman M, Martin-Orozco E, Goodman JS et al. Immunostimulatory DNA sequences function as T helper-1-promoting adjuvants. Nat Med 1997; 3:849-854.
104. Porgador A, Irvine KR, Iwasaki A et al. Predominant role for directly transfected dendritic cells in antigen presentation to CD8+ T cells after gene gun immunization. J Exp Med 1998; 188:1075-1082.
105. Greenberg PD, Reusser P, Goodrich JM et al. Development of a treatment regimen for human cytomegalovirus (CMV) infection in bone marrow transplantation recipients by adoptive transfer of donor-derived CMV-specific T cell clones expanded in vitro. Ann N Y Acad Sci 1991; 636:184-195.
106. Riddell SR, Watanabe KS, Goodrich JM et al. Restoration of viral immunity in immunodeficient humans by the adoptive transfer of T cell clones. Science 1992; 257:238-241.
107. Melief CJ. Tumor eradication by adoptive transfer of cytotoxic T lymphocytes. Adv Cancer Res 1992; 58:143-175.
108. Rosenberg SA, Yannelli JR, Yang JC et al. Treatment of patients with metastatic melanoma with autologous tumor-infiltrating lymphocytes and interleukin 2. J Natl Cancer Inst 1994; 86:1159-1166.
109. Yee C, Riddell SR, Greenberg PD. Prospects for adoptive T cell therapy. Curr Opin Immunol 1997; 9:702-708.
110. Sing AP, Ambinder RF, Hong DJ et al. Isolation of Epstein-Barr virus (EBV)-specific cytotoxic T lymphocytes that lyse Reed-Sternberg cells: Implications for immune-mediated therapy of EBV+ Hodgkin's disease. Blood 1997; 89:1978-1986.
111. Greenberg PD, Finch RJ, Gavin MA et al. Genetic modification of T-cell clones for therapy of human viral and malignant diseases. Cancer J Sci Am 1998; 4 Suppl 1:S100-105.
112. Gomi S, Nakao M, Niiya F et al. A cyclophilin B gene encodes antigenic epitopes recognized by HLA-A24-restricted and tumor-specific CTLs. J Immunol 1999; 163:4994-5004.
113. Walter EA, Greenberg PD, Gilbert MJ et al. Reconstitution of cellular immunity against cytomegalovirus in recipients of allogeneic bone marrow by transfer of T-cell clones from the donor. N Engl J Med 1995; 333:1038-1044.
114. Topalian SL, Gonzales MI, Parkhurst M et al. Melanoma-specific $CD4^+$ T cells recognize nonmutated HLA-DR-restricted tyrosinase epitopes. J Exp Med 1996; 183:1965-1971.
115. Pieper R, Christian RE, Gonzales MI et al. Biochemical identification of a mutated human melanoma antigen recognized by $CD4^+$ T cells. J Exp Med 1999; 189:757-766.
116. Wang RF, Wang X, Rosenberg SA. Identification of a novel major histocompatibility complex class II-restricted tumor antigen resulting from a chromosomal rearrangement recognized by $CD4^+$ T cells. J Exp Med 1999; 189:1659-1668.
117. Wang RF, Wang X, Atwood AC et al. Cloning genes encoding MHC class II-restricted antigens: mutated CDC27 as a tumor antigen. Science 1999; 284:1351-1354.
118. Chaux P, Vantomme V, Stroobant V et al. Identification of MAGE-3 epitopes presented by HLA-DR molecules to $CD4^+$ T lymphocytes. J Exp Med 1999; 189:767-778.
119. Schneider J, Gilbert SC, Blanchard TJ et al. Enhanced immunogenicity for $CD8^+$ T cell induction and complete protective efficacy of malaria DNA vaccination by boosting with modified vaccinia virus Ankara. Nat Med 1998; 4:397-402.

Chapter 13

Polysaccharide Vaccines

Stephen Freese

Polysaccharide Immunity

The immune response to polysaccharide antigens is qualitatively different from that to protein antigens. Considering only those aspects that bear upon immunity to polysaccharides, the differences may be roughly described as follows. Immune cells take up and digest foreign proteins, and the resulting peptide fragments are presented by MHC II molecules at the cell surface for recognition by T cells. Upon recognition of the bound peptide by a T cell it stimulates the B cell to multiply, to switch to the production of soluble immunoglobulins and causes memory B cells to be produced. Memory B cells are long lived and are able to undergo an additional round of genetic rearrangement resulting in an ability to produce antibodies of higher affinity than those produced initially. Because these responses are dependent upon stimulation by T cells, proteins are "T-dependent antigens". Their salient features are a memory response (i.e., priming), class switching, and affinity maturation.[1]

In contrast to T-dependent antigens, polysaccharides may be taken up by B cells, but can not be loaded into MHC II molecules and can not recruit T cell help. "T-independent antigens" do not produce a memory response, are largely IgM, and affinity of the antibodies does not increase with time.[2]

An additional feature of T-independent antigens, which is of primary importance for vaccine development, is the failure of children younger than about 18 months to mount an effective response to polysaccharide antigens. Although polysaccharide vaccines are efficacious in older people, they do not protect young children. It is primarily for this reason that conjugate vaccines were developed.[3]

Since the pioneering work of Goebel and Avery[4] it has been appreciated that covalently linking a saccharide to a protein carrier resulted in an immunogen which combined an antisaccharide response with the memory response characteristic of a protein. Essentially a polysaccharide-specific B cell is stimulated and simultaneously provided with a foreign T-cell epitope to garner the benefits T-cell stimulation.

Issues in Designing a Conjugate

There are several structural features which may be involved in conjugate immunogenicity. Among these are identity of carrier, molecular weight of saccharide, chemistry of the covalent bond between saccharide and protein, presence of a linker, attachment at a terminal position of the saccharide or at many points along the chain, and extent of cross-linking of polysaccharides. This is a long list and not surprisingly there are some answers for some constructs, but comprehensive answers are not available.

New Vaccine Technologies, edited by Ronald W. Ellis. ©2001 Eurekah.com.

Carrier Considerations

Although new carriers are continuously evaluated, those in common use are a rather small set. The carriers used in humans are usually one of the following, diphtheria toxoid (DT), a nontoxic DT mutant (cross reacting material, CRM197), tetanus toxoid (TT), and a outer membrane protein complex from *Neisseria meningitidis*. An appealing possibility is to use a carrier which is a functional antigen in its own right, such as pneumolysin toxin from pneumococcus.[5,6]

For human use immunogenicity is not the only issue involved in choosing a carrier. Because the three most common carriers (CRM197, TT and DT) are identical to or cross-reactive with the components in DTP vaccine there is a possibility of vaccine interference. Experience with *Haemophilus influenza*, type B polysaccharide (Hib) conjugate vaccines has shown that prior exposure to the carrier allows an enhanced response.[7,8] In one case simultaneous injection of carrier with a conjugate results in decreased anti-polysaccharide response.[9] However, when CRM197 was the carrier a combination of Hib conjugate and DTP gave higher titers for all the antigens.[10] As the number of coadministered conjugates built on the same carrier increases, the possibility of epitopic suppression becomes somewhat greater.[11]

One potentially useful approach to this problem is the use of peptide carriers.[12] If the immunological requirement for use as a carrier is the presence of T cell epitopes and if these epitopes can be known or predicted then it becomes possible to eliminate most of the protein and use a small peptide as the carrier. Bixler and coworkers did this and determined a minimal, 18 amino acid epitope from CRM197 required to stimulate T cells and showed that a Hib conjugate prepared from it could give an antibody response similar to the full Hib-CRM197 conjugate.[13] Also a peptide which contains both B and T cell epitopes was shown to be a good carrier while a peptide from the same protein which contained only B cell epitopes was inactive as a carrier.[14]

Effect of Chain Length on Immunogenicity

This aspect of conjugate design has attracted more effort than any other, but the answer seems to depend on the identity of the polysaccharide and upon other features of the design.

Seppala and Makela examined the effect of chain length on immunogenicity of dextran conjugates.[15] In order to eliminate the effect of cross-linking and multipoint attachment they conjugated by reductive amination through the reducing terminus either with or without linker. In either case they found that conjugates made from low MW dextran, either 1 or 4 kDa, gave Ab concentrations in mice several-fold higher than those of 40 kDa dextran. Both classes of conjugates boosted equally well, with the smaller preserving their higher response. When a linker was incorporated, the secondary IgG response was generally lower than without linker, but Seppala and Makela showed there was a clear trend of falling Ab concentration with rising saccharide MW.[15]

Working with Hib conjugates in which the saccharide varied from 4-12 RU Anderson et al found essentially no dependence on saccharide size when injected in infants.[16]

Finally, the most important data on saccharide size versus immunogenicity may be found among the clinical data for Hib vaccines. Several studies have shown that a 20 RU conjugate and polysaccharide conjugates all provide protective levels of antibody after three vaccinations although the exact values varied between studies.[17-19]

An important twist on the question of how immunogenicity varies with saccharide length is the possibility of conformational epitopes in a polysaccharide. For the most part the possibility of polysaccharides adopting preferred conformations has been assumed to be unimportant because of ready interconversion of hypothetical ordered structure with a "random" structure. Because the antibody combining site can accommodate about six monosaccharide residues in an extended conformation,[20] then oligosaccharides longer than this should have similar binding strengths in the absence of secondary structure. But Jennings et al observed that in order to inhibit binding of antibody to *N. meningitidis* group B polysaccharide at least 10 residues were required and inhibition continued to improve beyond 17 residues.[21] The authors argue that a conformational epitope is recognized. NMR and molecular modeling studies have supported the view that in solution this polysaccharide can exist as a stretched helix.[22,23] The requirement for an anomalously large oligosaccharide to inhibit binding is explained by the need for a minimum number of repeats to form a stable secondary structure.

This phenomenon is especially important in the case of group B polysaccharide because short oligomers of the same primary structure are found on the surface of certain cells of the nervous system, especially that of the fetus. This presumably accounts for the poor immunogenicity of this polysaccharide,[24,25] whereas the closely related group C polysaccharide is immunogenic.[26]

Inhibition of binding of antibody to polysaccharide only by similarly large oligosaccharides is also found in the case of type III group B streptococcus. In this case inhibition of antibody binding was increasingly inhibited by oligosaccharides up to and beyond 25 repeat units (RU).[27] Conjugates prepared from these oligosaccharides showed a trend toward higher antibody responses with increasing molecular weight. More striking, opsonophagocytic titers and passive protection were greatest with the antisera derived from conjugates made with the intermediate size oligosaccharide.[27] The authors speculated that the larger saccharide may better display the biologically relevant conformation, while smaller size favored a more T-dependent response with the attendant difference in isotype. These opposing trends could result in the medium size saccharide conjugates producing the most appropriate immune response.

Studies looking at other possible structural factors which influence conjugate efficacy are relatively scarce. Wessels et al examined the effect of polysaccharide size, conjugate size and degree of derivatization for group B streptococcal polysaccharides (GBS) type III conjugates made by random reductive amination. Antibody titers were highest for an intermediate size conjugate. Titers also increased strongly with increasing cross-linking, although at the highest degree of derivatization (89%) it seems that a nonnative epitope was primarily recognized.[28]

Issues in Making a Conjugate

Molecular Weight

Production of size-fractionated polysaccharide can be accomplished by several methods. Separation of native material by anion exchange columns is appropriate for oligosaccharides, but the resolution of adjacent N-mers fails after about 40 repeat units. Resolution of N-mers on a preparative scale has an even lower limit.

In order to reduce the size of native polysaccharide, hydrolysis is commonly employed when the backbone contains acid-labile groups such as phosphate or sialic acid (e.g., Hib or N. meningitidis groups B and C). Also periodate oxidation will cleave some polysaccharides (e.g., Hib and *N. meningitidis* group C).

Shearing by ultrasonication is a general technique which has been used for the Vi polysaccharide antigen of *Salmonella typhi*, Hib polysaccharide, dextran and pneumococcal polysaccharides. Based on mechanical rupturing of bonds, this technique is generally applicable. Also rupture tends to occur in the center of larger polysaccharides resulting in a fairly narrow distribution of molecular weight. Its primary drawback is an apparent lower limit on the size of polysaccharide that can be produced of about 50 kDa.[29]

Exceptionally, a glycosidase can be found to specifically cleave the polysaccharide backbone. Paoletti and coworkers generated oligosaccharides from type III GBS.[30]

Zou et al developed a chemical method to cleave glucosamine-containing polysaccharides. In this technique N-acetylglucosamines in the backbone are partially deacetylated, then reacted with nitrous acid. This results in an unstable diazonium intermediate which rearranges to eliminate the aglycon portion. The reducing end becomes an unnatural anhydromannofuranose bearing a free aldehyde.[31] They have used this to prepare oligosaccharides from one to over 200 repeat units. This technique should be applicable to samples in which aminosugars are present only in the backbone. Because the mechanism requires backside attack of the ring oxygen on the diazonium-bearing carbon and competing reactions are possible, the outcome is probably sensitive to conformation. For these reasons it may be restricted to particular glucosamine-containing polysaccharides, although galactosamine cleavage is also possible.

It has been shown that beta-D-aldosidic linkages can be oxidized to esters by ozone. Hydrolysis in mild base results in cleavage of the backbone without affecting acid sensitive sialic acid groups on the side chain. A great deal of specificity was obtained, apparently due to conformationally dependent differences in reactivity. This has been applied to Group B streptococcal polysaccharide but should be more generally applicable. It requires the preparation of per-acetyl polysaccharides and use of organic solvent which necessarily lengthen the procedure.[32]

Role of Labile Polysaccharide Epitopes

Preservation of immunologically important, labile epitopes is an essential goal during the isolation and conjugation of polysaccharides. Variable extents of O-acetylation are commonly encountered and these base-labile groups may or may not be immunologically important. Szu et al found that partial de-O-acetylation of Vi polysaccharide from *Salmonella typhi* had little effect on the antigenicity of the polysaccharide, presumably because a high concentration of O-acetyls are not critical for immunogenicity, but complete de-O-acetylation eliminated the response.[33]

In the case of type 9V pneumococcal polysaccharide, O-acetate response was present following vaccination with polysaccharide, but this specificity was not required for functional antibodies.[34] Konadu et al found that in order to obtain bactericidal antibodies from *S. paratyphi* O-specific polysaccharide O-acetyl groups were essential.[35]

Conjugation Chemistry

A large body of chemistry has been developed to achieve the conjugation of two macromolecules and is accordingly the subject of several reviews. An excellent and comprehensive review of the subject has been given by Dick et al,[36] so I will merely discuss certain aspects of some of the chemistry used commonly in the current literature. Several factors influence the choice of coupling chemistry. The linkage site should not induce antibodies specific for the linker. It must use groups available on the polysaccharide or that can be introduced. Reaction conditions must not destroy important epitopes.

Cyanogen bromide activation is still popular because it is quick, applicable to every polysaccharide, and has a long history of providing immunogenic conjugates. This reagent is attacked by hydroxyl groups in the polysaccharide to form a reactive cyanate ester which can then couple to amine groups on the protein leaving an isourea linkage.[37,38] Drawbacks are the noxious nature of the reagent and the high pH (~11) at which the initial step is performed. Both of these are addressed by use of the cyanylating reagent introduced by Kohn and Wilchek,[39] and later applied by Lees et al[40] This reagent, cyanodimethylaminopyridinium tetrafluoroborate (CDAP), is essentially analogous to the acylpridinium formed in pyridine-catalyzed acylation reactions. Because the cyanate ester intermediate is unstable it generally is not feasible to assay the intermediate. Also in some cases a linker is desirable to provide a better immune response.

For these reasons it is common to react the activated polysaccharide with adipic dihydrazide.[41-43] The hydrazide is easily assayed and very reactive.

Direct reductive amination was used in the first licensed conjugate,[44-46] and remains one of the best techniques. It does not introduce extraneous atoms and can be used on most polysaccharides and especially oligosaccharides. In this technique unstable Schiff bases formed between aldehydes and primary amines are specifically reduced by sodium cyanoborohydride.[47] This reagent has the appropriate activity to reduce the transiently formed Schiff base while leaving the less reactive aldehyde untouched. The Achilles' heel of the approach is the formation of aldehydes. In principle each polysaccharide molecule has one aldehyde at the reducing end of the chain. However, not only is the concentration of the free aldehyde vanishingly low, but the concentration of end groups in a polysaccharide is tiny. Taken together this means that even in a concentrated polysaccharide solution the aldehyde concentration is a few micromolar. This explains why reaction times can be very long. In one case, modest size dextrans required as much as 70 days to react.[15] This limitation of aldehyde concentration is commonly circumvented by periodate oxidation which may be used to introduce free aldehydes.[16]

Jennings and Lugowski used reductive amination to conjugate group A meningitidis polysaccharide.[26] Internal saccharides are resistant to oxidation because of the presence of O-acetates, but unlike group B the ends are also resistant. They circumvented this by first reducing the reducing end to destroy the pyranose ring structure leaving a straight chain diol. This could be oxidized to produce the free terminal aldehyde.

Ethyl(Dimethylaminopropyl)Carbodiimide (EDC) selectively activates carboxylic acids for nucleophillic attack by amines or hydrazides.[48] Most commonly, EDC is used with polysaccharide and an excess of ADH.[49] After removal of unbound ADH the polysaccharide is conjugated to carrier using more EDC. There are many variations possible on this theme. An active ester of N-hydroxysuccinimide (NHS) may be formed and promptly displaced by an amine on the other macromolecule.[50] The only limitations are the need for a carboxylic acid and the possibility of intramolecular cross-linking competing with intermolecular. This is especially the case when using sialic acid-containing polysaccharides. It has been shown that these will readily form lactones.[51]

Thioether bonds are introduced in two steps. Amines are converted to bromoacetamido groups by reaction with the corresponding reactive ester. Thiol groups can be introduced by EDC mediated reaction with cystamine followed by reduction. Thiol is an excellent nucleophile especially when bromide is the leaving group and mixing the components effects conjugation.[52,53]

A similar approach is the bigeneric linker approach of Marburg et al.[54] It is a general, if rather challenging method. Hydroxyl groups of the polysaccharide were derivatized by carbonyldiimidazole and used to attach amine groups. These were bromoacetylated by an active ester. Thiol groups were introduced in the protein and combination of the two components resulted in conjugation via a stable thioether bond.

In performing a coupling reaction many workers use linkers. There are several reasons, in principle, for doing so. From an immunological point of view it may be thought that separating the polysaccharide from the protein surface will give a more immunogenic construct. This sometimes is the case,[55] however, the excellent success of reductively aminated Hib, which does not use a linker, shows that this is not generally true.[45] Use of a linker may allow a reaction to proceed faster because of relief of the steric interaction attendant upon linking one macromolecule to another. Set against this is the possibility that the linker region itself will be immunodominant.[26]

Applications

Haemophilus Influenzae, Type b (Hib)

Haemophilus influenzae type b conjugates, first commercialized by Praxis Biologics (now part of Wyeth Lederle Vaccines), proved to be a splendid success and established the clinical importance of conjugates.[56] These vaccines have led to the virtual eradication of invasive Hib among vaccinated children.[57-59] Not only were these vaccines effective in preventing meningitis among toddlers, but they reduced carriage of the bacteria in the nasopharynx[60,61] and produced a high degree of herd immunity among the unvaccinated as well.[62]

The Hib conjugate vaccines introduced separately by Robbins[38] and Anderson[45] were different in construction, but similar in effect. Robbins' Hib conjugates were made by attaching adipic acid dihydrazide (ADH) to various proteins and then reacting with cyanogen bromide-activated polysaccharide. These were immunogenic, regardless of the carrier, and had the traits of t-dependent antigens. Anderson's conjugates were made from oligosaccharides of various sizes which were coupled to a diphtheria toxin derivative (cross-reacting material, CRM197) by reductive amination. They were well immunogenic and boostable as found for Robbins preparations.[63]

Among licensed Hib conjugates the chemistry varies, but all are protective after three doses. Note should be made of the Merck vaccine made by the Marburg method. In this instance there is a primary antibody response not seen in other vaccines.[19] A discussion of the conjugation chemistry used in licensed Hib vaccines has been published.[64]

Streptococcus Pneumoniae

Their are many capsule types associated with this species, but fortunately a few serotypes account for most disease.[65] In a large clinical trial, it recently was found that the mulitvalent conjugate vaccine developed by Wyeth Lederle Vaccines was 100% efficacious in preventing meningitis due to the vaccine serotypes.

Group B Streptococcus (GBS)

Five serotypes of GBS are important causes of neonatal meningitis.[66] It has been shown that conjugates against several of the GBS serotypes are immunogenic in humans and protective in animal models.[67]

Staphylococcus Aureus

Polysaccharide from the only major pathogenic serotypes, types 5 and 8,[68] have been conjugated and shown to be highly immunogenic.[69] Antibodies were opsonophagocytic and protective in a mouse i.p. challenge model.[70] The efficacy of anticapsular response in protection in an endocarditis model has been disputed. It has been argued that capsule was irrelevant to virulence,[71-73] or actually antivirulent.[74] Recently a consensus has developed that anticapsular antibodies are protective, at least in some models.[75-77] A trial in humans has not shown protection.[78] It may prove that a recently reported polysaccharide common antigen may be developed into a useful vaccine.[79]

Neisseria Meningitidis, Group B

Short oligosaccharides chemically identical to the group B polysaccharide of *Neisseria meningitidis* have been found in some human tissue.[24] Tolerance to these saccharides may explain the poor immunogenicity of group B meningococcal polysaccharide,[80] even as a conjugate.[26] A possible way around this difficulty is N-propionylated group B meningococcal polysaccharide in which the N-acetyl groups were removed and replaced with propionyl groups to obtain a novel material.[81,82] Conjugates of this new polysaccharide could induce antibodies, some of which cross-react with the native polysaccharide, but not with short oligosaccharides.[83,84] The explanation given for this is that the N-propionylated polysaccharide mimicked a conformational epitope found only in long stretches of native polysaccharide.

More recently some success has been achieved in raising antibodies using conventional conjugates of group B meningococcus polysaccharide.[85-87]

Neisseria Meningitidis, Other Groups

Tetravalent meningococcal polysaccharide (groups A, C, Y, W-135) has been available for use in adults and are natural targets of conjugate vaccine development. Conjugates of groups A and C polysaccharide have been reported.[88,89] Group C conjugates have been very successful in clinical trials has been in Great Britain soon. Recently there has been a notable increase in the incidence of group Y disease,[90] such that this and perhaps W-135 are now also promising targets.

LOS and LPS Conjugates

Gram negative bacteria produce, in addition to capsule, lipopolysaccharide (LPS, or if the repeating O saccharide is absent, LOS). These materials are attractive as vaccine candidates for two reasons. Much of the pathogenesis of gram negative infection is due not to the bacteria itself, but, released LPS.[91] Antibodies able to bind LPS and prevent its toxic effect might reduce the morbidity of the disease.[92] Also, because of the surface location of LPS, anti-LPS antibodies could be functional in clearing an infection. Caution is warranted because LPS is a subcapsular antigen; it is common that LPS is serologically invisible until a capsule is removed.[93] The attractiveness of LPS-based vaccines is especially great in the case of nonencapsulated bacteria (e.g., nontypeable H. influenzae) and bacteria for which the capsule is poorly immunogenic (e.g., N. meningitidis, group B).

The O-specific polysaccharide is heterogeneous within a strain; there are over 160 serotypes of *E. coli.*[93] Furthermore, within a strain there is considerable variation of substituents in the core.[94-96] Many different structures may be present in LOS isolated from a single culture. This raises the question of which if any of these molecules are present during an infection and which may be protective antigens. Sialidation, which results in poorly immunogenic structures, can occur.[97] Taken together this suggests caution in assuming that a vaccine derived from media-grown bacteria will contain the same epitopes as those present during infection.[98]

For the purpose of vaccine design, once the lipid A is removed from LPS the remaining polysaccharide can be dealt with using the same techniques as with capsular polysaccharides and conjugates of this nature are mentioned in separate sections above.

Lipid A is highly toxic; although conjugation reduces this to some extent, it is generally necessary to detoxify it. It has been found that removal of some of the O-linked lipid groups from lipid A substantially detoxify LPS. This has been accomplished by treatment with sodium hydroxide or with hydrazine.[99] An alternative to removing the lipid groups is to remove the lipid A altogether. This may be accomplished by mild acid hydrolysis taking advantage of the lability of the glycosidic bond of the adjacent KDO.

Gu and coworkers prepared conjugates from the LOS of nontypeable *H. influenzae,*[99] and *Moraxella catarrhalis*[100] by first detoxifying with hydrazine, then coupling through the carboxylic acid group of KDO. This provided vaccines which proved to be antigenic and induce opsonophagocytosis-promoting antibodies. In the case of NTHi it was also shown that the vaccines were significantly protective in the chinchilla model of otitis media.[101]

Pozsgay synthesized antigens corresponding to an O-specific polysaccharide of *Shigella dysenteria* in which 1-4 tetrasaccharide repeat units were single end coupled to albumin. Because the size of each saccharide was homogenous and relatively small, it was possible to use mass spectrometry to exactly characterize the number and size of saccharides on the resulting conjugates.[102] Using these conjugates Pozsgay et al demonstrated an optimal length of saccharide and optimal loading of saccharide molecules/protein molecule.[103]

Anti-Cancer Conjugate Vaccines

Several carbohydrate antigens are over-expressed on the surface of some types of cancerous cells. Although cell-mediated immunity is thought to be crucial for cancer immunity, higher antibody titers to these oligosaccharides tend to be correlated with an improved prognosis.[104] Therefore they have attracted attention as vaccine candidates. The target population would generally be adults, but the oligosaccharides under consideration are too small to be antigenic in their own right, so conjugation is being used.

In one case, GM3 ganglioside was "conjugated" by hydrophobic interaction to an outer membrane complex of proteins and lipids from *Neisseria meningitidis*.[105] This unusual vaccine construct was shown to have some efficacy in animal models, but its status as a T-dependent antigen was not established.

In a substantial body of work, several oligosaccharides, corresponding to gangliosides,[104] sialyl-TN,[106] Lewis-y,[107] globo H[108] and related saccharides from 2-7 residues in length have been chemically synthesized. They take advantage of the luxury synthetic chemists enjoy in designing linkage sites to order. Ceramides contain a carbon-carbon double bond which these authors echo by introducing a double bond-containing spacer in their synthetic product. This double bond is specifically cleaved with ozone and the resulting aldehyde coupled by reductive amination or to a maleimide-containing linker. Several of these vaccines are in clinical trials and are the only clinical applications of synthetic saccharide conjugates.

Abbreviations: KDO, ketodeoxyoctulosonic acid; RU, repeat unit; GBS, group B streptococcus; Hib, *Haemophilus influenzae* type b; CRM197, diphtheria toxin cross-reacting material; ADH, adipic acid dihydrazide; Ab, antibody

References

1. Goldblatt D. Recent developments in bacterial conjugate vaccines. J Med Microbiol 1998; 47(7):563-567.
2. Stein P, Dubois P, Greenblatt D et al. Induction of antigen-specific proliferation in affinity-purified small B lymphocytes: requirement for BSF-1 by type 2 but not type 1 thymus- independent antigens. J Immunol 1986; 136(6):2080-2089.
3. Becker RS. Conjugate vaccines: Practice and theory. Springer Semin Immunopathol 1993; 15(2-3):217-226.
4. Avery OT, Goebel WF. Chemo-immunological studies on conjugated carbohydrate-proteins. II. Immunological specificity of synthetic sugar-protein antigens. J Exp Med 1929; 50:533-550.
5. Michon F, Fusco PC, Minetti CA et al. Multivalent pneumococcal capsular polysaccharide conjugate vaccines employing genetically detoxified pneumolysin as a carrier protein. Vaccine 1998; 16(18):1732-1741.
6. Kuo J, Douglas M, Ree HK et al. Characterization of a recombinant pneumolysin and its use as a protein carrier for pneumococcal type 18C conjugate vaccines. Infect Immun 1995; 63(7):2706-2713.
7. Kurikka S. Priming with diphtheria-tetanus-pertussis vaccine enhances the response to the *Haemophilus influenzae* type b tetanus conjugate vaccine in infancy. Vaccine 1996; 14(13):1239-1242.
8. Granoff DM, Rathore MH, Holmes SJ et al. Effect of immunity to the carrier protein on antibody responses to *Haemophilus influenzae* type b conjugate vaccines. Vaccine 1993; 11 Suppl 1:S46-51.
9. Dagan R, Eskola J, Leclerc C et al. Reduced response to multiple vaccines sharing common protein epitopes that are administered simultaneously to infants. Infect Immun 1998; 66(5):2093-2098.
10. Paradiso P, Hogerman D, Madore D et al. Safety and Immunogenicity in infants of a Tetravalent vaccine composed of HbOC (HibTiter) and DTP (Triimmunol). Pediatric Research 1992; 31:174A.

11. Fattom A, Cho YH, Chu C et al. Epitopic overload at the site of injection may result in suppression of the immune response to combined capsular polysaccharide conjugate vaccines. Vaccine 1999; 17(2):126-133.
12. Paradiso PR, Dermody K, Pillai S. Novel Approaches to the Development of Glycoconjugate Vacccines with Synthetic Peptides as Carriers. Vaccine Research 1993; 2(4):239-248.
13. Bixler GS, Jr., Eby R, Dermody KM et al. Synthetic peptide representing a T-cell epitope of CRM197 substitutes as carrier molecule in a *Haemophilus influenzae* type B (Hib) conjugate vaccine. Adv Exp Med Biol 1989; 251:175-180.
14. Lett E, Gangloff S, Zimmermann M et al. Immunogenicity of polysaccharides conjugated to peptides containing T- and B-cell epitopes. Infect Immun 1994; 62(3):785-792.
15. Seppala I, Makela O. Antigenicity of dextran-protein conjugates in mice. Effect of molecular weight of the carbohydrate and comparison of two modes of coupling. J Immunol 1989; 143(4):1259-1264.
16. Anderson PW, Pichichero ME, Stein EC et al. Effect of oligosaccharide chain length, exposed terminal group, and hapten loading on the antibody response of human adults and infants to vaccines consisting of *Haemophilus influenzae* type b capsular antigen unterminally coupled to the diphtheria protein CRM197. J Immunol 1989; 142(7):2464-2468.
17. Decker MD, Edwards KM, Bradley R et al. Comparative trial in infants of four conjugate *Haemophilus influenzae* type b vaccines [see comments]. J Pediatr 1992; 120(2 Pt 1):184-189.
18. Holmes SJ, Murphy TV, Anderson RS et al. Immunogenicity of four *Haemophilus influenzae* type b conjugate vaccines in 17- to 19-month-old children. J Pediatr 1991; 118(3):364-371.
19. Granoff DM, Anderson EL, Osterholm MT et al. Differences in the immunogenicity of three *Haemophilus influenzae* type b conjugate vaccines in infants. J Pediatr 1992; 121(2):187-194.
20. Glaudemans CP. Seven structurally different murine monoclonal galactan-specific antibodies show identity in their galactosyl-binding subsite arrangements. Mol Immunol 1987; 24(4):371-377.
21. Jennings HJ, Roy R, Michon F. Determinant specificities of the groups B and C polysaccharides of Neisseria meningitidis. J Immunol 1985; 134(4):2651-2657.
22. Brisson JR, Baumann H, Imberty A et al. Helical epitope of the group B meningococcal alpha(2-8)-linked sialic acid polysaccharide. Biochemistry 1992; 31(21):4996-5004.
23. Evans SV, Sigurskjold BW, Jennings HJ et al. Evidence for the extended helical nature of polysaccharide epitopes. The 2.8 A resolution structure and thermodynamics of ligand binding of an antigen binding fragment specific for alpha-(2—>8)-polysialic acid. Biochemistry 1995; 34(20):6737-6744.
24. Weisgerber C, Husmann M, Frosch M et al. Embryonic neural cell adhesion molecule in cerebrospinal fluid of younger children: age-dependent decrease during the first year. J Neurochem 1990; 55(6):2063-2071.
25. Nedelec J, Boucraut J, Garnier JM et al. Evidence for autoimmune antibodies directed against embryonic neural cell adhesion molecules (N-CAM) in patients with group B meningitis. J Neuroimmunol 1990; 29(1-3):49-56.
26. Jennings HJ, Lugowski C. Immunochemistry of groups A, B, and C meningococcal polysaccharide- tetanus toxoid conjugates. J Immunol 1981; 127(3):1011-1018.
27. Paoletti LC, Kasper DL, Michon F et al. Effects of chain length on the immunogenicity in rabbits of group B Streptococcus type III oligosaccharide-tetanus toxoid conjugates. J Clin Invest 1992; 89(1):203-209.
28. Wessels MR, Paoletti LC, Guttormsen HK et al. Structural properties of group B streptococcal type III polysaccharide conjugate vaccines that influence immunogenicity and efficacy. Infect Immun 1998; 66(5):2186-2192.
29. Szu SC, Zon G, Schneerson R et al. Ultrasonic irradiation of bacterial polysaccharides. Characterization of the depolymerized products and some applications of the process. Carbohydr Res 1986; 152:7-20.
30. Paoletti LC, Kasper DL, Michon F et al. An oligosaccharide-tetanus toxoid conjugate vaccine against type III group B Streptococcus. J Biol Chem 1990; 265(30):18278-18283.
31. Zou W, Laferriere CA, Jennings HJ. Oligosaccharide fragments of the type III group B streptococcal polysaccharide derived from S. pneumoniae type 14 capsular polysaccharide by a chemoenzymatic method. Carbohydr Res 1998; 309(3):297-301.

32. Wang Y, Hollingsworth RI, Kasper DL. Ozonolysis for selectively depolymerizing polysaccharides containing beta-D-aldosidic linkages. Proc Natl Acad Sci U S A 1998; 95(12):6584-6589.
33. Szu SC, Li XR, Stone AL et al. Relation between structure and immunologic properties of the Vi capsular polysaccharide. Infect Immun 1991; 59(12):4555-4561.
34. McNeely TB, Staub JM, Rusk CM et al. Antibody responses to capsular polysaccharide backbone and O-acetate side groups of Streptococcus pneumoniae type 9V in humans and rhesus macaques. Infect Immun 1998; 66(8):3705-3710.
35. Konadu E, Shiloach J, Bryla DA et al. Synthesis, characterization, and immunological properties in mice of conjugates composed of detoxified lipopolysaccharide of Salmonella paratyphi A bound to tetanus toxoid with emphasis on the role of O acetyls. Infect Immun 1996; 64(7):2709-2715.
36. Dick WE, Jr., Beurret M. Glycoconjugates of bacterial carbohydrate antigens. A survey and consideration of design and preparation factors. Contrib Microbiol Immunol 1989; 10:48-114.
37. Jennissen HP. Cyanogen bromide and tresyl chloride chemistry revisited: the special reactivity of agarose as a chromatographic and biomaterial support for immobilizing novel chemical groups. J Mol Recognit 1995; 8(1-2):116-124.
38. Schneerson R, Barrera O, Sutton A et al. Preparation, characterization, and immunogenicity of *Haemophilus influenzae* type b polysaccharide-protein conjugates. J Exp Med 1980; 152(2):361-376.
39. Kohn J, Wilchek M. 1-Cyano-4-dimethylamino pyridinium tetrafluoroborate as a cyanylating agent for the covalent attachment of lignad to polysaccharide resins. FEBS Letters 1983; 154:209-210.
40. Lees A, Nelson BL, Mond JJ. Activation of soluble polysaccharides with 1-cyano-4- dimethylaminopyridinium tetrafluoroborate for use in protein- polysaccharide conjugate vaccines and immunological reagents. Vaccine 1996; 14(3):190-198.
41. Gupta RK, Taylor DN, Bryla DA et al. Phase 1 evaluation of Vibrio cholerae O1, serotype Inaba, polysaccharide-cholera toxin conjugates in adult volunteers. Infect Immun 1998; 66(7):3095-3099.
42. Takamatsu-Matsushita N, Yamaguchi N, Kawasaki M et al. Immunogenicity of Actinobacillus actinomycetemcomitans serotype b- specific polysaccharide-protein conjugate. Oral Microbiol Immunol 1996; 11(4):220-225.
43. Han Y, Ulrich MA, Cutler JE. Candida albicans mannan extract-protein conjugates induce a protective immune response against experimental candidiasis [In Process Citation]. J Infect Dis 1999; 179(6):1477-1484.
44. Smith DH, Madore DV, Eby RJ et al. Haemophilus b oligosaccharide-CRM197 and other Haemophilus b conjugate vaccines: a status report. Adv Exp Med Biol 1989; 251:65-82.
45. Anderson P. Antibody responses to *Haemophilus influenzae* type b and diphtheria toxin induced by conjugates of oligosaccharides of the type b capsule with the nontoxic protein CRM197. Infect Immun 1983; 39(1):233-238.
46. Anderson P, Pichichero M, Edwards K et al. Priming and induction of *Haemophilus influenzae* type b capsular antibodies in early infancy by Dpo20, an oligosaccharide-protein conjugate vaccine. J Pediatr 1987; 111(5):644-650.
47. Schwartz BA, Gray GR. Proteins containing reductively aminated disaccharides. Synthesis and chemical characterization. Arch Biochem Biophys 1977; 181(2):542-549.
48. Nakajima N, Ikada Y. Mechanism of amide formation by carbodiimide for bioconjugation in aqueous media. Bioconjug Chem 1995; 6(1):123-130.
49. Schifferle RE, Chen PB, Davern LB et al. Modification of experimental Porphyromonas gingivalis murine infection by immunization with a polysaccharide-protein conjugate. Oral Microbiol Immunol 1993; 8(5):266-271.
50. Grabarek Z, Gergely J. Zero-length crosslinking procedure with the use of active esters. Anal Biochem 1990; 185(1):131-135.
51. Lifely MR, Gilbert AS, Moreno C. Rate, mechanism, and immunochemical studies of lactonisation in serogroup B and C polysaccharides of Neisseria meningitidis. Carbohydr Res 1984; 134(2):229-243.
52. Bernatowicz MS, Matsueda GR. Preparation of peptide-protein immunogens using N-succinimidyl bromoacetate as a heterobifunctional crosslinking reagent. Anal Biochem 1986; 155(1):95-102.
53. DeWeers O, Beurret M, van Buren L et al. Application of cystamine and N,N'-Bis(glycyl)cystamine as linkers in polysaccharide-protein conjugation. Bioconjug Chem 1998; 9(3):309-315.

54. Marburg S, Jorn D, Tolman RL et al. Bimolecular Chemistry of Macromolecules: Synthesis of Bacterial Polysaccharide Conjugates with *Neisseria meningitidis* Membrane Protein. J Am Chem Soc. 1986; 108:5282-5287.
55. Fattom A, Shiloach J, Bryla D et al. Comparative immunogenicity of conjugates composed of the Staphylococcus aureus type 8 capsular polysaccharide bound to carrier proteins by adipic acid dihydrazide or N-succinimidyl-3-(2-pyridyldithio)propionate. Infect Immun 1992; 60(2):584-589.
56. Robbins JB, Schneerson R, Anderson P et al. The 1996 Albert Lasker Medical Research Awards. Prevention of systemic infections, especially meningitis, caused by *Haemophilus influenzae* type b. Impact on public health and implications for other polysaccharide-based vaccines. Jama 1996; 276(14):1181-1185.
57. Eskola J, Peltola H, Kayhty H et al. Finnish efficacy trials with *Haemophilus influenzae* type b vaccines. J Infect Dis 1992; 165 Suppl 1:S137-138.
58. Frasch CE, Hiner EE, Gross TP. Haemophilus b disease after vaccination with Haemophilus b polysaccharide or conjugate vaccine. Am J Dis Child 1991; 145(12):1379-1382.
59. Black SB, Shinefield HR. Immunization with oligosaccharide conjugate *Haemophilus influenzae* type b (HbOC) vaccine on a large health maintenance organization population: extended follow-up and impact on *Haemophilus influenzae* disease epidemiology. The Kaiser Permanente Pediatric Vaccine Study Group. Pediatr Infect Dis J 1992; 11(8):610-613.
60. Takala AK, Eskola J, Leinonen M et al. Reduction of oropharyngeal carriage of *Haemophilus influenzae* type b (Hib) in children immunized with an Hib conjugate vaccine. J Infect Dis 1991; 164(5):982-986.
61. Barbour ML, Mayon-White RT, Coles C et al. The impact of conjugate vaccine on carriage of *Haemophilus influenzae* type b. J Infect Dis 1995; 171(1):93-98.
62. Peltola H, Aavitsland P, Hansen KG et al. Perspective: a five-country analysis of the impact of four different *Haemophilus influenzae* type b conjugates and vaccination strategies in Scandinavia. J Infect Dis 1999; 179(1):223-229.
63. Anderson P, Pichichero ME, Insel RA. Immunogens consisting of oligosaccharides from the capsule of *Haemophilus influenzae* type b coupled to diphtheria toxoid or the toxin protein CRM197. J Clin Invest 1985; 76(1):52-59.
64. Powell MF, Newman MJ, eds. Vaccine Design. New York, London: Plenum Press 1995.
65. Butler JC, Breiman RF, Lipman HB et al. Serotype distribution of Streptococcus pneumoniae infections among preschool children in the United States, 1978-1994: Implications for development of a conjugate vaccine. J Infect Dis 1995; 171(4):885-889.
66. Hickman ME, Rench MA, Ferrieri P et al. Changing epidemiology of group B streptococcal colonization. Pediatrics 1999; 104(2 Pt 1):203-209.
67. Paoletti LC, Wessels MR, Michon F et al. Group B Streptococcus type II polysaccharide-tetanus toxoid conjugate vaccine. Infect Immun 1992; 60(10):4009-4014.
68. Sompolinsky D, Samra Z, Karakawa WW et al. Encapsulation and capsular types in isolates of Staphylococcus aureus from different sources and relationship to phage types. J Clin Microbiol 1985; 22(5):828-834.
69. Fattom A, Schneerson R, Szu SC et al. Synthesis and immunologic properties in mice of vaccines composed of Staphylococcus aureus type 5 and type 8 capsular polysaccharides conjugated to Pseudomonas aeruginosa exotoxin A. Infect Immun 1990; 58(7):2367-2374.
70. Fattom AI, Sarwar J, Ortiz A et al. A Staphylococcus aureus capsular polysaccharide (CP) vaccine and CP- specific antibodies protect mice against bacterial challenge. Infect Immun 1996; 64(5):1659-1665.
71. Albus A, Arbeit RD, Lee JC. Virulence of Staphylococcus aureus mutants altered in type 5 capsule production. Infect Immun 1991; 59(3):1008-1014.
72. Nemeth J, Lee JC. Antibodies to capsular polysaccharides are not protective against experimental Staphylococcus aureus endocarditis. Infect Immun 1995; 63(2):375-380.
73. Xu S, Arbeit RD, Lee JC. Phagocytic killing of encapsulated and microencapsulated Staphylococcus aureus by human polymorphonuclear leukocytes. Infect Immun 1992; 60(4):1358-1362.
74. Baddour LM, Lowrance C, Albus A et al. Staphylococcus aureus microcapsule expression attenuates bacterial virulence in a rat model of experimental endocarditis. J Infect Dis 1992; 165(4):749-753.

75. Thakker M, Park JS, Carey V et al. Staphylococcus aureus serotype 5 capsular polysaccharide is antiphagocytic and enhances bacterial virulence in a murine bacteremia model. Infect Immun 1998; 66(11):5183-5189.
76. Nilsson IM, Lee JC, Bremell T et al. The role of staphylococcal polysaccharide microcapsule expression in septicemia and septic arthritis. Infect Immun 1997; 65(10):4216-4221.
77. Lee JC, Park JS, Shepherd SE et al. Protective efficacy of antibodies to the Staphylococcus aureus type 5 capsular polysaccharide in a modified model of endocarditis in rats [In Process Citation]. Infect Immun 1997; 65(10):4146-4151.
78. Welch PG, Fattom A, Moore JJ et al. Safety and immunogenicity of Staphylococcus aureus type 5 capsular polysaccharide-Pseudomonas aeruginosa recombinant exoprotein A conjugate vaccine in patients on hemodialysis. J Am Soc Nephrol 1996; 7(2):247-253.
79. McKenney D, Pouliot KL, Wang Y et al. Broadly protective vaccine for Staphylococcus aureus based on an in vivo-expressed antigen. Science 1999; 284(5419):1523-1527.
80. Wyle FA, Artenstein MS, Brandt BL et al. Immunologic response of man to group B meningococcal polysaccharide vaccines. J Infect Dis 1972; 126(5):514-521.
81. Jennings HJ, Roy R, Gamian A. Induction of meningococcal group B polysaccharide-specific IgG antibodies in mice by using an N-propionylated B polysaccharide-tetanus toxoid conjugate vaccine. J Immunol 1986; 137(5):1708-1713.
82. Hayrinen J, Jennings H, Raff HV et al. Antibodies to polysialic acid and its N-propyl derivative: binding properties and interaction with human embryonal brain glycopeptides. J Infect Dis 1995; 171(6):1481-1490.
83. Pon RA, Lussier M, Yang QL et al. N-Propionylated group B meningococcal polysaccharide mimics a unique bactericidal capsular epitope in group B Neisseria meningitidis. J Exp Med 1997; 185(11):1929-1938.
84. Granoff DM, Bartoloni A, Ricci S et al. Bactericidal monoclonal antibodies that define unique meningococcal B polysaccharide epitopes that do not cross-react with human polysialic acid. J Immunol 1998; 160(10):5028-5036.
85. Devi SJ, Zollinger WD, Snoy PJ et al. Preclinical evaluation of group B Neisseria meningitidis and Escherichia coli K92 capsular polysaccharide-protein conjugate vaccines in juvenile rhesus monkeys. Infect Immun 1997; 65(3):1045-1052.
86. Bartoloni A, Norelli F, Ceccarini C et al. Immunogenicity of meningococcal B polysaccharide conjugated to tetanus toxoid or CRM197 via adipic acid dihydrazide. Vaccine 1995; 13(5):463-470.
87. Zollinger WD, Moran EE, Devi SJ et al. Bactericidal antibody responses of juvenile rhesus monkeys immunized with group B Neisseria meningitidis capsular polysaccharide-protein conjugate vaccines. Infect Immun 1997; 65(3):1053-1060.
88. Richmond P, Borrow R, Miller E et al. Meningococcal serogroup C conjugate vaccine is immunogenic in infancy and primes for memory [In Process Citation]. J Infect Dis 1999; 179(6):1569-1572.
89. Anderson EL, Bowers T, Mink CM et al. Safety and immunogenicity of meningococcal A and C polysaccharide conjugate vaccine in adults. Infect Immun 1994; 62(8):3391-3395.
90. Racoosin JA, Whitney CG, Conover CS et al. Serogroup Y meningococcal disease in Chicago, 1991-1997. Jama 1998; 280(24):2094-2098.
91. Zahringer U, Lindner B, Rietschel ET. Molecular structure of lipid A, the endotoxic center of bacterial lipopolysaccharides. Adv Carbohydr Chem Biochem 1994; 50:211-276.
92. Fletcher MA, Kloczewiak MA, Loiselle PM et al. A novel peptide-IgG conjugate, CAP18(106-138)-IgG, that binds and neutralizes endotoxin and kills gram-negative bacteria. J Infect Dis 1997; 175(3):621-632.
93. Orskov I, Orskov F, Jann B et al. Serology, chemistry, and genetics of O and K antigens of Escherichia coli. Bacteriol Rev 1977; 41(3):667-710.
94. Schweda EK, Hegedus OE, Borrelli S et al. Structural studies of the saccharide part of the cell envelope lipopolysaccharide from *Haemophilus influenzae* strain AH1-3 (lic3+). Carbohydr Res 1993; 246:319-330.
95. Phillips NJ, McLaughlin R, Miller TJ et al. Characterization of two transposon mutants from *Haemophilus influenzae* type b with altered lipooligosaccharide biosynthesis. Biochemistry 1996; 35(18):5937-5947.

96. Masoud H, Moxon ER, Martin A et al. Structure of the variable and conserved lipopolysaccharide oligosaccharide epitopes expressed by *Haemophilus influenzae* serotype b strain Eagan. Biochemistry 1997; 36(8):2091-2103.
97. Mandrell RE, McLaughlin R, Aba Kwaik Y et al. Lipooligosaccharides (LOS) of some Haemophilus species mimic human glycosphingolipids, and some LOS are sialylated. Infect Immun 1992; 60(4):1322-1328.
98. Rune Andersen S, Kolberg J, Hoiby EA et al. Lipopolysaccharide heterogeneity and escape mechanisms of Neisseria meningitidis: possible consequences for vaccine development. Microb Pathog 1997; 23(3):139-155.
99. Gu XX, Tsai CM, Ueyama T et al. Synthesis, characterization, and immunologic properties of detoxified lipooligosaccharide from nontypeable *Haemophilus influenzae* conjugated to proteins. Infect Immun 1996; 64(10):4047-4053.
100. Gu XX, Chen J, Barenkamp SJ et al. Synthesis and characterization of lipooligosaccharide-based conjugates as vaccine candidates for Moraxella (Branhamella) catarrhalis. Infect Immun 1998; 66(5):1891-1897.
101. Gu XX, Sun J, Jin S et al. Detoxified lipooligosaccharide from nontypeable *Haemophilus influenzae* conjugated to proteins confers protection against otitis media in chinchillas. Infect Immun 1997; 65(11):4488-4493.
102. Pozsgay V. Synthesis of glycoconjugate vaccines against *Shigella dysenteriae* type 1. J Org Chem 1998; 63:5983-5999.
103. Pozsgay V, Chu C, Pannell L et al. Protein conjugates of synthetic saccharides elicit higher levels of serum IgG lipopolysaccharide antibodies in mice than do those of the O- specific polysaccharide from Shigella dysenteriae type 1. Proc Natl Acad Sci U S A 1999; 96(9):5194-5197.
104. Zhang H, Zhang S, Cheung NK et al. Antibodies against GD2 ganglioside can eradicate syngeneic cancer micrometastases. Cancer Res 1998; 58(13):2844-2849.
105. Alonso DF, Gabri MR, Guthmann MD et al. A novel hydrophobized GM3 ganglioside/Neisseria meningitidis outer- membrane-protein complex vaccine induces tumor protection in B16 murine melanoma. Int J Oncol 1999; 15(1):59-66.
106. Sandmaier BM, Oparin DV, Holmberg LA et al. Evidence of a cellular immune response against sialyl-Tn in breast and ovarian cancer patients after high-dose chemotherapy, stem cell rescue, and immunization with Theratope STn-KLH cancer vaccine. J Immunother 1999; 22(1):54-66.
107. Kudryashov V, Kim HM, Ragupathi G et al. Immunogenicity of synthetic conjugates of Lewis(y) oligosaccharide with proteins in mice: towards the design of anticancer vaccines. Cancer Immunol Immunother 1998; 45(6):281-286.
108. Slovin SF, Ragupathi G, Adluri S et al. Carbohydrate vaccines in cancer: immunogenicity of a fully synthetic globo H hexasaccharide conjugate in man. Proc Natl Acad Sci USA 1999; 96(10):5710-5715.

Chapter 14

DNA Vaccines

Daniel E. McCallus, Catherine J. Pachuk, Shaw-guang Lee and C. Satishchandran

Introduction

Research into the use of DNA vaccines has shown that this mode of immunization has much potential for widescale application. The ability to elicit an immune response by injection of DNA encoding the gene for the protein-antigen of interest has been demonstrated by a variety of methods. The genes may be introduced into eukaryotic cells by viral or bacterial vectors or through uptake of naked or complexed plasmid DNA. All allow an exogenous gene to be transcribed inside the host (vaccinee's) cell and allow the presentation of the expressed protein to the immune system in a manner that will generate a potentially protective response. This review will describe the methods by which DNA vaccines are delivered to the cells and how the genes are expressed and presented to the immune system. Other important issues such as the safety and manufacturing of DNA vaccines will also be addressed.

Vaccination with DNA is a recent technology possessing distinct advantages over traditional vaccines (killed or attenuated organisms) and the more recently-derived subunit vaccines. DNA vaccines, like attenuated vaccines, are able to induce both cellular and humoral responses, while subunit vaccines typically elicit only humoral responses. The risk of revertants that is present with attenuated viral or bacterial vaccines are not an issue with DNA vaccines. In addition, plasmid DNA vaccines also may have advantageous stability properties and should be able to withstand worldwide transport to a greater extent as compared to their traditional counterparts. Plasmid DNA can also be easily manipulated using standard laboratory techniques, allowing for the creation of a "backbone" plasmid that can incorporate a potentially limitless number of different genes. The resulting plasmids can be readily scaled up to produce large quantities of vaccine stocks.

The ability of genes encoded on bacterially-derived plasmid DNA to express their products following injection into a live animal was demonstrated when beta-galactosidase activity was shown in mice injected intramuscularly with plasmid DNA containing the corresponding gene.[1] Elicitation of an immune response in mice to a foreign protein using this technique was shown a short time later.[2] At this point, the utility of this technique in producing potential vaccines against pathogenic organisms was apparent in the reports being published.[3,4] A wide variety of microorganisms (viral, bacterial, protozoal) have now been investigated as targets for the development of potential DNA vaccines. Several DNA vaccines have now progressed to the point of being approved for phase I clinical trials.[5,6] The potential for the use of DNA vaccines against cancer and for autoimmune conditions is also being studied.[7-9] While there have been no FDA-approved DNA vaccine products on the market as yet, and while the optimal DNA vaccine preparation or delivery method may not have been elucidated yet, the scale and scope

New Vaccine Technologies, edited by Ronald W. Ellis. ©2001 Eurekah.com.

by which this technology is being investigated increases the likelihood that DNA vaccines will become an approved alternative to standard forms of immunization.

Gene Expression

Transcription of DNA Vaccine Genes

Unless DNA vaccines make use of cytoplasmically-replicating viruses (such as alphaviruses,[10,11] or are expressed following direct injection of RNA, all DNA vaccines must gain entrance to the nucleus of the cell for transcription of the relevant gene by RNA polymerase II. Most existing DNA vaccines of this type use the human cytomegalovirus (CMV) immediate-early enhancer/promoter[12] to drive expression of the relevant gene. Muscle-specific promoters have been investigated but have not shown any enhancement of the immune response generated as compared to vectors containing the CMV promoter.[13,14] Intradermal immunization may benefit from the use of dendritic cell-specific promoters. However, the use of tissue-specific promoters offers uncertain prospects at this point, due to gaps in our knowledge regarding precise mechanisms and cell types involved in DNA vaccination. Many groups have chosen to include an intron in their expression cassette, either just 5' to the open reading frame (ORF) of their gene of interest or inserted into the gene itself. Inclusion of an intron has been shown to sometimes enhance the transcription of genes.[15,16] Transport of mRNA from the nucleus to the cytoplasm is critical for subsequent expression of antigen. Certain mRNAs, such as the *env* mRNA from the human immunodeficiency virus (HIV), contain unspliced introns and are precluded from leaving the nucleus unless transported by auxiliary proteins. Cis-acting sequences (the *rev* response element [RRE]) in the *env* message are recognized by the HIV *rev* protein and allow transport of this mRNA from the nucleus to the cytoplasm. Therefore, plasmid vaccines encoding certain mRNAs, like HIV *env* mRNA must contain the cis-acting sequences that allow for transport, along with the auxiliary genes or proteins (if they are not normally expressed by the host cell) that interact with them. Alternatively, the constitutive transport element (CTE) from the Mason-Pfizer Monkey Virus can be used to achieve Rev-independent expression.[17] The vector must also include a polyadenylation (polyA) site (such as the SV40 polyA sequence) necessary for proper 3' cleavage and polyadenylation of the mRNA.[18] Due to inefficiencies in transcription termination and the resulting phenomenon of promoter occlusion, it would appear advantageous to include a transcriptional termination sequence such as the alpha-globin pause site termination sequence.[19] This should enable higher expression from the plasmid promoter driving expression of the gene of interest. Some mRNAs are targets for nuclear editing enzymes.[20] The resultant edited mRNAs are different in primary sequence from those originally transcribed from the encoded gene. As the signals for RNA editing are not clearly defined, it is not possible to predict which sequences may be at risk. However, as RNA editing may prevent or alter antigen expression, it is important to consider this possibility if problems with expression arise.

Translation of the DNA Vaccine Message

Most mRNA transcribed from plasmid DNA will travel from the nucleus to the cytoplasm and will be translated in the same manner as host mRNA. For efficient translation, these messages need to be capped by host capping enzymes prior to exiting the nucleus (reviewed in ref. 21). Established rules of eukaryotic translation initiation dictate the optimal nucleotide sequence that is minimally required for acceptable levels of protein production.[22] Deviation from any of these sequence requirements can significantly decrease the amount of translation initiation from the message. However, problems with translation can still occur when all of the established rules are followed.

It is often desired to express more than one protein-antigen from a DNA vaccine. This can be accomplished by the use of either a multi-cistronic vector or through the incorporation of an internal ribosome entry site (IRES) into the mRNA. Multi-cistronic plasmid DNAs can be engineered through the addition of multiple expression cassettes (promoter, polyA site) into the vector. The genes for HIV *rev* and *env*, as well as the two genes for human interleukin-12 (p35 and p40) chains have been expressed from a single plasmid in our lab and have been used to generate immune responses in mice. IRES elements are highly structured RNA sequences present in certain viral and mammalian mRNAs that act in conjunction with other trans acting factors, allowing translation initiation to occur on uncapped mRNAs.[23,24] The recognition of IRES elements (as currently employed in plasmid systems) by the translational machinery is extremely inefficient, however, presumably due to the need for auxiliary viral or host cell factors. Nonetheless, these sequences have been used to direct the ribosome to initiate translation to a second ORF located 3' to an upstream ORF in both retroviral[25] and plasmid[26] DNA vectors.

One strategy for increasing the expression of heterologous proteins in mammalian cells is to alter the codon usage of the original mRNA such that codons specifying abundant tRNAs are used in place of codons specifying less frequently found tRNAs.[27,28] Codon changes affect not only mRNA translation but also the sequence and structure of the mRNA as well, potentially leading to changes in the stability and regulation of the RNA. For example, it has been demonstrated that alteration of the mRNA sequence and subsequent codon usage renders HIV *env* mRNA expression *rev*-independent.[27]

Mechanisms of Immunostimulation

Use of Immunomodulatory Genes as Adjuvants

The mammalian immune response has been found to be under the control of a number of different molecules that can direct or alter the type of response seen following infection or vaccination. These include cytokines and chemokines secreted by immune system cells, as well as adhesion molecules and costimulatory factors found on T-cells and antigen presenting cells (APCs). The presence or absence of these molecules can dramatically alter the magnitude and level of the immune response. A number of these have been examined with the intent of increasing or directing the immune response following injection of plasmid DNA containing the gene for a protein-antigen.

Due to the mechanism by which DNA vaccines are assumed to act – that is, by entering a cell, expressing the desired protein, and having it either processed by that cell or transferred to an APC—viral or intracellular bacterial pathogens are an attractive target for potential DNA vaccines. Protective immune responses to many viruses are dependent on a Th1-type immune response, characterized by elevated interferon-gamma (IFN-γ) levels and the presence of Th1-type $CD4^+$ (helper) and $CD8^+$ (cytotoxic) T cells. Therefore, augmenting the immune response to DNA vaccines by coimmunization with Th-1-inducing cytokines appears to be a logical avenue of investigation. The genes for interleukin-12 (IL-12) have been used in plasmid DNA form by a number of laboratories to induce IFN-γ production and specific cytotoxic T lymphocytes (CTL) production to desired antigens. This has often resulted in higher levels of immunity as compared to immunization with the plasmid containing the gene of interest alone[29-31] and has been shown to increase the level of protection in a mouse challenge model.[32] In some cases, injection of the IL-12 genes alone had beneficial effect. Mice infected with *M. tuberculosis* had decreased numbers of bacteria in the lung and spleen following injection of an IL-12-expressing plasmid as compared to vector controls,[33] although the possibility that CpG-induced nonspecific immune mechanisms were responsible for this decrease cannot be ruled out (see below). Plasmid expressing IL-12 was able to protect mice against a challenge with influenza A.[34] Injection of an IL-12 plasmid was able to prevent chronic graft-versus-host disease by suppressing a Th2 response, thought to be the cause of this phenomenon.[35] Therefore, plasmid-produced IL-12 may serve both to increase the Th1 response while simultaneously inhibiting the Th2 response. Coinjection of antigen-expressing plasmids with plasmids encoding the IFN-γ protein have shown either immune-enhancing[36] or inhibitory[37,38] effects depending on the antigen examined and promoters driving antigen expression. Specific T-cell proliferative responses were seen following coinjection with plasmids expressing IL-2 or tumor necrosis factor-α.[39] Levels of Th1 indicators were increased following coinjection of plasmids containing IL-15 and the gene for HIV envelope,[40] but these responses were not synergized with concurrent administration of plasmids containing the genes for IL-2 or IL-12.

Increases in specific humoral responses, indicative of a Th2 response, were seen when IL-4[36,39] or IL-10[39] were coinjected with the antigen-expressing plasmid. Administration of a plasmid expressing IL-10 was able to diminish antigen-induced delayed type hypersensitivity (DTH) responses in animals primed with an infection with herpes simplex virus,[41] implying that the immune response can be biased towards a Th2 response and away from a Th1 response.

Granulocyte-macrophage colony stimulating factor (GM-CSF) expressed from coinjected plasmids was able to enhance specific B- and T-helper cell responses to a rabies virus glycoprotein[37] and to increase CTL activity to a previously nonimmunogenic influenza nucleoprotein.[30] Coinjection of GM-CSF plasmids with plasmids containing HIV genes resulted in increases in specific antibody production and stimulation of T cell proliferation.[31] While coinjection of GM-CSF plasmids with plasmids expressing secreted proteins from *M. tuberculosis* led to enhanced IFN-γ levels, there was no decrease in the number of bacteria in the lung following challenge as compared to animals not receiving the GM-CSF plasmid.[42]

Data from experiments involving the coadministration of cytokine genes must be interpreted carefully, as many cytokines are potent regulators of the HCMV IE promoter (the most commonly used promoter to drive expression of an antigen in a DNA vaccine) and will lead to increases or decreases in antigen expression, depending on the coexpressed cytokine. Levels of antigen expression most likely also play a role in the nature of the immune response, and these effects must be distinguished from the immunoregulatory effects of cytokines.

Chemokines are modulators that activate and/or attract various immune system cells. Plasmids containing the genes for a number of chemokines were investigated for the ability to affect the immune response following coinjection with plasmids expressing HIV *env* or *gag/pol* proteins.[43] The various chemokines were able to influence the type of immune response generated, particularly the chemokines MIP-1a and RANTES that are strong CTL stimulators.

In addition to the binding of the T cell receptor to peptides presented on the surface of APCs, specific T-cell activation requires that APCs and T cells make intimate contact in a number of ways. Both T cells and APCs possess interacting receptors and ligands that facilitate engagement between the two cell types and costimulation of the T cell. Either expressing adhesion or costimulatory molecules on myocytes or increasing the number of these molecules present on APCs may assist in bringing specific T cells into contact with cells presenting the antigens expressed following DNA injection. Coimmunization of a plasmid containing the APC adhesion molecule ICAM-1 with an antigen-expressing plasmid increased specific $CD4^+$ and $CD8^+$ T-cell responses.[44] The APC costimulatory molecule B7-2 (CD86) expressed from a plasmid enhanced antigen-specific CTL levels in mice coinjected with an HIV *env*-expressing plasmid[45,46] or with a plasmid expressing an influenza nucleoprotein.[30] Expression of B7-2 protein was demonstrated in plasmid-injected mouse muscle cells,[29] indicating that this procedure may be used to allow muscle to potentially present antigen. While B7-2 has shown to be effective in these studies, B7-1 (CD80) was not able to induce CTLs.

The number of different molecules that have been shown to enhance the immune response to desired antigens following DNA vaccination makes a systematic comparison important. While it is fairly obvious which cytokines or chemokines will induce a Th1 or Th2 response, the optimal one(s) to choose is not yet known. Both arms of the immune system may be important in generating a protective response to a certain pathogen, therefore different immunomodulatory signals may be needed at priming and boosting. It may also be possible, if not likely, that the optimal immunoenhancer may be antigen-specific and may require empirical testing when developing each individual vaccine.

Immunostimulatory Activity of Bacterial DNA

While the specific immune response to encoded antigens on plasmid DNA is of obvious importance in vaccination, it has become increasingly apparent that bacterial DNA, in itself, possesses nonspecific, immunostimulatory activity. This activity is mediated by CpG motifs that are present in bacterial DNA with a frequency approximately 20-times greater than in mammalian DNA. In addition, the less abundant CpG motifs of mammalian DNA contain methylated cytosines, whereas the bacterial versions are unmethylated. This lack of methylation and the relative abundance of CpGs in bacterial DNA are responsible for the immunostimulatory activity.[47] It should be noted, however, that methylation of CpG motifs in mammals varies from 30-80% depending on species, cell type and level of cell differentiation. Optimal immunostimulatory activity of bacterial DNA was found when the CpGs were flanked by two 5' purines and two 3' pyrimidines. Synthetic oligodeoxynucleotides (ODNs) containing these immunostimulatory sequences (ISSs) were found to possess the same or greater immunostimulatory properties than bacterial DNA.[48] Other CpG motifs, found in "unfavorable" contexts, are actually able to downregulate the immune response.[49] Therefore, both immune stimulatory as well as immune suppressing CpGs must be taken into account when designing DNA vaccines.

The role of CpG motifs in DNA vaccination was demonstrated following the observation that plasmid vaccines containing an ampicillin-resistance gene with two ISS enhanced a Th1 response as compared to an equivalent vector which instead possessed a kanamycin resistance gene with no ISS.[50] The Th1 enhancing effect, measured by an increase in specific antibody and IFN-γ production, was replicated by insertion of two ISS elements to the flanking regions of the kanamycin-resistance gene. Further results have since corroborated the beneficial effects of CpG effects during DNA vaccination. The immunogenicity of a DNA vaccine containing the gene for the malarial circumsporozoite protein was greatly reduced by methylation of the plasmid CpG residues. However, immunization with suboptimal amounts of this DNA vaccine was significantly enhanced (as measured by specific antibody titers) by addition of DNA (plasmid or ODNs) containing ISSs.[51] While utilization of ISS may prove to be a powerful component of DNA vaccines, the observation that mice and fish recognize CpG motifs in different contexts indicates that the exact parameters necessary for human immunostimulation may need further investigation.[52]

The mechanism by which ISS enhance the immune response has been examined by the stimulation of cytokine production in various cell types using ISS-containing bacterial (chromosomal or plasmid) DNA or ODNs. In general, it has been found that ISS induce a Th1 response.[53-55] Bacterial DNA or ODNs containing ISSs can either directly or indirectly influence a number of cell types to respond. DNA containing these sequences have been shown to stimulate B-cell proliferation,[48] IL-6 production from B cells and $CD4^+$ T cells,[56] IL-12 production from dendritic cells[57] and IFN-γ from NK cells.[57,58] While macrophages are stimulated by bacterial DNA,[59] evidence has been provided that antigen processing may actually be downregulated by ISS-containing ODN.[60] The responses of immune cells to ISS appear to be an additional, possibly necessary, component of DNA vaccination. Nevertheless the extent to which ISS are required for a protective immune response in human vaccinees following DNA injection has yet to be determined. In addition, the precise mechanism by which these elements work remains to be elucidated.

Routes of Administration

The relative ease by which both humoral and cellular immune responses are generated in mice following DNA vaccination belies the fact that the actual mechanism by which these responses are being generated is still unknown. Furthermore, different modes of immune stimulation may be followed depending on the route or method of DNA delivery. Cell types taking up the DNA may vary depending on whether the DNA is injected intramuscularly, intradermally or delivered through a mucosal route such as the intranasal route. In addition, different methods of injection such as with a needle or with the use of a biolistic device such as a "gene gun" will probably expose the plasmid to different cell types even though both methods can be used to deliver DNA intradermally. A number of methods to deliver DNA to mucosal sites are being explored, many requiring different formulations. These will undoubtedly influence the trafficking of plasmid DNA and the cell types involved with DNA uptake. Therefore the search for any mechanism for the immune response generated by DNA vaccines must be tempered by the realization that more than one may actually be taking place in vivo.

Intramuscular Immunization

Most of the initial studies describing DNA vaccines were done using intramuscular injections. As described above, myocytes near the site of injection are able to take up plasmid and express the encoded protein.[1] However, the ability of myocytes to act as APCs is hampered by the relative lack of expression of costimulatory factors, such as B7-2 on these cells, which are essential for a successful T-cell response.[29] Immune responses following IM injection could result from direct uptake of plasmid DNA by resident APCs such as macrophages or dendritic cells or by migrating APCs. Alternatively, the myocytes containing the plasmid could express the protein and act as a reservoir releasing the protein, which could subsequently be taken up by nearby APCs. To address these questions, F1 ($H\text{-}2^{\beta\xi\delta}$) chimeric mice were either reconstituted with parental spleen or bone marrow-derived cells[61] or served as donors of these cells to $H\text{-}2^{\beta\xi\delta}$ recipients.[62,63] Immunization with plasmid DNA encoding proteins containing MHC-specific epitopes showed that the APC of the donor haplotype was responsible for presenting antigen to T cells. These data ruled out the somewhat unlikely possibility that myocytes of the injected mice were serving as APCs. In addition, removal of tissue at the site of injection in as little as 10 minutes following the injection did not decrease specific antibody responses[64] also indicating that myocytes are not critical for the subsequent immune response. Evidence has been presented that dendritic cells may be the APC following IM DNA immunization.[65] However, questions still remain as to how these APCs come to present the relevant antigens. These cells may have been directly transfected with injected plasmid DNA or may have taken up protein that had been expressed by transfected myocytes. If the APCs do take up protein expressed from other cells then the phenomenon of crosspriming, in which APCs are able to process exogenously produced proteins and then express these peptides with class I antigens to the appropriate T cells, must occur.[66] However, one series of experiments using a protein which does not appear to be released from cells suggests that, even during IM DNA immunization, the proteins must be expressed by the APC[67] for the elicitation of a specific immune response. The direct transfection of macrophages following IM injection of plasmid DNA has been suggested.[68] These transfected macrophages were found in the blood lymphocyte pool and in lymph nodes, indicating that they may serve as the APCs necessary for eliciting T-cell responses following IM DNA injection.

The ability to mount an antibody response to proteins expressed from DNA vaccines may follow a similar course. Foreign proteins must first bind to antigen-specific B cells through surface immunoglobulin. These proteins are also usually ingested by APCs that then process the proteins and present the peptides on class II receptors to MHC-compatible T cells. Proteins produced endogenously following uptake of plasmid DNA (either by myocytes, APCs or other cell types) must exit the cell and be taken up by APCs for proper presentation of the peptides. Some proteins not normally secreted by eukaryotic cells are still able to induce an immune response following DNA vaccination. The method by which these proteins are being released from the cell is still unknown, but perhaps they are released through apoptosis or through secretion via a nonclassical pathway. The possibility that CTLs, through a perforin-mediated phenomenon, lyse antigen expressing cells was investigated and found to be negative.[69]

Cutaneous Immunization

Tissue architecture as well as resident cell types makes delivery of DNA vaccines into or through the skin markedly different from IM injection. The increased presence of Langerhans and dendritic cells in this area makes intradermal (ID) immunization a potentially more efficacious method of eliciting a specific immune response. Following uptake of antigen in the dermal or epidermal region, these cells travel to the draining lymph nodes where antigen is then presented to T cells.[70] Biolistic delivery appears to be an efficient method of getting DNA directly into cells. Plasmid DNA is usually coated onto gold beads (< 1 mm) and then delivered through the skin using a "gene gun," although a recent report showed that specific immune responses could be elicited following a simple topical application of DNA to the skin.[71] In a study comparing different routes of immunization, it was found that gene gun-delivered DNA was superior to DNA given IM, intravenously, or intranasally (IN) as measured by specific antibody response.[72] Indeed, the amount of DNA required to generate an equivalent response was much less with gene gun immunization. Transfection of dendritic cells and generation of a specific anti-tumor response was shown following gene gun immunization, directly implicating this cell type as the probable APC following ID administration of DNA.[73] This study could not rule out the possibility that non-APCs were being transfected and transferring protein to resident dendritic cells. Removal of the skin at the site of the injection less than 24 hours following injection decreases the antibody and CTL responses, indicating that, in contrast in IM injection, cells at the site of injection are important in protein production and/or antigen presentation.[64] Transplantation of vaccinated skin to naive recipients generated a specific immune response in recipients.[74] However, in order for a memory response to be elicited, the cells had to be transferred within 12 hours of the injection, implying that migratory dendritic cells were responsible for immunologic memory. Both of these latter reports also indicate that the presence of nonmigratory keratinocytes enhance the specific immune response possibly due to cytokine production following transfection by plasmid DNA as an ISS-mediated phenomenon. In support of these findings, no cross-priming of dendritic cells by protein derived from keratinocytes was seen following immunization with a nonsecreted protein,[75] despite the production of a vigorous dendritic cell-mediated immune response.

Evidence has been presented that delivery of plasmid DNA in saline induces a Th1 response, while gene gun delivery induces a Th2 response.[76] This has given rise to the impression that IM and gene gun immunizations elicit Th1 or Th2 responses respectively. While this phenomenon may be true in a general sense, specific CTL responses have been induced following gene gun delivery of nanogram amounts of DNA.[77] It is possible that different antigens and their location following expression can have an influence on the type of immune response generated. The efficiency with which gene gun delivery of DNA allows for access to the interior of the cell makes it possible to use very small quantities of DNA. The increased amounts of DNA needed during IM injection may steer the immune response in the Th1 direction due to the nonspecific response to ISS sequences present in plasmid DNA,[78] although this hypothesis has been questioned.[76] The Th2 bias of gene gun delivery can be overcome by coimmunizing with plasmid containing the genes for IL-12 or IFN-α.[79] As more information is gathered regarding immune responses following IM or gene gun administration, it will become possible to tailor the response in a particular direction.

Mucosal Immunization

Because many of the pathogens infecting mammals enter the body through mucosal routes, it has been a goal of many laboratories to enhance the immune response at these sites. An increase in the humoral and/or cellular response in these areas may act as a formidable barrier preventing colonization by bacteria or viruses. However, selectively activating secretory immunoglobulin A (sIgA) in mucosal areas or generating CTLs in regional lymph nodes has not been an easy task. Delivery of plasmid vaccines to areas that will generate specific immune responses at mucosal sites has led researchers to explore different methods of administering the DNA. These include IN, oral and intravaginal (IVAG) delivery. The effects of complexing the DNA with various substances has usually been assessed in these experiments. In addition to the stimulation of mucosal immunity, it is hoped that these routes of delivery will be less traumatic than administration of the DNA through a gene gun or traditional needle and syringe. IN delivery of plasmid vaccines has been the most common route of delivery in these experiments. Administration of plasmids containing reporter genes showed expression of the reporter protein in nasal tissue,[80] lungs[72] as well as the spleen and draining lymph nodes.[81] Increases in intestinal IgA production as measured by levels of this antibody in feces have usually been observed following IN delivery of DNA vaccines. These levels have usually been increased by complexing the DNA with various lipidic substances or adjuvants. Increases in the levels of specific CTLs in the spleen and lymph nodes have also been seen in these studies as well as following IN administration of plasmids containing the genes for HIV *env* and *rev*.[82] Administration of plasmid DNA alone appears to stimulate a Th2 response,[82] although this can be skewed toward a Th1 response by complexing the DNA with monophosphoryl lipid A,[83] the lipid DMRIE/DOPE,[81] or by coimmunization (IN) with plasmids expressing IL-12 or GM-CSF.[82] Challenge studies following IN immunization have shown some protection following DNA in saline administration[72] or little to no protection, depending on the challenge dose[84] following coadministration with the adjuvant cholera toxin (CT). The addition of CT to the DNA vaccine induced a Th2 response (including vaginal IgA), which may have been inappropriate for successful protection following a vaginal challenge.

Oral delivery of DNA vaccines, while possessing certain obvious advantages, remains problematic due to the acidic environment of the stomach. Therefore, it has been necessary to find methods of protecting the DNA during its traversal through this organ. Plasmid DNA has been encapsulated into poly(DL-lactide-coglycolide) (PLG) microparticles which are absorbed into gut-associated lymphoid tissue following uptake by M cells. This technique has been shown to induce specific serum antibodies as well as mucosal IgA.[85] Plasmids containing the genes for rotavirus VP6 antigen given orally following PLG encapsidation induced specific serum antibodies and intestinal IgA. These animals were protected to some degree following rotavirus challenge, although not to the level of IM-immunized animals.[86] Later experiments demonstrated that similar results could be achieved following oral immunization with PLG-encapsidated plasmids containing the genes for rotavirus VP4 and VP7 genes.[87] It is possible that greater protection may be achieved following delivery of a combination of the rotavirus vaccine candidates. Other formulations of DNA have been tested in oral delivery of plasmid vaccines. Oral delivery of a plasmid complexed with chitosan (a biodegradable polymer) and which contained the gene for a peanut allergen was able to induce the production of specific IgA and reduce the severity of anaphylaxis.[88] While oral delivery of DNA vaccines have not yet given the same protective results in mice as have been seen with IM and gene gun delivery, the hurdles which need to be overcome with this technology are greater.[89] However, the projected benefits that oral DNA vaccines will bring imply that much more research in this area will undoubtedly be pursued.

Another potential site of immunization which would be especially important for early defense against sexually transmitted diseases is the vagina. IVAG administration of DNA containing genes for HIV *env* and *rev* elicited both IgG and IgA antibodies in vaginal washes.[90] Antibodies from vagina washes possessed some neutralizing activity, although the neutralizing isotype could not be determined. Using a plasmid encoding the gene for human growth hormone, rats immunized IVAG were shown to possess specific IgG and IgA in serum and the vagina.[91] A pregnant chimpanzee, immunized IVAG with plasmids containing genes from HIV, was able to produce specific IgA in the saliva and serum as well as specific IgG in the saliva and serum but not from vaginal washes.[92] Delivery and expression of plasmid DNA in genital mucosa is relatively unexplored territory, and optimal conditions remain to be determined. The development of a protective immune response in mucosal tissues is an important goal, but may only result from delivery of plasmid through multiple sites.

Intracellular Delivery of DNA Vaccines

DNA vaccines require entrance to the vaccinee's cells, where the relevant genes are transcribed and translated by endogenous enzymes. Therefore, a great effort has been put forth into understanding and increasing the efficiency by which plasmid DNA is taken up by host cells. Although naked DNA appears to be taken up by cells in vivo, the efficiency of the process is poor. No greater than 10^6 molecules are taken up by approximately as many cells following intramuscular inoculation of ~10^{14} molecules.[93]Increasing bioavailability, either by increasing the volume of inoculation or by injecting at multiple sites, has not resulted in the predicted increases in expression (unpublished results). Naked DNA is vulnerable to attack by nucleases, but this problem can be diminished by decreasing the time during which the DNA would potentially be in contact with serum or other extracellular enzymes. The use of gold beads to rapidly deliver DNA using a gene gun results in direct intracellular deposition of the transfecting molecule,[94] and ionotophoresis and electroporation use an electric current to transiently disrupt the cellular membrane and the electromotive force (emf) to cause inward migration of the DNA molecule.[95] All of these procedures result in increased uptake of DNA. In addition, treatment of mouse muscle with certain agents such as bupivacaine has allowed for the enhanced uptake of DNA following IM injection.[96] As an alternative to injection of naked DNA, the DNA can first be protected through condensation with different facilitating agents or through incorporation of the nucleic acid into viral or bacterial carriers. Some of these methods also allow for increased affinity of the DNA (or its carrier) for the host cell.

Synthetic Delivery Vehicles

The use of facilitators has been shown to increase the uptake of naked DNA, as monitored by the expression of the transgene or the immune response to the encoded antigen.[96] Passive transfection processes are predicted to require the administration of large amounts of plasmid molecules and/or plasmid molecules with increased stability. DNA molecules have been stabilized through entrapment in complexes that minimize interactions with nucleases[97] or through the formation of DNA salts that are not substrates for nuclease activity and that can be exchanged for cellular cations. Increasing stability by methods that use ionic interactions to form complexes with the nucleic acid is dependent on secondary interactions among the complexing molecules themselves, as it is the differences in the secondary interactions that afford various degrees of strength to these complexes. Stability can also be conferred to the nucleic acid by entrapment in microspheres.[98] The effectiveness of such delivery agents appears to be dependent on the size of the complexes or microspheres.[99]

Active transfection processes involve rapid interactions between the cells and the complexed nucleic acid, and therefore the stability of the DNA may not be as critical for cellular uptake. Active transfection can be achieved by directing interactions among nucleic acid molecules and the cellular membrane through the use of complexing agents containing ligands that interact with molecules on the cellular membranes. Mannosylated poly-lysine complexes of DNA appear to be efficiently targeted to peritoneal macrophages.[100] Complexes containing cholesteryl spermine-modified antibodies target the DNA molecule to cells containing the cognate antigen receptor on the cell surface.[101] Unfortunately, the endosomal location of the majority of the internalized DNA molecules renders the DNA useless in the acidic environment of the mature endosome. Utilization of endosomal-disrupting agents, such as certain viral-derived proteins and peptides[102] or synthetic zwitterionic lipids, appears to increase expression of internalized DNA, by allowing DNA release into the cytoplasm. A fraction of the released molecules localize to the nucleus where they are transcribed. Enhancement of nuclear localization may be achieved through the use of peptide sequences derived from proteins that localize to the nucleus.[103] These proteins, which contain positively-charged domains have been shown to be effective in transporting oligonucleotides and chimeric polypeptides into the nucleus. The use of cytoplasmic expression systems is predicted to obviate the need for nuclear targeting. The ultimate creation of synthetic virus-like particles that incorporate multiple abilities to target cell-surface receptors and to then exit endosomes and target the nucleus requires precise design features. Since the mechanics of viral entry and the biochemical features of the viral architecture are themselves poorly understood, the design of synthetic agents that simulate and mimic viral entry are somewhat empirical for the present time.

While IM injection results in the transfection of muscle and skin cells, plasmid molecules also track to distal sites in the animal. For nearly 24 hours following inoculation, a large majority of the plasmid molecules are found in the blood, most of which were degraded. However, plasmid molecules have been detected 30 days following injection, (presumably internalized), in gut tissues and in cells of the lymphoid tissues. It is not clear if plasmid molecules were transported to these distal sites by cells that were transfected at the inoculation sites or if intact plasmid DNA molecules migrated to these sites.[104]

Viral Delivery Vehicles

Virus expression vectors have been used to construct novel vaccines. The first systems to be developed were based on large DNA viruses such as baculoviruses and vaccinia virus, where foreign genes could be inserted into the viral genome and expressed along with viral genes.[105,106] Later, other DNA viruses such as adenovirus and herpesvirus were studied as vaccine vectors.[107,108] In addition, retroviruses and adeno-associated viruses have been used for gene therapy studies. Recently, much attention has been given to the development of alphavirus-based gene delivery systems for the cytoplasmic expression of transgenes.[109]

The alphavirus genome consists of a single-stranded, approximately 12-kilobase RNA (49S RNA) of positive polarity, which is capped at the 5'-end and polyadenylated at the 3'-end. The 5' two-thirds of the genome encodes the viral nonstructural protein replicases while the last one-third encodes the viral structural proteins.[110] Upon infection, the alphavirus 49S genomic RNA functions as a mRNA for the translation of the subsequently transcribed negative strand RNAs from the genomic RNA templates. The negative-strand RNAs then serve as templates for the synthesis of the 49S viral genomic RNAs and the shorter subgenomic 26S mRNAs, which encode the structural proteins. Robust transcription from the internal

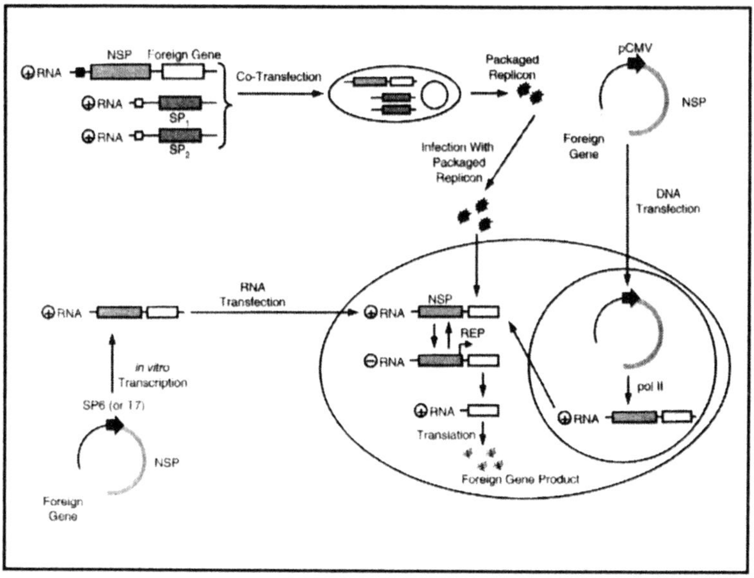

Fig. 14.1. Three strategies for genetic vaccines using alphavirus vectors. 1) Alphavirus replicons can be packaged by cotransfection of helper vector RNAs expressing the structural proteins. The packaged replicon particles are then used to infect cells for expression of the heterologous sequences. 2) Infectious RNA made from in vitro transfection and capping is used directly to transfect cells. 3) The alphavirus replicon is placed behind a polI promoter; the plasmid can directly be transfected into the cell where transcription occurs in the nucleus and further replication in the cell cytoplasm. In all three cases, the viral replicase uses the subgenomic promoter on the minus-strand RNA for the transcription of the subgenomic RNA that encodes the heterologous genome.

subgenomic promoter in the negative-strand results in the production of large amounts of subgenomic mRNAs. Replication of alphavirus RNAs occurs entirely in the cytoplasm of infected cells. (Fig. 14.1).

Full length cDNA clones of many alphaviruses have been constructed, of which the clones of Semliki Forest virus, Sindbis virus and Venezuelan equine encephalitis virus have been further developed into general expression vectors.[111] In these expression vectors, the region encoding the viral structural proteins is replaced with the transgene, while the viral replicase-coding region and all sequences required in *cis* for replication and packaging of the full-length recombinant RNA (replicon) are maintained. Transcription of the full-length replicon RNA can be achieved within a transfected cell by placing an RNA polymerase II-dependent promoter, such as the HCMV IE promoter, immediately upstream from the replicase genes and the alphavirus regulatory sequences. The replicon RNA is then transported to the cytoplasm, where it acts both as mRNA for the translation of the viral replicases and as template for replication. Transgene sequences are subsequently synthesized as subgenomic mRNA molecules from which the transgenes are translated. Alternatively, the RNA replicon can be made in vitro from a T7 or SP6 promoter and then directly transfected into cells or injected into animals.[112] In mice, intramuscular injection of replicon RNA has been shown to elicit both humoral and cellular immune responses to the encoded transgene antigen. In addition, immunization with the replicon RNA has been shown to protect mice from tumor challenge and to prolong the survival of mice with established tumors. When expressed in the presence of the structural proteins, the

RNA replicon is packaged into virus particles.[113] These particles can be harvested from cells and used to deliver the replicon to cells in vivo. The particles have a broad host range and can infect nondividing cells and thus may increase the in vivo delivery of the replicon relative to the delivery achieved when DNA or RNA is the delivery vehicle.

The self-replicating nature of the alphavirus genome has made alphavirus vectors useful tools for vaccine delivery. In a number of animal studies, strong immune responses have been elicited with low doses of alphavirus vectors, delivered either as packaged replicons, naked RNA, or DNA.[113] Alphavirus expression systems typically generate greater immune responses as compared to those elicited by traditional DNA vaccines. One mechanism underlying the enhanced immunogenicity may be the uptake by dendritic cells of transgene-encoded antigens that are released from cells that die apoptotically as a direct result of alphavirus RNA replication.[3] In addition, immunostimulatory signals may also be elicited by the presence of double-stranded RNA intermediates that are formed as a result of RNA replication.[114] Continued efforts in alphavirus vector development will illuminate the mechanism for cytoplasmic gene expression and the optimization of immune responses after vaccination.

Bacterial Delivery Vehicles

Recently, live bacteria have been used to deliver plasmid DNA for vaccination. In general, researchers have taken advantage of the ability of intracellular bacteria to gain entrance to the cytoplasm of eukaryotic cells. Transformation of these bacteria with plasmids containing a eukaryotic promoter driving the expression of the antigenic protein of interest appears to allow cellular expression of the antigen and efficient stimulation of a specific immune response. An added feature of this technique is that many of these bacterial species infect cells at mucosal sites. Thus, it may be possible to elicit a protective mucosal immune response. Due to the pathogenic nature of these bacteria however, it has been necessary to use attenuated mutants of these organisms. Attenuated versions of Shigella, Salmonella and Listeria have been used to deliver genes for reporter proteins and bacterial or viral virulence factors[115-119] with subsequent development of specific cellular and humoral immune responses. Direct expression of proteins from macrophages[117,118] and dendritic cells[119] following oral delivery has been shown. Protection against lethal bacterial[117] or tumor[119] challenge also has been demonstrated. While this method of plasmid delivery has certain advantages over administration of naked DNA, relatively high levels of plasmid DNA integration into the host cell chromosome[117] or host immune responses against the bacterial delivery vehicle will need to be addressed.

Safety of Nucleic Acid Vaccines

Several safety concerns exist with the use of DNA vaccines. Some concerns are related to the manufacturing process and formulation, while some are directly associated with the introduction of DNA molecules into a cell, including sequence-specific effects. The following section describes some of the safety issues involved in the use of DNA vaccines and potential strategies to address them.

Process Derived

The manufacturing process can affect the physical state of the plasmid DNA. At physiologic pH, DNA exists as a salt of a cation. Therefore, the potential for the introduction of cations to this strongly anionic molecule is fairly high. The periodicity of the anionic charge in DNA molecules can induce secondary cooperative interactions that have the potential to preclude desired interactions with carrier molecules, thus affecting consistency in formulations. In addition, strong secondary interactions between cationic species, such as triethyl amine and tetrabutyl amine, used in process development and formulation may result in the formation of stable complexes with DNA. Appropriate analytical methods will have to be developed to confirm the removal of residual substances.

Integration

Insertion of DNA into the host chromosome, often referred to as "integration," is presumed to be a random process, since illegitimate or random recombination is about 100-10000 times more frequent than homologous recombination in eukaryotic cells.[120] However, integration is dependent on several biochemical reactions, and it is predicted that certain sequences may be more preferred than others. The frequency of random integration of nonreplicating plasmid molecules into genomic sequences is $>10^{-4}$ in cell-culture models. Replicating plasmids appear to integrate at ~10 fold greater frequency than nonreplicating plasmids.[121] It is believed that the intermediates generated during plasmid replication are similar to those generated in the reaction pathways that lead to random insertion. In addition, random insertion of internalized DNA molecules has also been shown to invoke the host cell DNA repair reactions.[122,123] Open circular (OC) DNA invokes repair enzymes, and the plasmid DNA is therefore predicted to be replicated. However, it is not known if OC DNA integrates at a higher frequency than super-coiled DNA; this remains an important question to be answered. Also, since plasmid DNA isolated from *E.coli* contains methylated adenines (in the context of GATC sequences), and eukaryotic DNA contains no methylated adenines, the plasmid alkylated adenines are treated as mutagenic events which invoke cellular repair activities, at least after the methylated DNA has been integrated into the chromosome.[124,125]

The direct safety concern associated with random insertion of DNA is the permanent alteration of the host genome, particularly if germ cells are involved. Integration has been shown to disrupt cellular genes (resulting in altered metabolism), activate oncogenes and inactivate tumor suppressor genes in cell culture and in animal models. Invoking immunologic tolerance through the continued expression of the encoded genes from an integrated plasmid molecule is also a potential problem. However, several laboratories have been unable to detect the integration of administered plasmid DNA. The methods to evaluate inserted plasmid DNA have taken advantage of physical or chemical differences between free and integrated plasmid DNA sequences. We have used an assay that takes advantage of differences in methylation between inserted and free plasmid sequences and which detects one integration event in 10^5-10^7 genomes. Using this assay, no integration has been detected in several tissues (including lymphoid, muscle, skin, and gonads) in more than 60 injected animals.

Although theoretical considerations suggest that plasmid DNA sequences can integrate either as whole plasmids or as plasmid fragments, experimental evidence suggests that the integration frequency of plasmid fragments is at least two orders of magnitude lower.[122] Because integration is difficult to detect in animal models, extrapolation from cell culture experiments may be necessary. The integration frequency in cell lines is at least three logs greater than the frequency observed in animals following DNA administration (unpublished data). This is likely to be a reflection of the differences in the amounts of DNA taken up in tissue culture versus the amount of DNA taken up by cells in a living animal. As DNA delivery becomes more efficient, the frequency of DNA integration is predicted to also rise and approach the frequency observed in tissue culture. Nucleic acid vaccines that are RNA-derived are predicted to have no insertional abilities and therefore might be the ideal genetic vaccines.

Generation of Anti-Nucleic Acid and Anti-Chromatin Antibodies

Administration of large amounts of nucleic acid has been predicted to generate immune responses against nucleic acid. Coadministration of nucleic acids with certain adjuvants and DNA-protein complexes appear to elicit anti-nucleic acid (anti-chromatin) antibodies, similar to those observed in systemic lupus erythematosis. However, in preclinical animal models and in human clinical trials, administration of large amounts of purified DNA has not elicited a detectable anti-nucleic acid (anti-chromatin) response. Nevertheless, it must be noted that a DNA vaccine expressing the SV40 large-T antigen appeared to induce an anti-chromatin response in mice.[126] It is not clear if that result is a general finding related to DNA vaccines encoding nucleic acid-binding proteins or is an event uniquely associated with the expression of T antigen.

Oncogenesis and Biochemical Activities of the Encoded Gene Product

There does not exist any sequence-based analytical method to determine the oncogenic potential of an encoded expression cassette in a genetic vaccine. Given the complexity of the process and the poor understanding of precise biochemical events involved, the oncogenic potential of the encoded sequence can only be analyzed empirically.

Analysis of sequences for DNA binding motifs, for potential substrates of known protein kinases and for genes encoding proteins that participate in signal transduction pathways may be warranted. The protein products of these genes engage in reaction pathways that can potentially lead to oncogenesis. However, an accurate assessment of oncogenic potential of the encoded gene product will require careful observation of the growth characteristics of transfected cell types, and binding assays to determine if these proteins can interact with other cellular proteins and DNA.

Transfer of Plasmid Sequences to Endogenous Bacterial Flora

Another safety issue associated with the use of plasmid DNA vaccines, particularly following oral administration, is the potential for transformation of endogenous bacterial flora with plasmid DNA. Although analyses of fecal flora from animals, following oral administration of large quantities of plasmid DNA, have not yielded transformed bacteria, the risk of plasmid transfer will need to be further analyzed. The ColE1 replication origin used in current generations of DNA vaccines can autonomously replicate in bacteria that utilize gram-negative transcriptional elements. However, gram-negative bacteria require specialized mechanisms, such as F factors, to transfer nucleic acids internally. On the other hand, Gram-positive bacteria produce competence factors that allow easy assimilation of DNA and subsequent transformation. It is also possible that the fragments of plasmid DNA generated in the gut[127,128] can be recombined into existing chromosomal and extrachromosomal DNA molecules following cellular internalization. These recombined molecules have the potential to become stabilized under selective microenvironmental conditions. Of particular theoretical concern is the transfer of an antibiotic-resistance gene to endogenous microflora.

Future Directions of DNA Vaccines

The use of DNA to elicit an immune response to antigens encoded on that DNA has been a relatively recent area of exploration. Methods of increasing the immune response or specifically targeting certain types of immune responses are now being tested. There is little doubt that numerous other approaches will be undertaken in the near future to further exploit DNA vaccination.

Intracellular Trafficking and Presentation of Expressed Antigen

The intracellular production of antigen following the administration of DNA vaccines is the hallmark of DNA vaccination. Intracellular antigen synthesis allows MHC class I presentation of antigen-derived peptides that can prime cellular immune responses. The generation of peptides for MHC class I presentation is mediated in part through the ubiquitination of endogenously-produced proteins. Once bound to the substrate protein, ubiquitin itself becomes a target for ubiqiuitination, resulting in the polyubiquitination of the substrate protein. The polyubiquitinated substrate protein is then degraded by the 26S proteosome complex. A subset of the degradation products is comprised of peptides that can associate with MHC class I molecules.

The magnitude of a cellular response may be heightened, by increasing the efficiency with which an antigen is targeted for ubiqitination. Our knowledge of ubiquitination signals is limited however, and there appear to be multiple signals. In addition, ubiqutination is catalyzed by a myriad of ubiquitin carrier proteins that are specific for certain proteins. Since the expression of these carrier proteins is also regulated according to cell type and cell cycle, it does not seem likely that there is a universal signal for ubiquitination of all proteins in all cell types nor at all points in the cell cycle. Nonetheless, there have been some reported successes at targeting DNA vaccine-encoded antigens into the ubiquitin pathway. Rodriguez et al[129,130] demonstrated enhanced cytotoxic T-lymphocyte induction in mice immunized with DNA constructs encoding an N-terminal ubiquitin fusion protein. Enhanced responses were seen when ubiquitin was fused to either a lymphocytic choriomeningitis viral gene or minigene. The universality of this strategy will undoubtedly be tested over the next few years.

Another strategy for the increasing cellular immune responses involves the removal of signal peptides from antigens destined for secretion. This forces the cytosolic localization and the increased intracellular accumulation of these antigen proteins, which may allow for increased MHC class 1 presentation of antigen peptides.

The intracellular production of antigen that occurs following DNA vaccination results in a population of antigen molecules that become localized within antigen-presenting cells (APCs). It is unclear whether the antigen found in APCs merely reflects antigen taken up by an APC or whether some antigen is actually synthesized in an APC: most likely it is combination of these two events. Localization of antigen within an APC allows for the MHC class II presentation of antigenic peptides. The production of these peptides is achieved by antigen degradation in a lysosome-dependent pathway.

There has been much innovation in the design of DNA vaccines in an attempt to enable more efficient class II presentation of antigen. These modifications have included alterations to the design of the antigen molecule, expression of accessory molecules, and changes in regulatory elements such as the promoter.

In general, changes to the antigen are designed to result in the increased availability of antigen to APCs. These changes most often involve modifications that allow the antigen to be exported out of the cell. Comparisons of immune responses following DNA vaccination with plasmids expressing secreted or cell-associated antigen have demonstrated that it is possible to increase the magnitude of the immune response by targeting the antigen for secretion. For example, both Geissler and Inchauspe have reported higher seroconversion rates and higher antibody titers in BALB/c mice immunized with plasmid constructs expressing various forms of secreted Hepatitis C (HCV) core protein, as compared with those mice immunized with constructs expressing cell-associated core.[131,132]

As an alternative to secreting antigen from a cell via the use of secretory signals, the coexpression of proteins or peptides that induce cellular apoptosis may have merit. Not all proteins can be secreted from a cell due to the presence of domains that interact with intracellular membranes. In these instances, induced apoptosis following antigen synthesis would allow spillage of antigen from the cell, resulting in increased antigen availability to APCs. In addition, cells undergoing apoptosis may recruit inflammatory cells to the site and thereby increase antigen delivery to APCs. Consistent with this, a self-replicating RNA vaccine was recently demonstrated to be more efficacious than a traditional plasmid-based vaccine. The increased efficacy was postulated to be due to apoptosis of cells harboring the self-replicating RNA.

It has recently been demonstrated that secreted antigen can be targeted directly to APCs by a further modification of the antigen. Boyle et al have made C-terminal antigen fusions to CTLA4.[133] CTLA4, a ligand present on activated T cells, binds to B7 molecules expressed on the surface of APCs. The presence of CTLA4 on the antigen fusion protein should allow targeting of the antigen to B7-expressing cells, thereby increasing the efficiency with which antigen is taken up by APCs. Experimental data suggest that in fact was achieved. Mice immunized with plasmids expressing the B7 targeted antigen had much higher antibody levels than those mice immunized with the nontargeted antigen: at two weeks post-immunization, mice receiving the CTLA4 fusion protein plasmid had antigen-specific antibody titers 10,000-fold higher than those mice immunized with the native antigen plasmid. In addition, a 7000-fold increase in IgG1 levels was seen in mice immunized with the CTLA4 fusion construct as compared to mice immunized with the plasmid expressing the nontargeted antigen.

In addition to antigen modifications that allow targeting of antigen to specific cell types, modifications have also been incorporated that allow intracellular targeting of antigen. Expression of antigen as a fusion protein to lysosome-associated membrane protein (LAMP-1) has been shown to result in endosomal/lysosomal localization of antigen. LAMP-1 contains a localization signal that reroutes antigen into the lysosomal compartment and therefore into the MHC class II processing pathway. Immunization with constructs expressing LAMP-1-antigen fusion proteins has been shown to result in enhanced $CD4^+$ presentation of antigen. In addition, immunization of mice with a vector expressing a LAMP-1/tumor-antigen fusion protein was found to protect mice from tumor challenge, whereas mice vaccinated with a vector expressing the unmodified tumor antigen were not protected.[134] However, the use of LAMP-1 fusion proteins does not always result in increased immune response, and more research must be done in this area before the utility of this approach can be assessed.[135]

Other modifications can be made at the level of promoter choice. The use of dendritic cell and macrophage specific promoters would allow for the production of antigen in these APCs following uptake of plasmid DNA into these cells. It will be interesting to see if expression of antigen within APCs results in enhanced immune responses.

Prime-Boost Strategies

While injection of plasmid DNA has been shown to elicit immune responses in animals, the magnitude of these responses has not always been considered adequate for protection. As a remedy for this situation, researchers have explored methods to increase and broaden the immune responses seen in conjunction with DNA immunization. One way of increasing the antibody response has been to give a primary injection with plasmid DNA containing the gene of interest, followed by a boost with the recombinant protein itself. This has been particularly useful in cases where the protein has been poorly immunogenic when delivered by DNA. Plasmids containing the gene for HIV *env*, when given alone, do not usually generate high antibody levels, although CMI responses are better. When animals are subsequently boosted with recombinant *env* protein (as compared to boosting with DNA), the specific antibody titers are significantly heightened.[136] Protein boosting also increased the neutralizing activity and avidity of these antibodies. Boosting with a multicomponent peptide after DNA priming also increased a poor anti-*env* antibody response which complemented the CMI response generated following the initial DNA injection[137] A combination of DNA vaccine priming with a plasmid containing the HIV *env* gene followed by boosting with HIV *env* protein was able to induce both specific CTLs and antibodies in rhesus macaques and was able to protect these animals from a challenge with SIV containing the HIV *env* protein.[138] Similar increases in the antibody response were seen in mice primed with a DNA vaccine containing the gene for a *Plasmodium falciparum* antigen, followed by boosting with the analogous recombinant protein.[139] DNA priming followed by protein boosting may be an effective technique to generate both CMI (DNA) and humoral (protein) responses. An alternative method is to boost with a live virus containing the gene of interest. Mice primed with plasmid DNA containing the HIV *env* gene and boosted with a recombinant fowlpox virus also containing the HIV *env* gene protected mice from challenge with vaccinia virus expressing the HIV *env* gene.[140] This same immunization regimen also protected macaques from an intravenous HIV-1 challenge. Use of a multicomponent[141,142] (HIV and SIV) epitope-based sequence in both plasmid DNA and modified Ankara virus (MVA) constructs elicited specific CTLs in macaques primed with DNA and boosted with the MVA construct. Macaques immunized in this manner showed little protection (one out of three animals) following a challenge with live SIV, however.[143] Protection of mice from challenge with Plasmodium sporozoites was achieved by priming with DNA vaccines containing the genes for sporozoite antigens and boosting with vaccinia or MVA expressing the same protein.[144,145] Clearly, the DNA prime followed by recombinant protein or recombinant viral boost method possesses advantages over DNA priming and boosting. Extensive trials comparing the two methods in nonhuman primates or ideally human clinical trials will shed light on their respective effectiveness and safety.

Summary

While the age of DNA vaccines may not yet be upon us, the amount of research currently being done in this area, as well as the clinical trials which are in progress, imply that these types of vaccines may some day become part of the armamentarium by which mankind defends itself against infectious disease. Although the data that has emerged from these studies during the past decade is voluminous, many questions remain to be answered. The optimal route of delivery and type of formulation are just two of these areas that need further exploration. DNA vaccines may be less effective than preexisting vaccines unless accompanied by the coadministration of immune-enhancing genes or molecules. Additional safety studies may be needed before the general public feels secure in accepting these vaccines. It is possible that only through human clinical trials will many of these areas be adequately addressed. However, the potential attributes of DNA vaccines, as compared to existing vaccines, provides much reason for hope among the scientists researching them.

References

1. Wolff JA, Malone RW, Williams P et al. Direct gene transfer into mouse muscle in vivo. Science 1990; 247:1465-1468.
2. Tang D-C, DeVit M, Johnston SA. Genetic immunization is a simple method for eliciting an immune response. Nature 1992; 356:152-154.
3. Ulmer JB, Donnelly JJ, Parker SE et al. Heterologous protection against influenza by injection of DNA encoding a viral protein. Science 1993; 259:1745-1749.
4. Wang B, Ugen KE, Srikantan V et al. Gene inoculation generates immune responses against human immunodeficiency virus type 1. Proc Natl Acad Sci USA 1993: 90:4156-4160.
5. Ugen KE, Nyland SB, Boyer JB et al. DNA vaccination with HIV-1 expressing constructs elicits immune responses in humans. Vaccine 1998; 16:1818-1821.
6. Tacket CO, Roy MJ, Widera G et al. Phase 1 safety and immune response studies of a DNA vaccine encoding hepatitis B surface antigen delivered by a gene delivery device. Vaccine 1999; 17:2826-2829.
7. Ying H, Zaks TZ, Wang RF et al. Cancer therapy using a self-replicating RNA vaccine. Nat Med 1999; 5:823-827.
8. Syrengelas AD, Chen TT, Levy R. DNA immunization induces protective immunity against B-cell lymphoma. Nat Med 1996; 2:1038-1041.
9. Waisman A, Ruiz PJ, Hirschberg DL et al. Suppressive vaccination with DNA encoding a variable region gene of the T-cell receptor prevents autoimmune encephalomyelitis and activates Th2 immunity. Nat Med 1996; 2:899-905.

10. Davis NL, Brown KW, Johnston RE. A viral vaccine vector that expresses foreign genes in lymph nodes and protects against mucosal challenge. J Virol 1996; 70:3781-3787.
11. Zhou X, Berglund P, Zhao H et al. Generation of cytotoxic and humoral immune responses by nonreplicative recombinant Semliki Forest virus. Proc Natl Acad Sci USA 1995; 92:3009-3013.
12. Boshart M, Weber F, Jahn G et al. A very strong enhancer is located upstream of an immediate early gene of human cytomegalovirus. Cell 1985; 41:521-530.
13. Ambrovic A, Adam M, Monteil M et al. Efficacy of replication-defective adenovirus-vectored vaccines: protection following intramuscular injection is linked to promoter efficiency in muscle representative cells. Virology 1997; 238:327-335.
14. Loriat D, Li Z, Mancini M, Tiollais P et al. Muscle-specific expression of Hepatitis B surface antigen: no effect on DNA-raised immune responses. Virology 1999; 260:74-83.
15. Brinster RL, Allen JM, Behringer RR et al. Introns increase transcriptional efficiency in transgenic mice. Proc Natl Acad Sci USA 1988; 85: 836-840.
16. Buchman AR, Berg P. Comparison of intron-dependent and intron-independent gene expression. Mol Cell Biol 1988; 8:4395-4405.
17. Bray M, Prasad S, Dubay JW et al. A small element from the Mason-Pfizer monkey virus genome makes human immunodeficiency virus type 1 expression and replication Rev-independent. Proc Natl Acad Sci 1994; 91:1256-1260.
18. Keller W, Bienroth S, Lang KM et al. Cleavage and polyadenylation factor CPF specifically interacts with the premRNA 3' processing signal AAUAAA. EMBO J 1991; 10:4241-4219.
19. Enriquez-Harris P, Levitt N, Briggs D et al. A pause site for RNA polymerase II is associated with termination of transcription. EMBO J 1991; 10:1833-1842.
20. Polson AG, Ley HL 3rd, Bass BL et al. Hepatitis delta virus RNA editing is highly specific for the amber/W site and is suppressed by hepatitis delta antigen. Mol Cell Biol 1998; 18:1919-1926.
21. Sachs AB, Sarnow P, Hentze MW. Starting at the beginning, middle, and end: translation initiation in eukaryotes. Cell 1997; 89:831-838.
22. Kozak M. Features in the 5' noncoding sequences of rabbit alpha and beta-globin mRNAs that affect translational efficiency. J Mol Biol 1994; 235:95-110.
23. Pelletier J, Sonenberg N. Internal initiation of translation of eukaryotic mRNA directed by a sequence derived from poliovirus mRNA. Nature 1998; 334:320-325.
24. Macejak DG, Sarnow P. Internal initiation of translation mediated by the 5' leader of a cellular mRNA. Nature 1991; 353:90-94.
25. Berlioz C, Torrent C, Darlix JL. An internal ribosome entry signal in the rat VL30 region of the Harvey murine sarcoma virus leader and its use in dicistronic retroviral vectors. J Virol 1995; 69:6400-6407.
26. Clarke NJ, Hissey P, Buchan K et al. PPV: A novel IRES-containing vector to facilitate plasmid immunization and antibody response characterization. Immunotechnology 1997; 3:145-153.
27. Andre S, Seed B, Eberle J et al. Increased immune response elicited by DNA vaccination with a synthetic gp120 sequence with optimized codon usage. J Virol 1998; 72:1497-1503.
28. Vinner L, Nielsen HV, Bryder K et al. Gene gun DNA vaccination with Rev-independent synthetic HIV-1 gp160 envelope gene using mammalian codons. Vaccine 1999; 17:2166-2175.
29. Tsuji T, Hamajima K, Fukushima J et al. Enhancement of cell-mediated immunity against HIV-1 induced by coinoculation of plasmid-encoded HIV-1 antigen with plasmid expressing IL-12. J Immunol 1997; 158:4008-4013.
30. Iwasaki A, Steirnholm BJ, Chan AK et al. Enhanced CTL responses mediated by plasmid DNA immunogens encoding costimulatory molecules and cytokines. J Immunol 1997; 158:4591-4601.
31. Kim JJ, Ayyavoo V, Bagarazzi ML et al. In vivo engineering of a cellular immune response by coadministration of IL-12 expression vector with a DNA immunogen. J Immunol 1997; 158:816-826.
32. Sin JI, Kim JJ, Arnold RL et al. IL-12 gene as a DNA vaccine adjuvant in a Herpes mouse model: IL-12 enhances Th1-type $CD4^+$ T cell-mediated protective immunity against Herpes simplex virus-2 challenge. J Immunol 1999; 162:2912-2921.
33. Lowrie DB, Tascon RE, Bonato VLD et al. Therapy of tuberculosis in mice by DNA vaccination. Nature 1999; 400:269-271.

34. Operschall E, Schuh T, Heinzerling L et al. Enhanced protection against viral infection by coadministration of plasmid DNA coding for viral antigen and cytokines in mice. J Clin Virol 1999; 13:17-27.
35. Okubo T, Hagiwara E, Ohno S et al. Administration of an IL-12-encoding plasmid prevents the development of chronic graft-versus-host disease (GVHD). J Immunol 1999; 162:4013-4017.
36. Ishii KJ, Weiss WR, Ichino M et al. Activity and safety of DNA plasmids encoding IL-4 and IFN gamma. Gene Ther 1999; 6:237-244.
37. Xiang Z, Ertl HC. Manipulation of the immune response to a plasmid-encoded viral antigen by coinoculation with plasmids expressing cytokines. Immunity 1995; 2:129-135.
38. Xiang ZQ, He Z, Wang Y et al. The effect of interferon-γ on genetic immunization. Vaccine 1997; 15:896-898.
39. Kim JJ, Trivedi NN, Nottingham LK et al. Modulation of amplitude and direction of in vivo immune responses by coadministration of cytokine gene expression cassettes with DNA immunogens. Eur J Immunol 1998; 28:1089-1103.
40. Xin KQ, Hamajima K, Sasaki S et al. IL-15 expression plasmid enhances cell-mediated immunity induced by an HIV-1 DNA vaccine. Vaccine 1999; 17:858-866.
41. Chun S, Daheshia M, Lee S, et al. Distribution fate and mechanism of immune modulation following mucosal delivery of plasmid DNA encoding IL-10. J Immunol 1999; 163:2393-2402.
42. Kamath AT, Hanke T, Briscoe H et al. Coimmunization with DNA vaccines expressing granulocyte-macrophage colony-stimulating factor and mycobacterial secreted proteins enhances T-cell immunity, but not protective efficacy against *Mycobacterium tuberculosis*. Immunology 1999; 96:511-516.
43. Kim JJ, Nottingham LK, Sin JI et al. CD8 positive T cells influence antigen-specific immune responses through the expression of chemokines. J Clin Invest 1998; 102:1112-1124.
44. Kim JJ, Tsai A, Nottingham LK et al. Intracellular adhesion molecule-1 modulates b-chemokines and directly costimulates T cells in vivo. J Clin Invest 1999; 103:869-877.
45. Tsuji T, Hamajima K, Ishii N et al. Immunomodulatory effects of a plasmid expressing B7-2 on human immunodeficiency virus-1-specific cell-mediated immunity induced by a plasmid encoding the viral antigen. Eur J Immunol 1997; 27:782-787.
46. Kim JJ, Bagarazzi ML, Trivedi N et al. Engineering of in vivo immune responses to DNA immunization via codelivery of costimulatory molecule genes. Nat Biotechnol 1997; 15:641-646.
47. Pisetsky DS. Immune activation by bacterial DNA: A new genetic code. Immunity 1996; 5:303-310.
48. Kreig AM, Yi AK, Matson S et al. CpG motifs in bacterial DNA trigger direct B-cell activation. Nature 1995; 374:546-549.
49. Kreig AM, Wu T, Weeratna R et al. Sequence motifs in adenoviral DNA block immune activation by stimulatory CpG motifs. Proc Natl Acad Sci USA 1998; 95:12631-12636.
50. Sato Y, Roman M, Tighe H et al. Immunostimulatory DNA sequences necessary for effective intradermal gene immunization. Science 1996; 273:352-354.
51. Klinman DM, Yamshchikov G, Ishigatsubo Y. Contribution of CpG motifs to the immunogenicity of DNA vaccines. J Immunol 1997; 158:3635-3639.
52. Kanellos TS, Sylvester ID, Butler VL et al. Mammalian granulocyte-macrophage colony-stimulating factor and some CpG motifs have an effect on the immunogenicity of DNA and subunit vaccines in fish. Immunology 1999; 96:507-510.
53. Chu RS, Targoni OS, Kreig AM et al. CpG oligodeoxynucleotides act as adjuvants that switch on T helper 1 (Th1) immunity. J Exp Med 1997; 186:1623-1631.
54. Roman M, Martin-Orozco E, Goodman JS et al. Immunostimulatory DNA sequences function as T helper-1-promoting adjuvants. Nat Med 1997; 3:849-854.
55. Kline JN, Waldschmidt TJ, Businga TR et al. Modulation of airway inflammation by CpG oligodeoxynucleotides in a murine model of asthma. J Immunol 1998; 160:2555-2559.
56. Klinman DM, Yi AK, Beaucage SL et al. CpG motifs present in bacterial DNA rapidly induce lymphocytes to secrete interleukin 6, interleukin 12, and interferon g. Proc Natl Acad Sci USA 1996; 93:2879-2883.
57. Jakob T, Walker PS, Krieg AM et al. Activation of cutaneous dendritic cells by CpG-containing oligodeoxynucleotides: a role for dendritic cells in the augmentation of Th1 responses by immunostimulatory DNA. J Immunol 1998; 161:3042-3049.

58. Cowdery JS, Chace JH, Yi AK et al. Bacterial DNA induces NK cells to produce IFN-γ in vivo and increases the toxicity of lipopolysaccharides. J Immunol 1996; 156:4570-4575.
59. Stacey KJ, Sweet MJ, Hume DA. Macrophages ingest and are activated by bacterial DNA. J Immunol 1996; 157:2116-2122.
60. Chu RS, Askew D, Noss EH et al. CpG oligodeoxynucleotides down-regulate macrophage class II MHC antigen processing. J Immunol 1999; 163:1188-1194.
61. Corr M, Lee DJ, Carson DA et al. Gene vaccination with naked plasmid DNA: Mechanism of CTL priming. J Exp Med 1996; 184:1555-1560.
62. Doe B, Selby M, Barnett S et al. Induction of cytotoxic T lymphocytes by intramuscular immunization with plasmid DNA is facilitated by bone marrow-derived cells. Proc Natl Acad Sci USA 1996; 93:8578-8583.
63. Fu TM, Ulmer JB, Caulfield MJ et al. Priming of cytotoxic T lymphocytes by DNA vaccines: requirement for professional antigen presenting cells and evidence for antigen transfer from myocytes. Mol Med 1997; 3:362-371.
64. Torres CA, Iwasaki A, Barber BH et al. Differential dependence on target site tissue for gene gun and intramuscular DNA immunizations. J Immunol 1997; 158:4529-4532.
65. Casares S, Inaba K, Brumeanu TD et al. Antigen presentation by dendritic cells after immunization with DNA encoding a major histocompatibility complex class II-restricted viral epitope. J Exp Med 1997; 186:1481-1486.
66. Bevan MJ. Antigen presentation to cytotoxic T lymphocytes in vivo. J Exp Med 1995; 182:639-641.
67. Whitton JL, Rodriguez F, Zhang J et al. DNA immunization: mechanistic studies. Vaccine 1999; 17:1612-1619.
68. Chattergoon MA, Robinson TM, Boyer JD et al. Specific immune induction following DNA-based immunization through in vivo transfection and activation of macrophages/antigen presenting cells. J Immunol 1998; 160:5707-5718.
69. Hassett DE, Zhang J, Whitton JL. Induction of antiviral antibodies by DNA immunization requires neither perforin-mediated nor $CD8^+$-T-cell-mediated lysis of antigen-expressing cells. J Virol 1999; 73:7870-7873.
70. Kaiserlian D, Etchart N. Epicutaneous and transcutaneous immunization using DNA or proteins. Eur J Dermatol 1999; 9:169-176.
71. Fan H, Lin Q, Morrissey GR et al. Immunization via hair follicles by topical application of naked DNA to normal skin. Nat Biotechnol 1999; 17:870-872.
72. Fynan EF, Webster RG, Fuller DH et al. DNA vaccines: protective immunizations by parenteral, mucosal, and gene-gun inoculations. Proc Natl Acad Sci USA 1993; 90:11478-11482.
73. Condon C, Watkins SC, Celluzzi CM et al. DNA-based immunization by in vivo transfection of dendritic cells. Nat Med 1996; 2:1127.
74. Klinman DM, Sechler JM, Conover J et al. Contribution of cells at the site of DNA vaccination to the generation of antigen-specific immunity and memory. J Immunol 1998; 160:2388-2392.
75. Akbari O, Panjwani N, Garcia S et al. DNA vaccination: transfection and activation of dendritic cells as key events for immunity. J Exp Med 1999; 189:169-177.
76. Feltquake DM, Heaney S, Webster RG et al. Different T helper cell types and antibody isotypes generated by saline and gene gun DNA immunization. J Immunol 1997; 158:2278-2284.
77. Pertmer TM, Eisenbraun MD, McCabe D et al. Gene gun-based nucleic acid immunization: elicitation of humoral and cytotoxic T lymphocyte responses following epidermal delivery of nanogram quantities of DNA. Vaccine 1995; 13:1427-1430.
78. Barry MA, Johnston SA. Biological features of genetic immunization. Vaccine 1997; 15:788-791.
79. Tuting T, Gambotto A, Robbins PD et al. Codelivery of T helper 1-biasing cytokine genes enhances the efficacy of gene gun immunization of mice: studies with the model tumor antigen beta-galactosidase and the BALB/c meth A p53 tumor-specific antigen. Gene Ther 1999; 6:629-636.
80. Klavinskis LS, Gao L, Barnfield C et al. Mucosal immunization with DNA-liposome complexes. Vaccine 1997; 15:818-820.
81. Klavinskis LS, Barnfield C, Gao L et al. Intranasal immunization with plasmid DNA-lipid complexes elicits immunity in the female genital and rectal tracts. J Immunol 1999; 162:254-262.
82. Okada E, Sasaki S, Ishii N et al. Intranasal immunization of a DNA vaccine with IL-12- and granulocyte-macrophage colony-stimulating factor (GM-CSF)-expressing plasmids in liposomes

induces strong mucosal and cell-mediated immune responses against HIV-1 antigens. J Immunol 1997; 159:3638-3647.

83. Sasaki S, Hamajima K, Fukushima J et al. Comparison of intranasal and intramuscular immunization against human immunodeficiency virus type 1 with a DNA-monophosphoryl lipid A adjuvant vaccine. Infect Immun. 1998; 66:823-826.
84. Kuklin N, Daheshia M, Karem K et al. Induction of mucosal immunity against herpes simples virus by plasmid DNA immunization. J Virol 1997; 71:3138-3145.
85. Jones DH, Corris S, McDonald S et al. Poly (DL-lactide-coglycolide)-encapsulated plasmid DNA elicits systemic and mucosal antibody responses to encoded protein after oral administration. Vaccine 1997; 15:814-817.
86. Chen SC, Jones DH, Fynan EF et al. Protective immunity induced by oral immunization with a rotavirus DNA vaccine encapsulated in microparticles. J Virol 1998; 72:5757-5761.
87. Herrmann JE, Chen SC, Jones DH et al. Immune responses and protection obtained by oral immunization with rotavirus VP4 and VP7 DNA vaccines encapsulated in microparticles. Virology 1999; 259:148-153.
88. Roy K, Mao HQ, Huang SK et al. Oral gene delivery with chitosan-DNA nanoparticles generates immunologic protection in a murine model of peanut allergy. Nat Med 1999; 5:387-391.
89. Lundholm P, Asakura Y, Hinkula J et al. Induction of mucosal IgA by a novel jet delivery technique for HIV-1 DNA. Vaccine 1999; 17:2036-2042.
90. Wang B, Dang K, Agadjanyan MG et al. Mucosal immunization with a DNA vaccine induces immune responses against HIV-1 at a mucosal site. Vaccine 1997; 15:821-825.
91. Livingston JB, Lu S, Robinson H et al. Immunization of the female genital tract with a DNA-based vaccine. Infect Immun 1998; 66:322-329.
92. Bagarazzi ML, Boyer JD, Javadian MA et al. Systemic and mucosal immunity is elicited after both intramuscular and intravaginal delivery of human immunodeficiency virus type 1 DNA plasmid vaccines to pregnant chimpanzees. J Infect Dis 1999; 180:1351-1355.
93. Winegar RA, Monforte JA, Suing KD et al. Determination of tissue distribution of an intramuscular plasmid vaccine using PCR and in situ hybridization. Hum Gene Ther 1996; 10:2185-2194.
94. Haynes JR, McCabe DE, Swain WF et al. Particle-mediated nucleic acid immunization. J Biotechnol 1996; 44:37-42.
95. Goto T, Nishi T, Tamura T et al. Highly efficient electro-gene therapy of solid tumor by using an expression plasmid for the herpes simplex virus thymidine kinase gene. Proc Natl Acad Sci USA 2000; 97:354-359.
96. Danko I, Fritz JD, Jiao S et al. Pharmacological enhancement of in vivo gene expresion in muscle. Gene Ther 1994; 1:114-121.
97. Ferrari ME, Nguygen CM, Zelphati O et al. Analytical methods for the characterization of cationic-lipid nucleic acid complexes. Hum Gene Ther 1998; 9:341-351.
98. Capan Y, Woo BH, Cebrekidan S et al. Stability of poly(L-lysine)-complexed plasmid DNA during mechanical stress and DNase I treatment. Pharm Dev Technol 1999; 4:491-498.
99. Singh M, Briones M, Ott G et al. Cationic microparticles: a potent delivery system for DNA vaccines. Proc Natl Acad Sci USA 2000; 97:811-816.
100. Ferkol T, Perales JC, Mularo F et al. Receptor-mediated gene transfer into macrophages. Proc Natl Acad Sci USA 1996; 93:101-105.
101. Mohr L, Schauer JI, Boutin RH et al. Targeted gene transfer to hepatocellular carcinoma cells in vitro using a novel monoclonal antibody-based gene delivery system. Hepatology 1999; 29:82-89.
102. Moradpour D, Schauer JI, Zurawski VR Jr, et al. Efficient gene transfer into mammalian cells with cholesteryl-spermidine. Biochem Biophys Res Commun 1996; 221:82-88.
103. Liang MR, Alestrom P, Collas P. Glowing zebrafish: integration, transmission, and expression of a single luciferase transgene promoted by noncovalent DNA-nuclear transport peptide complexes. Mol Reprod Dev 2000; 55:8-13.
104. Ciccarelli RB, Pachuk CJ, Samuel M et al. Preclinical safety of DNA vaccines. In: Lowrie DB, Whalen RG, eds. Methods in Molecular Medicine, vol 29, DNA Vaccines: Methods and Protocols. Totowa, NJ: Humana Press Inc., 1998:473-488.

105. Wyatt LS, Moss B, Rozenblatt S. Replication-deficient vaccinia virus encoding bacteriophage T7 RNA polymerase for transient gene expression in mammalian cells. Virology 1995; 210:202-205.
106. O'Reilly DR, Miller LK, Luckow VA. In: Baculovirus Expression Vectors. Oxford: Oxford University Press 1994.
107. Bramson JL, Graham FL, Gauldie J. The use of adenovirus vectors for gene therapy and gene transfer in vivo. Curr Opin Biotechnol 1995; 6:590-595.
108. Brooks AI, Halterman MW, Federoff HJ. Use of herpes amplicon system as a vehicle for somatic gene transfer. Advanced Drug Delivery Reviews 1997; 27:29-39.
109. Garoff H, Li KJ. Recent advances in gene expression using alphavirus vectors. Curr Opin Biotechnol 1998; 9:464-469.
110. Johnston RE and Peters CJ. Alphaviruses. In: Fields BN, ed. Virology. 3rd edition. Philadelphia: Lipincott-Raven Publishers, 1996:843-898.
111. Frolov I, Hoffman TA, Pragai BM et al. Alphavirus-based expression vectors: strategies and applications. Proc Natl Acad Sci USA 1996; 93:11371-11377.
112. Johanning FW, Conry RM, LoBuglio AF et al. A sindbis virus mRNA polynucleotide vector achieves prolonged and high level heterologous gene expression in vivo. Nucleic Acids Res 1995; 23:1495-1501.
113. Rice CM. Alphavirus-based expression systems. Adv Exp Med Biol 1996; 397:31-40.
114. Berglund P, Smerdou C, Fleeton MN et al. Enhance immune responses using suicidal DNA vaccines. Nat Biotechnol 1998; 16:562-565.
115. Sizemore DR, Branstrom AA, Sadoff JC. Attenuated Shigella as a DNA delivery vehicle for DNA-mediated immunization. Science 1995; 270:299-302.
116. Darji A, Guzman CA, Gerstel B et al. Oral somatic transgene vaccination using attenuated *S. typhimurium*. Cell 1997; 91:765-775.
117. Dietrich G, Bubert A, Gentschev I et al. Delivery of antigen-encoding plasmid DNA into the cytosol of macrophages by attenuated suicide *Listeria monocytogenes*. Nat Biotechnol 1998; 16:181-185.
118. Fennelly GJ, Khan SA, Abadi MA et al. Mucosal DNA vaccine immunization against measles with a highly attenuated *Shigella flexneri* vector. J Immunol 1999; 162:1603-1610.
119. Paglia P, Medina E, Arioli I et al. Gene transfer in dendritic cells, induced by oral DNA vaccination with *Salmonella typhimurium*, results in protective immunity against a murine fibrosarcoma. Blood 1998; 9:3172-3176.
120. Roth D, Wilson J. Illegitimate recombination in mammalian cells. In: Kucherlapati R, Smith CR, eds. Genetic Recombination. Washington DC: American Society for Microbiology Press, 1988:621-654.
121. Subramani S, Berg P. Homologous and nonhomologous recombination in monkey cells. Mol Cell Biol 1983; 3:1040-1052.
122. Sancar A, Sancar GB. DNA repair enzymes. Annu Rev Biochem 1988; 57:29-67.
123. Priebe S, Westmoreland J, Nillson-Tillgren T et al. Induction of recombination between homologous and diverged DNAs by double-strand gaps and breaks and role of mismatch repair. Mol Cell Biol 1994; 14:4802-4814.
124. Hare JT, Taylor JH. One role for DNA methylation in vertebrate cells is strand discrimination in mismatch repair. Proc Natl Acad Sci USA 1985; 82:7350-7354.
125. Huang JC, Hsu DS, Kazantsev A et al. Substrate spectrum of human exinuclease: repair of abasic sites, methylated bases, mismatches, and bulky adducts. Proc Natl Acad Sci USA 1994; 91:12213-12217.
126. Moens U, Seternes OM, Hey AW et al. In vivo expression of a single viral DNA-binding protein generates systemic lupus erythematosus-related autoimmunity to double-stranded DNA and histones. Proc Natl Acad Sci USA 1995; 92:12393-12397.
127. Doerfler W, Schubbert R. Uptake of foreign DNA from the environment: the gastrointestinal tract and the placenta as portals of entry. Wein Klin Wochenschr 1998; 110:40-44.
128. Schubbert R, Renz D, Schmitz B et al. Foreign (M13) DNA ingested by mice reaches peripheral leukocytes, spleen, and liver via the intestinal wall mucosa and can be covalently linked to mouse DNA. Proc Natl Acad Sci USA 1997; 94:961-966.

129. Rodriguez F, Zhang J, Whitton JL. DNA immunization: ubiquitination of a viral protein enhances cytotoxic T-lymphocyte induction and antiviral protection but abrogates antibody induction. J Virol 1997; 71:8497-8503.
130. Rodriguez F, An LL, Harkins S et al. DNA immunization with minigenes: low frequency of memory cytotoxic T lymphocytes and inefficient antiviral protection are rectified by ubiquitination. J Virol 1998; 72:5174-5181.
131. Geissler M, Tokushige K, Wakita T et al. Differential cellular and humoral immune responses to HCV core and HBV envelope proteins after genetic immunizations using chimeric constructs. Vaccine 1998; 16:857-867.
132. Inchauspe G, Vitvitski L, Major ME et al. Plasmid DNA expressing a secreted or a nonsecreted form of hepatitis C virus nucleocapsid: Comparative studies of antibody and T-helper responses following genetic immunization. DNA Cell Biol 1997; 16:185-195.
133. Boyle JS, Brady JL, Lew AM. Enhanced responses to a DNA vaccine encoding a fusion antigen that is directed to sites of immune induction. Nature 1998; 392:408-411.
134. Ji H, Wang TL, Chen CH et al. Targeting human papillomavirus type 16 E7 to the endosomal/lysosomal compartment enhances the antitumor immunity of DNA vaccines against murine human papillomavirus type 16 E7-expressing tumors. Hum Gene Ther 1999; 10:2727-2740.
135. Vidalin O, Tanaka E, Spengler U, Trepo C et al. Targeting of hepatitis C virus core protein for MHC I or MCH II presentation does not enhance induction of immune responses to DNA vaccination. DNA Cell Biol 1999; 18:611-621.
136. Richmond JFL, Lu S, Santoro JC et al. Studies of the neutralizing activity and avidity of anti-human immunodeficiency virus type 1 *env* antibody elicited by DNA priming and protein boosting. J Virol 1998; 72:9092-9100.
137. Okuda K, Xin KQ, Tsuji T et al. DNA vaccination followed by macromolecular muticomponent peptide vaccination against HIV-1 induces strong antigen-specific immunity. Vaccine 1997; 15:1049-1056.
138. Letvin NL, Montefiori DC, Yasutomi Y et al. Potent, protective anti-HIV immune responses generated by bimodal HIV envelope DNA plus protein vaccination. Proc Natl Acad Sci USA 1997; 94:9378-9383.
139. Haddad D, Liljeqvist S, Stahl S et al. Characterization of antibody responses to a *Plasmodium falciparum* blood-stage antigen induced by a DNA prime/protein boost immunization protocol. Scand J Immunol 1999; 49:506-514.
140. Kent SJ, Zhao A, Best SJ et al. Enhanced T-cell immunogenicity and protective efficacy of a human immunodeficiency virus type 1 vaccine regimen consisting of consecutive priming with DNA and boosting with recombinant fowlpox virus. J Virol 1998; 72:10180-10188.
141. Hanke T, McMichael A. Preclinical development of a multi-CTL epitope-based DNA prime MVA boost vaccine for AIDS. Immunol Lett 1999; 66:177-181.
142. Hanke T, Neumann VC, Blanchard TJ et al. Effective induction of HIV-specific CTL by multi-epitope using gene gun in a combined vaccination regime. Vaccine 1999; 17:589-596.
143. Hanke T, Samuel RV, Blanchard TJ et al. Effective induction of simian immunodeficiency virus-specific cytotoxic T lymphocytes in macaques by using a multiepitope gene and DNA prime-modified vaccinia virus Ankara boost vaccination regimen. J Virol 1999; 73:7524-7532.
144. Sedegah M, Jones TR, Kaur M et al. Boosting with recombinant vaccinia increases immunogenicity and protective efficacy of malaria DNA vaccine. Proc Natl Acad Sci USA 1998; 95:7648-7653.
145. Schneider J, Gilbert SC, Blanchard TJ et al. Enhanced immunogenicity for CD8+ T cell induction and complete protective efficacy of malaria DNA vaccination by boosting with modified vaccinia virus Ankara. Nat Med 1998; 4:397-402.

CHAPTER 15

Plant-Derived Vaccines

Amanda M. Walmsely and Charles J. Arntzen

Introduction

Vaccination has traveled a long road since the contents of smallpox pustules were used to inoculate individuals.[1] Whilst vaccines consisting of attenuated or inactivated whole organisms are still in use, advances in recombinant DNA technology have unleashed the potential of subunit vaccines. Subunit vaccines contain specific antigenic substances that have been expressed and purified using recombinant DNA technology. The production and purification of recombinant antigens may be time consuming and relatively expensive; however the reduced exposure to the pathogen decreases the patient complication rate. This aspect has persuaded health administrations and product manufacturers to favor the subunit approach over other vaccine strategies.[2]

It is thought humans have used medicinal plants since the beginning of civilization. The low cost and availability of medicinal plants ensure their continued use worldwide, particularly in developing countries. During the past decade, advances in plant molecular biology have provided another avenue through which plants can be used for medicinal purposes: use of transgenic plants for production of pharmaceutical proteins. Transgenic plants are arguably the ideal means for production and delivery of subunit vaccines. Like traditional medicinal plants, transgenic plants expressing subunit vaccines are low capital means for large-scale production of therapeutic proteins. Simple technology is involved in the growth and harvest of transgenic plants and increase in production merely requires more area put to seed. Plant-derived subunit vaccines lack contamination with animal pathogens; may be delivered orally; may be engineered to contain multiple antigens; and require minimal processing to produce a stable, heat tolerant formulation. These qualities combine to decrease cost and simplify dispersal of the vaccine, hence increasing the chance of success of a vaccination program, particularly in the resource-limited health systems of developing countries.

Mucosal Vaccines

Many human and animal pathogens initiate infection at the mucosal surfaces lining the digestive tract, respiratory tract and urino-reproductive tract. The primary defense of these tissues is the mucosal immune system (MIS). Whilst parenteral vaccines are effective at eliciting the systemic immune system, they are not effective at eliciting the MIS. Instead the MIS is best induced by mucosal immunization via oral or nasal delivery. The oral and nasal routes evoke less pain and discomfort; have high patient acceptance and compliance; and low costs of administration. Unlike systemic immunity, mucosal immunity does not appear to be subject to age-associated dysfunction and develops earlier than systemic immunity.[3]

Induction of a Mucosal Immune Response

Induction of the MIS begins with antigen recognition and capture by specialized epithelial cells, called M cells, located within the mucus membranes. The antigens are channeled through the epithelial layer to antigen presenting cells that process and present the antigen to B-cells. The now activated B-cells migrate to the mesenteric lymph nodes where they mature into plasma cells before moving into circulation. Upon arrival at a mucosal tissue the mature B-cells, named plasma cells, produce immunoglobulin A (IgA). In moving through the epithelial layer towards the lumen, IgA complexes with membrane bound secretary components to form secretary IgA (sIgA). Upon transportation of the sIgA into the lumen, it interacts with specific antigenic epitopes and neutralizes the invading pathogen.

Production of Plant-Derived Vaccines

Transformation can be classed as stable or transient depending upon the fate of the foreign gene. If the foreign gene is integrated into a genome and passed on to the following cell generations, stable transformation has occurred. If the foreign gene is not incorporated into the targeted genome it is not maintained in following cell generations and is said to be transiently expressed in the recipient cell. Both transformation types are used for production of plant-derived vaccines.

Production and Delivery of Vaccines by Plants

This strategy principally uses stable transformation of plants. Transgenic plants for delivery of subunit vaccines are cost efficient, easy to deliver and do not require the degree of refrigeration that purified recombinant proteins require. However the process of stable plant transformation is time consuming, labor intensive and typically gives rise to erratic expression of the recombinant protein. Although many techniques have been developed and used for stable plant transformation, only *Agrobacterium*-mediated gene transfer has been used for stable expression of subunit vaccines in plants.

Agrobacterium-mediated transformation uses a plant pathogen to transfer and integrate a specific segment of the bacterium's DNA (the transfer- or T-DNA) into the genome of the host plant. After insertion of the candidate gene into the T-DNA, the *Agrobacterium* is incubated with plant tissues. Plant wound-associated compounds produced by either the plant tissue or added to the media, induce the transfer of T-DNA into the plant cells. Transgenic plants are produced through selection and regeneration of transformed cells in tissue culture.

It has been demonstrated that expression and stability of a foreign protein may be increased by directing protein expression to a specific compartment in the plant cell.[4] This is usually achieved through fusing an appropriate signal or transit peptide to the amino terminus of the candidate protein. It may also be advantageous to use promoter elements to express the candidate protein in a specific tissue or stage of development. For example the candidate antigen could be expressed in most tissues all the time; localized in fruit or seeds; or induced at a convenient point in production.

Plants as Bioreactors of Vaccines

Plants are classified as bioreactors when a plant-derived recombinant protein is purified to some degree before use. The protein purification may be due to the plant tissues being unpalatable, not practical for delivery (for example cell suspensions), or containing toxins. Transient expression of recombinant proteins using plant viruses often involves plants as bioreactors. Disadvantages of using recombinant plant viruses for expression of candidate proteins include the possible loss of genetic stability after several passages and the chance of the recombinant virus escaping. However the ease of genetic manipulation and ability to quickly produce large amounts of recombinant protein (up to 2 g/kg of plant tissue in less than two weeks)[5] make plant viruses attractive alternatives for vaccine production.

There are a number of ways to use recombinant plant viruses for over-expression of candidate proteins including viral gene replacement, gene insertion, epitope presentation and complementation.[6] To date however, only epitope presentation has been used to express a plant-derived vaccine. Epitope presentation involves the translational fusion of a small peptide to a viral protein so that it is presented on the exterior of the virus particle. This may be achieved via a fusion within the coat protein or by a "leaky" stop codon at 3' end of the coat protein.

Plant-Derived Vaccines

Plant-derived pharmaceuticals may elicit active immunization (vaccination), passive immunization or auto-tolerance. Plant-derived pharmaceuticals for passive immunization and auto-tolerance have proven immunologically active[7-9] and effective in animals[10-13] and humans[14] but the majority of studies have investigated plant-derived vaccines.

Human Vaccine Antigens

Escherichia coli Heat-Labile Enterotoxin/Cholera Toxin

At present there is no available vaccine for diarrhea, the highest cause of death in children under five in the developing world.[15] Enterotoxigenic *Escherichia coli* (ETEC) and *Vibrio cholerae* are two causal agents of bacterial diarrhea. These organisms enter the body through contaminated food or water and initiate disease by adhering to the surface of the gut epithelial cells and secreting enterotoxins. Two examples of enterotoxins are the ETEC heat labile toxin (LT) and the *V. cholerae* cholera toxin (CT). LT and CT consist of two subunits, the A subunit (LTA or CTA) and the B subunit (LTB or CTB). The B subunit autonomously forms a doughnut-shaped pentamer that has binding affinity with G_{M1} gangliosides, the sugar moieties present on the epithelial cells of the gut. This affinity targets the B subunit and therefore the holotoxin to the gut wall. Diarrhea ensues when the A subunit traverses the gut epithelial cells and causes the loss of electrolytes and water.[16]

The expression of LTB in plants provided proof of concept that plant-derived oral vaccines could be effective in humans.[17] Expression and correct assembly of LTB by tobacco and potato tissues was demonstrated before use of the plant tissues in animal feed trials.[4] After ingestion of the plant-derived recombinant LTB (rLTB) mice produced LTB specific serum IgG and mucosal IgA.[4]

Assembly of a synthetic LTB gene substantially increased rLTB expression in plant tissues.[18] The synthetic LTB gene was designed to use amino acid codons best processed by plants and remove signals that might cause premature termination of gene processing or the degradation of its products. Mason et al[18] demonstrated that mice fed potato tissues expressing the synthetic "plant friendly" rLTB gene developed appropriate IgG and IgA responses sufficient to provide significant protective immunity when challenged with the LT holotoxin.

In 1997, the first human clinical trial with a transgenic, plant-derived, antigen was planned, approved (US Food and Drug Administration) and performed.[17] Eleven "test" participants consumed three doses of 50-100 g of raw, transgenic potato and three "control" participants consumed three doses of 50 g of raw, untransformed potato. Serum and fecal samples were collected and analyzed for LTB-specific antibodies at multiple time points before, during and after the trial. A significant rise (minimum 4-fold increase) in LTB antibodies was displayed by 10 of the 11 test participants while no LTB specific antibodies were detected in the control participants. The serum responses after consumption of transgenic potatoes were comparable to those measured when volunteers were challenged with 10^6 virulent ETEC organisms.[19]

Although expression of CTA and CTB was demonstrated in separate tobacco lines in 1995,[20] report of retention of its native structure and immunological properties did not occur until 1997.[21] Potato-derived CTB was subsequently shown to induce both mucosal and systemic CTB-specific antibodies in orally immunized mice.[22] When challenged by intraileal injection with native CT, the plant-immunized mice showed up to a 60% reduction in diarrheal fluid accumulation. The transgenic plant delivered CTB was therefore able to generate partial protective immunity in mice against the bacterial enterotoxin.

Hepatitis B Virus Surface Antigen

Hepatitis B virus (HBV) is a highly contagious, blood borne virus capable of causing chronic liver disease. It is the leading cause of liver cancer and persistent viremia in humans. Current estimates establish that there are approximately 300 million carriers of HBV world-wide with approximately 1.25 million chronic HBV carriers in the United States.[15] Hepatitis B vaccines were first introduced in the early 1980s as either heat or chemically inactivated particles comprised of the hepatitis B surface antigen (HBsAg) derived from the plasma of chronic hepatitis B patients. Due to the risks involved with the use of serum-derived products, several subunit vaccines were produced in recombinant yeast and are now licensed for clinical use. However the cost of the resulting vaccine is prohibitive for immunization programs in developing countries.[23] Research programs have therefore been initiated to express a hepatitis vaccine in transgenic plants.

In 1992 Mason et al[23] expressed the HBsAg in tobacco and demonstrated the plant-derived antigen formed virus-like particles (VLPs). The particles were physically similar to those derived from human serum and retained antigenicity. Mice immunized with partially purified, tobacco-derived hepatitis B VLPs responded in a manner similar to those immunized with yeast-derived rHBsAg (commercial vaccine).[24] Potato-derived HBsAg is currently being tested in human clinical trials using the orally delivered antigen as a booster for the commercial vaccine. Three additional groups have reported expression of hepatitis B antigens in plant tissues but formation of VLPs or antigenicity have not been described.[25-27]

Norwalk Virus Capsid Protein

Studies have indicated that Norwalk virus (NV) and Norwalk-like viruses are very important causes of diarrhea in the US causing an estimated 42% of severe gastroenteritis outbreaks.[28] Recent advances in Norwalk molecular biology have facilitated the expression of the Norwalk virus capsid protein (NVCP) in an insect expression system. The recombinant particles have proven morphologically and antigenically similar to authentic virus particles; stable upon storage at 4°C and after lyophilization; resistant to pH 3.0 treatment; and capable of inducing the production of serum and mucosal antibodies against NVCP.[29,30] Investigation into the feasibility of the insect-derived recombinant NV particles as a vaccine against Norwalk virus is currently in progress (Mary Estes, Baylor College of Medicine, Houston, TX: personal communication).

Mason et al demonstrated tobacco-expressed NV capsid proteins formed virus-like particles indistinguishable from the insect recombinant VLPs.[31] When gavaged or fed to mice, the plant-derived Norwalk VLPs proved orally immunogenic, inducing both systemic and mucosal immune responses. Recent human clinical trials investigating oral delivery of NV VLPs in transgenic potato tubers resulted in 19 of 20 volunteers developing an immune response against NV.[31a] Although rises in serum antibody levels were modest, 95% of the volunteers who ingested the transgenic potato experienced significant rise in number of specific IgA antibody secreting cells.

Human Rhinovirus 14 Epitope

Cowpea Mosaic Virus has been modified to express an epitope from the human rhinovirus 14 (HRV) in black-eyed bean.[32,33] Large quantities of purified virus particles (1.2-1.5 mg of virus per gram of fresh tissue) expressing the HRV epitope were extracted from leaf tissues and injected intramuscularly and subcutaneously into rabbits. Immunogenicity of the inserted epitope was demonstrated when the sera collected from rCPMV treated rabbits detected denatured HRV-14 virus in Western analysis.

Human Immunodeficiency Virus Type 1 Epitopes

Thousands of people continue to be infected with human immunodeficiency virus (HIV) everyday. In 1996, an estimated 8500 new HIV infections occurred worldwide per day.[15] To date several recombinant HIV antigens expressed by yeast, insect or mammal cells have proven safe and immunogenic in human trials. The antigens include the envelope proteins gp120 and 160, a transmembrane protein gp41, and protease genes.[15]

Plant cell expression of an epitope of the HIV-1 virus was first demonstrated in 1995.[34] A translational fusion of a gp120 epitope to the CP of Tobacco Mosaic Virus (TMV) resulted in expression of virus particles consisting of the CP and CP fusion protein. Immunogenicity of plant-expressed HIV-1 epitopes was later demonstrated in mice using parenteral injections of the purified rCPMV-expressed gp41,[5,33,35] rAIMV-expressed V3 loop of gp120,[36] and recombinant tomato bushy stunt virus-expressed V3 loop.[37] Durrani et al[38] induced the mucosal immune system with a black-eyed bean-expressed, recombinant gp41 CPMV. Purified rCPMV combined with the mucosal adjuvant CT was administered intranasally or through gastric gavage. Two intranasal doses were sufficient to induce both gp41-specific and CPMV-specific IgG in serum and IgA in feces. Oral immunization was less effective than intranasal immunization. Although anti-CPMV antibodies were detected in the serum, feces and intestine, the levels were much lower in comparison to titers received after intranasal immunization. No gp41-specific IgA could be detected in either the feces or intestine of orally immunized animals, and only two mice produced low levels of gp41-specific serum antibodies. The lack of responsiveness was believed to be due to degradation of the HIV-1 peptide or the recombinant virus within the stomach or intestine.

Epitopes from Opportunistic Pathogens

Pseudomonas aeruginosa and *Staphylococcus aureus* are opportunistic pathogens that cause disease in the immunosuppressed, burn victims and chronic pulmonary infections in children with cystic fibrosis. Immunization with the outer membrane F protein has provided protection from *P. aeruginosa* in a number of animal models[39-41] while the D2 domain of the fibronectin-binding protein B (FnBP) has proven effective with *S. aureus.*[42,43]

In 1999, black-eyed bean plants expressed a synthetic peptide containing two epitopes in tandem from the outer membrane F protein from *P. aeruginosa.*[44] The coding region for the synthetic peptide was inserted into two coat proteins of CPMV (S and L) and the rCPMV used to inoculate black-eyed bean plants. Purified rCPMV was subcutaneously injected into mice in Freund's or QuilA adjuvant. Both S and L fusions induced F protein-specific antibodies but were found to induce antibodies predominantly or exclusively against one epitope of the synthetic peptide. It was concluded that the site of peptide expression on CPMV influences its immune recognition.

Black-eyed bean also expressed the D2 domain of the fibronectin-binding protein B (FnBP) of *S. aureus.*[45] The D2 domain was inserted within the coding region of the S coat protein of CPMV and the recombinant virus used to over-express the epitope of interest in plants. Purified recombinant virus was administered intranasally or by gavage to mice either alone or in the presence of the mucosal adjuvant ISCOM matrix. All intranasal treatments (with and without adjuvant) generated high titers for CPMV- and FnBP-specific IgG in sera and IgA and IgG of bronchial, intestinal and vaginal lavage fluids. This confirmed the ability of recombinant plant viruses to induce immune responses at distant mucosal sites and without the presence of a mucosal adjuvant. Assays also demonstrated the sera to completely inhibit the binding of human serum fibronectin to *S. aureus* FnBP.

As found in the studies performed by Durrani et al,[38] oral immunization with recombinant plant viruses was not as successful as intranasal administration. Although oral immunization resulted in production of CPMV- and FnBP-specific serum IgG, the titers were significantly lower and more variable than those generated by the intranasal treatments. FnBP-specific intestinal IgA was not detected.

Animal Vaccine Antigens

Rabies Virus Glycoprotein and Nucleoprotein

Rabies continues to be a significant international health problem. Approximately 12 human fatalities occur annually in the United States and the number of postexposure treatments is rapidly increasing.[15] Unfortunately postexposure treatment is not anticipated in developing countries; instead mass immunization is the most feasible course. This goal however is hampered by the relatively high cost of vaccine production and distribution. The currently available vaccine is recombinant, using the rabies virus glycoprotein (G protein) derived from an insect expression system.

McGarvey et al[46] used stable transformation to demonstrate the ability of plants to express the G protein. However demonstration of immunological activity of a plant-derived rabies antigen employed a plant virus expression system.[36] Yusibov et al[36] expressed a B-cell epitope from the G protein and a T-cell epitope from the nucleoprotein through fusion to the coat protein (CP) of the alfalfa mosaic virus (AIMV) and subsequent inoculation of tobacco. Injection of mice with isolated recombinant virus particles induced virus-neutralizing antibodies. The ability of the recombinant AIMV to be delivered orally was tested through gastric intubation of plant extracts or feeding on virus-infected spinach leaves.[47] Both treatments induced local and systemic immune responses that proved sufficient to improve the clinical signs of the virus upon challenge with an attenuated rabies strain. These results were the first indication that an orally delivered, modified virus could induce a mucosal immune reponse.

Foot and Mouth Virus VP1 Coat Protein

Foot and Mouth Disease Virus (FMDV) affects all cloven-hoofed animals producing vesicles in the mouth, on the teats, and on the skin between and above the hoofs. Since foot and mouth disease is extremely infectious an outbreak quickly closes export markets of livestock and livestock products. The basis of eradication of the disease is destruction of afflicted animals and vaccination of susceptible hosts using inactivated virus.[48] Recent studies have demonstrated the structural protein VP1 to carry critical epitopes responsible for the induction of neutralizing antibodies.[49-52] These epitopes have been expressed within a diverse range of prokaryotic and eukaryotic systems.

A plant virus expression system was the first reported means of expressing the VP1 epitope in plants.[32] Black-eyed bean protoplasts were inoculated with a recombinant Cowpea Mosaic Virus (CPMV) containing a sequence from VP1 cloned into the coding region of the S coat protein. Immunosorbent electron microscopy and Western analysis demonstrated the presence of recombinant virions in the plant cell extracts. In 1999 Wigdorovitz et al reported the use of transgenic alfalfa to express and deliver the FMDV VP1 epitope.[53] Immunization of mice through injection of leaf extracts or feeding freshly harvested leaves induced virus-specific immune responses sufficient for protection upon FMDV challenge. After 36 hours, 77-80% of the parenteral immunizations and 66-75% of the oral immunizations exhibited absence of viremia.

Murine Zona Pellucida 3 Protein

A novel use of plant-derived vaccines is the delivery of immunocontraceptive vaccines for the control of problem animal populations. For many years, methods such as shooting, poisoning or trapping have been used to control problem animal populations. However, changing social beliefs and increasing urbanization have prompted many to advocate the use of more humane, contraceptive approaches to control these populations. Several investigations have found peptides of zona pellucida 3 (ZP3), a protein involved in the docking of sperm to eggs, capable of inducing antibody-mediated contraception. However, these vaccines have proven difficult and expensive to deliver parenterally to free-ranging animals. Delivery of immunocontraceptive vaccines through baiting with edible plant tissues expressing an appropriate immunogen would decrease the cost of production and labor and increase the ease of delivery.

A plant virus expression system was used by Fitchen et al[54] to express an epitope from the murine ZP3 within the coat protein of tobacco mosaic virus (TMV). The recombinant coat protein was found to accumulate to high concentrations and form particles in inoculated tobacco plants. Parenteral immunization of mice with tobacco extracts containing the recombinant CP resulted in the production of serum antibodies recognizing the ZP3 epitope and the recruitment in vivo of antibodies to the ZP in female mice. Unfortunately the effect of the murine ZP3 epitope on fertility was not reported.

VP2 Epitope from the Mink Enteritis Virus

Mink enteritis virus (MEV) is characterized by severe inflammation of the intestine, resulting in profuse diarrhea. Existing vaccines are either chemically inactivated or live attenuated viruses propagated from cell cultures.[55] VP2 is the major structural protein required for MEV particle formation and is capable of self-assembling into virus-like particles when expressed in an insect cell expression system. Peptide vaccines based on VP2 sequences have been shown to induce neutralizing antibodies in guinea pigs and rabbits as well as solid protection against disease in the natural hosts after heavy challenge.[56]

Upon insertion of a VP2 epitope into the S protein of CPMV Dalsgaard et al[55] demonstrated the ability of black-eyed bean plants to express the VP2 epitope on the surface of CPMV. The immunogenicity of the rCPMV was demonstrated after one subcutaneous injection of the purified recombinant virus into mink conferred protection against clinical disease and virtually abolished shedding of virus after challenge with virulent MEV. This was the first demonstration of a plant virus expressed vaccine providing full protection against a pathogen.

Murine Hepatitis Virus Epitope

Murine hepatitis virus (MHV) is responsible for a variety of acute and chronic diseases in mice. MHV contains at least three dominant structural proteins, the membrane glycoprotein (M protein), the nucleocapsid protein (N protein) and the spike glycoprotein (S protein).[57] Many studies have focused upon the S protein since it is a critical determinant of viral pathogenicity and contains major immunodominant neutralization domains. Protective neutralizing antibodies have been induced in mice by immunization with purified S protein and synthetic peptides.[58-60]

The epitope from the S protein was expressed in inoculated tobacco plants after insertion within the coding region of the TMV CP.[61] The purified recombinant virus was administered subcutaneously with an adjuvant or intranasally. Intranasal treatments developed serum IgG and IgA specific for the S protein epitope and for the TMV coat protein. Subcutaneous treatment only induced S epitope- and TMV CP-specific serum IgG. Both treatments induced antibody levels that enabled survival when challenged with a lethal dose of MHV.

Antigens that have been expressed by plants but not tested for immunogenicity include epitopes from malaria expressed on the surface of TMV[62] and the immunodominant glycoprotein B complex (gB, UL55) of human cytomegalovirus (HCMV) in seeds of transgenic tobacco.[63]

Summary

Evidence has quickly accumulated in support of plant-derived antigens as effective, safe and inexpensive means of acquiring or delivering a vaccine. The success of human trials with edible transgenic plant vaccines[17,31a] and successful challenge in animal studies[53,56,61] have demonstrated the potential of plant-derived vaccines to expand the armament of preventative medicine. Vaccines produced using plants as bioreactors have been successful when parenterally or intranasally administered. However oral immunization has only been successful when the vaccine was delivered in planta. This suggests the additional protection afforded by plant tissues permits successful oral immunization.

Future Use of Plant-Delivered Vaccines

Before global distribution of plant-delivered vaccines, several issues need to be considered. These include dosage of antigens delivered through edible vaccines and analysis of the possible development of oral tolerance to edible vaccines. Batch processing of edible transgenic tissues expressing a candidate antigen circumvents the anticipated problem of variable expression levels within and between transgenic plants. In accordance with traditional vaccines the distribution of edible vaccines to humans is envisioned to be regulated by a health-care system. The vaccine would hence not be delivered as normal constituents of food but allocated in prescribed amounts.

Although the picture of the development of oral tolerance is incomplete, known factors include solubility of the antigen, frequent exposure and extreme doses. Extreme doses would be prevented through the health-care system employing an appropriate vaccination regime. In addition, candidate antigens for oral vaccines likely mimic disease-causing agents, for example virus like particles, pathogen surface antigens and toxin analogs. It is therefore unlikely that these antigens would induce tolerance but careful testing would be required on a case by case basis.

As with any area of study, progress requires overcoming challenging obstacles. The demonstrated potential of plants for production and/or delivery of vaccines and the rapid advance in the related technologies lend promise to swift response and solution of difficulties. The prospect of developing effective plant-derived vaccines is real.

References

1. Fenner F, Henderson DA, Arita L et al. Smallpox and its eradication. Geneva: World Health Organization, 1988.
2. Mackett M, Williamson JD. Developing New Vaccines: Inactivated and Subunit Approaches. In: Mackett M, Williamson JD, eds. Human Vaccines and Vaccination. Oxford: BIOSIS Scientific Publishers Limited, 1995:99-119.
3. Fasano A. Innovative strategies for the oral delivery of drugs and peptides. Trends Biotechnol 1998; 16(4):152-157.
4. Haq TA, Mason HS, Clements JD et al. Oral immunization with a recombinant bacterial antigen produced in transgenic plants. Science 1995; 268:714-716.
5. Porta C, Spall VE, Lin T et al. The development of cowpea mosaic virus as a potential source of novel vaccines. Intervirol 1996;39:79-84.

6. Scholthof HB, Scholthof K-BG, Jackson AO. Plant virus gene vectors for transient expression of foreign proteins in plants. Ann Rev Phytopathol 1996; 34:299-323.
7. During K, Hippe S, Kreuzaler F et al. Synthesis and self-assembly of a functional monoclonal antibody in transgenic *Nicotiana tabacum*. Plant Mol Biol 1990; 15:281-293.
8. Ma JK, Hiatt A, Hein M et al. Generation and assembly of secretory antibodies in plants. Science 1995; 268:716-719.
9. Verch T, Yusibov V, Kowprowski H. Expression and assembly of a full-length monoclonal antibody in plants using a plant virus vector. J Immunol Meth 1998; 220:69-75.
10. Ma S-W, Zhao D-L, Yin Z-Q et al. Transgenic plants expressing autoantigens fed to mice to induce oral immune tolerance. Nature Med 1997; 3(7):793-796.
11. Zeitlin L, Olmsted SS, Moench TR et al. A humanized monoclonal antibody produced in transgenic plants for immunoprotection of the vagina against genital herpes. Nature Biotechnol 1998; 16:1361-1364.
12. Arakawa T, Yu J, Chong DKX et al. A plant-based cholera toxin B subunit—insulin fusion protein protects against the development of autoimmune diabetes. Nature Biotechnol 1998; 16:934-938.
13. McCormick AA, Kumagai MH, Hanley K et al. Rapid production of specific vaccines for lymphoma by expression of the tumor-derived single-chain Fv epitopes in tobacco plants. Proc Nat Acad Sci USA 1999; 96:703-708.
14. Ma JK-C, Hikmat BY, Wycoff K et al. Characterization of a recombinant plant monoclonal secretory antibody and preventative immunotherapy in humans. Nature Med 1998;4:601-606.
15. The Jordon Report: Accelerated Development of Vaccines: National Institute of Allergy and Infectious Diseases. 1998.
16. Sixma TK, Pronk SE, Kalk KH et al. Crystal structure of a cholera toxin-related heat-labile enterotoxin from *E. coli* Nature 1991; 351:371-373.
17. Tacket CO, Mason HS, Losonsky G et al. Immunogenicity in humans of a recombinant bacterial antigen delivered in a transgenic potato. Nature Med 1998; 4(5):607-609.
18. Mason HS, Haq TA, Clements JD. Edible vaccine protects mice against *Escherichia coli* heat-labile enterotoxin (LT): potatoes expressing a synthetic LT-B gene. Vaccine 1998; 16(13):1336-1343.
19. Tacket CO, Reid RH, Boedeker EC et al. Enternal immunisation and challenge of volunteers given enterotoxigenic *E. coli* CFA/II encapsulated in biodegradable microspheres. Vaccine 1994; 12:1270-1274.
20. Hein MB, Yeo T-C, Wang F et al. Expression of cholera toxin subunits in plants. Ann NY Acad Sci 1996; 792:50-56.
21. Arakawa T, Chong DKX, Merritt JL et al. Expression of cholera toxin B subunit oligomers in transgenic potato plants. Trans Res 1997; 6:403-413.
22. Arakawa T, Chong DKX, Langridge WHR. Efficacy of a food plant-based oral cholera toxin B subunit vaccine. Nature Biotechnol 1998; 16:292-297.
23. Mason HS, Lam DM-K, Arntzen CJ. Expression of hepatitis B surface antigen in transgenic plants. Proc Nat Acad Sci USA 1992; 89:11745-11749.
24. Thanavakam Y, Yang Y-F, Lyons P et al. Immunogenicity of transgenic plant-derived hepatitis B surface antigen. Proc Nat Acad Sci USA 1995; 92:3358-3361.
25. Liu YL, Wang JF, Qui BS et al. Expression of human hepatitis B surface antigen gene in transgenic tobacco. Sci China B 1994; 37:27-40.
26. Domansky N, Ehsani P, Salmanian A-H. Organ-specific expression of hepatitis B surface antigen in potato. Biotechnol Letts 1995; 17:863-866.
27. Ehsani P, Khabiri A, Domansky N. Polypeptides of hepatitis B surface antigen produced in transgenic potato. Gene 1997; 190:107-111.
28. Kaplan JE, Feldman R, Campbell DS et al. The frequency of a Norwalk-like pattern of illness in oubreak of acute gastroenteritis. Am J Pub Health 1982; 72:1329-1332.
29. Jiang X, Wang M, Graham DY et al. Expression, self-assembly, and antigenicity of the Norwalk virus capsid protein. J Virol 1992; 66:6527-6532.
30. Ball JM, Estes MK, Hardy ME et al. Recombinant Norwalk virus-like particles as an oral vaccine. Arch Virol 1996; 12:243-249.
31. Mason HS, Ball JM, Shi J-J et al. Expression of Norwalk virus capsid protein in transgenic tobacco and potato and its oral immunogenicity in mice. Proc Nat Acad Sci USA 1996; 93:5353-5340.

31a. Tacket CO, Mason HS, Lansonsky G et al. Human immune responses to a novel norwalk virus vaccine delivered in transgenic potatoes. J Infect Dis 2000; 182(1):302-305.
32. Usha R, Rohll JB, Spall VE et al. Expression of an animal virus antigenic site on the surface of a plant virus particle. Virology 1993; 197:366-374.
33. Porta C, Spall VE, Loveland J et al. Development of cowpea mosaic virus as a high-yielding system for the presentation of foreign peptides. Virology 1994; 202:949-955.
34. Sugiyama Y, Hamamoto H, Takemoto S et al. Systemic production of foreign peptides on the particle surface of tobacco mosaic virus. FEBS Letts 1995; 359:247-250.
35. McLain L, Durrani Z, Wisniewski LA et al. Stimulation of neutralizing antibodies to human immunodeficiency virus type 1 in three strains of mice immunized with a 22 amino acid peptide of gp41 expressed on the surface of a plant virus. Vaccine 1996; 14(8):799-810.
36. Yusibov V, Modelska A, Steplewski K et al. Antigens produced in plants by infection with chimeric plant viruses immunize against rabies virus and HIV-1. Proc Nat Acad Sci USA 1997; 94:5784-5788.
37. Joelson T, Akerblom L, Oxelfelt P et al. Presentation of a foreign peptide on the surface of tomato bushy stunt virus. J Gen Virol 1997; 78:1213-1217.
38. Durrani Z, McInerney TL, McLain L et al. Intranasal immunization with a plant virus expressing a peptide from HIV-1 gp41 stimulates better mucosal and systemic HIV-1-specific IgA and IgG than oral immunization. J Immunol Meths 1998; 220:93-103.
39. Gilleland HE, Parker MG, Matthews JM et al. Use of purified outer membrane protein F (porin) preparation of *Pseudomonas aeruginosa* as a protective vaccine in mice. Infec Immunol 1984; 44:49-54.
40. Matthews-Greer JM, Robertson DE, Gilleland LB et al. *Pseudomonas aeruginosa* outer membrane protein F produced in *Escherichia coli* retains vaccine efficacy. Curr Microbiol 1990; 20:171-175.
41. Gilleland HE, Gilleland LB, Hughes EE et al. Recombinant outer membrane protein F of *Pseudomonas aeruginosa* elicits antibodies that mediate opsonophagocytic killing but not complement-mediated bacteriolysis of various strains of *P. aeruginosa.* Curr Microbiol 1992; 24:1-8.
42. Flock J-I, Froman G, Jonsson K et al. Cloning and expression of the gene for a fibronectin-binding protein of *Staphylococcus aureus.* EMBO 1987; 6:2351-2357.
43. Signas C, Raucci G, Jonsson K et al. Nucleotide sequence of the gene for a fibronectin-binding protein from *Staphylococcus aureus*: Use of this peptide sequence in the synthesis of biologically active peptides. Proc Nat Acad Sci USA 1989; 86:699-703.
44. Brennan FR, Jones TD, Gilleland LB et al. *Pseudomonas aeruginosa* outer-membrane protein F epitopes are highly immunogenic in mice when expressed on a plant virus. Microbiol 1999; 145:211-220.
45. Brennan FR, Bellaby T, Helliwell SM et al. Chimeric plant virus particles administered nasally or orally induce systemic and mucosal immune responses in mice. J Virol 1999: 73(2):930-938.
46. McGarvey PB, Hammond J, Dienelt MM et al. Expression of the rabies virus glycoprotein in transgenic tomatoes. Bio/Technol 1995; 13:1484-1487.
47. Modelska A, Dietzschold B, Sleysh N et al. Immunization against rabies with a plant-derived antigen. Proc Nat Acad Sci USA 1998; 95:2481-2485.
48. Brown F. Vaccination against foot and mouth disease virus. Vaccine 1992;10:1022-1026.
49. Kleid DG, Yansura D, Small B et al. Cloned viral protein vaccine for foot and mouth disease: Responses in cattle and swine. Science 1981; 214:1125-1129.
50. DiMarchi R, Brooke G, Gale C et al. Protection in cattle against foot and mouth disease by a synthetic peptide. Science 1986; 232:639-641.
51. Morgan DO, Moore DM. Protection in cattle and swine against foot and mouth disease (FMD) with a biosynthetic peptide vaccine. Am J Vet Res 1990; 51:40-45.
52. Zamorano P, Wigdorovitz A, Perez-Filgueira M et al. A 10-amino acid linear sequence of VP1 of foot and moth disease virus (FMDV) containing B- and T-cell epitopes induces protection in mice. Virology 1995; 212:614-621.
53. Wigdorovitz AW, Carrillo C, Dus Santos MJ et al. Induction of a protective antibody response to foot and mouth disease virus in mice following oral or parenteral immunization with alfalfa transgenic plants expressing the viral structural protein VP1. Virology 1999; 255:347-353.

54. Fitchen J, Beachy RN, Hein MB. Plant virus expressing hybrid coat protein with added murine epitope elicits autoantibody response. Vaccine 1995; 13(12):1051-1057.
55. Dalsgaard K, Uttenthal A, Jones TD et al. Plant-derived vaccine protects target animals against a viral disease. Nature Biotechnol 1997; 15:248-252.
56. Langeveld JPM, Kamstrup S, Uttenthal A et al. Full protection in mink against mink enteritis virus with new generation canine parvovirus vaccines based on synthetic peptide or recombinant protein. Vaccine 1995; 13:1033-1037.
57. Cavanagh D, Brian DA, Enjuanes L et al. Recommendations to the Coronavirus study group for the nomenclature of the structural proteins, mRNA and genes of Coronaviruses. Virology 1990; 176:306-307.
58. Talbot PJ, Dione G, Lacroix M. Vaccination against lethal coronavirus-induced encephalitis with a synthetic decapeptide homologous to the domain in the predicted pelomer stalk. J Virol 1988; 62:3032-3036.
59. Daniel C, Talbot PJ. Protection from lethal coronavirus infection by affinity-purified spike glycoprotein of murine hepatitis virus, strain A59. Virology 1990; 174:87-94.
60. Koolen MJM, Borst MAJ, Horzinek MC et al. Immunogenic peptide comprising a mouse hepatitis virus A59 B-cell epitope and an influenza virus T-cell epitope protects against lethal infection. J Virol 1990; 64:6270-6273.
61. Koo M, Bendahmane M, Lettieri GA et al. Protective immunity against murine hepatitis virus (MHV) induced by intranasal or subcutaneous administration of hybrids of tobacco mosaic virus that carries an MHV epitope. Proc Nat Acad Sci USA 1999; 96:7774-7779.
62. Turpen TH, Reinl SJ, Charoenvit Y et al. Malarial epitopes expressed on the surface of recombinant tobacco mosaic virus. Bio/Technol 1995; 13:53-57.
63. Tackaberry ES, Dudani AK, Prior F et al. Development of biopharmaceuticals in plant expression systems: Cloning, expression and immunological reactivity of human cytomegalovirus glycoprotein B (UL55) in seeds of transgenic tobacco. Vaccine 1999; 17(23-24):3020-3029.

CHAPTER 16

Biological Aspects and Prospects for Adjuvants and Delivery Systems

Bror Morein and Ke-Fei Hu

Introduction

There is no all-encompassing definition for an **adjuvant**, but an old definition is: a substance that enhances the immunogenicity of coadministered antigens without inducing an immune response to itself. Allison and Byars[1] introduced functional structures into the terminology by defining an **adjuvant** as an agent that augments specific immune stimulation to antigens, a **vehicle** as the substance used for the delivery of antigen, and an **adjuvant formulation** as the combination of an adjuvant with a suitable vehicle. An even broader definition was advanced by Cox and Coulter[2] who defined adjuvants as "any substance or procedure that results in a specific increase in the immunogenicity of a vaccine component". Recently Morein and Lövgren-Bengtsson[3] defined three major functionally important areas for adjuvant, without claiming to be all-encompassing: 1) Physical presentation in the form of particles with multimeric antigens on the surface. This form mimics a microorganism, which the host evolutionarily considers dangerous if the antigens on the surface are not self. Examples of such particulate vehicles are protein micelles, liposomes, immune stimulating complexes (ISCOMs), or virus-like particles (VLPs). 2) Targeting properties are required to guide the vaccines to the antigen-presenting cells (APC) as well as into cellular compartments of the APC, i.e., to the endosomal pathway for MHC class II presentation, while MHC class I responses require delivery of antigen to the cytosol. Targeting to the lymphatic organs and cells for further tissue distribution is important. 3) Immune modulation encompasses the type of immunity including both T- and B-cell responses as well as the level of immune response.

For an adjuvant to succeed in guiding an immune response, we have to intervene in the development of the immune response along the lines shaped by nature. Thus, a main target for adjuvants is the innate immune system which recognizes molecules with structures divergent from those in the host, thereby leading to inflammatory responses and eventually acquired immune responses. Some vaccine antigens from microorganisms have immunomodulating activity, e.g., in respiratory syncytial virus (RSV) the G envelope protein promotes a Th2 type of response that may exacerbate disease, while the F envelope protein promotes a Th1 type of response.[4] Therefore the G antigen in its native form should be avoided as an immunogen; despite harboring protective properties, it may enhance disease or subsequent natural infection.[5] Likewise experimental vaccination of mice with *Trypanosoma cruzi* whole-cell antigens exacerbated disease following subsequent challenge of mice, while deletion of one antigen resulted in a protective experimental vaccine.[6] Thus the parasite has learned evolutionarily to use the reaction of the host to evade a protective immune response. Adjuvants are often of bacterial origin; well-known examples are lipopolysaccharide (LPS), muramyldipeptide (MDP), various

New Vaccine Technologies, edited by Ronald W. Ellis. ©2001 Eurekah.com.

Table 16.1. Recognition and activation of the innate immune system for initiating acquired immune responses

Features		**Characterized by**	**Examples of inducer (adjuvant)**
Cognitive phase	Recognition of physical forms	The particulate form of viruses and microorganism with multimeric formation of antigens	VLPs, micelles, ISCOMs, liposomes, SAF-1, nanoparticles, PLGA microspheres, emulsions
	Recognition of biochemical compositions	Microbial carbohydrate, glycolipids, polypeptides, DNA	LPS (and derivatives), carbohydrates, polymers (e.g., mannose), cell wall skeleton and CpG motifs
Effector phase	Inflammation	Production of chemokines and cytokines, e.g., β-chemokines, IL-1, IL-6, IL-8, GM-CSF, TNF-α	LPS, saponins, Freund's complete adjuvant, block polymers
	Costimulation	The requirement of B7.1 (CD80) and B7.2 (CD86) and their common receptors CD28 and CTLA-4	ISCOMs
	Immune regulation	Mainly regulatory cytokines over MHC class II, IL-12, IL-15, IL-18, IFN-γ (Th1) or IL-4 and IL-10 (Th2)	See Table 16.4
		or over MHC class I mainly by delivery of Ag to the cytosol	See Table 16.5

bacterial cell wall components, and CpG motifs of prokaryotic DNA. Virus components such as double-stranded RNA may act as adjuvants by inducing interferon (IFN)-α/β. Adjuvant-active substances also are found in plants (e.g., among saponins and oils), among minerals (aluminum hydroxide and phosphate or calcium phosphate) and can be synthetic (e.g., block polymers and polyacryls in various formations) (ref. 7, Table 16.2). All these molecules harbor structures divergent from those intrinsic to the host, as is the case with most vaccine antigens. It is therefore obvious that adjuvant activity has to be evaluated in the total formulation of antigen, carbohydrates, lipids and delivery system. Any adjuvant, from the point of view of immune protection, may influence the immune response negatively or positively depending on the biological properties of the pathogen that are targeted for the vaccine. Several excellent articles dealing with adjuvants have been published recently. Thus, we are using a complementary approach for covering biological aspects and the prospects for adjuvants to intervene in the immune response.

Table 16.2. Main adjuvant sources and their key properties

Origin		Properties
Microbial	LPS and derivatives such as MPL and Lipid A-detoxified derivatives of lipopolysaccharide derived from bacteria	Induction of co-stimulatory signals on APCs and favors Th1 response
	CT (B) and LT (B)	Mucosal targeting with a Th2 profile
	Non-replicative vectors, e.g., recombinant adenylate cyclase toxins of *B. pertussis* and recombinant parvovirus-like particles	Deliver Ag to cytosol inducting CTL response
	Mycobacteria whole cells (FCA) and derivatives such as MDP, MTP and GMDP and TDM	Th1 response
	CWS (cell wall skeleton)	Th1 response
	Prokaryotic DNA CpG motifs	Th1 response
Plants	Saponins	Th1 and Th2 responses and IgA via mucosal administration
	Vegetable oil in emulsion	degradable
	Vitamin D3	Th2, secretory IgA
	γ-Inulin	Th1 response and complement activation
Mineral	Calcium salts	Depot
	Aluminum salts	Depot and Th2 response
	Mineral oil (IFA)	Th2
Synthetic	Block co-polymers	Th1
	Microspheres	Depot and Th2 response
	Stearyl tyrosine	Th1 and Th2 responses
Combined sources	w/o *M. tuberculosis* + mineral oil (FCA)	Th1
	o/w Block polymers, MDP (SAF-1)	Th1 response
	CWS, TDM, MPL	Th1
	LPS derivatives with $Al(OH)_3$	Th1/Th2
	ISCOMs formulated with any hydrophobic antigens, various s saponins and lipids	Strong Th1 and Th2 with good antigen presentation, targeting and CTL induction
	Liposomes as for ISCOMs	Moderate antigen presentation, targeting and also CTL induction properties

Innate Immunity: The Gateway to an Acquired Immune Response

Innate immunity has been considered to provide fast and nonspecific protection against infections or tumors in a surveillance action. Acquired immunity evolved about 400 million years ago from a phylogenetically very old system, i.e., innate immunity which provides an instructive role for the ensuing acquired immune response. However, it should not be forgotten that the innate immune system still is of primary importance, e.g., in protection against infections or tumor development. For example, the expectant mother has a partially suppressed immune status, particularly with respect to Th1 responses, but in compensation there is an upregulation of the innate system.[9]

Innate immunity is a primary target for adjuvants and delivery systems and includes at least four phases: recognition, inflammation, costimulation and immunomodulation (Table 16.1). All these phases should be considered for tailor-making vaccines with defined immunological properties. The primary event for activation of the innate system against an infection is the recognition of divergent foreign physical forms of antigens. This particulate form of foreign intruder has been mimicked in a number of vaccines or experimental vaccine formulations. Examples of delivery systems encompassing antigens and various adjuvant functions such as targeting and immune modulation include micelles, VLPs, liposomes, SAF-1, biodegradable microspheres[10] and ISCOMs.[7,11]

Prokaryotic microbes (Table 16.2) are a rich source of substances that act as host recognition signals. Many types of bacterial molecules have been used as adjuvants, among which the most common for recognition are microbial carbohydrate structures in the form of outer membrane glycolipids and cell wall carbohydrates. Carbohydrate structures activate complement (C′) through the classical pathway by binding to collectin (a molecule having both lectin and collagen domains)[12] or through the alternative pathway by interacting with sialic acid-free carbohydrates. Carbohydrates often cover bacterial surfaces, thus presenting a wide range of molecular structures and patterns serving as fingerprints for host recognition. The host's soluble mannose-binding protein (MBP) and the cell surface mannose-binding receptor (MR, not limited to mannose) are able to recognize these patterns to discern self from nonself and to activate innate immunity accordingly.

The MBP is synthesized by liver cells as a monomer that has a carboxy-terminal carbohydrate recognition (lectin) domain (CRD), a neck region, and a collagen region; these qualify MBP as a collectin.[13] The helical collagen structure provides increased flexibility to MBP, thus contributing to a broadening of the freedom of ligand binding. MBP is also classified as an acute-phase protein. Multimer forms of MBP can substitute C1q in the classical and alternative C′ cascade. More important is the activation mechanism of the C′ cascade by MBP-associated serine proteases (MASP-1 and MASP-2) discovered in man and mouse.[14,15] These proteins can be traced evolutionarily (together with lectin-like proteins) to tunicates, suggesting an evolution for C′ activation. In view of the link between C′ activation and its major role in generating an antibody response,[16] it is of interest to look for adjuvant-activating C′ in early life forms. The opsonin activity of MBP[17] suggests a role as a broad-spectrum "antibody" and a promoter for antigen uptake by APCs.

Explanations for the selective binding of MBP to carbohydrate structures of pathogens are given in the review of Frazer et al[18] and the pioneering work of Weis et al.[19] The latter showed, by cocrystallization of the carbohydrate domain of rat MBP with a defined oligomannose ligand-binding site, that an equatorial orientation of the C_3 and C_4-OH groups of the hexoses is required for the binding. This orientation of hydroxyl groups is found not only in mannose but also in glucose, N-acetylglucosamine and fucose[19] in the cell walls of $Gram^+$ and $Gram^-$ bacteria, yeasts, and certain parasites. These structures in O-linked polysaccharides of $Gram^-$ bacteria and in lipoteichoic acid of $Gram^+$ bacteria are certainly important for activation of the innate immune response. The analyses of key amino acids in the rat MBP-binding region also clarifies the requirements for galactose binding to the C-type lectin-binding domains. In contrast, the ultimate and penultimate sialic acid and galactose sugars of mammalian cells, probably constituting a marker of self for the innate immune system, are not bound to the MBP-binding site. Thus, the MBP is a target for adjuvant molecules with its recognition and opsonin functions. Carbohydrates that bind to MBP have prospects for use in complex adjuvant and delivery systems. Viral carbohydrates also are bound to MBP, as exemplified by herpesviruses and HIV-1. Conceptually, innate immunity recognizing adjuvants is similar to its recognizing microorganisms; this event initiates the danger signal[20] or mimics the pathogen-associated molecular pattern (PAMP) to target the corresponding pattern recognition receptors (PRR) and to provoke a cascade of events eventually leading to protective immunity.

The cellular MR, a 180-kilodalton (kD) molecule, is probably the most important PRR found on resident macrophages and dendritic cells (DC). MR has a unique capacity to bind and mediate the internalization both of soluble ligands by endocytosis and of particulate ligands by phagocytosis. The MR has five domains* that appear to recognize different PAMPs. The receptor originally was named after mannose; its binding properties are mediated by a CRD tandem repeat that also binds fucose, N-acetylglucosamine and glucose (but not galactose) in a calcium-dependent manner. The individual CRDs of the receptor contribute differently to overall avidity[19,21] This CRD-mediated pattern contributes to differentiating self from pathogen surface carbohydrates that are not found on mammalian glycoconjugates and cells. Additional motifs are recognized by the cysteine-rich MR region that is upregulated after immunization of mice. Oligosaccharides terminating with $GalNac_4$-SO_4 independently bind calcium to a site different from that of mannosylated ligands. This motif is present on glycoprotein hormones and may be involved in determining serum half-life. Sulfate-derivitized polysaccharide adjuvants used for animals and proposed for humans[7,28-30] may be active through a similar mechanism. High-molecular-weight sulfated dextran was used in water/oil (w/o) emulsions; its mechanism of action is probably complex.[30] The ability to bind and internalize microorganisms via the MR in the absence of opsonizing antibodies is highly interesting for protection against infection and also represents an adjuvant concept for vaccine delivery to target APC.

The endocytic and phagocytic activities mediated by the MR are independently regulated. IFNγ enhances MR-mediated phagocytic activity while inhibiting MR expression and IL-4-induced endocytosis is leading to a Th2 response as demonstrated by microbes and their products.[31,32]

MR expression is regulated in vivo by immunoglobulins, corticosteroids and prostaglandins.[33] Direct evidence for the potential of adjuvants to target the MRs through carbohydrates is that mannosylation of antigens enhances the MHC class II-restricted T-cell response over DCs[34,35] for targeting captured antigens to the MHC class II loading compartment.[33] Polymers of mannose and (1-3 glucose have been proposed as adjuvants for human vaccines, either mixed[2,7] or conjugated to an immunogen in order to target the MR or DEC receptor on macrophages or DCs; such formulations upregulate Th1 responses.

In the last few years, it has been revealed that lipid antigens could induce immune responses by antigen presentation to lymphocytes by members of the CD1 family of molecules.[36,37] The CD1 pathway of antigen presentation, which functions similarly to that of MHC class II

*The five domains are 1, amino-terminal, extracellular, cysteine-rich region; 2, a fibronectin type-II repeat-containing domain; 3, eight tandem lectin-like carbohydrate recognition domains (CRD); 4, a transmembrane domain; and 5. an intracellular carboxy-terminal tail.[22-24] This domain-structure defines a family of multilectin mannose receptors having 8-10 CRDs with other members including phopholipase A_2 receptor (PLA_2),[25] DEC-205, and an unnamed lectin.[26]

molecules, opens a new channel for the presentation of a novel chemical universe of lipids. Studies suggest that CD1 might not be primarily involved with presenting foreign ligands, but rather would function by triggering regulatory cells to respond to danger signals (stress or damage) through exerting or recruiting appropriate protective responses (reviewed in ref. 38). It also is well established that mycolic acids and ceramides, which can be presented by CD1, are strong adjuvants of adaptive immune responses[39,40] implicating their importance in glycolipid vaccine formulations (e.g., *Mycobacterium tuberculosis* vaccine).

LPS of Gram$^-$ bacteria (reviewed in 41) has long been used as an adjuvant. During the last 10 years, substantial understanding has been gained in the mechanisms controlling cellular responses to LPS. The biologically-active endotoxic structure within this complex glycolipid is the lipid A moiety (a phosphorylated glucosamine), with multiple fatty acids anchoring the structure in the outer membrane of the Gram$^-$ bacteria. LPS binds to soluble LPS-binding serum proteins (LBP)[42] or the homologous bacterial/permeability-increasing protein (BPI).[43] Binding by LBP promotes delivery of LPS to the CD14 receptor of monocytes, which constitutes a LBP/CD14 pathway leading to cellular activation and TNF-α production. No other serum proteins are able to replace LBP as an opsonin for LPS that promotes transfer to CD14 and subsequent phagocytosis. However, CD14 only increases the sensitivity of macrophages to LPS, because it does not have a cytoplasmic signaling domain. Recent studies by Janeway's group revealed a mammalian homologue of a *Drosophila* protein known as toll.[44] More than five such proteins have been identified in humans and are called toll-like receptors (TLRs). All these are transmembrane proteins containing repeated leucine-rich motifs in their extracellular portions, similar to such motifs in other pattern-recognition proteins. TLRs contain a cytoplasmic portion that is homologous to the interleukin-1 receptor (IL1R) and hence could trigger signaling pathways. Evidence suggests that CD14 recruits LPS to TLR proteins, thereby facilitating signal transduction and leading to 1. the release of cytokines and expression of costimulatory molecules from macrophages; 2. the recruitment of IL1R-associated kinase-2 and TNF receptor-associated factor-6; 3. activation of nuclear factor-kB; and 4. subsequent gene transcription.[45,46] Agonists of TLRs in the form of vaccines could prove useful as adjuvants for stimulating innate immunity, thereby enhancing adaptive immune responses.[47] This pathway is down-regulated by the BPI-LPS complex and may control LPS aggregation and delivery to monocytes. The inflammatory reaction follows this recognition, as seen with Gram$^-$ bacteria and the production of chemokines and proinflammatory cytokines like IL-1, IL-6, IL-8 GM-CSF and TNF-α. Subsequent to inflammation, regulatory cytokines are produced, with IL-12 (synergistically with IL-15 and IL-18) and IL-4 as the main actors for Th1 and Th2 responses, respectively.

Besides carbohydrate structures of microorganisms, the innate immune system recognizes lipids (e.g., in LPS) and prokaryote peptides or their derivatives (e.g., MDP, GMDP and MTP), bacterial cell walls components (e.g., trehalose dimycolate from *M. tuberculosis*), and prokaryotic DNA (e.g., nonmethylated DNA). Such recognition events activate the innate system. These products, found in various delivery systems and formulations, also provoke cells other than DCs and macrophages in the innate immune system, e.g., NK cells, γ/δ T cells and neutrophils.[8] Cellular elements that evoke Th2 responses include "natural" T cells producing IL-4.[8] Mast cells and basophils also produce IL-4 and therefore have a regulatory role in promoting a Th2 response. Freund's complete adjuvant (FCA), containing whole mycobacterial cells, promotes a Th1 response, while incomplete Freund's adjuvant (IFA) favors a Th2 type response. MDP, a mycobacterial peptide, is the source of various derivatives. Lipophilic derivatives (MDP, MTP, GMDP) generally induce a Th1 response, hydrophilic derivatives a Th2 response (Table 16.4), and DNA CpG motifs a Th1 response as recently shown by the pathway of a TLR, all being used as adjuvants and in adjuvant formulations. Effector cells of the innate

Table 16.3. Cells and cytokines of the innate immune system which modulate the acquired immune responses

Cell type		Cytokine produced	T-helper response	B-cell response
APC#	Macrophages	TGF-β		IgA
	Macrophages	IL-4	Th2	IgG1, IgE
	Macrophages, dentritic cells	IL-10	Th2	IgG1, IgA , IgE
	Macrophages, dentritic cells	IL-12, IL-15	Th1	IgG2a
	Macrophages	IL-18	Th1	IgG2a
Accessory cells	NK cells	IFN-γ	Th1	IgG2a
	Mast cells, Basophils	IL-4	Th2	IgG1, IgE
	Eosinophils	IL-5	Th2	IgE
	Epithelial cells	TGF-β		IgA
	NK T cells	IL-4	Th2	IgG1, IgE

B cells have about 1000 times stronger antigen presenting capacity. But with little or no immunomodulating capacity, they do not belong to the innate immune system.

immune system being recruited by activated Th1 cells are macrophages and neutrophils producing several inflammatory products and respiratory burst.

Saponins and γ-inulin are examples of plant-derived adjuvants. The saponin from the tree *Quillaja saponia Molina*, in the form of ISCOMs, is efficiently taken up by macrophages, DCs and B cells. Many fold-higher uptakes were observed in these APCs when glycosylated influenza virus envelope proteins were incorporated into ISCOMs. Macrophages, DCs, and B cells incubated with ISCOMs stimulated committed and naive T cells in vitro to proliferation and cytokine production.[48] A major advantage of ISCOMs, in contrast to emulsions and aluminum, is their suitability for in vitro studies.

APCs Instruct the Acquired Immune System

There are no distinct boundaries between the innate and acquired immunities. The APCs include members of the innate system (i.e., DCs and macrophages), while B cells belong to the acquired immune system. Adjuvants often interact with both systems, with initial recognition at one end and immune memory at the other end, but with effector functions distributed in both systems.

APCs have a key role in inducting and controlling T- and B-cell responses as well as immunity as a whole (Table 16.3). DCs are involved in optimizing the clonal selection of low-frequency $CD4^+$ and $CD8^+$ T cells and recruiting and activating quiescent naive and memory B and T cells. An essential sequence for antigen presentation is the development of immature DCs into mature DCs. Immature DCs have a prominent capacity for antigen uptake by pinocytosis or phagocytosis, but lack CD86 and CD80 required for interaction with CD28 and CTLA-4 on T cells. With maturation comes a loss of capacity for antigen uptake but increased capacity to process and present antigens to T cells. The maturation is induced, e.g., by infectious agents and their products (LPS) and mediated by GM-CSF and IL-4. These processes give DCs the role of mobile sentinels bringing antigens to T cells after expression of

T-cell ligands (CD40 and CD56) and costimulators (CD80 and CD86). DCs upregulated during infection also respond to signals for trafficking that are guided by expression of adhesion molecules (e.g., LFA-3, ICAM-1 and -3, and CD24), chemokine receptors, and chemokine secretion (e.g., IL-8, MIP-1α, β and IL-12) (reviewed in ref. 49). This process is a target for adjuvant action but is biologically very complex, which explains why many adjuvant-active molecules are found in microorganisms (Table 16.2) or their products (e.g., LPS, MDP, and other bacterial cell wall components containing a multitude of carbohydrates). Adjuvants interact with various receptors on the DCs, MRs, and receptors for LPS, TLRs and DEC-205.[50] This interaction probably occurs in the case of delivery systems (e.g., ISCOMs containing a complex carbohydrate structure), SAF-1 or liposomes formulated with proper immunomodulators. The production of IL-12 leads to Th1 responses and of IL-4 and IL-10 to Th2 responses.[51] Cytokines such as TGF-β and IL-10 will inhibit maturation to Th1 responses, while IFN-γ will down-regulate Th2 responses (Fig. 16.1). However, DCs might not be readily manipulatable with adjuvants, in particular the steering of maturation and migration to the lymphatics and the T- and B-cell areas. Importantly, we have adjuvants with the capacity to induce immune modulatory secretory products like IL-12 (e.g., by LPS derivatives ISCOMs and CpG driving the Th1 response and IFN-γ production[48,52,53]), which in most cases are likely regulated through DCs. It is becoming more evident that DCs have an important role for antigen presentation toward CTL induction. Macrophages are also important targets for immunomodulators (Table 16.3) and particularly for Th1 responses[8] that involve phagocytosis and intracellular killing of microorganisms. There appears also to be a synergistic effect between macrophages and DCs as APCs.[48,52,53] Adjuvants that promote macrophage presentation are often of bacterial origin and are partly recognized through the MR. Adjuvants in the form of sulfated polymers are endocytosed (soluble antigen) or phagocytosed (particulate antigens) by macrophages[54-56] and DCs, possibly through the MR. It has been suggested that such molecules reduce the scavenger activity of macrophages, thus promoting APC activity. Particles generally drive macrophages to phagocytosis and Th1 responses. There are several particulate formulations with and without carbohydrate moieties that target macrophages (e.g., microspheres, liposomes, VLPs and ISCOMs). Th2 responses have been considered independent of macrophages and are mediated by noncytotoxic antibodies, mast cells, eosinophils and "natural" T cells. However, subsets of macrophages also mediate Th2 responses (Table 16.3). Soluble antigens incorporated by endocytosis tend to induce Th2 responses.

Immune Modulation Is Based on Cross-Talk Between Innate Immunity and Helper T Cells

Immune modulation towards Th1 and Th2 responses, often set as a goal for adjuvants, is highly dependent on APC activities. As discussed above, the Th1 response requires IL-12 and is enhanced by IL-15[57] and IL-18,[58] which are products of innate immunity. IL-2 and IFN-γ (regulatory Th1 cytokines) are produced by both innate and acquired systems, as are IL-4, IL-10 and TGF-β (regulatory Th2 cytokines). Since Th1 cells express the Fas ligand, they are targets for contact-apoptosis by Fas+ cells.[59] These cytokines maintain immune responses in autocrine cycles and lack the down-regulation seen in hormone systems. Therefore, these cycles depend on cross-regulation of respective T-helper responses preventing exaggerated immune reactions (Fig. 16.1). Nature often formulates a balanced response, but under pathological conditions this balance is upset, which also may be an issue for the use of adjuvants. The regulatory mechanism of IL-10 and TGF-β is physiological, to avoid overproduction of IL-12 and to down-regulate Th1 leading to pathological problems (e.g., toxic shock syndrome).[60] Hepatitis B infection may induce strong reactions and tissue destruction due to a strong CTL activity enhanced by IL-12 and IFN-γ.[61] Adjuvants inducing IL-10 are suitable for down-regulation of DCs by preventing such maturation;[62] this is also a tool for the persistence

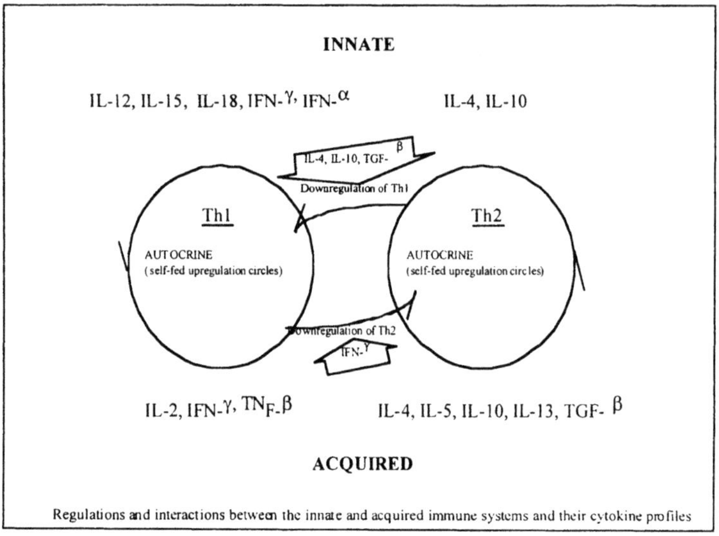

Fig. 16.1. Regulations and interactions between the innate and acquired immune systems and their cytokine profiles.

of microorganisms by evading the immune system (as suggested for Epstein-Barr virus).[63] Thus, cross-talk between innate and acquired immunity and cross-regulation between Th1 and Th2 are important for avoiding either tissue damage by a Th1 response[64] or allergy and autoimmune disorders by a Th2 response. Consequently, adjuvants promoting Th2 responses are of interest for regulating Th1 responses; they also provide for protective activities against parasites in which mast cells and eosinophils are involved. IFN-γ can down-regulate the Th2 response, as can adjuvants.[60] Examples of inducers or adjuvants and their cellular targets promoting Th1 or Th2 responses or both are listed in Table 16.4. Despite these numerous examples, there is a limited choice of such adjuvants for vaccine use.

The Collaboration Between the Complement System and B Cells: Roles for Adjuvants

The C′ system is the major humoral effector mechanism for innate immunity. Its recruitment is based on recognition of microbial carbohydrates along two pathways. The classical pathway is initiated by binding of collectin to certain carbohydrates,[12] while the alternative pathway is induced by carbohydrates lacking sialic acid.[65] Thus, carbohydrates (such as glycolipids) activating the C′ system have strong enhancing effects on the immunogenicity of the antigens to which they are attached (reviewed in 66), including instruction towards an immune response. Well-known adjuvants of microbial origin that enhance C′ cooperation are LPS and its derivatives, various cell wall elements and outer membrane molecules, synthetic molecules such as block polymers,[67] γ-inulin, or plant-derived adjuvants.[68] The delivery system SAF-1 was designed to activate the C′ system by incorporation of MTP,[69] which contains cell wall elements.[7](Table 16.2)

The involvement of C′ in stimulating B-cell responses is becoming increasingly recognized as important. C3d components of C′ attach covalently to microbial antigens and bind to the CR2 C' receptor (CD21) on follicular DCs. This binding results in an enhanced immune response to the linked antigen (reviewed in ref. 70). The complex of CR2-CD19 linked to IgM activates the B lymphocyte to proceed along one of two pathways. Development of B-cell

Table 16.4. Principal actions of immunomodulatory compounds

Th1 induction	hydrophobic MDP and derivatives (e.g., MTP-PE), LPS-lipid A and derivatives (e.g., MPL) avridine, DDA (amines toxic) TDM (from *M. tuberculosis* used in emulsions) CWS (cell wall skeletons) DHEA (dehydroepi-androsterone)
Th2 induction	hydrophilic MDP (e.g., GMDP) vitamin D3 poly A:U aluminum salts vitamin D3
Th1/Th2 induction	saponins poly I:C and poly ICLC stearyl tyrosine

memory and its localization to a germinal center is mediated by IL-4 and by the interaction between CD40 on B cells and the CD40 ligand on activated T cells. The development of plasma cells is facilitated by the cytokines IL-3, IL-6 and IL-10 and the expression of the OX40 ligand on B cells and of OX40 on activated T cells. Two down-regulating systems are involved in C′-dependent B-cell activation (i.e., CD22 and Fcγ RIIB), the latter providing memory for a previous immune response.[70] Basically this knowledge should facilitate the creation of adjuvants and vaccines inducing B-cell memory and/or plasma-cell responses.

Immune Modulation For CTL

The MHC class I-restricted immune response, in connection with vaccine development, is today mainly focused on CTL, although $CD8^+$ cells with other functions may be important for protection against infection, in particular viral. For example, protection against HIV infection has been associated with the production of the β-chemokines MIP-1α and β and RANTES, produced by $CD8^+$ cells and which block the second receptor essential for infection by facilitating the fusion of the virion to host cells.[71] It was shown recently that ISCOMs with HIV-1 antigens induced macaque $CD8^+$ cells to produce these β-chemokines;[72] the immunized macaques were protected from infection by a chimeric HIV-SIV. Other functions of $CD8^+$ cells may be of future vaccine interest, e.g., for down-regulation of the immune response.

Until recently it was thought that the MHC class I-restricted CTL response was only efficiently evoked by endogeneously produced antigens, e.g., after virus infection or as tumor antigens. However, intranasal immunization with low doses of influenza virus antigens in ISCOMs elicited CTL.[73] Takahashi et al[74] showed that parenteral immunization with ISCOMs containing influenza virus envelope antigens and ISCOMs with HIV-1 gp160 induced CTL in a mouse model. Deres et al[75] showed that lipopeptides with tripalmitic acid linked via serine to influenza virus peptides induced CTL. Other systems that efficiently induce CTL are acid-sensitive liposomes[76] and fusogenic liposomes (e.g., Sendai virus F protein of).[77] Various adjuvant formulations that evoke CTL are listed in Table 16.5. In the case of ISCOMs, the incorporation of the particle into the cell membranes exposes the antigens to cytosolic proteases and delivers the antigens to the cytosol. The antigens are evenly delivered to the cytosol and to cellular vesicles.[78] The lipophilic center of lipopeptides similarly may incorporate them in the membrane, then proteases clip the peptide into the cytosol. Liposomes can integrate into

Table 16.5. Adjuvants that induce CTL responses

Property	Adjuvant examples	Remarks
Surfactants	Saponins and ISCOM-matrix	Low to moderate
	Block co-polymers	Low
Lipophilic compounds	Synthetic lipopeptide	Efficient
Emulsions	MF59	Low
	SAF-1 with MTP-PE o/w emulsion	Efficient only with short peptides
Combination of emulsion and lipiphilic properties	o/w emulsion + MPL + QS21	Low
Delivery systems	Liposomes (fusogenic or acid sensitive)	Efficient
	ISCOMs	Efficient after both mucosal and parenteral administrations
Receptor mediated	Nonreplicative vectors for CTL activation (e.g., recombinant adeylate cyclase toxins of *B. pertussis*[103,104] and recombinant parvovirus-like[105,106]	Efficient

cell membranes by means of fusion in the acidic compartment of endolysosomes or by means of fusion proteins, thereby exposing the antigens to cytosolic proteases for processing and transport to the ER. Thus, the key for CTL induction is delivery of antigens to the cytosol and up-regulation of IL-2 and IFN-γ,[79-81] which is the case with ISCOMs and with certain liposomes discussed above.[82] It has been convincingly demonstrated that CTL immunity can be induced naturally by exogenously delivered antigens;[83] this required professional APC and especially DCs. In the case of viruses, when APCs are not infected (i.e., no endogenous virus antigen is produced), the DCs efficiently take up exogenous viral antigens for processing and presentation from other cells in the infected mouse, resulting in the development of CTL responses. This information points to more ways for manipulating MHC class I responses, which is especially required under conditions whereby the APCs are not being infected by nonreplicating delivery systems.

Delivery Systems

It can be argued whether the term **delivery system** should include adjuvants that influence the delivery of antigen over a period of time {$Al(OH)_3$ and gel emulsions such as CFA or IFA). Their effect is mainly the release of antigen from a depot. The benefit of that effect can be questioned, since surgical removal of the injection site after three days did not hamper the antibody response.[84] A recent concept of such a depot is slow sustained-release and controlled-release formulations, with the objective of giving two or more doses in one injection. There are several formulations that continuously release antigens, e.g. synthetic polymers and nonionic block copolymers surfactants[85] composed of hydrophilic polyoxyethylene (POE) and hydrophobic polyoxypropylene (POP). Certain proportions of POE and POP correlate with high adjuvant activity and are more efficient in emulsions than by themselves. Methylmethacrylate polymers with entrapped antigens formed into spheres of < 1 μm diameter are named nanoparticles.[86] Microencapsulation exists in several forms. Perhaps most successful is the biodegradable polyester poly-D,L-lactide-coglycolide (PLGA). Formulation of antigens into 1- to 10-μm PLGA microspheres[87] enhances antibody- and cell-mediated immune responses following parenteral or mucosal administration.[88] However, polymerization by organic solvents may create practical problems, and acid degradation products may compromise antigenicity. In contrast, crosslinking of water-soluble polyphosphorus and alginate polymers exclude the use of organic solvents. By crosslinking of the carboxyl groups of the polyphosphorus polymers with divalent cations, hydrogel microspheres are formed with properties for mucosal administration. Antigens in these polymers generally enhance antibody responses compared to corresponding aqueous solutions of the antigen. Some block copolymers in emulsion (e.g., in SAF-1) have strong immunomodulatory effects towards a Th1 response. To analyze the effect of sustained antigen release from polymers, we have studied the effect of ISCOMs released from alginate gel and from w/o emulsions and also mimicked pulsatile release by weekly injection of ISCOMs. In all formulations or procedures where ISCOMs were released repeatedly or release was protracted, the antibody response was lower and the proportion of IgG2a was significantly lower than after two injections of ISCOMs given six weeks apart. It should be noted that ISCOMs are rapidly removed from the site of injection. Thus, it can be discussed how the release should be effected in order to give optimal priming and boost (Johansson et al, in press).

Delivery systems for vaccine antigens generally are not immunomodulating,[7] and the targeting is limited by the particulate form, with improved uptake by macrophages. Externally exposed carbohydrates as antigens may contribute to uptake by macrophages and DCs. Conceptually, delivery systems for vaccine antigens should contain antigen, target delivery to the lymphatic system, and preferably include an immunomodulator. Examples of such delivery systems are liposomes, SAF-1 and ISCOMs. Liposomes and ISCOMs are versatile, allowing the incorporation of target molecules and immunomodulators. These delivery systems can be formulated to induce prominent Th1 responses; liposomes and ISCOMs also have been formulated for mucosal administrations.

Vaccines for Newborns and Elderly Require Suitable Strong Adjuvants

We lack some vaccines for both newborns and the elderly due to difficulties in eliciting protective immunity. Here cease the similarities between the two age groups. While the immature immune system has components for mounting a long-lasting response, the aged immune system has decreased capacity and the induced immune response is of shorter duration.[89] After the pioneering work of Medawar,[90] it was generally accepted that neonates are difficult to immunize and may be tolerized. More recent work has shown that the immaturity of the neonate immune system is quantitative to a great extent, due to few T cells and few APCs compared to the antigen dose given,[91,92] which promotes high-dose tolerance.[93] Sarzotti and colleagues[94] showed that the immune system of newborns contains a >100-fold lower number of immune cells than that of adults. High doses of virus promoted a Th2 response, while decreasing the virus dose promoted a Th1 response. It is becoming clearer that adjuvants may enable the development of some vaccines for newborns.[95] The Th2 bias of neonates may be overcome with strong Th1 adjuvants[92] or a delivery system such as ISCOMs containing Sendai virus (SV) membrane proteins, which induced a Th1 response in 2-day-old mice compared to an alum-adjuvanted vaccine that induced a Th2 response (submitted). Interestingly the ISCOMs seem to boost and convert a Th2 response primed by alum-adjuvanted SV vaccine to a Th1 response (Blomquist et al., manuscript in preparation). The presence of maternal-derived blocking antibodies may cause problems for vaccinating newborns. Osterhaus' group has shown in primates[97] that passively-transferred antibodies to measles virus blocked the commercial measles vaccine from inducing antibody- and cell-mediated responses, but ISCOMs containing the H and F proteins overcome the effect of blocking antibodies and induced potent antibody and cell-mediated immune responses. ISCOMs containing equine herpes virus II envelope antigens induced protective immunity by a vaccination protocol with priming at 10 days of age followed by a boost at 1 month.[98]

While early work claimed that human neonates produce IgM but do not switch efficiently to the production of other immunoglobulin isotypes, recent work has shown that when T-cell help is provided in vitro[99] or in vivo[100] in an adoptive host, there is a B-cell switch to other classes. Neonate mice, in contrast to humans, respond with a switch to the various isotypes and IgG subclasses, i.e., alum promotes Th2 responses, while ISCOMs, block polymers or DNA vaccines promote Th1 responses.[91,92,101]

Aging is associated with the decline of various immunological functions. Vaccination is still carried out against certain infection in elderly (e.g., influenza), although they are less responsive and often develop antibodies with low affinity and of shorter duration. Suboptimal costimulation is a distinct factor for the decreased responsiveness of the elderly, which was overcome by an experimental ISCOM influenza virus vaccine in an aged mouse model. Sambhara et al[102] showed that the enhanced immune response and protection to challenge infection in the elderly compared to current influenza vaccine was correlated with a restored capacity of APCs to express CD86 and to a lesser extent CD80.

The Present Situation and Future Aspects of Adjuvants and Delivery Systems

There is no universal adjuvant because different antigens require different adjuvant activities for inducing protective immunity against the target pathogen. Aluminum hydroxide, which induces Th2 responses, is effective at adjuvanting some antigens, e.g., bacterial toxoids. It is used in several viral and bacterial vaccines despite being suboptimal for inducing the inappropriate type of immune response in some cases. Even closely related microorganisms may require different type of adjuvants, e.g., protective immunity against the South American *Trypanosoma cruzi* requires Th1 responses, while that against African trypanosomes require Th2 responses. If inappropriate immune responses are induced, then disease might be enhanced. Classical w/o emulsions often cause strong local reactions. However, o/w or double emulsions cause fewer side effects and have stronger immune modulatory effects. Liposomes may be versatile. They are produced by a number of different protocols that may be complicated, and they may also be fragile. In contrast, ISCOMs, which also are versatile, are very stable and are used in animal vaccines. The use of adjuvants is generally empirical. The selected antigen and the pathogenesis of the microorganisms for which the antigen is intended should be factors for determining the choice of adjuvant. However, the big problem is that, at least for human vaccines, there are few approved adjuvants.

Slow-release systems often require organic solvents for preparing formulations. They continuously release antigens after administration in a way which may be suboptimal. Furthermore, some of them produce polymer breakdown products in the form of organic acids which may negatively influence the structural stability of antigens.

During the next decade, we may anticipate that adjuvants would be constructed with defined natural properties, starting with signals of recognition including a particulate format for activating the innate immune system. This step is followed by selected enhanced recruitment of the proinflammatory response and stimulation for expression of the costimulatory molecules and their receptors. Eventually the adjuvant activity would have an immunomodulating action. This may involve a delivery system where the components driving the described actions can be exchanged for tailor-made activities. The delivery system also would include targeting properties for organs, cells and intracellular distribution of antigens and other components. Particular considerations for targeting properties are required for mucosal administration. For that purpose devices addressing local as well as remote mucosal sites will be constructed using mechanisms of adhesions, integrins and homing. Such devices also are of great interest for drug delivery. WHO has long called for one injection delivering two or more doses. Such a delivery system awaits the development of controlled pulsed-release formulations for replacing continuous release.

References

1. Allison AC, Byars NE. Adjuvant formulations and their mode of action. Semin Immunol 1990; 2(5):369-74.
2. Cox JC, Coulter AR. Advances in adjuvant technology and application. In: Animal parasite control utilizing biotechnology. CRC Press, Inc; 1992:51.
3. Morein B, Lovgren-Bengsson K. Functional aspects of iscoms. Immunology and Cell Biology 1998; 76:295-9.
4. Jackson M, Scott R. Different patterns of cytokine induction in cultures of respiratory syncytial (RS) virus-specific human TH cell lines following stimulation with RS virus and RS virus proteins. J Med Virol 1996; 49(3):161-9.
5. Johnson TR, Johnson JE, Roberts SR et al. Priming with secreted glycoprotein G of respiratory syncytial virus (RSV) augments interleukin-5 production and tissue eosinophilia after RSV challenge. J Virol 1998; 72(4):2871-80.
6. Hansen DS, Alievi G, Segura EL et al. The flagellar fraction of Trypanosoma cruzi depleted of an immunosuppressive antigen enhances protection to infection and elicits spontaneous T cell responses. Parasite Immunol 1996; 18(12):607-15.
7. Cox JC, Coulter AR. Adjuvants--a classification and review of their modes of action. Vaccine 1997; 15(3):248-56.
8. Fearon DT, Locksley RM. The instructive role of innate immunity in the acquired immune response. Science 1996; 272:50-4.
9. Sacks G, Sargent I, Redman C. An innate view of human pregnancy. Immunol Today 1999; 20(3):114-8.
10. Bergers JJ, ten Hagen TL, van Etten EW et al. Liposomes as delivery systems in the prevention and treatment of infectious diseases. Pharm World Sci 1995; 17(1):1-11.
11. Morein B, Lovgren-Bengtsson K, Villacres-Eriksson M. What to expect of future vaccines? Acta Vet Scand Suppl 1996; 90:101-6.
12. Holmskov U, Malhotra R, Sim RB et al. Collectins: Collagenous C-type lectins of the innate immune defense system. Immunol Today 1994; 15(2):67-74.
13. Malhotra R, Willis AC, Jensenius JC et al. Structure and homology of human C1q receptor (collectin receptor). Immunology 1993; 78(3):341-8.
14. Matsushita M, Fujita T. Activation of the classical complement pathway by mannose-binding protein in association with a novel C1s-like serine protease. J Exp Med 1992; 176(6):1497-502.
15. Thiel S, Vorup-Jensen T, Stover CM et al. A second serine protease associated with mannan binding lectin that activates complement. Nature 1997; 386(6624):506-10.
16. Carroll MC, Prodeus AP. Linkages of innate and adaptive immunity. Curr Opin Immunol 1998; 10(1):36-40.
17. Schweinle JE, Ezekowitz RA, Tenner AJ et al. Human mannose-binding protein activates the alternative complement pathway and enhances serum bactericidal activity on a mannose-rich isolate of Salmonella. J Clin Invest 1989; 84(6):1821-9.
18. Frazer A, Daws LC. Serotonin transporter function in vivo: Assessment by chronoamperometry. Ann N Y Acad Sci 1998; 861:217-29.
19. Weis WI, Drickamer K, Hendrickson WA. Structure of a C-type mannose-binding protein complexed with an oligosaccharide. Nature 1992; 360(6400):127-34.
20. Matzinger P. An innate sense of danger. Semin Immunol 1998; 10(5):399-415.
21. Weis WI, Kahn R, Fourme R et al. Structure of the calcium-dependent lectin domain from a rat mannose-binding protein determined by MAD phasing. Science 1991; 254(5038):1608-15.

22. Ezekowitz RA, Sastry K, Bailly P et al. Molecular characterization of the human macrophage mannose receptor: Demonstration of multiple carbohydrate recognition-like domains and phagocytosis of yeasts in Cos-1 cells. J Exp Med 1990; 172(6):1785-94.
23. Taylor ME, Conary JT, Lennartz MR et al. Primary structure of the mannose receptor contains multiple motifs resembling carbohydrate-recognition domains. J Biol Chem 1990; 265(21):12156-62.
24. Harris N, Super M, Rits M et al. Characterization of the murine macrophage mannose receptor: Demonstration that the downregulation of receptor expression mediated by interferon-gamma occurs at the level of transcription. Blood 1992; 80(9):2363-73.
25. Ancian P, Lambeau G, Lazdunski M. Multifunctional activity of the extracellular domain of the M-type (180 kDa) membrane receptor for secretory phospholipases A2. Biochemistry 1995; 34(40):13146-51.
26. Wu K, Yuan J, Lasky LA. Characterization of a novel member of the macrophage mannose receptor type C lectin family. J Biol Chem 1996; 271(35):21323-30.
27. Martinez-Pomares L, Kosco-Vilbois M, Darley E et al. Fc chimeric protein containing the cysteine-rich domain of the murine mannose receptor binds to macrophages from splenic marginal zone and lymph node subcapsular sinus and to germinal centers. J Exp Med 1996; 184(5):1927-37.
28. Finnerty M, Enright WJ, Morrison CA et al. Immunization of bull calves with a GnRH analogue-human serum albumin conjugate: effect of conjugate dose, type of adjuvant and booster interval on immune, endocrine, testicular and growth responses. J Reprod Fertil 1994; 101(2):333-43.
29. Diamantstein T, Wagner B, Beyse I et al. Stimulation of humoral antibody formation by polyanions. II. The influence of sulfate esters of polymers on the immune response in mice. Eur J Immunol 1971; 1(5):340-3.
30. Hoskinson RM, Rigby RD, Mattner PE et al. Vaxstrate: an anti-reproductive vaccine for cattle. Aust J Biotechnol 1990; 4(3):166-70, 76.
31. Lefkowitz DL, Lincoln JA, Lefkowitz SS et al. Enhancement of macrophage-mediated bactericidal activity by macrophage- mannose receptor-ligand interaction. Immunol Cell Biol 1997; 75(2):136-41.
32. Shepherd VL, Lane KB, Abdolrasulnia R. Ingestion of Candida albicans down-regulates mannose receptor expression on rat macrophages. Arch Biochem Biophys 1997; 344(2):350-6.
33. Fraser IP, Koziel H, Ezekowitz RA. The serum mannose-binding protein and the macrophage mannose receptor are pattern recognition molecules that link innate and adaptive immunity. Semin Immunol 1998; 10(5):363-72.
34. Engering AJ, Cella M, Fluitsma D et al. The mannose receptor functions as a high capacity and broad specificity antigen receptor in human dendritic cells. Eur J Immunol 1997; 27(9):2417-25.
35. Tan MC, Mommaas AM, Drijfhout JW et al. Mannose receptor-mediated uptake of antigens strongly enhances HLA class II-restricted antigen presentation by cultured dendritic cells. Eur J Immunol 1997; 27(9):2426-35.
36. Porcelli S, Morita CT, Brenner MB. CD1b restricts the response of human CD4-8- T lymphocytes to a microbial antigen. Nature 1992; 360(6404):593-7.
37. Porcelli SA. The CD1 family: A third lineage of antigen-presenting molecules. Adv Immunol 1995; 59(6404):1-98.
38. Park SH, Chiu YH, Jayawardena J et al. Innate and adaptive functions of the CD1 pathway of antigen presentation. Semin Immunol 1998; 10(5):391-8.
39. Bendelac A. CD1: Presenting unusual antigens to unusual T lymphocytes. Science 1995; 269(5221):185-6.
40. Sallusto F, Nicolo C, De Maria R et al. Ceramide inhibits antigen uptake and presentation by dendritic cells. J Exp Med 1996; 184(6):2411-6.
41. Ulevitch RJ, Tobias PS. Recognition of gram-negative bacteria and endotoxin by the innate immune system. Curr Opin Immunol 1999; 11(1):19-22.
42. Lamping N, Hoess A, Yu B et al. Effects of site-directed mutagenesis of basic residues (Arg 94, Lys 95, Lys 99) of lipopolysaccharide (LPS)-binding protein on binding and transfer of LPS and subsequent immune cell activation. J Immunol 1996; 157(10):4648-56.
43. Tobias PS, Soldau K, Iovine NM et al. Lipopolysaccharide (LPS)-binding proteins BPI and LBP form different types of complexes with LPS. J Biol Chem 1997; 272(30):18682-5.
44. Medzhitov R, Preston-Hurlburt P, Janeway CA, Jr. A human homologue of the Drosophila Toll protein signals activation of adaptive immunity. Nature 1997; 388(6640):394-7.

45. Muzio M, Natoli G, Saccani S et al. The human toll signaling pathway: Divergence of nuclear factor kappaB and JNK/SAPK activation upstream of tumor necrosis factor receptor-associated factor 6 (TRAF6). J Exp Med 1998; 187(12):2097-101.
46. Kirschning CJ, Wesche H, Merrill Ayres T et al. Human toll-like receptor 2 confers responsiveness to bacterial lipopolysaccharide. J Exp Med 1998; 188(11):2091-7.
47. Modlin RL, Brightbill HD, Godowski PJ. The toll of innate immunity on microbial pathogens. N Engl J Med 1999; 340(23):1834-5.
48. Villacres-Eriksson M. Antigen presentation by naive macrophages, dendritic cells and B cells to primed T lymphocytes and their cytokine production following exposure to immunostimulating complexes. Clin Exp Immunol 1995; 102(1):46-52.
49. Banchereau J, Steinman RM. Dendritic cells and the control of immunity. Nature 1998; 392(6673):245-52.
50. Jiang W, Swiggard WJ, Heufler C, Peng M et al. The receptor DEC-205 expressed by dendritic cells and thymic epithelial cells is involved in antigen processing. Nature 1995; 375(6527):151-5.
51. Nelson PJ, Krensky AM. Chemokines, lymphocytes and viruses: what goes around, comes around. Curr Opin Immunol 1998; 10(3):265-70.
52. Smith RE, Donachie AM, Grdic D et al. Immune-stimulating complexes induce an IL-12-dependent cascade of innate immune responses. J Immunol 1999; 162(9):5536-46.
53. Grdic D, Smith R, Donachie A et al. The mucosal adjuvant effects of cholera toxin and immune-stimulating complexes differ in their requirement for IL-12, indicating different pathways of action. Eur J Immunol 1999; 29(6):1774-84.
54. de Haan A, Groen G, Prop J et al. Mucosal immunoadjuvant activity of liposomes: Role of alveolar macrophages. Immunology 1996; 89(4):488-93.
55. Parish CR, Rylatt DB, Snowden JM. Demonstration of lymphocyte surface lectins that recognize sulphated polysaccharides. J Cell Sci 1984; 67:145-58.
56. Parish CR, Hogarth PM, McKenzie IF. Evidence that Thy-1 and Ly-5 (T-200) antigens interact with sulphated carbohydrates. Immunol Cell Biol 1988; 66(Pt 3)):221-30.
57. Carson WE, Caligiuri MA. Interleukin 15: A potential player during the innate immune response to infection. Exp Parasitol 1996; 84(2):291-4.
58. Okamura H, Kashiwamura S, Tsutsui H et al. Regulation of interferon-gamma production by IL-12 and IL-18. Curr Opin Immunol 1998; 10(3):259-64.
59. Ramsdell F, Seaman MS, Miller RE et al. Differential ability of Th1 and Th2 T cells to express Fas ligand and to undergo activation-induced cell death. Int Immunol 1994; 6(10):1545-53.
60. Trinchieri G, Kubin M, Bellone G et al. Cytokine cross-talk between phagocytic cells and lymphocytes: Relevance for differentiation/activation of phagocytic cells and regulation of adaptive immunity. J Cell Biochem 1993; 53(4):301-8.
61. Bertoletti A, Ferrari C, Fiaccadori F. Role of the cell-mediated immune response in the pathogenesis of hepatitis B virus infection: possible immune-therapeutic strategies. Acta Biomed Ateneo Parmense 1996; 67(3-4):87-93.
62. Kitajima T, Ariizumi K, Bergstresser PR et al. A novel mechanism of glucocorticoid-induced immune suppression: The inhibiton of T cell-mediated terminal maturation of a murine dendritic cell line. J Clin Invest 1996; 98(1):142-7.
63. Moore KW, O'Garra A, de Waal Malefyt R et al. Interleukin-10. Annu Rev Immunol 1993; 11:165-90.
64. Scott B, Liblau R, Degermann S et al. A role for non-MHC genetic polymorphism in susceptibility to spontaneous autoimmunity. Immunity 1994; 1(1):73-83.
65. Fearon DT, Austen KF. Current concepts in immunology: The alternative pathway of complement—a system for host resistance to microbial infection. N Engl J Med 1980; 303(5):259-63.
66. Allison AC. Immunological adjuvants and their modes of action. Arch Immunol Ther Exp (Warsz) 1997; 45(2-3):141-7.
67. Hunter RL, McNicholl J, Lal AA. Mechanisms of action of nonionic block copolymer adjuvants. AIDS Res Hum Retroviruses 1994; 10 Suppl 2(16):S95-8.
68. Cooper PD. Vaccine adjuvants based on gamma inulin. Pharm Biotechnol 1995; 6(16):559-80.

69. Lefrancier P, Derrien M, Lederman I et al. Synthesis of some new analogs of the immunoadjuvant glycopeptide MDP (N- acetyl-muramyl-L-alanyl-D-isoglutamine). Int J Pept Protein Res 1978; 11(4):289-96.
70. Fearon DT. The complement system and adaptive immunity. Semin Immunol 1998;10(5):355-61.
71. Paxton WA, Martin SR, Tse D et al. Relative resistance to HIV-1 infection of CD4 lymphocytes from persons who remain uninfected despite multiple high-risk sexual exposure. Nat Med 1996; 2(4):412-7.
72. Heeney JL, Teeuwsen VJ, van Gils M et al. beta-chemokines and neutralizing antibody titers correlate with sterilizing immunity generated in HIV-1 vaccinated macaques. Proc Natl Acad Sci U S A 1998; 95(18):10803-8.
73. Jones PD, Tha Hla R, Morein B et al. Cellular immune responses in the murine lung to local immunization with influenza A virus glycoproteins in micelles and immunostimulatory complexes (iscoms). Scand J Immunol 1988; 27(6):645-52.
74. Takahashi H, Takeshita T, Morein B et al. Induction of CD8+ cytotoxic T cells by immunization with purified HIV-1 envelope protein in ISCOMs. Nature 1990; 344(6269):873-5.
75. Deres K, Schild H, Wiesmuller KH et al. In vivo priming of virus-specific cytotoxic T lymphocytes with synthetic lipopeptide vaccine. Nature 1989; 342(6249):561-4.
76. Zhou F, Huang L. Liposome-mediated cytoplasmic delivery of proteins: An effective means of accessing the MHC class I-restricted antigen presentation pathway. Immunomethods 1994; 4(3):229-35.
77. Hale AH, Lyles DS, Fan DP. Elicitation of anti-Sendai virus cytotoxic T lymphocytes by viral and H- 2 antigens incorporated into the same lipid bilayer by membrane fusion and by reconstitution into liposomes. J Immunol 1980; 124(2):724-31.
78. Villacres MC, Behboudi S, Nikkila T et al. Internalization of iscom-borne antigens and presentation under MHC class I or class II restriction. Cell Immunol 1998; 185(1):30-8.
79. Bernhard H, Maeurer MJ, Jager E et al. Recognition of human renal cell carcinoma and melanoma by HLA-A2- restricted cytotoxic T lymphocytes is mediated by shared peptide epitopes and up-regulated by interferon-gamma. Scand J Immunol 1996; 44(3):285-92.
80. Yang Y, Xiang Z, Ertl HC et al. Upregulation of class I major histocompatibility complex antigens by interferon gamma is necessary for T-cell-mediated elimination of recombinant adenovirus-infected hepatocytes in vivo. Proc Natl Acad Sci U S A 1995; 92(16):7257-61.
81. Mehrotra PT, Grant AJ, Siegel JP. Synergistic effects of IL-7 and IL-12 on human T cell activation. J Immunol 1995; 154(10):5093-102.
82. Lipford GB, Wagner H, Heeg K. Vaccination with immunodominant peptides encapsulated in Quil A- containing liposomes induces peptide-specific primary CD8+ cytotoxic T cells. Vaccine 1994; 12(1):73-80.
83. Sigal LJ, Crotty S, Andino R et al. Cytotoxic T-cell immunity to virus-infected nonhaematopoietic cells requires presentation of exogenous antigen. Nature 1999; 398(6722):77-80.
84. White RG. The adjuvant effect of microbial products on the immune response. Annu Rev Microbiol 1976; 30:579-600.
85. Hunter RL. Nonionic block copolymers: New preparations and review of the mechanism of action. In: Spriggs DRaK WC eds. Topics in Vaccine Adjuvant Research. Boca Raton: CRC Press; 1991:89-97.
86. Kreuter J. Microcapsules and nanoparticles. In: Donbrow ME, ed. Medicine and Pharmacology. Boca Raton: CRC Press; 1992:125-48.
87. Eldridge JH, Gilley RM, Staas JK et al. Biodegradable microspheres: Vaccine delivery system for oral immunization. Curr Top Microbiol Immunol 1989; 146:59-66.
88. O'Hagan DT, Jeffery H, Maloy KJ et al. Biodegradable microparticles as oral vaccines. Adv Exp Med Biol 1995; 371B:1463-7.
89. Lesourd B. Immune response during disease and recovery in the elderly. Proc Nutr Soc 1999; 58(1):85-98.
90. Billingham RE, Brent L, Medawar PB. Actively acquired tolerance of foreign cells. Nature 1953; 172:603-6.

91. Barrios C, Brawand P, Berney M et al. Neonatal and early life immune responses to various forms of vaccine antigens qualitatively differ from adult responses: Predominance of a Th2-biased pattern which persists after adult boosting. Eur J Immunol 1996; 26(7):1489-96.
92. Barrios C, Brandt C, Berney M et al. Partial correction of the TH2/TH1 imbalance in neonatal murine responses to vaccine antigens through selective adjuvant effects. Eur J Immunol 1996; 26(11):2666-70.
93. Okamura H, Tsutsi H, Komatsu T et al. Cloning of a new cytokine that induces IFN-gamma production by T cells. Nature 1995; 378(6552):88-91.
94. Sarzotti M, Robbins DS, Hoffman PM. Induction of protective CTL responses in newborn mice by a murine retrovirus. Science 1996; 271(5256):1726-8.
95. Adkins B. T-cell function in newborn mice and humans. Immunol Today 1999; 20(7):330-5.
96. Morein B, Villacres-Eriksson M, Ekstrom J et al. ISCOM: A delivery system for neonates and for mucosal administration. Adv Vet Med 1999; 41(16):405-13.
97. Osterhaus A, van Amerongen G, van Binnendijk R. Vaccine strategies to overcome maternal antibody mediated inhibition of measles vaccine. Vaccine 1998; 16(14-15):1479-81.
98. Nordengrahn A, Rusvai M, Merza M et al. Equine herpesvirus type 2 (EHV-2) as a predisposing factor for Rhodococcus equi pneumonia in foals: Prevention of the bifactorial disease with EHV-2 immunostimulating complexes. Vet Microbiol 1996; 51(1-2):55-68.
99. Pirenne-Ansart H, Paillard F, De Groote D et al. Defective cytokine expression but adult-type T-cell receptor, CD8, and p56lck modulation in CD3- or CD2-activated T cells from neonates. Pediatr Res 1995; 37(1):64-9.
100. Ueno Y, Ichihara T, Hasui M et al. T-cell-dependent production of IgG by human cord blood B cells in reconstituted SCID mice. Scand J Immunol 1992; 35(4):415-9.
101. Adkins B, Du RQ. Newborn mice develop balanced Th1/Th2 primary effector responses in vivo but are biased to Th2 secondary responses. J Immunol 1998; 160(9):4217-24.
102. Sambhara S, Woods S, Arpino R et al. Influenza (H1N1)-ISCOMs enhance immune responses and protection in aged mice. Mech Ageing Dev 1997; 96(1-3):157-69.
103. Fayolle C, Ladant D, Karimova G et al. Therapy of murine tumors with recombinant Bordetella pertussis adenylate cyclase carrying a cytotoxic T cell epitope. J Immunol 1999; 162(7):4157-62.
104. Fayolle C, Sebo P, Ladant D et al. In vivo induction of CTL responses by recombinant adenylate cyclase of Bordetella pertussis carrying viral CD8+ T cell epitopes. J Immunol 1996; 156(12):4697-706.
105. Sedlik C, Dridi A, Deriaud E et al. Intranasal delivery of recombinant parvovirus-like particles elicits cytotoxic T-cell and neutralizing antibody responses. J Virol 1999; 73(4):2739-44.
106. Sedlik C, Saron M, Sarraseca J et al. Recombinant parvovirus-like particles as an antigen carrier: a novel nonreplicative exogenous antigen to elicit protective antiviral cytotoxic T cells. Proc Natl Acad Sci USA 1997; 94(14):7503-8.

CHAPTER 17

Transcutaneous Immunization

Gregory M. Glenn

Introduction

The skin is a highly complex yet well-orchestrated system committed to its protective barrier function. Human integument not only keeps moisture in and foreign material out, but is also designed for protection against the hostile microbial world which frequently gains entry through micro-trauma or other barrier disruptive events. Scientists in the past few decades have begun to more thoroughly describe skin immunobiology and vaccinologists have subsequently appreciated that the skin immune system is a desirable and potent target for immunization.[1-4] Topical applications and immune responses have traditionally been associated by clinicians only with immunopathologies of the skin. Intradermal immunization has been appreciated for the potency of subsequent immune responses, but until recently has not been appreciated for its connection with the skin immune system. The premise of transcutaneous immunization (TCI) is that vaccinologists can capitalize on the potency and accessiblity of the skin immune system and utilize this refined protective system for immunoprotection and immunotherapy by topical application of the immunizing formulation.

TCI is based on an elegant but simple amalgam of established scientific insights, each with an extensive background of literature that can be utilized for development of topical immunization techniques: skin penetration techniques, the presence and activities of antigen presenting cells in the epidermis, and adjuvants as enhancers and modulators of the immune response. Thus, although TCI is a relatively new observation, the understanding of skin penetration, skin immunobiology and adjuvant use underlying this technique provides an enormous fund of knowledge for realization of this technique in the clinic. This chapter reviews the working hypothesis of TCI, surveys the relevant literature and current data, and outlines potential future directions for research that aims at stimulating further research and development for this and related vaccine technologies.

Barriers and Targets for TCI

The skin is composed of three principal layers: the stratum corneum, the epidermis and the dermis. The stratum corneum, the outer protective layer of the epidermis, is composed principally of quiescent, keratin-filled epidermal cells encased in a mortar of surrounding lipids that were secreted by maturing keratinocytes. The stratum corneum is widely accepted as the principle barrier to penetration.[5] The living epidermis that underlies the stratum corneum (SC) is composed primarily (95%) of epidermal keratinocytes. However, 25% of the total skin surface area in humans is undergirded by antigen presenting cells (Langerhans cells) distributed among the viable keratinocytes[6] and thus represents an extensive, highly superficial network

New Vaccine Technologies, edited by Ronald W. Ellis. ©2001 Eurekah.com.

barrier of immune system elements. In contrast to transdermal drug delivery, which presumes transport of a drug through the epidermis into the vasculature found in the deeper layers of the skin (dermis), TCI appears to target only these superficial layers of the skin, the epidermis, where Langerhans cells (LCs) can initiate systemic immune responses.[7] The extensive dedication of biological resources in the form of antigen presenting cells to immunoprotection in the superficial layers of the skin indicates that the stratum corneum is often penetrated by microbes and suggests that it can be penetrated for purposes of immunization. The fact that TCI has been used in mice, rabbits, guinea pigs, cats, dogs, sheep and humans supports this premise.[8,9]

The transit pathways utilized by antigens to traverse the stratum corneum are unknown at this time. The stratum corneum is the principal barrier to delivery of drugs and antigens through the skin. Transdermal drug delivery of polar drugs is widely held to occur through aqueous intercellular channels formed between the keratinocytes.[10] Although the SC is the limiting barrier for penetration, it is breached by hair follicles and sweat ducts. Whether antigens penetrate directly through the SC or via the epidermal appendages may depend on a host of factors. These appendages are thought to play only a minor role in transdermal drug delivery.[11] Despite some evidence in mice that topical immunization using DNA may utilize hair follicles as the pathway for skin penetration,[12] it is more likely that robust immune responses will require utilization of more of the skin surface area. Disruption of the SC barrier can be accomplished by simple interventions such as hydration of the skin which is considered to be the most effective skin penetration enhancer technique.[5] Hydration of the SC induced by occlusion is widely used for enabling penetration of drugs with transdermal patches and has been the principle technique employed to date for TCI.[13]

LCs are bone marrow derived antigen presenting cells that reside in the epidermis where they sample antigen and have a baseline level of traffic to the draining lymph nodes. LCs are known to increase their rate of migration out of the skin in response to activating stimuli, such as contact sensitizers or intradermally injected LPS or TNF-α and travel to the lymph node where antigen presentation occurs to T-cells.[14] In immune pathologies such as topically induced delayed-type hypersensitivity (DTH), sensitized T-cells generated from previous antigen exposure migrate back to the skin and produce local inflammation.[15] We have hypothesized that LCs phagocytose antigen, are activated by adjuvants, and the activation enhances migration out of the skin and to the draining lymph nodes, where antigen presentation and activation of T-cells occurs (Fig. 17.1). The efferent of contact sensitization which involves T-cell migration back to the skin is not required for TCI.

The role of LCs and mechanisms engaged for TCI are undoubtedly complex yet early observations are intriguing. Observations of activated LCs in response to topical administration of CT or CTB, as indicated by morphological changes, have proven to be reproducible. This is illustrated in Figure 17.2. Split epidermis from mouse ears immunized with cholera toxin (CT), harvested at 26 hours post-immunization, and FITC stained for class II demonstrate rounding of the cells and loss of dendritic processes typical of LC activation by contact sensitizers and intradermal injection of LPS or TNF-α.[16,17] TNF-α and IL-1β are the primary cytokines responsible for LC migration out of the skin[16] and the role of these cytokines in TCI is under study.

In TCI, inflammation of the skin in human trials is notably absent and histologic sections taken from the site of immunization are devoid of lymphocytic infiltration.[18] Thus, although TCI may share some cellular pathways with DTH responses, TCI is not DTH. Furthermore, the absence of skin inflammation and the induction of robust immune responses is highly suggestive of the involvement of LCs in TCI as they are the only antigen presenting cells found in uninflamed skin.[19] Further studies are clearly needed to clarify the role of LCs in topical immunization and elucidate the mechanisms underlying TCI. Clearly, the targeting of dendritic

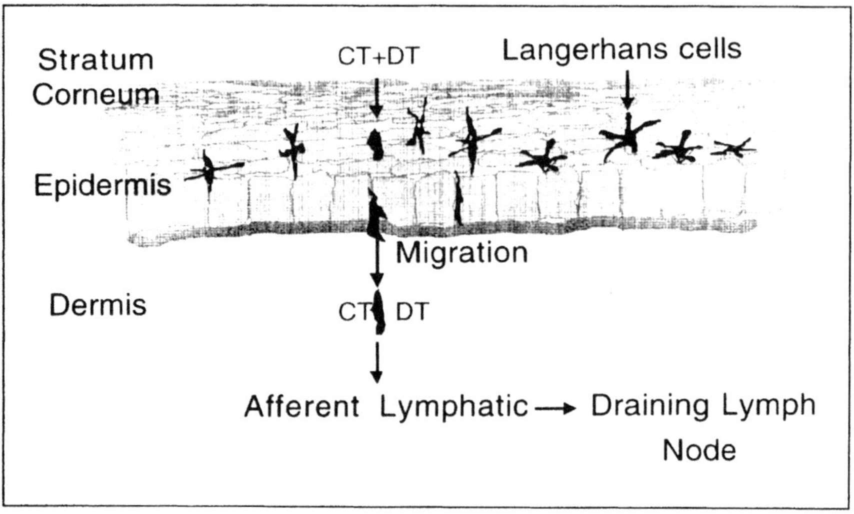

Fig. 17.1. Illustration of the operating hypothesis for transcutaneous immunization. Antigen and adjuvant placed on the skin penetrate into the superficial layers of the epidermis through simple hydration. Langerhans cells phagocytose antigen (DT) and are activated by CT. Activated Langerhans cells migrate out of the skin and into the draining lymph nodes, where antigen presentation of DT to T-cells occurs, resulting in a subsequent systemic immune response to DT. Reprinted with permission from: Glenn GM, Scharton-Kersten T, Alving CR. Exp Opin Invest Drugs 1999;8(6):797-805. ©1999 Ashley Publications LTD.

cells such as LCs has great potential for immunotherapies and thus elucidation of the mechanisms underlying TCI will be of general interest.

Adjuvants and TCI

Adjuvants are used to augment both the immune response to vaccine antigens as well as direct the qualitative response, and are amply discussed elsewhere.[20] The induction of robust immune responses following TCI appears to be dependent on the presence of an adjuvant in the formulation.[13] The use of adjuvants targeting potent APCs opens a host of possibilities for manipulation of the immune response and enhancing efficacy through augmentation of the immune response.

The greatest experience using adjuvants for TCI has been with the bacterial ADP-ribosylating exotoxins (bAREs) which include CT, heat labile enterotoxin from *E. coli* (LT)[21] and their mutants.[22] bAREs have had extensive use as adjuvants via intranasal and oral routes which has provided a wealth of applicable experience for their use on the skin.[21] TCI is similar to intranasal or oral immunization as the simple admixture of CT or LT with a coadministered antigen such as tetanus toxoid or influenza hemagluttinin results in far higher antibody levels compared to the administration of antigen alone.[13] Similarly, use of bAREs either by intranasal or by TCI can induce cell-mediated immunity to the coadministered antigens such as $CD4^{+}$[7,23] or $CD8^{+}$ T cells.[24]

Among the bAREs, there are several adjuvant choices for TCI. Point mutations and other techniques have resulted in mutant toxins that appear to have adjuvant activities similar to the native toxins but are less prone to induce potential gastrointestinal side effects. The use of these mutant toxins as TCI adjuvants may allay the possible concern regarding toxicity that is associated with the use of native toxins[22] which cause diarrhea upon ingestion in fasting subjects in which the gastric acid has been neutralized.[25] Purified cholera toxin B-subunit has also been

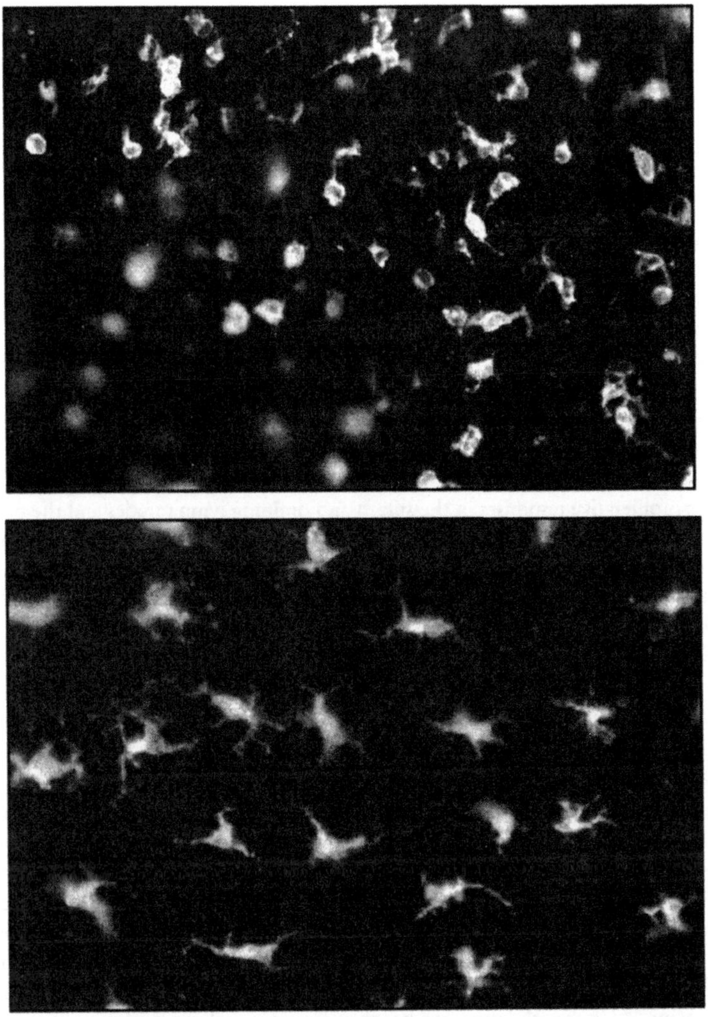

Fig. 17.2. After wetting the skin with water, the ventral ear surfaces of C57BL/6 mice ears were treated with CT (250 mg) on one ear and PBS on the contralateral ear. Mice were sacrificed 26 hours after treatment. Epidermal sheets were prepared as described[17] and stained for the murine epidermal LC marker MHC class II using hybridoma supernatant (top panel: Y3P, mouse IgG2a, ATCC, Manassas, VA), or irrelevant isotype control supernatant (lower panel) and FITC-labeled goat anti-mouse Ig secondary Ab (Biosource International, Camarillo, CA).

found to be a potent adjuvant on the skin in contrast to rCTB which is less potent but readily induces anti-CTB antibodies.[26] Other adjuvants including mutant toxins, bacterial DNA, cytokines and LPS have been shown to act as adjuvants but their potency and qualitative effects compared to the bAREs remains to be determined.[26]

Immune Responses to Transcutaneous Immunization

Initial work has shown that CT acts as an adjuvant when applied to the skin, resulting in classic priming and secondary antibody responses to coadministered antigens.[13] Mice immunized with CT or LT alone and subsequently boosted demonstrate both strong priming and a typical secondary antibody response to the toxins. By contrast, when CT is coadministered with the vaccine antigens tetanus or diphtheria toxoid (DT), minimal priming antibody responses to the coadministered antigen are seen, yet after boosting immunizations, classic secondary antibody responses are observed.[13] Antigens given without adjuvant can, after several doses, induce a measurable antibody response which is consistently several $\log_{10}$ titers below the response induced using an adjuvant.[26,27]

Most vaccines acquire their protective levels of antibodies through boosting regimens; similarly, it appears that TCI elicits immune responses that can be readily boosted.[13] Importantly, the humoral response to the adjuvant has been shown not to interfere with the response to the coadministered antigen, and multiple immunizations with different antigens using the same adjuvant may be conducted.[13] Additionally, the anti-toxin response to LT or CT as adjuvants, which also shows secondary antibody response kinetics, may also have a protective effect to prevent toxin-mediated diarrheas.[28-30]

In general, antigen-specific T-cells underlie the secondary antibody responses seen in response to vaccination and would be expected to be induced using TCI. Consistent with this concept, mice immunized on the skin with DT using CT as adjuvant have been found to have DT-specific proliferative responses in the spleen and draining lymph nodes and the proliferative responses to DT induced using TCI are due to $CD4^+$ cells.[7] T-cell responses to several antigens have been seen and may represent the most sensitive immunological marker for detecting a response to TCI. While the induction of secondary antibody and T-cell responses in spleen and draining lymph nodes may seem in retrospect to be an obvious expectation of TCI, the unique nature of this topical response and the historical association of topical applications almost exclusively with immunopathologies of the skin made demonstration of secondary systemic immune responses an essential task.

Mucosal Responses

Intramuscular and mucosal immunization appear to induce immune responses in different immune compartments. Mucosal immune responses are desirable for their ability to block pathogens at the point of entry at the mucosal lining. Intramuscular and subcutaneous immunization generally result in systemic immune responses with little or no mucosal component, although this may depend on a host of factors including the nature of the antigen or adjuvant.[31] Intramuscular immunization, however, can be effective against 'mucosal' pathogens such as *Hemophilus influenzae B* that has nearly been eradicated by the use of an intramuscularly injected vaccine.[32] Clearly, the divisions between systemic and mucosal immune responses to immunization are not simple. Similarly, mucosal responses are diverse and extend well beyond locally produced secretory IgA.[35]

We have postulated that skin immune system elements may be predisposed to mucosal responses. The mucosa and skin share common elements including Langerhans cells and secretory organs and the presence of immunoglobulins. More specifically, the sweat glands are known to secrete IgA[34] and immunohistochemical staining reveals both IgA and secretory chain in the glandular lumina.[35] Furthermore, microorganisms found on the skin demonstrate coating with immunoglobulins, including IgA and secretory chain.[36,37] Thus, it appears that the skin immune system has inherent mechanisms for responding to microbes with mucosal-like immune responses.[38] It may also be possible that immune responses in lymph nodes that drain to nodes receiving lymph from mucosal sites result in mucosal responses.[39,40]

The topical application of bAREs such as CT and coadministered antigens induces antibodies that can be detected at the mucosa[13,28] as well as mucosal cellular responses (unpublished observations). In Figure 17.3, mice immunized transcutaneously with cholera toxin produce anti-CT IgG and IgA antibodies in the stool and pulmonary secretions. These antibodies

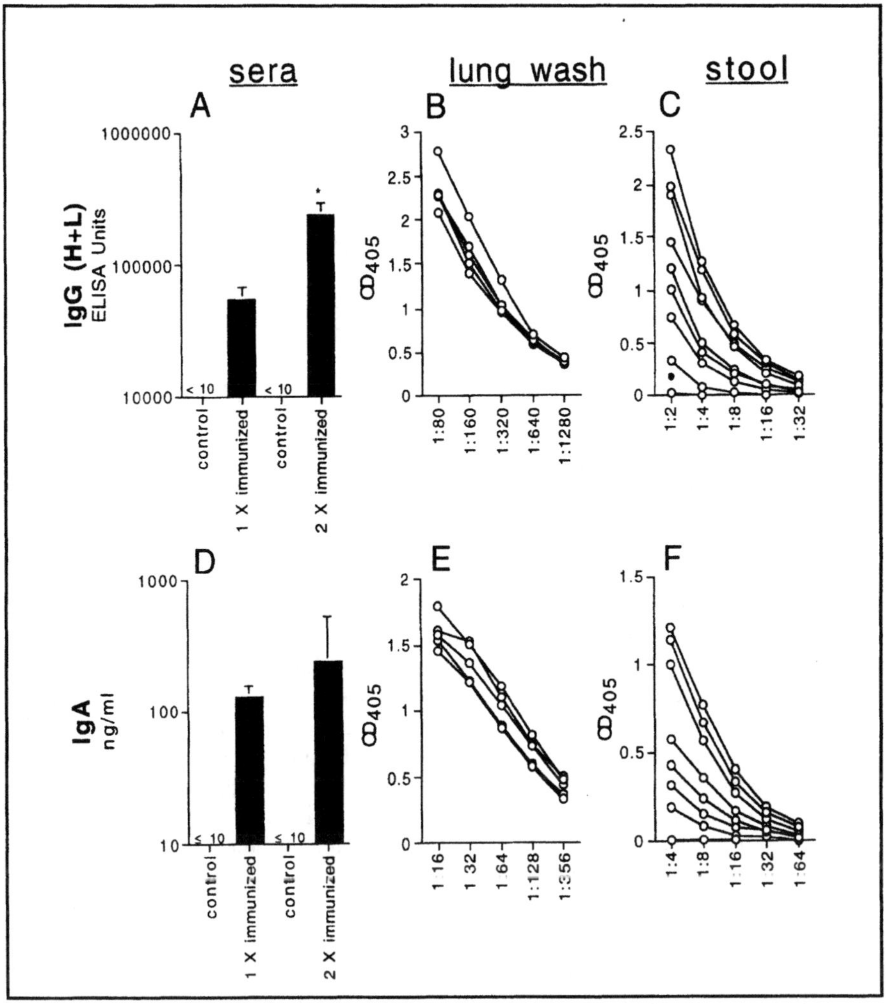

Fig. 17.3. Sera (A and D), mucosal lung (B and E), and stool (C and F) antibody responses to CT after transcutaneous immunization. C57BL/6 mice (17-22 animals per group) were immunized transcutaneously at 0 and 3 weeks with 100 mg CT. Sera was collected at 3 and 6 weeks and the CT specific Ig (H+L) and IgA levels assessed by ELISA. Data shown are the geometric mean ± SEM for measurements from 5 individual animals. An asterisk denotes a statistically significant ($p < 0.05$) difference between the titers measured in the 1X and 2X immunization groups (A and B). C57BL/6 mice were immunized transcutaneously at 0 weeks. Lung washes were performed on representative mice (n=5) after sacrifice on the day of challenge (3 weeks) by tracheal transection. Ig (H+L) and IgA levels were assessed by ELISA and the titers (optical density, 405 nm) from individual animals are shown. Neither IgG nor IgA were detected in lung washes from unimmunized animals (B and E). C57BL/6 mice were immunized transcutaneously at 0 weeks. Single stool pellets were collected immediately after defecation on the day before toxin challenge (6 weeks). Antibodies were extracted from fecal homogenates as described in the Materials and Methods. Ig (H+L) and IgA levels were assessed by ELISA and the titers (optical density 405 nm) from 8 (F) or 9 (C) individual animals are shown. CT specific IgA was not detected in stool samples from unimmunized mice. A solid circle denotes the maximal level of anti-CT Ig antibody detected in 1:2 dilutions of sera from unimmunized mice (background). Reprinted with permission from: Glenn GM, Scharton-Kersten T, Vassell R, Mallett C, Hale TL, Alving CR. J Immunol 1998;161:3211-3214. ©1998. The American Association of Immunologists.

Table 17.1. Antibody responses to CT, CT-subunits and other bAREs

Immunizing Antigen	Antibody Specificity	Serum IgG (ELSA Units)	SEM
CT	Anti-CT	39,828	(17,826-44,838)
CTB	Anti-CTB	7,480	(3,756-14,896)
rCTB	Anti-CTB	9,324	(5,271-13,372)
CTA	Anti-CTA	0	0
LT	Anti-LT	22,461	(20,262-27,167)
ETA	Anti-ETA	3,758	(1,951-77,240)
BSA	Anti-BSA	0	0

Balb/C Mice (n=5) were immunized with 100 mg of antigen at 0 and 3 weeks. Antibodies were measured by ELISA at 6 weeks. The results are reported as the geometric mean ± SEM of individually assayed sera in ELISA.
Reprinted with permission from: Glenn GM, Scharton-Kersten T, Vassell R, Matyas GR, Alving CR. Infect Immun 1999;67(3):1100-1106. © 1999 American Society for Microbiology.

can neutralize toxin-induced pathology at both the pulmonary[28] and intestinal mucosa (unpublished observation). IgG antibodies to the coadministered antigen have been detected in the mucosa in several settings. Mice immunized with CT as adjuvant and DT produce high levels of serum anti-DT antibodies. When the mucosal secretions of mice immunized in this fashion are assayed for the presence of antibodies by ELISA, anti-DT IgG can be found in both lung washes and stool pellets.[13] Stool and lung wash studies are not quantitative or sensitive assessments of mucosal antibodies, but these results are compelling due to the complete absence of antibodies in 0-week control stool specimens and control lung washes. These early findings raise many mechanistic questions, yet suggest that induction of responses by TCI may be used to enhance vaccine efficacy through induction of both mucosal and systemic responses.

Diversity of Antigens

Early studies with topical immunization utilized bAREs and toxoids as test antigens. As shown in Table 17.1, antibodies to a variety of bAREs and the B subunit of CT can be readily induced. CT and LT are organized as A:B5 structures that are essentially proenzymes with the ADP-ribosyltransferase activity contained in the A subunit and its target cell binding region located on the B subunit which binds to the ubiquitous cell membrane ganglioside G_{M1}.[41,42] The hallmark of their activities at the cellular level is the rise in intracellular cAMP upon binding of CT to the ganglioside G_{M1} by the B-subunit and insertion of the A-subunit into the cytoplasm of epithelial cells. In the intestine, this leads to fluid loss and diarrhea. However, the mechanism of adjuvant effects by CT or LT on the immune system is not fully understood.[23] They also act as antigens themselves, inducing, for example, anti-CT IgG and IgA.[28] The bacterial ADP-ribosylating exotoxins (bAREs) family includes CT, LT, *Bordetella pertussis*-derived pertussis toxin (PT), *Pseudomonas aeruginosa* exotoxin A (ETA) and DT.[42]

As shown in Table 17.1, a structurally dissimilar bARE, ETA, was tested to determine if the phenomena of TCI was independent of ganglioside G_{M1} binding. ETA, derived from *Pseudomonas aeroginosa*, is a single 613 amino acid peptide with A and B domains on the same peptide, unlike CT or LT which are composed of large (86 kD) noncovalently linked subunits with A:B5 stoichiometry.[42] ETA binds to an entirely different receptor, the α_2-macroglobulin receptor/low density lipoprotein receptor-related protein.[43] Despite the dissimilarities in size, structure, and binding target between ETA and CT, ETA also induced an anti-ETA antibody response when topically administered (Table 17.1). Thus, TCI is not entirely dependent on CT, LT, or binding of ganglioside G_{M1} although in the case of CT or LT such binding may play an essential role as suggested by the poor response to CTA alone. Consistent with later studies, the presence of adjuvant activity seemed to be essential to the induction of antibodies to coadministered antigen (Table 17.1).

Antibodies to a wide variety of coadministered antigens can be induced using TCI. Antigens successfully used include 600 kD recombinants, 32 mer synthetic peptides, killed viruses, live viruses, split virus preparations, VLPs, lysates, and polysaccharide conjugates (unpublished observations). Although responses to antigens administered without adjuvants have been seen, they are uniformly not robust.

To illustrate the breadth of antigens that may be used by TCI, a live viral vector encoding a foreign gene was administered topically with and without adjuvant and compared to intramuscular delivery of virus.[8] The recombinant viral vector utilized in this study was a replication competent mengovirus vector encoding the F1 capsular protein from the plague-associated bacterium *Y. pestis*. Mengovirus is a picornavirus consisting of a 30nm nonenvelope spherical virion and containing a single plus stranded RNA genome. The vector used contains the F1 coding region and signal peptide coding sequence inserted into its genome. Mengovirus protease cleavage sites located on both ends of the F1 protein sequence result in the production of the F1 antigen by the vector. F1-specfic IgG responses were detected, as illustrated in Figure 17.4. Mice receiving recombinant mengovirus via TCI without coadministered CT had F1-specific antibody at preimmune levels. Upon TCI of recombinant mengovirus coadministered with CT, F1-specific antibody was readily detected and did not significantly differ from those elicited in the positive control group that received the virus intramuscularly. The ability to generate F1 specific antibody titers despite poor growth characteristics of the recombinant virus vector confirms other studies that have shown that immunization with particulate antigens is feasible using TCI. Even with the delivery of the complex system, the role of the adjuvant appears critical for the induction of robust immune responses and has been a constant feature of our studies.[3,7,13,44]

Delivery Options Using Transcutaneous Immunization

TCI appears to provide a new level of flexibility for delivery strategies. Although clinical studies using patches are underway, topical delivery of vaccines may not be restricted to patches but may involve gels, creams or ointments. Other practical strategies target multiple draining lymph nodes, delivery of multivalent vaccines, increasing the frequency of boosting, anatomical targeting and boosting subjects primed by other routes (intramuscular, intranasal) or by other immunization strategies (e.g., DNA immunization).

Optimization for Enhancement of the Immune Response

The earliest experiments conducted using TCI were performed using empiric antigen doses and no skin manipulation. The working hypothesis for TCI suggests that potent immune responses comparable or superior to immune responses induced by standard routes should be feasible as TCI uses 'gold standard' mucosal adjuvants to target 'gold standard' antigen presenting cells, also known as 'nature's adjuvants' (Langerhans cells).[16] In line with this expectation, optimization experiments have suggested that TCI can be enhanced by simple manipulations and can elicit responses comparable in magnitude to those seen in response to established routes of immunization. The model antigens tetanus toxoid and diphtheria toxoid have been

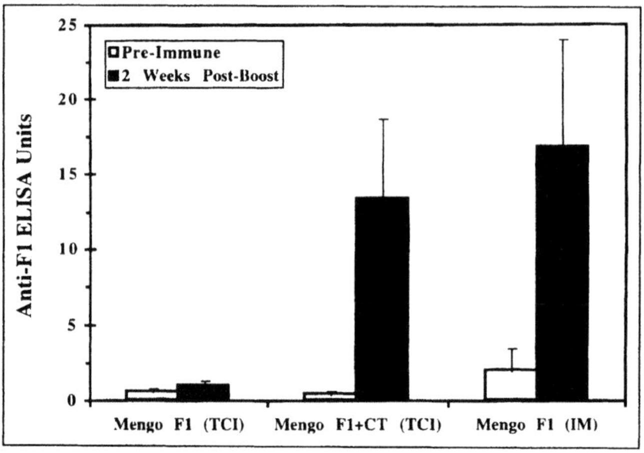

Fig. 17.4. A recombinant F1 mengovirus vector can elicit an antigen specific humoral response in transcutaneously immunized mice. Three groups of mice (n=5) were immunized either parentally or transcutaneously with 7 x 10^4 PFU on day 0 and 1 x 10^7 PFU on day 35 of recombinant F1 mengovirus. Mean antibody responses presented as ELISA Units were measured before immunization (pre-immune; open columns) and two weeks following the second inoculation (2 weeks post-boost; filled columns). Standard deviations of the sample mean for each set of data was plotted with Y-error bars. Reprinted with permission from: Hammond SA, Tsonis C, Sellins K, Rushlow K, Scharton-Kersten T, Colditz I, Glenn GM. Transcutaneous immunization of domestic animals: opportunities and challenges. Advanced Drug Delivery Reviews. ©2000. Elsevier Science.

used initially to define some of the optimization principles but penetration enhancement may be required for a broader variety of antigens delivered by topical immunization.

Alcohol swabbing prior to hydration can enhance the immune response elicited by TCI.[44] Antibody titers against CT were improved when the area for immunization was swabbed with isopropanol prior to application of the antigen. This procedure may delipidate the affected skin and appears to enhance penetration as the immunizing solution is absorbed more rapidly compared to the standard procedure which involves hydration with water prior to immunization. The process of swabbing itself may also remove corneocytes and enhance penetration through physical means. In mice in which the skin was hydrated and alcohol swabbed prior to immunization with CT and DT, anti-DT antibody responses were comparable to anti-DT responses elicited by intramuscular injection with alum and intranasal immunization using LT as adjuvant.[7]

Other studies have shown that lower doses of adjuvant and antigen than initially described[3] can be used for TCI. Initially, doses of adjuvant and antigen used in TCI studies were chosen empirically. Using DT as an antigen and CT as adjuvant, further studies suggested that similar immune responses could be induced using lower doses. Mice that received 10 μg of adjuvant and 100 μg of antigen had anti-DT antibody responses that were equivalent to the mice receiving 100 μg of antigen and adjuvant.[7] In another study, a range of adjuvant (LT) doses were used with a range of tetanus toxoid doses.[26] As shown in Figure 17.6, the adjuvant effect was seen at 1 μg of LT. In comparison to the response to TT with alum delivered i.m., 10 μg of LT and 10 μg of TT yielded anti-TT antibodies of a similar magnitude. These studies suggest that doses of adjuvant and antigen similar to those used by other routes may be used in the induction of immune responses using TCI.

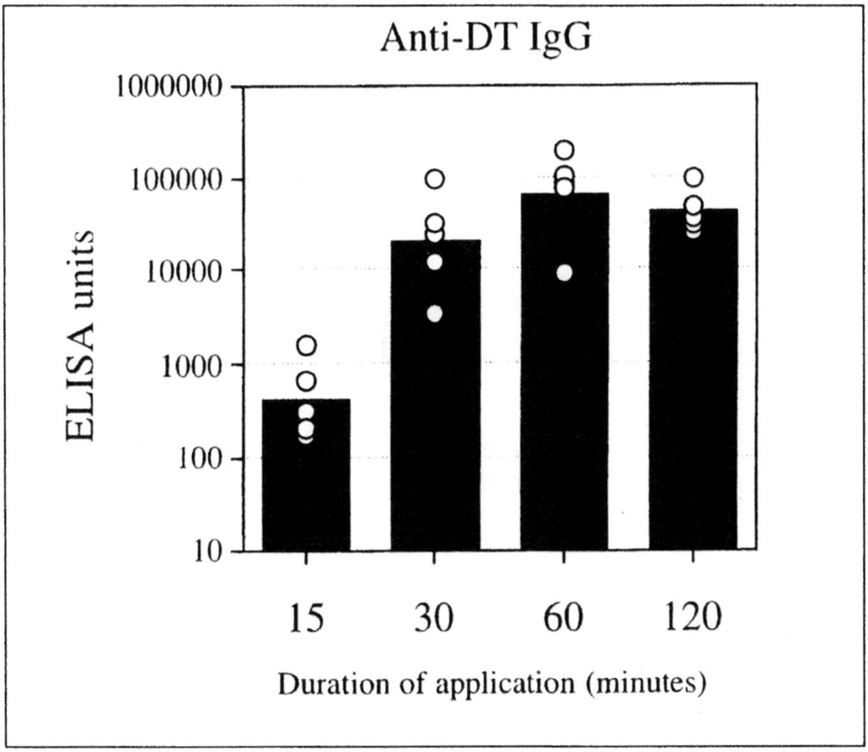

Fig. 17.5. Achievement of maximal anti-DT titers in animals immunized for 30 min. C57BL/6 mice (n = 5) were immunized with 100 μg of CT and 100 μg of DT on the skin at 0, 4, and 8 weeks. The duration of immunization was 15, 30, 60 or 120 min. Anti-DT IgG titers were assessed by ELISA on serum collected 13 weeks after the primary immunization. Titers in pre-bleed serum from the same batch of animals were = 20 EU. Results are reported in ELISA Units, which are defined as the inverse dilution of the sera that yields an OD of 1.0 at 405 nm. Reprinted with permission from: Glenn, Scharton-Kersten and Alving. Expert Opinion on Investigational Drugs 1999; 8(6):797-805. ©1999. Ashley Publications LTD.

In combination with simple skin manipulation, or GRAS (Generally Recognized As Safe) penetration enhancers, further efficiency of antigen use may be expected. This may be especially important for veterinary applications or vaccines with high cost of goods. Chemical removal of the stratum corneum with a high pH base emulsion has also been utilized to enable topical immunization.[45] Use of physical and chemical penetration enhancement in concert with TCI will clearly play a role in future developments but will have to achieve a balance between irritation, reactogenicity and robust immune responses. The use of an adjuvant offers a method for augmenting the immune response without the risk of severe chemical or mechanical skin irritation.

The duration of time required for induction of an immune response via TCI is not clear, but if hydration of the skin and passive diffusion of the antigen into the epidermis are the underlying physical phenomena required for immunization, then a short period of immunization may be feasible. The time required for TCI was studied by applying the immunizing solution for various times and after 15 minutes of application, the animals were well immunized (Fig. 17.5). By 30-60 minutes, maximal immune responses were obtained.[7] These early studies suggest that simple optimization strategies may greatly enhance the immune response to TCI,

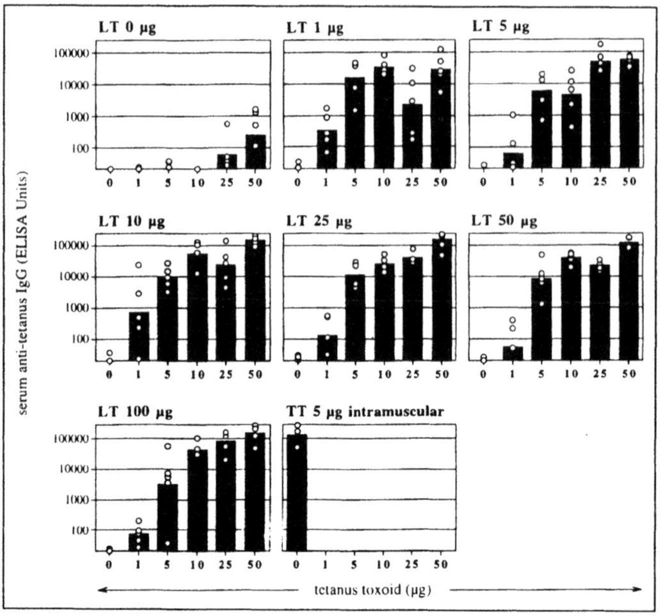

Fig. 17.6. Serum anti-TTx titers from C57/B16 mice immunized with different doses of LT and TTx. Mice were immunized at 0, 3 and 6 weeks and serum collected 2 weeks after the third immunization . Anti-TTx titers were analyzed by ELISA. Data shown as bars represent the geometric mean of 5 mice with individual titers represented by open circles. Reprinted with permission from Glenn, Scharton-Kersten. Infection and Immunity. ©2000. American Society for Microbiology

and that clinical strategies may anticipate low doses of adjuvant and antigen and brief application periods that may fit into a clinic visit.

Human Studies

Several phase I clinical trials are under way using TCI. The objective of the first study was to demonstrate that a large antigen such as LT (86 kD) could be safely delivered and induce an immune response also. LT is an ideal candidate in that it is immunogenic and can produce anti-LT antibodies. It can also act as an adjuvant for future studies, and it is relatively large, exceeding the usual size of molecules that can be delivered through the skin (0.5kD). Using a simple patch, we tested the hypothesis that the skin of humans sould be similarly be utilized for immunization. After 3 doses of LT, a greater than 14-fold rise in anti-LT IgG was seen.[17] The magnitude of anti-LT responses using oral immunization can be used as a comparison to assess the robustness of the immune response.[30] This study demonstrated that TCI can safely induce a systemic immune response and validated the animal findings that a large antigen could pass through the skin and induce an immune response.

Conclusions

The use of topically applied vaccines may address the urgent need for needle-free vaccine delivery,[46] decrease the barriers to immunization and allow flexible delivery for multiple boosting and multivalent vaccines. TCI appears to offer a new method for the delivery of vaccines with practical and immunological advantages. The exploitation of the skin immune system may hold promise for improved efficacy for established vaccines through both mucosal and systemic responses and thus open new possibilities for vaccines under development. Clearly, there are significant challenges ahead for the full development of TCI, but this technique has entered into the array of immunization and vaccine delivery strategies.

References

1. Condon C, Watkins SC, Celluzzi CM et al. DNA-based immunization by in vivo transfection of dendritic cells. Nat Med 1996; 2(10):1122-1128.
2. Kaiserlian D, Etchart N. Epicutaneous and transcutaneous immunization using DNA or proteins. Eur J Dermatol 1999; 9(3):169-176.
3. Glenn GM, Rao M, Matyas GR et al. Skin immunization made possible by cholera toxin. Nature 1998; 391(6670):851.
4. Fletcher MA, Saliou P. Vaccines and infectious disease. EXS 2000; 89:69-88.
5. Roberts MS, Walker M. Water, the most natural penetration enhancer. In: Walters KA, Hadgraft J, ed(s). Pharmaceutical Skin Penetration Enhancement. New York: Marcel Dekker, 1993:33.
6. Yu RC, Abrams DC, Alaibac M et al. Morphological and quantitative analyses of normal epidermal Langerhans cells using confocal scanning laser microscopy. Brit J Dermatol 1994; 131(6):843-848.
7. Glenn GM, Scharton-Kersten T, Alving CR. Advances in vaccine delivery: transcutaneous immunisation. Exp Op Invest Drugs 1999; 8(6):797-805.
8. Hammond SA, Tsonis C, Rushlow, K et al. Transcutaneous immunization of animals: promises and challenges. Advanced Drug Delivery Reviews 2000; 43(1):45-55.
9. Cope RB, Colditz IG. Effect on systemic antibody concentrations of topical application of cholera toxin to skin of sheep. Aust Vet J 2000; 78(2):121-123.
10. Ghosh TK, Pfister WR. Transdermal and topical delivery systems: an overview and future trends. In: Ghosh TK, Pfister WR, Yum SI, ed(s). Transdermal and Topical Drug Delivery Systems. Buffalo Grove:Interpharm Press, Inc., 1997:6.
11. Barry BW. Mode of action of penetration enhancers in human skin. J Control Rel 1987; 6(Spec. Iss.):85-97.
12. Fan H, Lin Q, Morrissey GR, Khavari PA. Immunization via hair follicles by topical application of naked DNA to normal skin. Nat Biotechnol 1999; 17(9):870-872.
13. Glenn GM, Scharton-Kersten T, Vassell R et al. Transcutaneous immunization with bacterial ADP-ribosylating exotoxins as antigens and adjuvants. Infect Immun 1999; 67(3):1100-1106.
14. Kripke ML, Munn CG, Jeevan A et al. Evidence that cutaneous antigen-presenting cells migrate to regional lymph nodes during contact sensitization. J Immunol 1990; 145(9):2833-2838.
15. Macatonia SE, Knight SC, Edwards AJ et al. Localization of antigen on lymph node dendritic cells after exposure to the contact sensitizer fluorscein isothiocyanate. Funcitonal and morphological studies. J Exp Med 1987; 166(6):1654-1667.
16. Jakob T, Udey MC. Langerhans cells: From neurons to nature's adjuvants. Adv Dermatol 1999; 14:209-258.
17. Aiba S, Katz S. Phenotypic and functional charcteristics of in vivo activated Langerhans cells. J Immunol 1990; 145(9):2791-2796.
18. Glenn GM, Taylor DN, Li X et al. Transcutaneous immunization: A human vaccine delivery strategy using a patch. Nat Med 2000; in press.
19. Udey, MC. Cadherins and Langerhans cell immunobiology. Clin Exp Immunol 1997; 107 (Suppl 1):6-8.
20. Singh M, O'Hagan D. Advances in vaccine adjuvants. Nat Biotechnol 1999; 17(11):1075-1081.
21. Snider DP. The mucosal adjuvant activities of ADP-ribosylating bacterial enterotoxins. Crit Rev Immunol 1995; 15(3-4):317-348.
22. Giuliani MM, Del Giudice G, Giannelli V et al. Mucosal adjuvanticity and immunogenicity of LTR72, a novel mutant of Escherichia coli heat-labile enterotoxin with partical knockout of ADP-ribosyltransferase activity. J Exp Med 1998; 187(7):1123-1132.
23. Lycke N. The mechanism of cholera toxin adjuvanticity. Res Immunol 1997; 148(8-9):504-520.
24. Porgador A, Staats HF, Faiola B et al. Intranasal immunization with CTL epitope peptides from HIV-1 or ovalbumin and the mucosal adjuvant cholera toxin induces peptide-specific CTLs and protection against tumor development in vivo. J Immunol 1997; 158(2):834-841.

25. Levine MM, Kaper JB, Black RE et al. New knowledge on pathogenesis of bacterial enteric infections as applied to vaccine development. Microbiol Rev 1983; 47(4):510-550.
26. Scharton-Kersten T, Yu J, Vassell R et al. Adjuvants for topical immunization on the skin: beyond the bacterial ADP-ribosylating exotoxin family. Infect Immun 2000; 68:5306-5313.
27. Paul A, Cevc G, Bachhawat BK. Transdermal immunisation with an integral membrane component, gap junction protein, by means of ultradeformable drug carriers, transfersomes. Vaccine 1998; 16(2-3):188-195.
28. Glenn GM, Scharton-Kersten T, Vassell R et al. Transcutaneous immunization with cholera toxin protects mice against lethal mucosal toxin challenge. J Immunol 1998; 161(7):3211-3214.
29. Clemens JD, Sack DA, Harris JR et al. Cross-protection by B subunit-whole cell cholera vaccine against diarrhea associated with heat-labile toxin-producing enterotoxigenic Escherichia coli: results of a large-scale field trial. J Infect Dis 1988; 158(2):372-377.
30. Michetti P, Kreiss C, Kotloff KL et al. Oral immunization with urease and Escherichia coli heat-labile enterotoxin is safe and immunogenic in Helicobacter pylori-infected adults. Gastroenterology 1999; 116(4):804-812.
31. Guy B, Fourage S, Hessler C et al. Effects of the nature of adjuvant and site of parenteral immunization on the serum and mucosal immune responses induced by a nasal boost with a vaccine alone. Clin Diagn Lab Immunol 1998; 5(5):732-736.
23. The Jordan Report: Accelerated Development of Vaccines. Division of Microbiology and Infectious Diseases, National Institute of Allergy and Infectious Diseases, NIH, 1998, p. iii.
33. Bouvet JP, Fischetti VA. Diversity of antibody-mediated immunity at the mucosal barrier. Infect Immun 1999; 67(6):2687-2691.
34. Okada T, Konishi H, Ito M et al. Identification of secretory immunoglobulin A in human sweat and sweat glands. J Invest Dermatol 1988; 90(5):648-651.
35. Gebhart W, Metze D, Jurecka W et al. IgA in human skin appendages. In: Caputo R, ed. Immunodermatology. CIC Edizioni Internationali, 1987:185.
36. Metze D, Kersten A, Jurecka W et al. Immunoglobulins coat microorganisms of skin surface: a comparative immunohistochemical and ultrastructural study of cutaneous and oral microbial symbionts. J Invest Dermatol 1991; 96:439-445.
37. Hard, GC. Electron microscopic examination of Corynebacterium ovis. J Bacteriol 1969; 97(3):1480-1485.
38. Mestecky J, McGhee JR. Prospects for human mucosal vaccines. Adv Exp Med Biol 1992; 327:13-23.
39. Gao, Y, Daley, MJ, Splitter, GA. BHV-1 glycoprotein 1 and recombinant interleukin 1ß efficiently elicit mucosal IgA response. Vaccine 1995; 13(9):871-877.
40. Babiuk S, Baca-Estrada M, Babiuk LA et al. Cutaneous vaccination: The skin as an immunological active tissue and the challenge of antigen delivery. J Control Release 2000; 66(2-3)199-214.
41. Holmgren J, Svennerholm AM. Mechanisms of disease and immunity in cholera: a review. J Infect Dis 1977; 136:S105-S112.
42. Krueger KM, Barbieri JT. The family of bacterial ADP-ribosylating exotoxins. Clin Microbiol Rev 1995; 8(1):34-47.
43. Kounnas MZ, Morris RE, Thompson MR et al. The a_2-macroglobulin receptor/low density lipoprotein receptor-related protein binds and internalizes Pseudomonas exotoxin A. J Biol Chem 1992; 267(18):12420-12423.
44. Scharton-Kersten T, Glenn GM, Vassell R et al. Principles of transcutaneous immunization using cholera toxin as an adjuvant. V[illegible] 1999; 17 (Suppl 2):37-43.
45. Tang DC, Shi Z, Curiel DT. [illegible]on onto bare skin. Nature 1997; 388(6644):729-730.
46. World Health Organization. Reducing the risk of unsafe injections in immunization programmes. The role of injection equipment. Expanded Programme on Immunization. Geneva: WHO; 1996.

Index

F

G

H

I

J

K

L

M

T

V

Y

For Product Safety Concerns and Information please contact our EU representative GPSR@taylorandfrancis.com Taylor & Francis Verlag GmbH, Kaufingerstraße 24, 80331 München, Germany

Batch number: 10725935

Printed by Printforce, the Netherlands